STUDENT RESOURCE MANUAL

MODERN DIFFERENTIAL EQUATIONS

SECOND EDITION

LORRAINE M. BRASELTON
MARTHA L. ABELL
JAMES P. BRASELTON
GEORGIA SOUTHERN UNIVERSITY

Australia • Canada • Mexico • Singapore • Spain • United Kingdom • United States

Printed in the United States of America

3 4 5 6 7 05 04 03 02 01

0-03-029751-6

For more information about our products, contact us at:
Thomson Learning Academic Resource Center
1-800-423-0563

For permission to use material from this text, contact us by:
Phone: 1-800-730-2214
Fax: 1-800-731-2215
Web: www.thomsonrights.com

Asia
Thomson Learning
60 Albert Complex, #15-01
Alpert Complex
Singapore 189969

Australia
Nelson Thomson Learning
102 Dodds Street
South Street
South Melbourne, Victoria 3205
Australia

Canada
Nelson Thomson Learning
1120 Birchmount Road
Toronto, Ontario M1K 5G4
Canada

Europe/Middle East/South Africa
Thomson Learning
Berkshire House
168-173 High Holborn
London WC1 V7AA
United Kingdom

Latin America
Thomson Learning
Seneca, 53
Colonia Polanco
11560 Mexico D.F.
Mexico

Spain
Paraninfo Thomson Learning
Calle/Magallanes, 25
28015 Madrid, Spain

Preface

What is here?

The *Student Resource Manual* to accompany *Modern Differential Equations 2e* contains solutions, partial solutions and hints, including *Mathematica* and *Maple V* code, to asterisked exercises in both the section and chapter ending review exercises. Additionally, selected solutions to problems within the *Differential Equations at Work* sections are also included.

The Solutions

There is a wide range of detail in the solutions and hints provided in this manual. Most solutions carefully explain the key steps that are encountered in the calculations although answers to *very* complicated problems were constructed with the help of a computer algebra system.

The Computer Code

For the most part, the code supplied will generate output from which reasonable conclusions can be made. Due to space limitations, the code presented in not a course on how to use *Mathematica* or *Maple V* to solve differential equations; the code is not thoroughly explained (see **Other Resources** below.) Throughout the *Student Resource Manual*, whenever we present computer code, we use the convention that *Mathematical* appears in the left column in **`bold Courier`** and *Maple V* code appears in the right column in `Monaco`. If the code for only one computer algebra system is shown, the aforementioned fonts are used.

```
Here is a sample of Mathematica code.
```

```
Here is a sample of Maple V code.
```

Other Resources

You can find substantial guidance in learning how to use computer algebra systems like *Mathematica* and *Maple V* from a wide variety of introductory texts, some of which are available below:

Abell, Martha L. and Braselton, James P., *Differential Equations with Maple V 2e*, Academic Press, 1999.

Abell, Martha L. and Braselton, James P., *Differential Equations with Mathematica 2e*, Academic Press, 1997.

Abell, Martha L. and Braselton, James P., *Maple V By Example 2e*, Academic Press, 1998.

Abell, Martha L. and Braselton, James P., *Mathematica By Example 2e*, Academic Press, 1997.

Lorraine M. Braselton
Martha L. Abell
James P. Braselton
Georgia Southern University
Statesboro, Georgia
August 2000

Table of Contents

Introduction to Differential Equations

1

EXERCISES 1.1

5. (a) ordinary; (b) first-order; (c) nonlinear; derivative is squared
9. (a) ordinary; (b) second-order; (c) nonlinear; second derivative is raised to a power
11. (a) partial; (c) nonlinear
13. If y is the dependent variable, we write the equation as $dy/dx = 2x - y$ and see that it is (a) ordinary; (b) first-order; and (c) linear. If x is the dependent variable, write it as $dx/dy = 1/(2x - y)$ and see that it is (a) ordinary; (b) first-order; and (c) nonlinear.
15. If y is the dependent variable, we write the equation as $dy/dx = (2x - y)/y$ to see that it is (a) ordinary; (b) first-order; and (c) nonlinear. If x is the dependent variable, we write the equation as $dx/dy = y/(2x - y)$ to see that it is (a) ordinary; (b) first-order; and (c) nonlinear.
16. (a) $\begin{cases} x' = y \\ y' = y + 6x \end{cases}$; (b) Let $x' = y$. Then, $y' = (x')' = x''$ and substitution into $4x'' + 4x' + 37x = 0$ yields $4y' + 4y + 37x = 0$. Solving for y' gives us $\begin{cases} x' = y \\ y' = \frac{1}{4}(-4y - 37x) \end{cases}$; (c) $\begin{cases} x' = y \\ y' = -\frac{g}{L}\sin x \end{cases}$; (d) Let $x' = y$. Then, $y' = (x')' = x''$ and substitution into $x'' - \mu(1 - x^2)x' + x = 0$ yields $y' - \mu(1 - x^2)y + x = 0$. Solving for y' gives us $\begin{cases} x' = y \\ y' = \mu(1 - x^2)y - x \end{cases}$; (e) $\begin{cases} x' = y \\ y' = \frac{1}{t}[(t - b)y + ax] \end{cases}$

19. $y(x) = e^{-x} - \frac{1}{2}\cos x + \frac{1}{2}\sin x \Rightarrow y'(x) = -e^{-x} + \frac{1}{2}\cos x + \frac{1}{2}\sin x$. Substituting gives us

$$y' + y = -e^{-x} + \tfrac{1}{2}\cos x + \tfrac{1}{2}\sin x + e^{-x} - \tfrac{1}{2}\cos x + \tfrac{1}{2}\sin x = \sin x.$$

23. $x = A\cos t + B\sin t + \frac{1}{4}t^2 \sin t - \frac{1}{2}t\sin t + \frac{1}{4}t\cos t \Rightarrow x' = \left(B + \frac{1}{4} - \frac{1}{2}t + \frac{1}{4}t^2\right)\cos t + \left(-A - \frac{1}{2} + \frac{1}{4}t\right)\sin t$ and $x'' = \left(-A - 1 + \frac{3}{4}t\right)\cos t + \left(-B + \frac{1}{2}t - \frac{1}{4}t^2\right)\sin t$. Substitution into the equation gives us

$$x'' + x = \left(-A - 1 + \tfrac{3}{4}t\right)\cos t + \left(-B + \tfrac{1}{2}t - \tfrac{1}{4}t^2\right)\sin t + \left(A + \tfrac{1}{4}t\right)\cos t + \left(B - \tfrac{1}{2}t + \tfrac{1}{4}t^2\right)\sin t = (-1 + t)\cos t.$$

29.

$$2x + 2y\frac{dy}{dx} = 0$$

$$\frac{dy}{dx} = -\frac{x}{y}$$

$$0^2 + y^2 = 16 \Rightarrow y = \pm 4; (0, \pm 4)$$

```
ContourPlot[x^2 + y^2, {x, -5, 5},
 {y, -5, 5}, Contours → {16},
 ContourShading → False, Frame → False,
 Axes → Automatic, AxesOrigin → {0, 0},
 PlotPoints → 200]
```

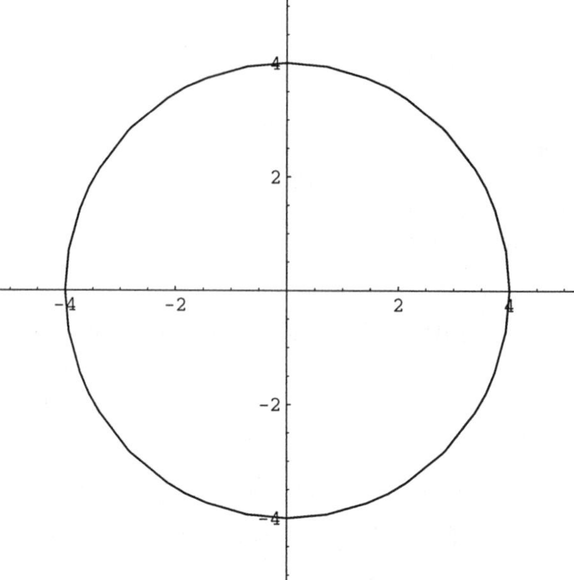

31.

$$3x^2 + 2xy + x^2\frac{dy}{dx} = 0$$

$$\frac{dy}{dx} = -\frac{3x^2 + 2xy}{x^2};$$

$$1^3 + y = 100 \Rightarrow y = 99; (1, 99)$$

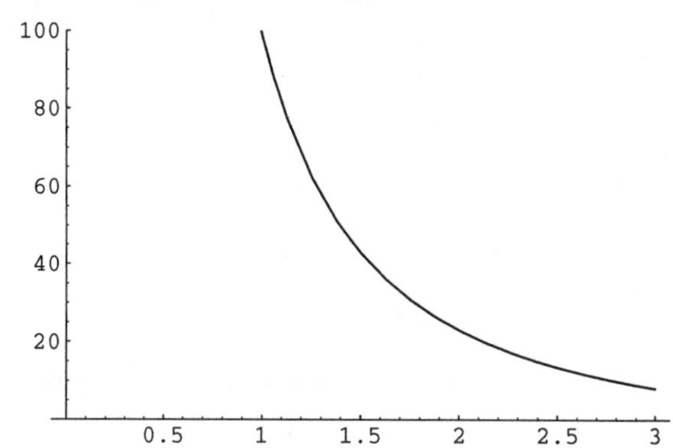

37. Use a u-substitution with $u = \ln x \Rightarrow du = 1/x\, dx$:

$$y = \int \frac{1}{x \ln x}\, dx = \int \frac{1}{u}\, du = \ln|u| + C = \ln|\ln x| + C.$$

41. Use partial fractions:

$$y = \int \frac{x - x^2}{(x+1)(x^2+1)}\, dx = \int \left(-\frac{1}{x+1} + \frac{1}{x^2+1} \right) dx = -\ln|x+1| + \tan^{-1} x + C.$$

47. $y = Ae^{4x} + Be^{-3x} \Rightarrow y' = 4Ae^{4x} - 3Be^{-3x}$ and applying the initial conditions yields

$$\begin{cases} A + B = 0 \\ 4A - 4B = 1 \end{cases} \Rightarrow A = 1/7,\ B = -1/7$$

so $y = \frac{1}{7}\left(e^{4x} - e^{-3x}\right)$.

51. $y = A + Bx + Ce^{2x} \Rightarrow y' = B + 2Ce^{2x}$ and $y'' = 4Ce^{2x}$. Applying the initial conditions yields

$$\begin{cases} A+C=0 \\ B+2C=1 \Rightarrow A=-3/4,\ B=-1/2,\ C=3/4 \\ 4C=3 \end{cases}$$

so $y=-\frac{3}{4}-\frac{1}{2}x+\frac{3}{4}e^{2x}$.

57. Integration yields

$$y(x)=\int\frac{1}{x^2}\cos\left(\frac{1}{x}\right)dx \underset{u=1/x\Rightarrow -du=1/x^2\,dx}{=} -\int\cos u\,du=-\sin u+C=-\sin\left(\frac{1}{x}\right)+C.$$

To find C, we apply the initial condition which yields $-\sin(\pi/2)+C=1\Rightarrow -1+C=1\Rightarrow C=2$ so $y(x)=-\sin\left(\frac{1}{x}\right)+2$.

```
Plot[-Sin[1 / x] + 2, {x, 0, 2 Pi},
 AspectRatio -> 1, PlotRange -> {-1,
```

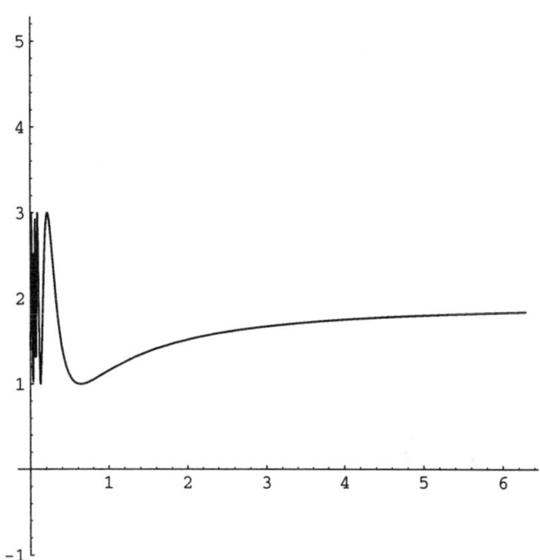

61. (a) $P(0)=\dfrac{r}{a+Ce^{-r\cdot 0}}=P_0\Rightarrow C=\dfrac{r-aP_0}{P_0}\Rightarrow P(t)=\dfrac{r}{a+\dfrac{r-aP_0}{P_0}e^{-rt}}=\dfrac{rP_0e^{rt}}{r+aP_0\left(e^{rt}-1\right)}$

(b) If $r>0$, $\lim\limits_{t\to\infty}P(t)=\lim\limits_{t\to\infty}\dfrac{r}{a+\dfrac{r-aP_0}{P_0}e^{-rt}}=\dfrac{r}{a+\dfrac{r-aP_0}{P_0}\cdot 0}=\dfrac{r}{a}$

65. $u_t=16ke^{-16t}\cos 4x, u_{xx}=16e^{-16t}\cos 4x; u(\pi,0)=2;\ \lim\limits_{t\to\infty}u(x,t)=3$

67. If $y=x^m$, $y'=mx^{m-1}$ and $y''=m(m-1)x^{m-2}$. Substitution into the equation yields

$$\begin{aligned} x^2y''-2xy'+2y&=0\\ x^2\bullet m(m-1)x^{m-2}-2x\bullet mx^{m-1}+2x^m&=0\\ m(m-1)x^m-2mx^m+2x^m&=0\\ x^m\left[m(m-1)-2m+2\right]&=0\\ m^2-3m+2&=0\\ (m-2)(m-1)&=0\end{aligned}$$

so $m=1, m=2$.

69.

$$e^{2x}\frac{dy}{dx}+2e^{2x}y=e^x$$
$$\frac{d}{dx}\left(e^{2x}y\right)=e^x$$
$$e^{2x}y=\int e^x dx=e^x+C$$
$$y=\frac{e^x+C}{e^{2x}}=e^{-x}+Ce^{-2x}$$

77. **Mathematica**

```
Clear[x, y]
y[x_] = x^(-1) Sin[3 Log[x]];
y'[x] // Together
y''[x] // Together
y'''[x] // Together
```

$$\frac{3\,\mathrm{Cos}[3\,\mathrm{Log}[x]] - \mathrm{Sin}[3\,\mathrm{Log}[x]]}{x^2}$$

$$\frac{-9\,\mathrm{Cos}[3\,\mathrm{Log}[x]] - 7\,\mathrm{Sin}[3\,\mathrm{Log}[x]]}{x^3}$$

$$\frac{6\,(\mathrm{Cos}[3\,\mathrm{Log}[x]] + 8\,\mathrm{Sin}[3\,\mathrm{Log}[x]]}{x^4}$$

```
Plot[y[x], {x, 0, Pi}]
```

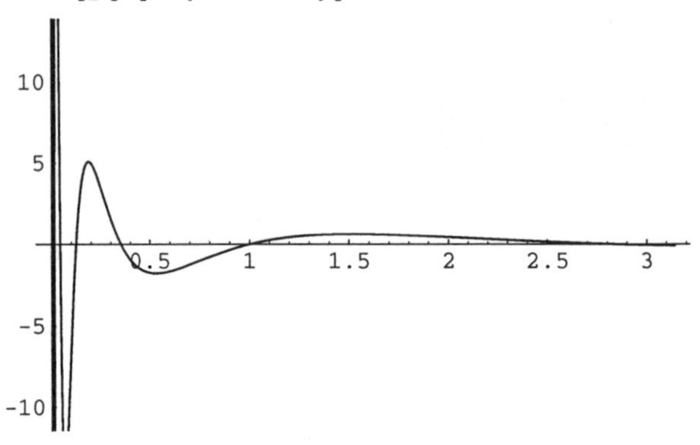


```
x^3 y'''[x] + x^2 y''[x] + x y'[x] -
    40 y[x] // Expand
```

0

79.

Mathematica

```
Clear[x,y,eq]
eq=(x^2+y^2)^2==5x  y;
step1=Dt[eq]
step2=step1  /.  {Dt[x]->1,Dt[y]->dydx}
step3=Solve[step2,dydx]
```

Maple

```
x:='x':y:='y':
step1:=D((x^2+y^2)^2=5*x*y);
step2:=subs({D(x)=1,D(y)=dydx},
   step1);
solve(step2,dydx);
```

```
plot1=ContourPlot[(x^2+y^2)^2-5x  y,
   {x,-2,2},{y,-2,2},
   Contours->{0},Frame->False,
   ContourShading->False,Axes-
>Automatic,
   AxesOrigin->{0,0},PlotPoints->60]
plot2=Graphics[{Dashing[{0.02}],
   Line[{{1,-2},{1,2}}],
   Line[{{2,-0.319},{-2,-0.319}}]}];
Show[plot1,plot2]
```

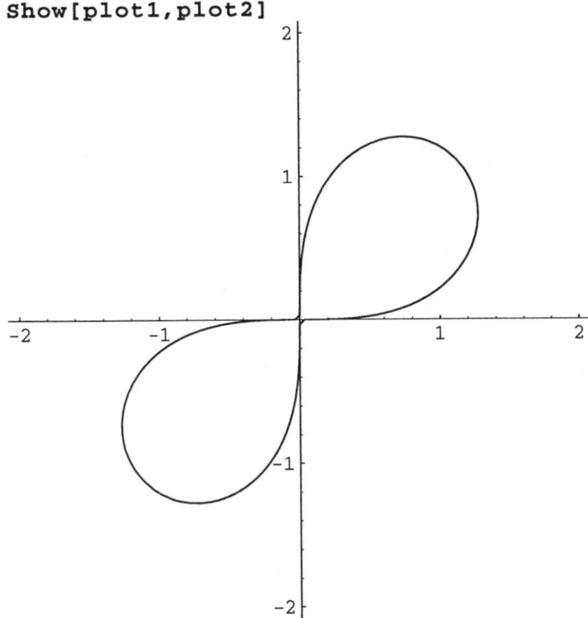

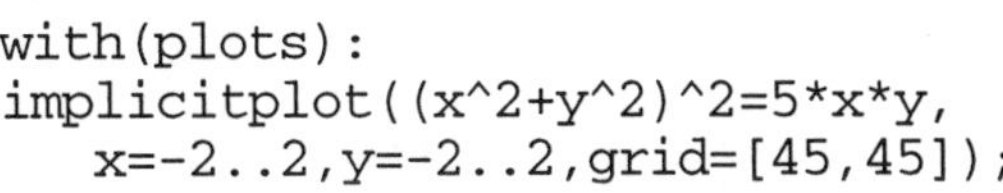

```
with(plots):
implicitplot((x^2+y^2)^2=5*x*y,
   x=-2..2,y=-2..2,grid=[45,45]);
```

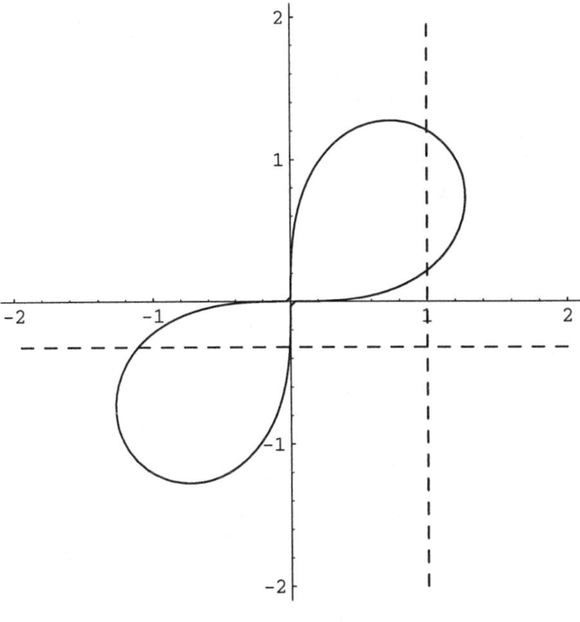

```
partc=eq  /.  x->1
Solve[partc]//N
partd=eq  /.  y->-0.319
Solve[partd]//N
```

```
fsolve((1+y^2)^2=5*y);
fsolve((x^2+(-0.319)^2)^2=
   -5*x*0.319);
```

80. **Mathematica**

```
sol = DSolve[{y[x] == Sin[x] ^ 4, y[0] == 0},
  y[x], x]
```

$\{\{y[x] \to \mathrm{Sin}[x]^4\}\}$

```
Plot[y[x] /. sol, {x, 0, 4 Pi}]
```

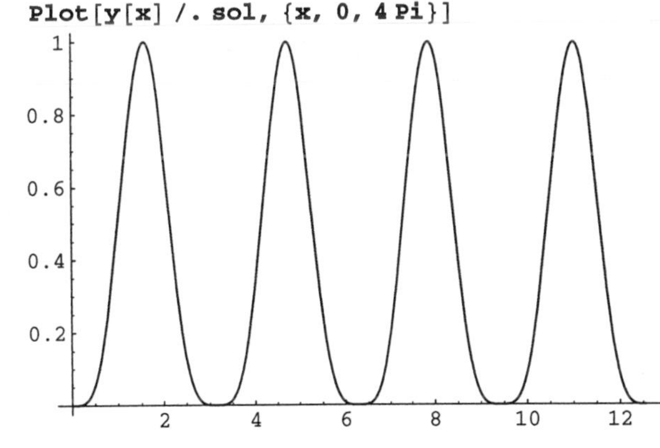

83. **Mathematica**

```
Clear[x, y]
sol =
 DSolve[{x'[t] == -5 x[t] + 4 y[t],
   y'[t] == 2 x[t] + 2 y[t], x[0] == 4,
   y[0] == 0}, {x[t], y[t]}, t]
```

$$\left\{\left\{x[t] \to \frac{4}{9} e^{-6t} \left(8 + e^{9t}\right), \; y[t] \to \frac{8}{9} e^{-6t} \left(-1 + e^{9t}\right)\right\}\right\}$$

```
Plot[Evaluate[{x[t], y[t]} /. sol],
 {t, 0, 1},
 PlotStyle ->
  {GrayLevel[0], GrayLevel[.5]}]
```

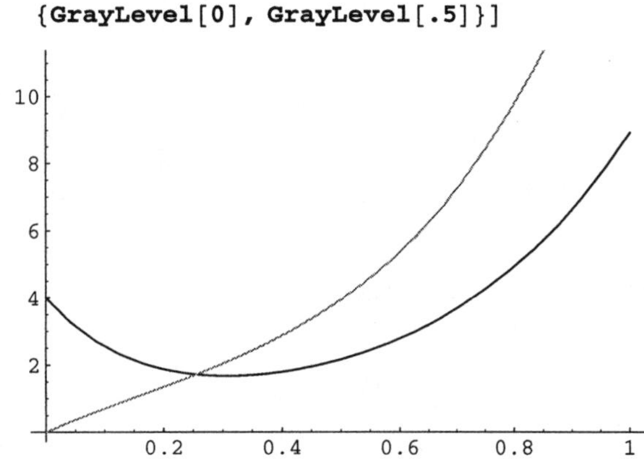


```
ParametricPlot[
 Evaluate[{x[t], y[t]} /. sol], {t, 0, 1}
 AspectRatio -> Automatic]
```

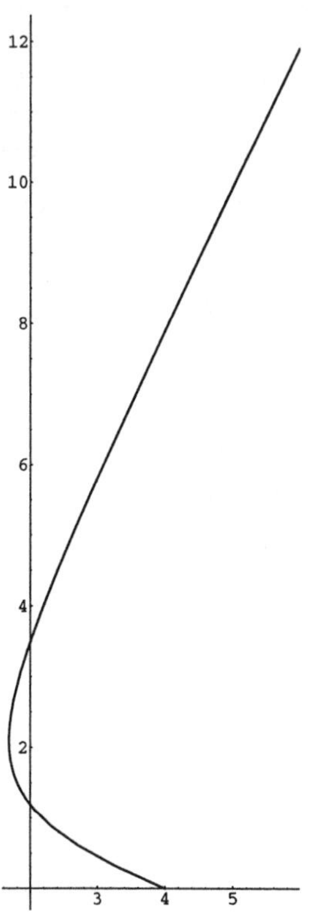

EXERCISES 1.2

5. **Mathematica**

```
<< Graphics`PlotField`
p1 = PlotVectorField[{1, x / y}, {x, -2, 2},
    {y, -2, 2}, Axes → Automatic,
    ScaleFunction → (1 &),
    AxesOrigin → {0, 0}, HeadLength -> 0,
    PlotPoints -> 30,
    DisplayFunction -> Identity];
```

```
sols =
 Map[
  DSolve[{y'[x] == x / y[x],
     y[#[[1]]] == #[[2]]}, y[x], x] &,
  {{-3 / 2, 0}, {0, -1}, {1, 1}}]
```

$\{\{\{y[x] \to -\sqrt{-\frac{9}{4} + x^2}\},$

$\{y[x] \to \sqrt{-\frac{9}{4} + x^2}\}\},$

```
pts = ListPlot[{{-3 / 2, 0}, {0, -1},
    PlotStyle -> PointSize[0.03],
    DisplayFunction -> Identity];
```

```
p2 = Plot[Evaluate[y[x] /. sols],
   {x, -2, 2},
   PlotStyle ->
    {{GrayLevel[.5], Thickness[0.01]}},
   DisplayFunction -> Identity];
```

```
Show[p1, pts, p2,
 DisplayFunction -> $DisplayFunction]
```

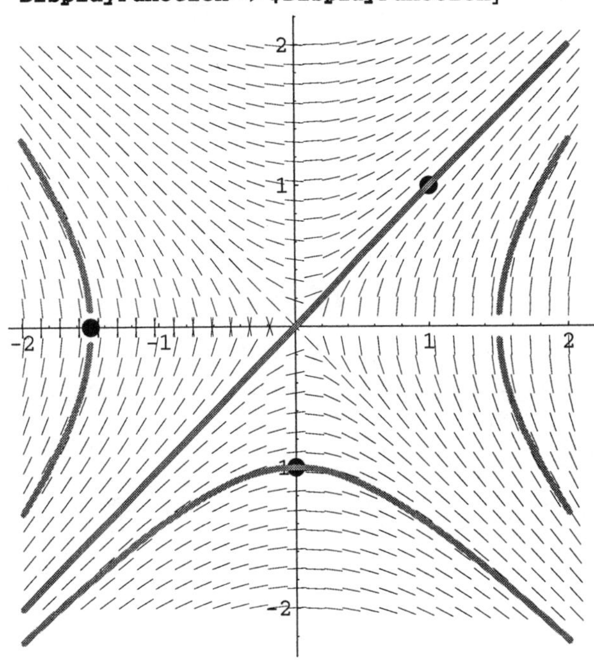

11. Integrating gives $y = -e^{-x} + C$ and applying the initial condition $y(0) = y_0$ gives

$$-e^{-0} + C = y_0 \Rightarrow C = y_0 + 1 \Rightarrow y = 1 + y_0 - e^{-x}.$$

$\lim_{x\to\infty}\left(1 + y_0 - e^{-x}\right) = 1 + y_0$. This agrees with the graphical result but is hard to see. (See below.)

Mathematica

```
<< Graphics`PlotField`
p1 = PlotVectorField[{1, Exp[-x]},
    {x, 0, 2}, {y, -1, 1}, Axes → Automatic,
    ScaleFunction → (1 &), AxesOrigin → {0, 0},
    HeadLength -> 0, PlotPoints -> 30];
```

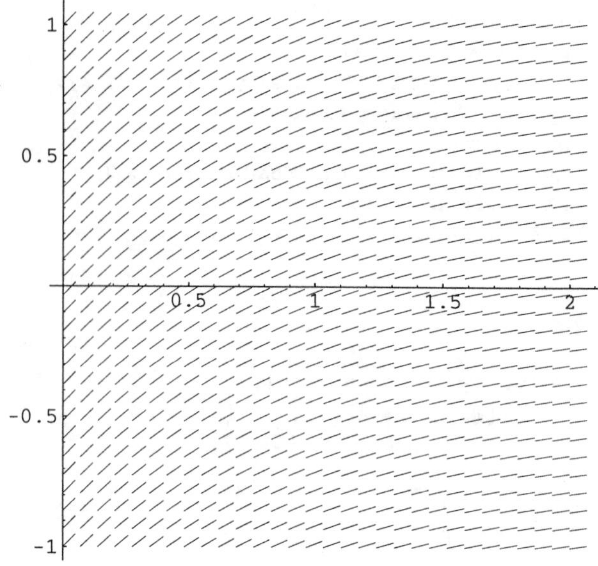

17. Let $x' = y \Rightarrow y' = (x')' = x'' = -4x' - 13x = -4y - 13x$ so $\begin{cases} x' = y \\ y' = -13x - 4y \end{cases}$.

21. **Mathematica**

```
<< Graphics`PlotField`
Clear[x, y]
DSolve[y'[x] == Exp[-x^2], y[x], x]
```

$\{\{y[x] \to C[1] + \frac{1}{2}\sqrt{\pi}\,\text{Erf}[x]\}\}$

```
sol = DSolve[{y'[x] == Exp[-x^2], y[0] == a
  y[x], x]
```

$\{\{y[x] \to \frac{1}{2}(2\,a + \sqrt{\pi}\,\text{Erf}[x])\}\}$

```
toplot = Table[sol[[1, 1, 2]], {a, -2, 2}]
```

$\{\frac{1}{2}(-4 + \sqrt{\pi}\,\text{Erf}[x]),\ \frac{1}{2}(-2 + \sqrt{\pi}\,\text{Erf}[x]),\ \frac{1}{2}\sqrt{\pi}\,\text{Erf}[x],\ \frac{1}{2}(2 + \sqrt{\pi}\,\text{Erf}[x]),\ \frac{1}{2}(4 + \sqrt{\pi}\,\text{Erf}[x])\}$

```
p1 = Plot[Evaluate[toplot], {x, 0, 6},
    PlotRange -> {-3, 3},
    AspectRatio -> Automatic,
    PlotStyle ->
     {{GrayLevel[.5], Thickness[0.01]}},
    DisplayFunction -> Identity];
```

```
p2 = PlotVectorField[{1, Exp[-x^2]},
    {x, 0, 6}, {y, -3, 3}, Axes → Automatic,
    ScaleFunction → (1 &), AxesOrigin → {0, 0},
    HeadLength -> 0, PlotPoints -> 30,
    DisplayFunction -> Identity];
```

```
Show[p2, p1, DisplayFunction ->
  $DisplayFunction]
```

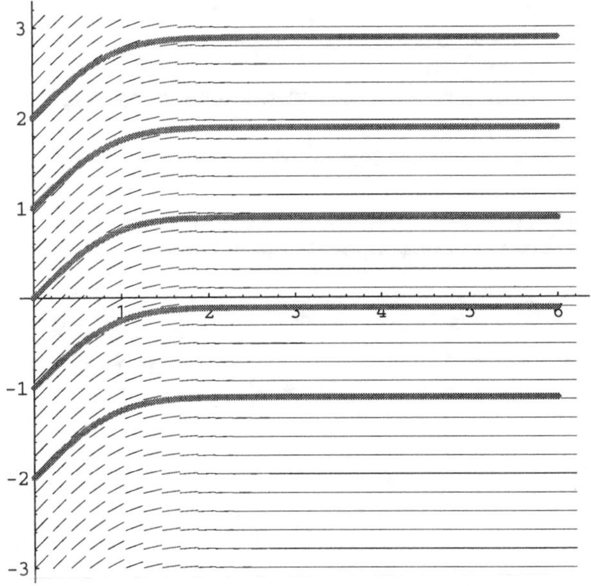

23. **Mathematica**

```
<< Graphics`PlotField`
p1 = PlotVectorField[{-y, x}, {x, -1, 1},
    {y, -1, 1}, Axes → Automatic,
    ScaleFunction → (1 &),
    AxesOrigin → {0, 0}, PlotPoints -> 20,
    DisplayFunction -> Identity];


sols =
  Map[
    DSolve[{x'[t] == -y[t], y'[t] == x[t],
        x[0] == #[[1]], y[0] == #[[2]]},
      {x[t], y[t]}, t] &,
    {{1/2, 0}, {-1/4, 0}, {0, 3/4},
     {0, -1/2}}];

p2 = ParametricPlot[
    Evaluate[{x[t], y[t]} /. sols],
    {t, -5, 5},
    PlotStyle ->
     {{GrayLevel[.5], Thickness[0.01]}},
    Compiled -> False,
    DisplayFunction -> Identity];

Show[p1, p2, PlotRange -> {{-1, 1}, {-1, 1}
 DisplayFunction -> $DisplayFunction]
```

```
p1 = PlotVectorField[{-y, -x}, {x, -1, 1},
    {y, -1, 1}, Axes → Automatic,
    ScaleFunction → (1 &),
    AxesOrigin → {0, 0}, PlotPoints -> 20,
    DisplayFunction -> Identity];


sols =
  Map[
    DSolve[{x'[t] == -y[t], y'[t] == -x[t],
        x[0] == #[[1]], y[0] == #[[2]]},
      {x[t], y[t]}, t] &,
    {{1/2, 0}, {-1/4, 0}, {0, 3/4},
     {0, -1/2}}];

p2 = ParametricPlot[
    Evaluate[{x[t], y[t]} /. sols],
    {t, -5, 5},
    PlotStyle ->
     {{GrayLevel[.5], Thickness[0.01]}},
    Compiled -> False,
    DisplayFunction -> Identity];

Show[p1, p2, PlotRange -> {{-1, 1}, {-1, 1}
 DisplayFunction -> $DisplayFunction]
```

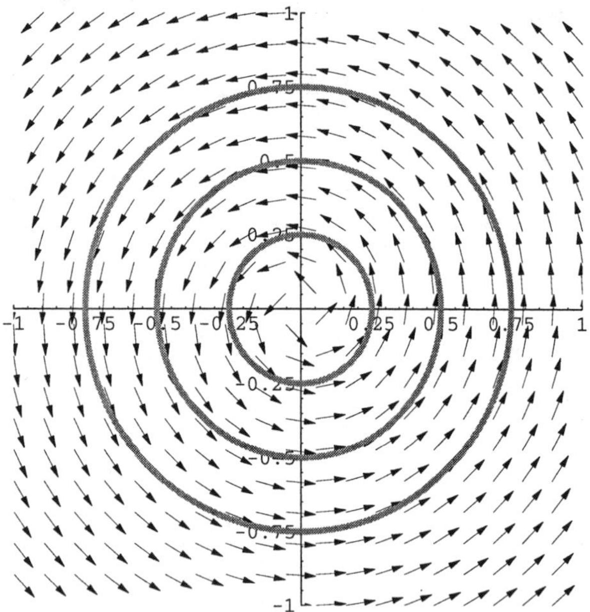

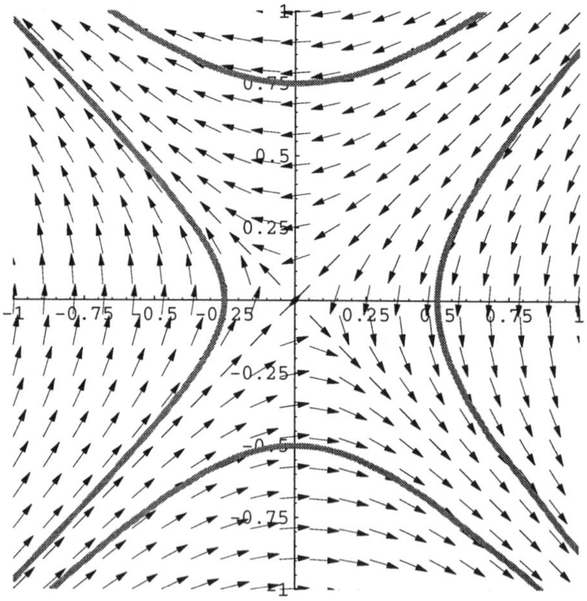

2

First-Order Ordinary Differential Equations

EXERCISES 2.1

3. Separating variables and integrating yields

$$\frac{dy}{dx}=\frac{3y^7}{x^8}$$

$$y^{-7}dy=3x^{-8}dx$$

$$-\frac{1}{6}y^{-6}=-\frac{3}{7}x^{-7}+C$$

$$\frac{1}{y^6}=\frac{18}{7x^7}+C.$$

7. $4\left(\frac{1}{4}\cosh 4y\right)=6\left(\frac{1}{3}\sinh 3x\right)+C$,

$\cosh 4y=2\sinh 3x+C$

11. $3\sin x dx=4\cos y dy$, $-3\cos x=4\sin y+C$

15. $20\sinh y dy=-(\cosh 6x+5\sinh 4x)dx$,

$20\cosh y=-\frac{1}{6}\sinh 6x-\frac{5}{4}\cosh 4x+C$

19. $-3\cos x+\frac{1}{3}\cos 3x=\frac{1}{4}\sin 4y-4\sin y+C$

23. We first rewrite the equation:

$$\tan y\sec^2 y\,dy+\cos^3 2x\sin 2x\,dx=0$$

$$\tan y\sec^2 y\,dy=-\cos^3 2x\sin 2x\,dx.$$

Then,

$$\int\tan y\sec^2 y\,dy \underset{\substack{u=\tan y\Rightarrow\\ du=\sec^2 y dy}}{=} \int u\,du=\frac{1}{2}u^2+C_1=\frac{1}{2}\tan^2 y+C_1$$

and

$$\int-\cos^3 2x\sin 2x\,dx \underset{\substack{u=\cos 2x\Rightarrow\\ -\frac{1}{2}du=\sin 2x\,dx}}{=} \frac{1}{2}\int u^3du=\frac{1}{8}u^4+C_2=\frac{1}{8}\cos^4 2x+C_2$$

so

$$\tfrac{1}{2}\tan^2 y=\tfrac{1}{8}\cos^4 2x+C \quad\text{or}\quad \tan^2 y=\tfrac{1}{4}\cos^4 2x+C.$$

27. We integrate the left and right-hand sides of the equation to obtain:

$$\int\frac{\cos y}{(1-\sin y)^2}dy \underset{\substack{u=1-\sin y\Rightarrow\\ -du=\cos y\,dy}}{=} -\int u^{-2}du=\frac{1}{u}+C_1=\frac{1}{1-\sin y}+C_1$$

and

$$\int\sin^3 x\cos x\,dx \underset{\substack{u=\sin x\Rightarrow\\ du=\cos x\,dx}}{=} \int u^3du=\frac{1}{4}u^4+C_2=\frac{1}{4}\sin^4 x+C_2.$$

Then,

$$\frac{1}{1-\sin y}=\frac{1}{4}\sin^4 x+C$$

33. First, separate variables and integrate to obtain

$$dy=\sec x\,dx$$

$$y=\ln|\sec x+\tan x|+C.$$

Application of the initial conditions yields

$$2=\ln|\sec 0+\tan 0|+C$$

$$2=\ln|1+0|+C$$

$$2=C$$

so the solution to the initial-value problem is

$y(x) = \ln|\sec x + \tan x| + 2$.

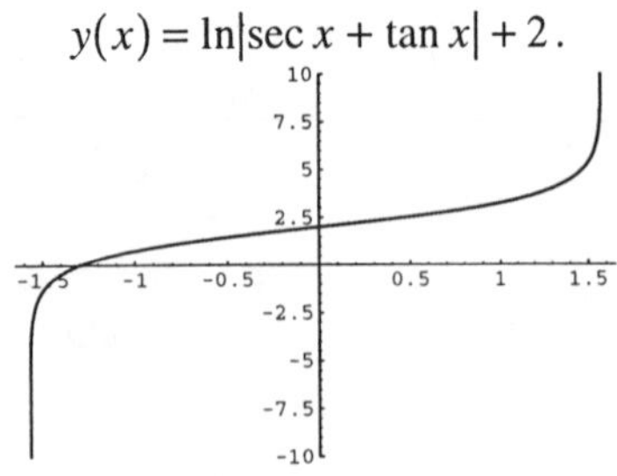

37. Integration and application of the initial condition yields $\frac{1}{2}y^2 + y = e^x - 1$. We use the quadratic formula to solve for y:

$$y = \frac{-1 \pm \sqrt{1 - 4 \cdot \frac{1}{2} \cdot \left(1 - e^x\right)}}{2 \cdot \frac{1}{2}} = -1 \pm \sqrt{2e^x - 1}.$$

Because $y(0) = -2$, the solution is $y = -1 - \sqrt{2e^x - 1}$.

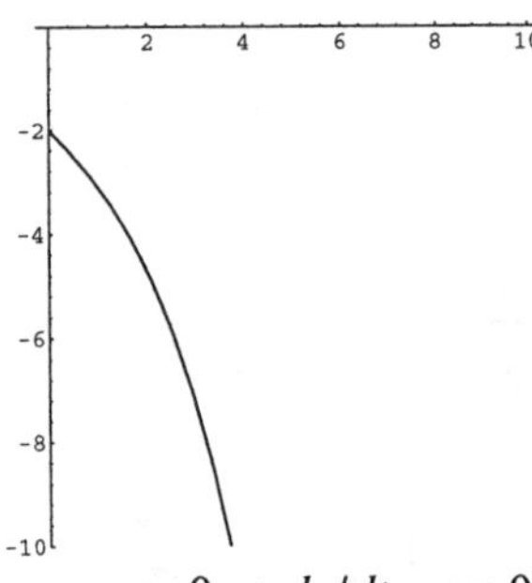

41. $y(x) = \tan^{-1} x + 1$

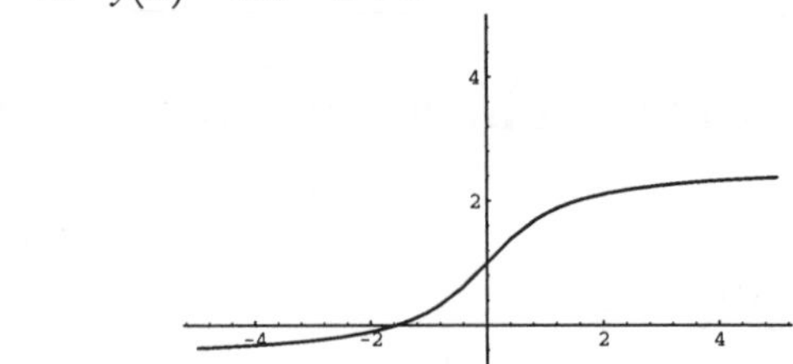

45. (a) $x = r\cos\theta \Rightarrow dx/dt = \cos\theta\, dr/dt - r\sin\theta\, d\theta/dt$ and $y = r\sin\theta \Rightarrow dy/dt = \sin\theta\, dr/dt + r\cos\theta\, d\theta/dt$. Substituting into the first equation, we have

$$\frac{dx}{dt} = \cos\theta\frac{dr}{dt} - r\sin\theta\frac{d\theta}{dt} = r\cos\theta\,(1-r) - \omega\, r\sin\theta$$

and into the second

$$\frac{dy}{dt} = \sin\theta\frac{dr}{dt} + r\cos\theta\frac{d\theta}{dt} = r\sin\theta\,(1-r) + \omega\, r\cos\theta.$$

Now, multiply the first of these two equations by $\cos\theta$ and the second by $\sin\theta$, add, and simplify the result to obtain

$$\frac{dr}{dt} = r(1-r)$$

and then multiply the first by $-\sin\theta$, the second by $\cos\theta$, add and simplify the result to obtain

$$\frac{d\theta}{dt} = \omega.$$

(b) and (c) follow directly from (a).

53. $y = 0$ is unstable; $y = 2$ is asymptotically stable.

```
p1 = PlotVectorField[{1, 16 y - 8 y^2},
    {x, -3, 3}, {y, -3, 3}, Axes → Automatic,
    ScaleFunction → (1 &),
    AxesOrigin → {0, 0}, HeadLength -> 0,
    PlotPoints -> 20];
```

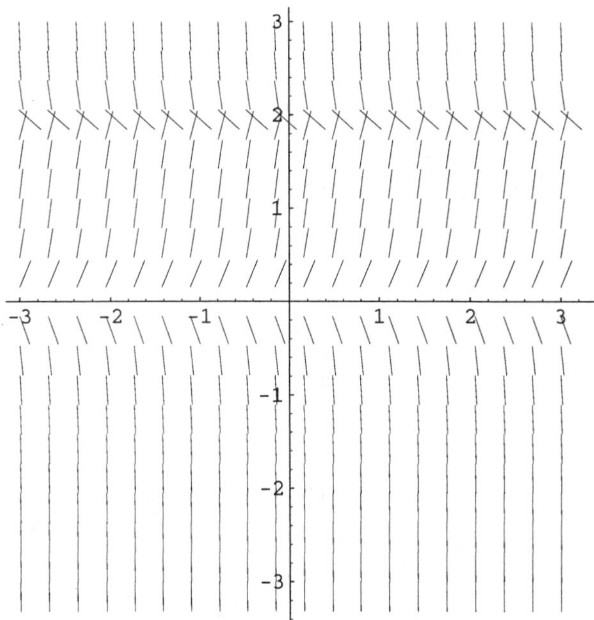

57. First, $\frac{dN}{dt} = -k\, N \Rightarrow \frac{1}{N} dN = -k\, dt \Rightarrow \ln N = -kt + C_1 \Rightarrow N = Ce^{-kt}$, where $C = e^{C_1}$. Then, $N(0) = N_0 \Rightarrow N(t) = N_0\, e^{-kt}$ and $N(D) = \frac{1}{10} N_0 = N_0 e^{-k\, D} \Rightarrow e^{-k\, D} = \frac{1}{10} \Rightarrow D = \frac{\ln 10}{k} \approx 2.30259\, k^{-1}$.

63. First, an implicit solution to the equation is found.

```
gensol=DSolve[Exp[y[x]]Cos[y[x]]y'[x]
    ==x^2/Sqrt[9-x^2],
    y[x],x]
gensol[[1,1]]
```

```
gensol:=dsolve(exp(y)*cos(y)*
    diff(y(x),x)=x^2/
    sqrt(9-x^2),y(x));
```

We then graph the solution by graphing various level curves of

$$f(x, y) = \frac{1}{2} x\sqrt{9 - x^2} - \frac{9}{2}\sin^{-1}(x/3) + \frac{1}{2} e^y (\cos y + \sin y).$$

```
toplot=gensol[[1,1]] /. y[x]->y
ContourPlot[toplot,{x,-3,3},{y,-3,3},
    Frame->False,
    ContourShading->False,
    Axes->Automatic,
    AxesOrigin->{0,0},
    PlotPoints->120]
```

```
with(plots):
contourplot(lhs(gensol),x=-3..3,
    y=-3..3,grid=[60,60],axes=NORMAL);
```

64. In this case, we are unable to compute the solution to the initial-value problem directly.

```
partsol=DSolve[{y'[x]==x^2/
    (Sqrt[9-x^2] Exp[y[x]] Cos[y[x]]),
    y[0]==0},y[x],x]
```

```
partsol:=dsolve({diff(y(x),x)=x^2/
    (sqrt(9-x^2)*exp(y)*cos(y)),
    y(0)=0},y(x));
```

However, we are able to take advantage of the general solution obtained in (63) to determine the value of the arbitrary constant so that the initial condition is satisfied, substitute into the general solution, and graph the resulting equation.

```
step1=gensol[[1]] /. {y[x]->0,x->0}
cval=Solve[step1,C[1]]
toplot=gensol[[1]] /.
    {cval[[1,1]],y[x]->y}
<<Graphics`ImplicitPlot`
ImplicitPlot[toplot,{x,-3,3},
    {y,-3,3}]
```

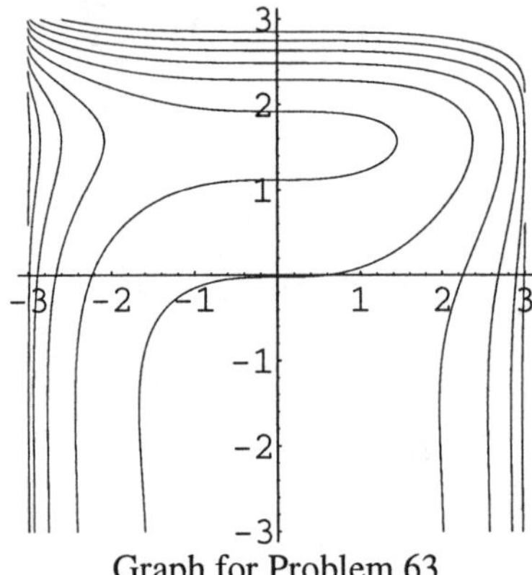

Graph for Problem 63

Remember that it is not necessary to reload the **plots** *package if you have already loaded the* **plots** *package during your current Maple session.*

```
step_1:=subs({y(x)=0,x=0},gensol);
step_2:=eval(step_1);
c_val:=solve(step_2,_C1);
toplot:=subs(_C1=c_val,gensol);
with(plots):
implicitplot(toplot,x=-3..3,
   y=-3..3,grid=[50,50]);
```

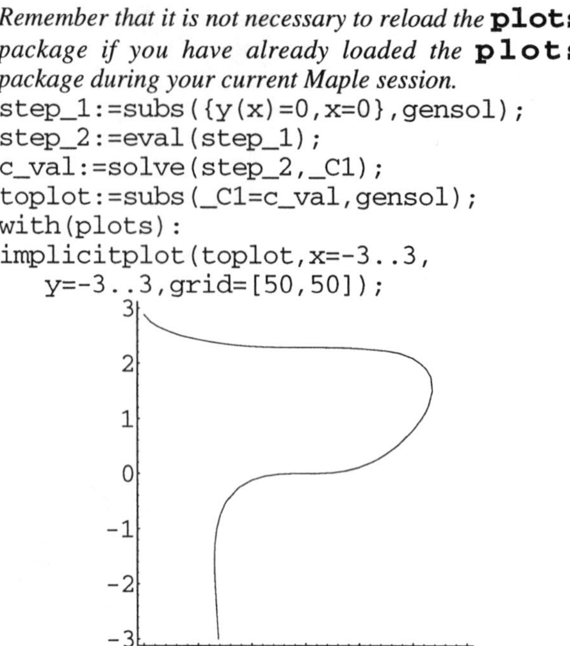

Graph for Problem 64

65. First, we determine that the solution to the initial-value problem is $y = e^{c-c/x}$.

```
Clear[x,y,c]
sol=DSolve[{y'[x]==c y[x]/x^2,
    y[1]==1},y[x],x]
```

```
y:='y':c:='c':
sol:=dsolve({diff(y(x),x)=c*y/x^2,
   y(1)=1},y(x));
```

We then graph the solution using $c = -2, -5/3, -4/3, \ldots, 5/3, 2$ for $0 < x \le 4$.

```
Clear[x,y,c]
sol=DSolve[{y'[x]==c y[x]/x^2,
    y[1]==1},y[x],x]
toplot=Table[sol[[1,1,2]],
    {c,-2,2,1/3}]
grays=Table[GrayLevel[i],
    {i,0,.7,.7/12}];
Plot[Evaluate[toplot],{x,0.01,4},
    PlotStyle->grays,
    PlotRange->{0,4},AspectRatio->1]
```

```
cvals:=seq(-2+1/3*k,k=0..12);
toplot:=seq(subs(c=k,rhs(sol)),
   k=cvals);
plot({toplot},x=0.1..4,
   view=[0..4,0..4]);
```

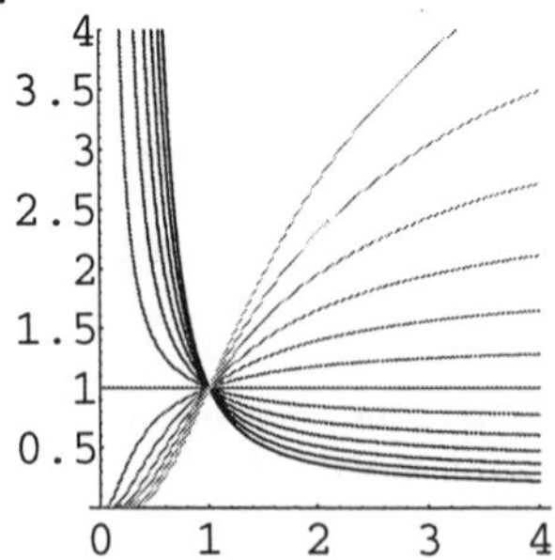

66. (a) We verify that $y(t) = 2\,e^{\int_1^t f(u)\,du}$ is the solution to the initial-value problem.

```
Clear[y]
y[t_]=2 Exp[Integrate[f[u],{u,1,t}]]
y[0]
y'[t]-y[t]f[t]
```

```
y:='y':
y:=t->2*exp(Int(f(u),u=1..t));
y(0);
diff(y(t),t)-y(t)*f(t);
```

(b) For (i) we choose $f(t)=\sin t$, (ii) we choose $f(t)=-1/t$, and (iii) we choose $f(t)=1$.

```
f[t_]=Sin[t];
y[t]
Plot[y[t],{t,0,4Pi}]
f[t_]=-1/t;
y[t]
Plot[y[t],{t,0,4Pi}]
f[t_]=1;
y[t]
Plot[y[t],{t,0,4Pi}]
```

```
f:=t->sin(t):
plot(y(t),t=0..4*Pi);
f:=t->-1/t:
plot(y(t),t=0..4*Pi);
f:=t->1:
plot(y(t),t=0..4*Pi);
```

8
6
4
2
2 4 6 8 10 12

30
25
20
15
10
5
2 4 6 8 10 12

60000
50000
40000
30000
20000
10000
2 4 6 8 10 12

(c) Let $c>0$. To have $\lim_{t\to\infty} y(t)=c$ we must have

$$\lim_{t\to\infty} 2\,e^{\int_1^t f(u)du}=c$$

$$\lim_{t\to\infty} e^{\int_1^t f(u)du}=c/2$$

$$\lim_{t\to\infty}\int_1^t f(u)\,du=\ln(c/2).$$

Thus, if F is an antiderivative of f,

$$\lim_{t\to\infty}\int_1^t f(u)\,du=\lim_{t\to\infty}\left(F(t)-F(1)\right)=\ln(c/2)$$

$$\lim_{t\to\infty} F(t)=F(1)+\ln(c/2).$$

Consider $F(t)=\ln(c/2)\dfrac{t-1}{t+1}$. Then, $\lim_{t\to\infty} F(t)=F(1)+\ln(c/2)$ and $F'(t)=\dfrac{2\ln(c/2)}{(t+1)^2}$. If we let $f(t)=\dfrac{2\ln(c/2)}{(t+1)^2}$, $\lim_{t\to\infty} y(t)=c$.

```
Clear[f,t,y]
partsol=DSolve[{y'[t]==
    y[t] 2 Log[c/2]/(t+1)^2,
    y[1]==2},y[t],t]
Limit[partsol[[1,1,2]],t->Infinity]
```

```
y:='y':
partsol:=dsolve({diff(y(t),t)=
    y*2*ln(c/2)/(t+1)^2,y(1)=2},y(t));
assign(partsol):
limit(y(t),t=infinity);
```

EXERCISES 2.2

3. An integrating factor is $\mu(x) = e^{\int 1/x\, dx} = x$ so

$$\frac{dy}{dx} + \frac{1}{x}y = e^x$$

$$x\frac{dy}{dx} + y = xe^x$$

$$\frac{d}{dx}(xy) = xe^x$$

$$xy = \int xe^x dx \underbrace{=}_{\substack{\text{integration by parts} \\ \text{with } u=x \Rightarrow du=dx \\ \text{and} \\ dv=e^x dx \Rightarrow v=e^x}} xe^x - \int e^x dx$$

$$= xe^x - e^x + C$$

$$y = e^x - \frac{e^x}{x} + \frac{C}{x}.$$

7. First, we rewrite the equation

$$dy = \left(2x + \frac{xy}{x^2 - 1}\right)dx$$

$$\frac{dy}{dx} - \frac{x}{x^2 - 1}y = 2x.$$

Then, an integrating factor is given by

$$\mu(x) = e^{\int -x/(x^2-1)\, dx} = e^{-\ln|x^2-1|/2} = 1/\sqrt{x^2 - 1}$$

so

$$\frac{dy}{dx} - \frac{x}{x^2 - 1}y = 2x$$

$$\frac{1}{\left(x^2 - 1\right)^{1/2}}\frac{dy}{dx} - \frac{x}{\left(x^2 - 1\right)^{3/2}}y = \frac{2x}{\left(x^2 - 1\right)^{1/2}}$$

$$\frac{d}{dx}\left[\frac{1}{\left(x^2 - 1\right)^{1/2}}y\right] = \frac{2x}{\left(x^2 - 1\right)^{1/2}}$$

$$\frac{1}{\left(x^2 - 1\right)^{1/2}}y = 2\left(x^2 - 1\right)^{1/2} + C$$

$$y = \sqrt{x^2 - 1}\left(2\sqrt{x^2 - 1} + C\right).$$

11. An integrating factor is given by

$$\mu(x) = e^{\int -3x/(x^2-4)\, dx} = e^{-3\ln|x^2-4|/2} = 1/\left(x^2 - 4\right)^{3/2}.$$

Then,

$$\frac{dy}{dx} - \frac{3x}{x^2-4}y = x^2$$

$$\frac{1}{\left(x^2-4\right)^{3/2}}\frac{dy}{dx} - \frac{3x}{\left(x^2-4\right)^{5/2}}y = \frac{x^2}{\left(x^2-4\right)^{3/2}}$$

$$\frac{dy}{dx}\left[\frac{1}{\left(x^2-4\right)^{3/2}}y\right] = \frac{x^2}{\left(x^2-4\right)^{3/2}}.$$

To evaluate $\int \frac{x^2}{\left(x^2-4\right)^{3/2}}dx$ we use a trigonometric substitution with $x = 2\sec\theta$:

$$\int \frac{x^2}{\left(x^2-4\right)^{3/2}}dx = \int \frac{4\sec^2\theta}{\left(4\sec^2\theta - 4\right)^{3/2}}2\sec\theta\tan\theta\, d\theta = \int \frac{8\sec^3\theta\tan\theta}{\left(4\tan^2\theta\right)^{3/2}}d\theta$$

$$= \int \frac{\sec^3\theta}{\tan^2\theta}d\theta = \int \frac{\left(\tan^2\theta+1\right)\sec\theta}{\tan^2\theta}d\theta = \int\left(\sec\theta + \frac{\sec\theta}{\tan^2\theta}\right)d\theta$$

$$= \int\left(\sec\theta + \frac{\cos\theta}{\sin^2\theta}\right)d\theta = \ln\left|\sec\theta+\tan\theta\right| - \frac{1}{\sin\theta} + C.$$

Because $x = 2\sec\theta$, $\tan\theta = \sqrt{\sec^2\theta - 1} = \sqrt{x^2-4}/2$ and $\sin\theta = \tan\theta\cos\theta = \frac{\tan\theta}{\sec\theta} = \frac{\sqrt{x^2-4}}{x}$. Therefore,

$$\int \frac{x^2}{\left(x^2-4\right)^{3/2}}dx = \ln\left|\sec\theta+\tan\theta\right| - \frac{1}{\sin\theta} + C = \ln\left|\frac{x+\sqrt{x^2-4}}{2}\right| - \frac{x}{\sqrt{x^2-4}} + C$$

$$= \ln\left|x+\sqrt{x^2-4}\right| - \frac{x}{\sqrt{x^2-4}} + C,$$

where we take advantage of $\ln\left|\frac{x+\sqrt{x^2-4}}{2}\right| = \ln\left|x+\sqrt{x^2-4}\right| - \underbrace{\ln 2}_{\text{this is a constant}}$. Thus,

$$\frac{1}{\left(x^2-4\right)^{3/2}}y = \ln\left|x+\sqrt{x^2-4}\right| - \frac{x}{\sqrt{x^2-4}} + C$$

$$y = \left(x^2-4\right)^{3/2}\ln\left|x+\sqrt{x^2-4}\right| - x\left(x^2-4\right) + C\left(x^2-4\right)^{3/2}.$$

15. An integrating factor is given by

$$\mu(x) = e^{\int -9x/\left(9x^2+49\right)dx} = 1/\sqrt{9x^2+49}.$$

Then,

$$\frac{dy}{dx}-\frac{9x}{9x^2+49}y=x$$

$$\frac{1}{\left(9x^2+49\right)^{1/2}}\frac{dy}{dx}-\frac{9x}{\left(9x^2+49\right)^{3/2}}y=\frac{x}{\left(9x^2+49\right)^{1/2}}$$

$$\frac{d}{dx}\left[\frac{1}{\left(9x^2+49\right)^{1/2}}y\right]=\frac{x}{\left(9x^2+49\right)^{1/2}}$$

$$\frac{1}{\left(9x^2+49\right)^{1/2}}y=\frac{1}{9}\sqrt{9x^2+49}+C$$

$$y=\frac{1}{9}\left(9x^2+49+C\right)\sqrt{9x^2+49}.$$

19. Rewrite the equation so that it is linear in x:

$$\frac{dy}{dx}=\frac{1}{y^2+x}$$

$$\frac{dx}{dy}=y^2+x$$

$$\frac{dx}{dy}-x=y^2.$$

Then, an integrating factor is given by $\mu(y)=e^{\int -1\cdot dy}=e^{-y}$ so

$$\frac{dx}{dy}-x=y^2$$

$$e^{-y}\frac{dx}{dy}-e^{-y}x=y^2e^{-y}$$

$$\frac{d}{dy}\left(e^{-y}x\right)=y^2e^{-y}$$

$$e^{-y}x=\int y^2e^{-y}\,dy \underbrace{=}_{\substack{\text{integration by parts with}\\ u=y^2\Rightarrow du=2y\,dy\\ \text{and}\\ dv=e^{-y}dy\Rightarrow v=-e^{-y}}} =-y^2e^{-y}+\underbrace{2\int ye^{-y}\,dy}_{\substack{\text{use integration}\\ \text{by parts with}\\ u=y\Rightarrow du=dy\\ \text{and}\\ dv=e^{-y}dy\Rightarrow v=-e^{-y}}}$$

$$=-y^2e^{-y}+2\left(-ye^{-y}+\int e^{-y}dy\right)=e^{-y}\left(-y^2-2y-2\right)+C$$

$$x=-y^2-2y-2+C\,e^{y}.$$

23. First, we rewrite the equation in the form $\frac{dp}{dt}=t^3+\frac{p}{t}\Rightarrow\frac{dp}{dt}-\frac{1}{t}p=t^3$.

Then, an integrating factor is given by $\mu(t) = e^{\int -1/t\, dt} = 1/t$ so

$$\frac{dp}{dt} - \frac{1}{t}p = t^3$$

$$\frac{1}{t}\frac{dp}{dt} - \frac{1}{t^2}p = t^2$$

$$\frac{d}{dt}\left(\frac{1}{t}p\right) = t^2$$

$$\frac{1}{t}p = \frac{1}{3}t^3 + C$$

$$p = \frac{1}{3}t^4 + Ct.$$

27. An integrating factor is

$$\mu(x) = e^{\int 3x^2 dx} = e^{x^3}$$

so a general solution is

$$\frac{dy}{dx} + 3x^2 y = e^{-x^3}$$

$$e^{x^3}\frac{dy}{dx} + 3x^2 e^{x^3} y = 1$$

$$\frac{d}{dx}\left(e^{x^3} y\right) = 1$$

$$e^{x^3} y = x + C$$

$$y = xe^{-x^3} + Ce^{-x^3}.$$

Application of the initial condition results in

$$2 = 0 \bullet e^{-0^3} + Ce^{-0^3}$$

$$C = 2$$

so

$$y = 2e^{-x^3} + x\,e^{-x^3}.$$

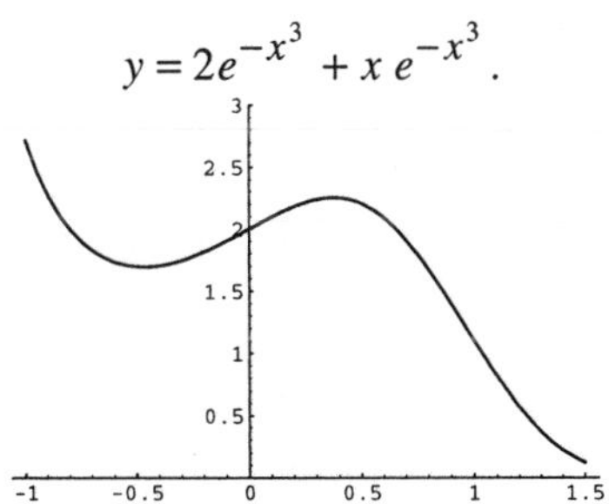

31. An integrating factor is

$$\mu(x) = e^{\int e^x/(e^x+1)\, dx} = e^x + 1$$

so a general solution of the differential equation is

$$\begin{aligned}\frac{dy}{dx}+\frac{e^x}{e^x+1}y&=\frac{x}{e^x+1}\\ \left(e^x+1\right)\frac{dy}{dx}+e^xy&=x\\ \frac{d}{dx}\left[\left(e^x+1\right)y\right]&=x\\ \left(e^x+1\right)y&=\frac{1}{2}x^2+C\\ y&=\left(e^x+1\right)^{-1}\left(\frac{1}{2}x^2+C\right).\end{aligned}$$

Application of the initial condition results in

$$\begin{aligned}1&=\left(e^0+1\right)^{-1}\left(\frac{1}{2}0^2+C\right)\\ 1&=\frac{1}{2}C\\ C&=2\end{aligned}$$

so

$$y=\frac{x^2}{2\left(e^x+1\right)}+\frac{2}{e^x+1}.$$

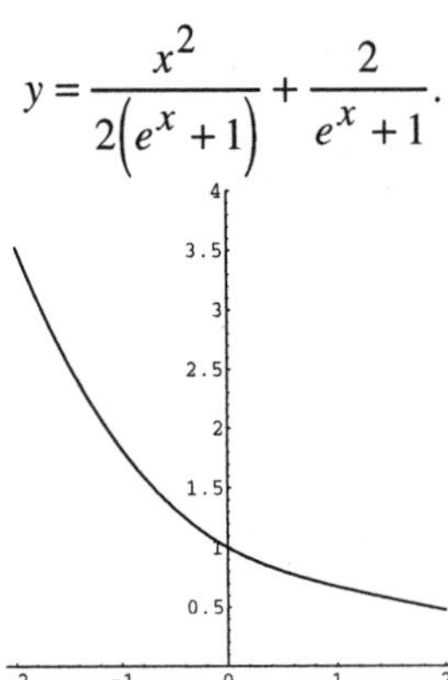

37.

$$\begin{aligned}\frac{d}{dx}\left(y_1(x)+y_2(x)\right)+p(x)\left(y_1(x)+y_2(x)\right)&=y_1'(x)+y_2'(x)+p(x)y_1(x)+p(x)y_2(x)\\ &=\underbrace{y_1'(x)+p(x)y_1(x)}_{=r(x)}+\underbrace{y_2'(x)+p(x)y_2(x)}_{=q(x)}\\ &=r(x)+q(x)\end{aligned}$$

45. First, we solve $\begin{cases}dy/dx+2y=0\\ y(0)=1\end{cases}$:

$$\frac{d}{dx}\left(e^{2x}y\right)=0$$

$$y=Ce^{-2x};\ y(0)=1\Rightarrow C=1.$$

Because $y(1)=e^{-2}$, we now solve $\begin{cases}dy/dx+4y=0\\ y(1)=e^{-2}\end{cases}$:

$$\frac{d}{dx}\left(e^{4x}y\right)=0$$

$$y=Ce^{-4x};\ y(1)=e^{-2}\Rightarrow C=e^2.$$

Thus, the solution to the initial-value problem is

$$y=\begin{cases}e^{-2x}, 0\le x\le 1\\ e^2e^{-4x}, x>1\end{cases}.$$

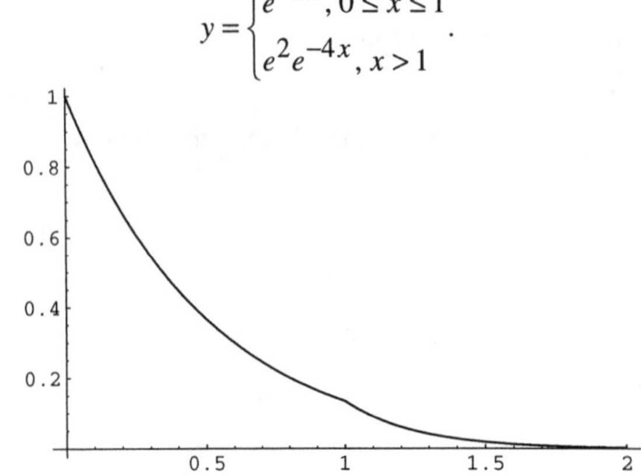

49. The corresponding homogeneous equation is $y' + y = 0$, which is separable:

$$\frac{dy}{dx} = -y$$
$$\frac{1}{y}dy = -dx$$
$$\ln|y| = -x + C$$
$$y_h = Ce^{-x}.$$

$y_p = Axe^{-x} \Rightarrow y_p' = Ae^{-x} - Axe^{-x}$ and substituting into the nonhomogeneous equation yields

$$y_p' + y_p = Ae^{-x} - Axe^{-x} + Axe^{-x}$$
$$= Ae^{-x} = e^{-x} \Rightarrow A = 1$$

so $y_p = xe^{-x}$ and a general solution to the nonhomogeneous equation is $y = y_h + y_p = Ce^{-x} + xe^{-x}$.

Because Ae^{-x} is a solution to the corresponding homogeneous equation.

53. A general solution of the corresponding homogeneous equation is $y_h = Ce^{x/2}$. $y_p = A\cos x + B\sin x + Ce^x \Rightarrow y_p' = B\cos x - A\sin x + Ce^x$ and substituting into the nohomogeneous equation yields

$$y_p' - \frac{1}{2}y_p = A\cos x + B\sin x + Ce^x - \frac{1}{2}\left(B\cos x - A\sin x + Ce^x\right)$$
$$= \left(-\frac{1}{2}A + B\right)\cos x + \left(-A - \frac{1}{2}B\right)\sin x + \frac{1}{2}Ce^x$$
$$= 5\cos x + 2e^x.$$

Equating coefficients gives us

$$\begin{cases} -\frac{1}{2}A + B = 5 \\ -A - \frac{1}{2}B = 0 \Rightarrow A = -2,\ B = 4,\ C = 2 \\ \frac{1}{2}C = 1 \end{cases}$$

so $y_p = -2\cos x + 4\sin x + 2e^x$ and a general solution to the nonhomogeneous equation is

$$y = Ce^{x/2} - 2\cos x + 4\sin x + 2e^x.$$

55. By inspection, we see that a general solution to the corresponding homogeneous equation is $y_h = Ce^{-10x}$ so we search for a particular solution to the nonhomogeneous equation of the form $y_p = Ae^x \Rightarrow y_p' = Ae^x$. Substituting into the nonhomogeneous equation and equating coefficients gives us

$$y_p' + 10y_p = Ae^x + 11Ae^x = 12Ae^x = 2e^x$$
$$\Rightarrow 12A = 2 \Rightarrow A = 1/6$$

so a particular solution is $y_p = \frac{1}{6}e^x$ and a general solution to the nonhomogeneous equation is $y = Ce^{-10x} + \frac{1}{6}e^x$.

59. $y_h = Ce^{-x}$; $y_p = A\cos x + B\sin x + Cx + D \Rightarrow y_p' = B\cos x - A\sin x + C \Rightarrow$

$$\begin{aligned} y_p' + y_p &= B\cos x - A\sin x + C + A\cos x + B\sin x + Cx + D \\ &= (A+B)\cos x + (-A+B)\sin x + Cx + (C+D) \\ &= 2\cos x + x \\ &\Rightarrow \begin{cases} A+B=2 \\ -A+B=0 \\ C=1 \\ C+D=0 \end{cases} \Rightarrow A=1,\ B=1,\ C=1,\ D=-1 \end{aligned}$$

so $y_p = \cos x + \sin x + x - 1$ and $y = y_h + y_p = Ce^{-x} + \cos x + \sin x + x - 1$.

62. First, we find a general solution and then graph it for various values of the arbitrary constant.

```
gensol=DSolve[x y'[x]+y[x]==
    x Cos[x],y[x],x]
toplot=Table[gensol[[1,1,2]] /. C[1]->a,
    {a,-7,7}];
grays=Table[GrayLevel[i],{i,0,.7,.7/14}];
Plot[Evaluate[toplot],{x,0,2Pi},
    PlotRange->{-10,10},PlotStyle->grays]
```

```
y:='y':
gensol:=dsolve(x*diff(y(x),x)+y=
   x*cos(x),y(x));
toplot:=seq(rhs(gensol),_C1=-7..7);
plot({toplot},x=0..2*Pi,
   view=[0..2*Pi,-10..10]);
```

63. We use the built-in commands (`DSolve` for Mathematica and `dsolve` for Maple V) to solve all four equations at the same time.

```
solutions=
    Map[DSolve[{y'[x]+y[x]==#,
        y[0]==0},y[x],x][[1,1,2]]&,
    {x,Sin[x],Cos[x],Exp[x]}]
grays=Table[GrayLevel[i],{i,0,.7,.7/3}];
Plot[Evaluate[solutions],{x,-2,6},
    PlotRange->{-4,4},
    PlotStyle->grays,AspectRatio->1]
```

```
x:='x':y:='y':
eqs:={diff(y(x),x)+y=x,
   diff(y(x),x)+y=sin(x),
   diff(y(x),x)+y=cos(x),
   diff(y(x),x)+y=exp(x)};
sol:=eqn->rhs(dsolve({eqn,y(0)=0},
   y(x))):
solutions:=map(sol,eqs);
plot(solutions,x=-2..6,-4..4);
```

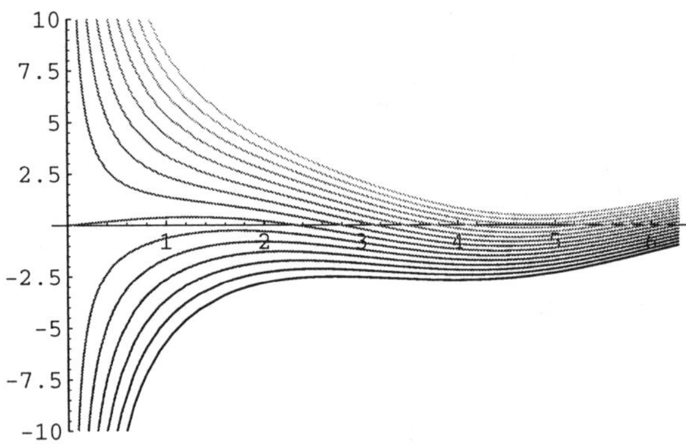

Graph for Exercise 62

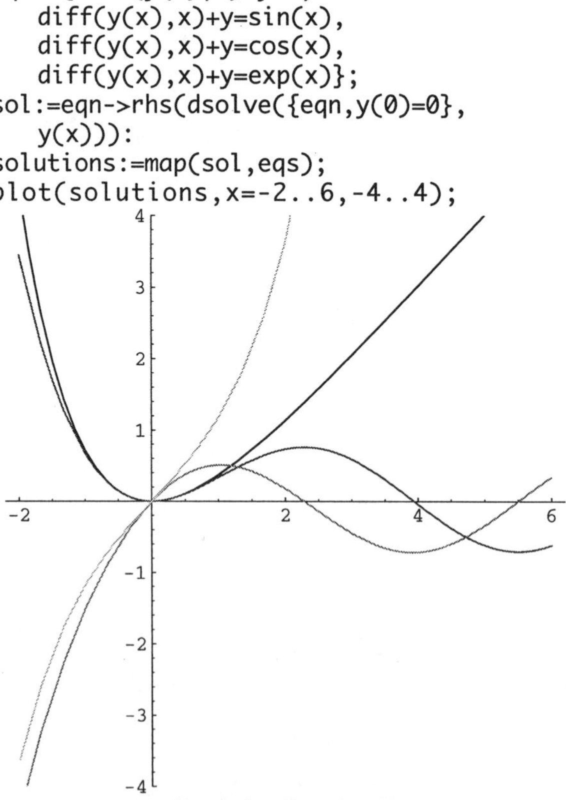

Graph for Exercise 63

64.

```
solutions=Map[DSolve[{y'[x]+# y[x]==x,
    y[0]==1},y[x],x][[1,1,2]]&,
    {-2,-1,0,1,2}]
grays=Table[GrayLevel[i],{i,0,.7,.7/4}];
Plot[Evaluate[solutions],{x,-3,3},
    PlotRange->{0,6},AspectRatio->1,
    PlotStyle->grays]
```

```
eqs:={seq(diff(y(x),x)+k*y=x,
    k=-2..2)};
sol:=eqn->rhs(dsolve({eqn,y(0)=1},
    y(x))):
solutions:=map(sol,eqs);
plot(solutions,x=-3..3,0..6);
```

65.

```
solutions=Map[DSolve[{y'[x]+y[x]==x,
    y[0]==#},y[x],x][[1,1,2]]&,
    {-2,-1,0,1,2}]
grays=Table[GrayLevel[i],{i,0,.7,.7/4}];
Plot[Evaluate[solutions],{x,-4,4},
    PlotRange->{-4,4},AspectRatio->1,
    PlotStyle->grays]
```

```
solutions:={seq(
    rhs(dsolve({diff(y(x),x)+y=x,
        y(0)=k},y(x))),k=-2..2)};
plot(solutions,x=-4..4,-4..4);
```

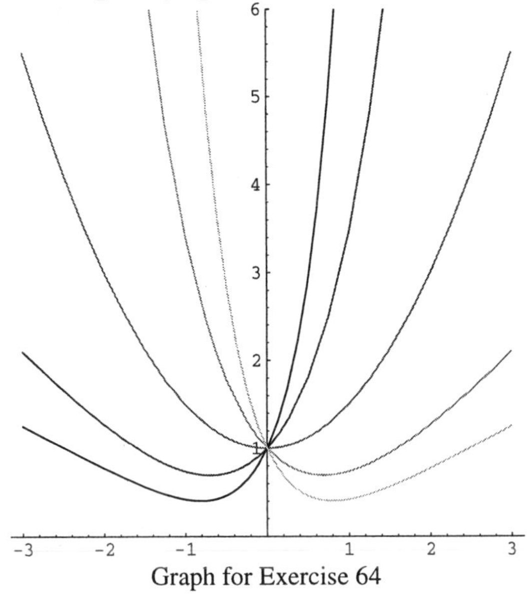

Graph for Exercise 64

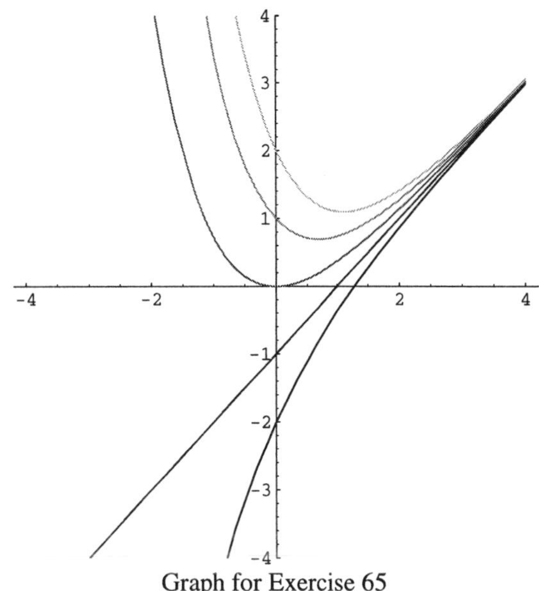

Graph for Exercise 65

66.

```
<< Graphics'PlotField'
p1 = PlotVectorField[{1, y - x}, {x, -2, 2
  {y, -1, 3},
    Axes -> Automatic, AxesOrigin -> {0,
    HeadLength -> 0, HeadWidth -> 0,
    ScaleFunction -> (1 &),
  AspectRatio -> 1]
```

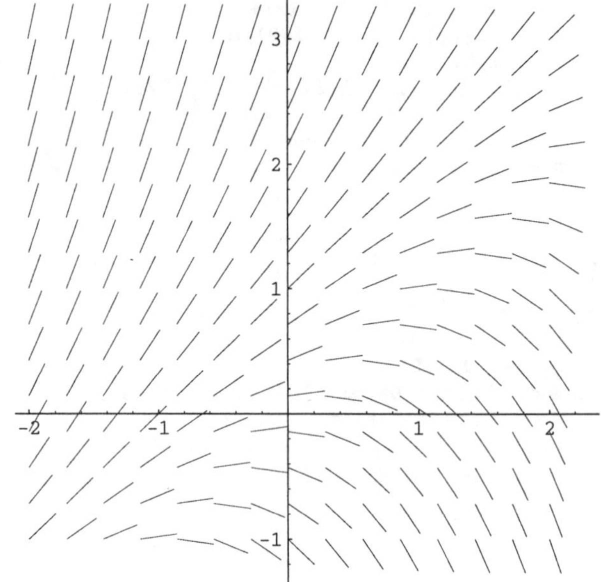

```
solutions =
  (DSolve[{y'[x] == y[x] - x, y[0] == #1},
       y[x], x][[1, 1, 2]] &) /@ {1, 1.1, 0.9}
p2 = Plot[Evaluate[solutions], {x, -2, 2},
   PlotRange → {-1, 3},
   PlotStyle → {Thickness[0.01]},
   AspectRatio → 1]
```

$\{1 + x,\ 1. + 0.1\, e^{x} + x,\ 1. - 0.1\, e^{x} + x\}$

```
Show[p1, p2, PlotRange -> {{-2, 2},
```

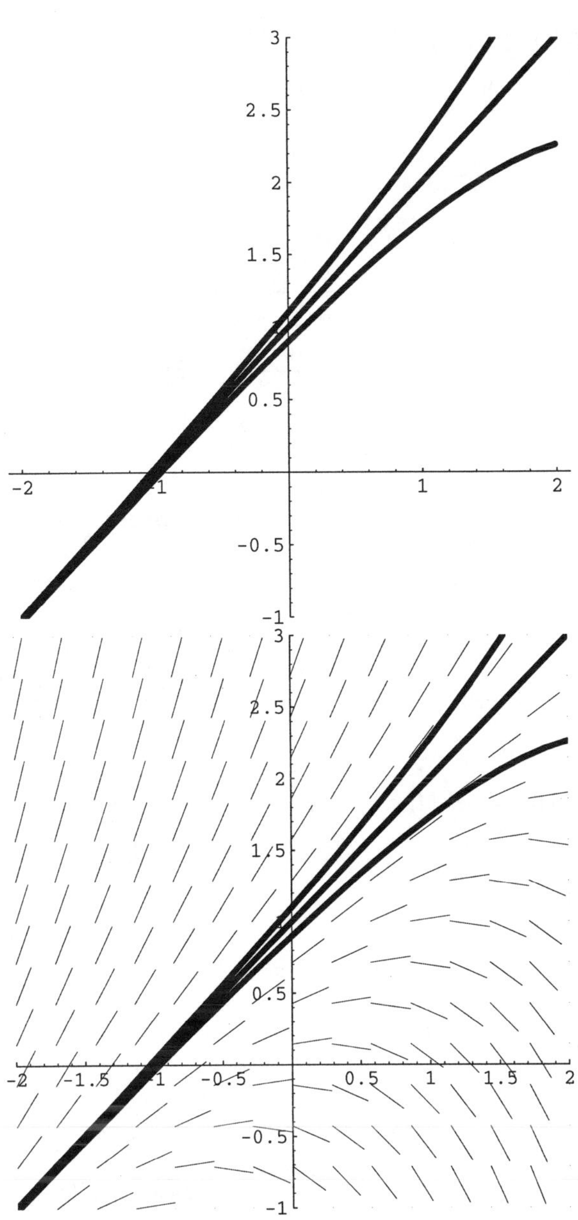

EXERCISES 2.3

3. We let $w = y^{1-3} = y^{-2}$. Then, $dw/dx = -2y^{-3}\,dy/dx \Rightarrow dy/dx = -\frac{1}{2}y^3\,dw/dx$ so

$$y' - \frac{1}{2x}y = y^3\cos x$$
$$-\tfrac{1}{2}y^3\frac{dw}{dx} - \frac{1}{2x}y = y^3\cos x$$
$$\frac{dw}{dx} + \frac{1}{x}w = -2\cos x$$
$$x\frac{dw}{dx} + w = -2x\cos x \qquad \left(\mu(x) = e^{\int \frac{1}{x}dx} = e^{\ln x} = x,\ x>0\right)$$
$$\frac{d}{dx}(x\,w) = -2x\cos x$$
$$x\,w = \int -2x\cos x\,dx \underbrace{=}_{\substack{\text{integration by parts}\\ \text{with } u=x\Rightarrow du=dx \text{ and}\\ dv=\cos x\,dx \Rightarrow v=\sin x}} -2x\sin x - 2\cos x + C$$
$$w = -2\sin x - 2x^{-1}\cos x + C\,x^{-1}.$$

Substituting $w = 1/y^2$ yields $\dfrac{1}{y^2} = -\dfrac{2\cos x + 2x\sin x + C}{x}$.

9. In this case, $M(x,y) = \sqrt{x^2 + xy}$ and $N(x,y) = -xy$. Because

$$M(tx,ty) = \sqrt{(tx)^2 + tx\cdot ty} = t\sqrt{x^2+xy} = t\,M(x,y)$$

and

$$N(tx,ty) = -tx\cdot ty = t^2\cdot -xy = t^2N(x,y)$$

the equation is not homogeneous.

13. No

19. Letting $y = ux$ we have, $dy = xdu + udx$. Then,

$$\left(xy - y^2\right)dx + x(x-3y)dy = 0$$
$$x^2u(1-u)dx + x(x-3xu)(x\,du + u\,dx) = 0$$
$$2x^2u(1-2u)dx + x^3(1-3u)du = 0$$
$$2x^2u(1-2u)dx = x^3(3u-1)du$$
$$\frac{2}{x}dx = \frac{3u-1}{u(1-2u)}du = \left(\frac{1}{1-2u} - \frac{1}{u}\right)du$$
$$2\ln|x| = -\frac{1}{2}\ln|1-2u| - \ln|u| + C.$$

Because $u = y/x$ we can rewrite the solution as

$$2\ln|x| = -\frac{1}{2}\ln|1-2y/x| - \ln|y/x| + C$$

$$\ln x^2 = \ln\left|\frac{1}{\sqrt{1-2y/x}\cdot\frac{y}{x}}\right| + C$$

$$\ln x^2 = \ln\left|\frac{x}{y\sqrt{1-2y/x}}\right| + C$$

$$x^2 = C\frac{x}{y\sqrt{1-2y/x}}$$

$$x^2 = C\frac{x}{xy^2-2y^3}$$

$$x^2y^2 - 2xy^3 = C \qquad or \qquad \frac{1}{2}x^2y^2 - xy^3 = C.$$

We use Mathematica to graph $x^2y^2 - 2xy^3 = C$ for various values of C.

```
ContourPlot[x^2 y^2 - 2 x y^3, {x, -10, 10},
 {y, -10, 10}, PlotPoints -> 300,
 Contours -> 30, ContourShading -> False,
 Frame -> False, Axes -> Automatic,
 AxesOrigin -> {0, 0}]
```

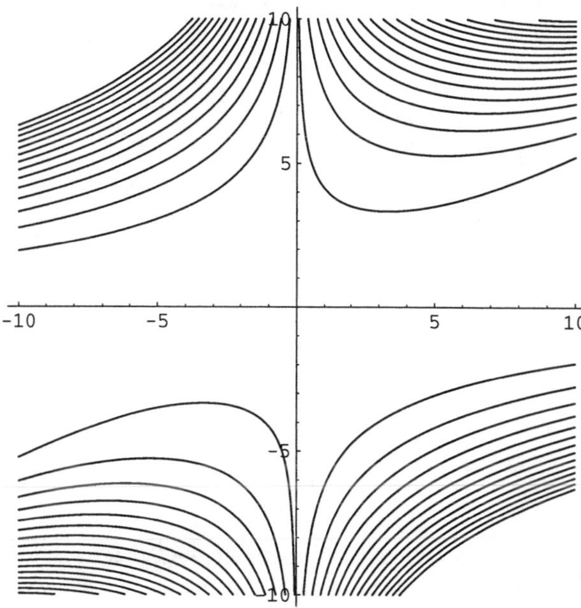

23. This is homogeneous of degree 1. Letting $y = ux \Rightarrow dy = xdu + udx$ and substituting into the equation yields

$$(x-ux)dx + x(xdu+udx) = 0$$

$$(1-u)dx + (xdu+udx) = 0$$

$$dx = -xdu$$

$$\frac{1}{x}dx = -du$$

$$\ln|x| = -u + C$$

$$\ln|x| = -\frac{y}{x} + C$$

$$y = -x\ln|x| + Cx.$$

Observe that the equation is also linear:

$$(x-y)dx + xdy = 0$$
$$x\frac{dy}{dx} - y = -x$$
$$\frac{dy}{dx} - \frac{1}{x}y = -1.$$

An integrating factor is $e^{\int -1/x\,dx} = 1/x$. Then,

$$\frac{1}{x}\frac{dy}{dx} - \frac{1}{x^2}y = -\frac{1}{x}$$
$$\frac{d}{dx}\left(\frac{1}{x}y\right) = -\frac{1}{x}$$
$$\frac{1}{x}y = -\ln|x| + C$$
$$y = -x\ln|x| + Cx.$$

Notice that we obtain the same result when we solve the equation by viewing it as a linear equation as we obtained when solving it as a homogeneous equation.

27. Let $x = v\,y$ so $dx = v\,dy + y\,dv$. Then,

$$y^2dx - \left(xy - 4x^2\right)dy = 0$$
$$y^2(y\,dv + v\,dy) - y^2v(1-4v)dy = 0$$
$$y^3dv + 4v^2y^2dy = 0$$
$$y^3dv = -4v^2y^2dy$$
$$\frac{1}{v^2}dv = -\frac{4}{y}dy$$
$$-\frac{1}{v} = -4\ln|y| + C$$
$$-\frac{y}{x} = -4\ln|y| + C,$$

which can be written as $y = C\,e^{y/(4x)}$ or $ye^{-y/(4x)} = C$.

29. Dividing both sides of the equation by $x+y$ yields $(x-y)\,dy + y\,dx = 0$. This also shows that $y = -x$ is a singular solution of the original equation. Let $x = v\,y$ so $dx = v\,dy + y\,dv$. Then,

$$(x-y)dy + y\,dx = 0$$
$$y(v-1)dy + y(v\,dy + y\,dv) = 0$$
$$y^2dv = y(1-2v)dy$$
$$\frac{1}{1-2v}dv = \frac{1}{y}dy$$
$$-\tfrac{1}{2}\ln|1-2v| = \ln|y| + C_1$$
$$\ln\sqrt{\frac{y}{y-2x}} = \ln|y| + C_1$$
$$\sqrt{\frac{y}{y-2x}} = C_2y \Rightarrow \frac{1}{y-2x} = C_3y \Rightarrow y^2 - 2xy = C.$$

```
ContourPlot[y^2 - 2 x y, {x, -10, 10},
  {y, -10, 10}, PlotPoints → 300,
  Contours → 30, ContourShading → False,
  Frame → False, Axes → Automatic,
  AxesOrigin → {0, 0}]
```

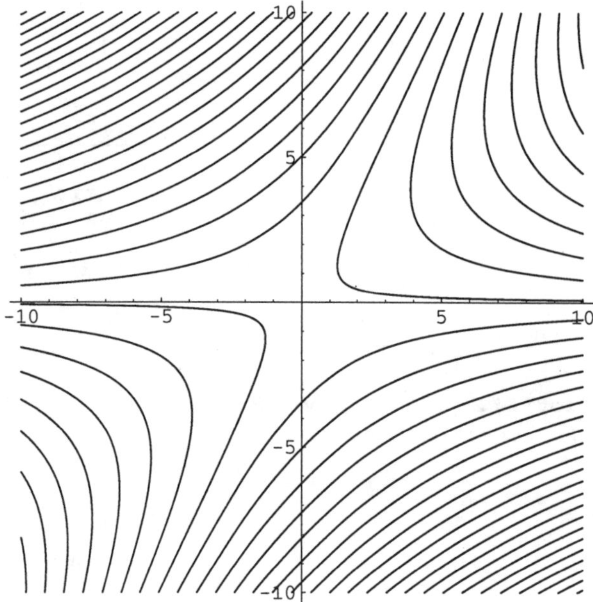

31.

$$x = vy \Rightarrow dx = vdy + ydv \Rightarrow \frac{dx}{dy} = v + y\frac{dv}{dy}$$

$$v + y\frac{dv}{dy} = \frac{2}{v}e^{-v} + v \Rightarrow ve^{v}dv = \frac{2dy}{y} \Rightarrow ve^{v} - e^{v} = 2\ln|y| + C$$

$$e^{x/y}\left(\frac{x}{y} - 1\right) = 2\ln|y| + C$$

35. Let $y = ux \Rightarrow dy = xdu + udx$. Substituting into the equation yields

$$x(xdu + udx) - \left(ux + \sqrt{x^2 + (ux)^2}\right)dx = 0$$

$$(xdu + udx) - \left(u + \sqrt{1 + u^2}\right)dx = 0$$

$$\frac{1}{x}dx = \frac{1}{\sqrt{1 + u^2}}du$$

$$\ln|x| = \ln\left|u + \sqrt{1 + u^2}\right| + C$$

$$\ln|x| = \ln\left|\frac{y}{x} + \sqrt{1 + \left(\frac{y}{x}\right)^2}\right| + C.$$

Application of the initial condition gives us $\ln|1| = \ln|1| + C \Rightarrow C = 0$

Thus,

$$\frac{y}{x}+\sqrt{1+\left(\frac{y}{x}\right)^2}=x$$

$$\frac{y+\sqrt{x^2+y^2}}{x}=x$$

$$\sqrt{x^2+y^2}=x^2-y$$

$$x^2+y^2=x^4-2x^2y+y^2$$

$$y=\frac{1}{2}\left(x^2-1\right).$$

```
Plot[1 / 2 (x^2 - 1), {x, -2, 2},
 AspectRatio -> Automatic]
```

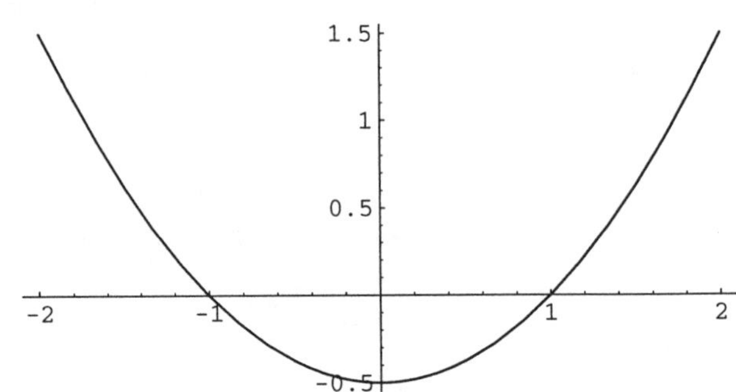

39.

$$y=u\,x\Rightarrow y^4dx+\left(x^4-xy^3\right)dx=0\Leftrightarrow(u\,x)^4\,dx+\left(x^4-x\,u^3x^3\right)(u\,dx+x\,du)=0$$

$$\Rightarrow x^4u\,dx+x^5\left(1-u^3\right)du=0\Rightarrow\frac{1}{x}dx=\frac{u^3-1}{u}du\Rightarrow\ln|x|=\frac{1}{3}u^3-\ln|u|+C$$

$$\Rightarrow\ln|x|=\frac{1}{3}(y/x)^3-\ln|y/x|+C$$

$$y(1)=2\Rightarrow 0=\frac{1}{3}\cdot 8-\ln 2+C\Rightarrow C=\frac{1}{3}(3\ln 2-8)\approx -1.07352$$

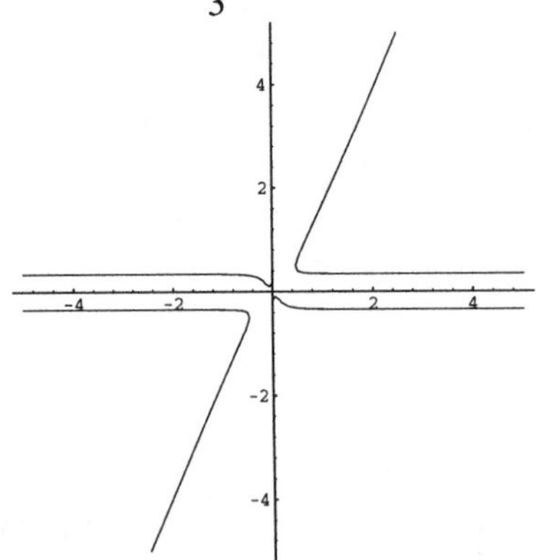

40. (a) First we solve

$$\begin{cases} h-2k+1=0 \\ 4h-3k-6=0 \end{cases}\Rightarrow(h,k)=(3,2).$$

Let $x=X+3\Rightarrow dx=dX$ and $y=Y+2\Rightarrow dy=dY$. Then,

$$(x-2y+1)dx+(4x-3y-6)dy=0$$
$$\big(X+3-2(Y+2)+1\big)dX+\big(4(X+3)-3(Y+2)-6\big)dY=0$$
$$(X-2Y)dX+(4X-3Y)dY=0$$
$$(X-2uX)dX+(4X-3uX)(Xdu+udX)=0 \qquad (Y=uX \Rightarrow dY=udX+Xdu)$$
$$(1-2u)dX+(4-3u)(Xdu+udX)=0$$
$$\left(1+2u-3u^2\right)dX+(4-3u)Xdu=0$$
$$\frac{1}{X}dX=\frac{4-3u}{3u^2-2u-1}du$$
$$\frac{1}{X}dX=\left(\frac{1}{4}\frac{1}{u-1}-\frac{15}{4}\frac{1}{3u+1}\right)du$$
$$\ln|X|=\frac{1}{4}\ln|u-1|-\frac{5}{4}\ln|3u+1|+C.$$

Now we combine logarithms, substitute $u=Y/X$, and simplify the result:

$$\ln|X|=\frac{1}{4}\ln|u-1|-\frac{5}{4}\ln|3u+1|+C.$$
$$4\ln|X|=\ln\left|\frac{u-1}{(3u+1)^5}\right|+C$$
$$X^4=C\frac{Y/X-1}{(3Y/X+1)^5}$$
$$X^4=C\frac{X^4(Y-X)}{(X+3Y)^5}$$
$$\frac{(X+3Y)^5}{(Y-X)}=C.$$

But, $X=x-3$ and $Y=y-2$ so $(x+3y-9)^5/(y-x+1)=C$, which we graph for various values of C with Mathematica.

```
ContourPlot[(x + 3 y - 9) ^ 5 / (y - x + 1),
  {x, -10, 10}, {y, -10, 10},
  PlotPoints → 300, Contours → 30,
  ContourShading → False, Frame → False,
  Axes → Automatic, AxesOrigin → {0, 0}]
```

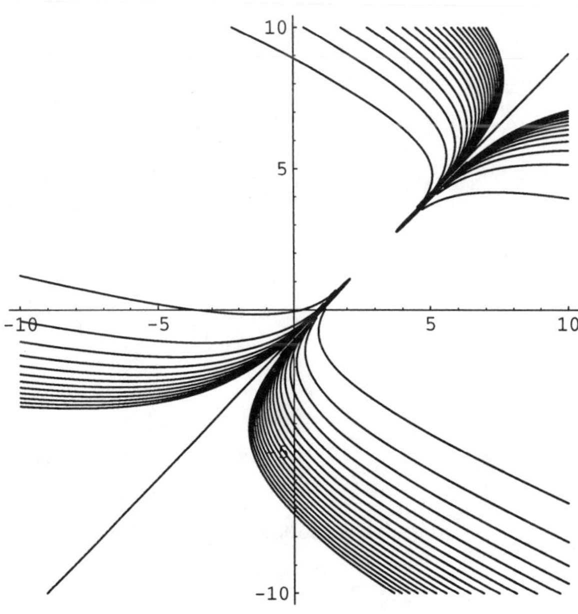

42. (b)
43. (d)
44. (a)
45. (c)

```
<< Graphics`PlotField`
s =
 Map[
  PlotVectorField[#, {x, -3,
    {y, -3, 3},
    Axes -> Automatic,
    AxesOrigin -> {0, 0},
    HeadLength -> 0,
    HeadWidth -> 0,
    ScaleFunction -> (1 &),
    AspectRatio -> 1,
    DisplayFunction ->
     Identity] &,
  {{1, x Exp[y / x] / y},
   {1, y Exp[y / x] / x},
   {1, y Exp[x / y] / x},
   {1, x Exp[x / y] / y}}]


Show[GraphicsArray[
  Partition[s, 2]]]
```

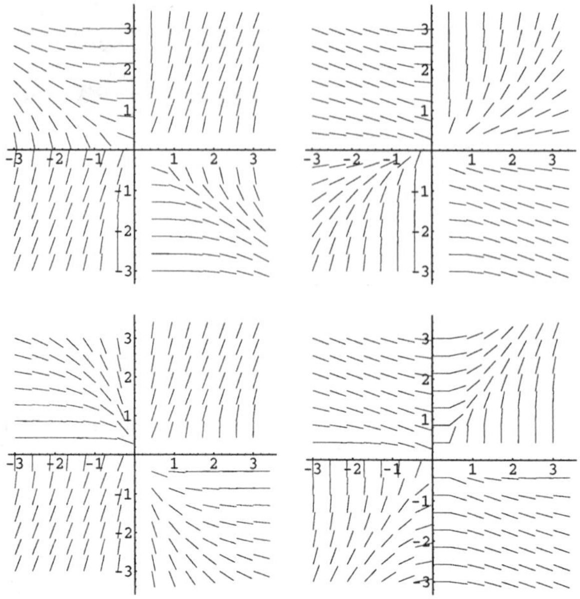

47. Because $M(x,y)dx + N(x,y)dy = 0$ is homogeneous, we can write the equation in the form $dy/dx = F(y/x)$. If $x = r\cos\theta$ and $y = r\sin\theta$, $dx = \cos\theta\, dr - r\sin\theta\, d\theta$ and $dy = \sin\theta\, dr + r\cos\theta\, d\theta$. Substituting into the equation gives us

$$dy/dx = F(y/x)$$

$$\frac{\sin\theta\, dr + r\cos\theta\, d\theta}{\cos\theta\, dr - r\sin\theta\, d\theta} = F\left(\frac{r\sin\theta}{r\cos\theta}\right) = F\left(\frac{\sin\theta}{\cos\theta}\right)$$

$$\sin\theta\, dr + r\cos\theta\, d\theta = F\left(\frac{\sin\theta}{\cos\theta}\right)\cos\theta\, dr - F\left(\frac{\sin\theta}{\cos\theta}\right) r\sin\theta\, d\theta$$

$$F\left(\frac{\sin\theta}{\cos\theta}\right) r\sin\theta\, d\theta + r\cos\theta\, d\theta = F\left(\frac{\sin\theta}{\cos\theta}\right)\cos\theta\, dr - \sin\theta\, dr$$

$$r\left(F\left(\frac{\sin\theta}{\cos\theta}\right)\sin\theta + \cos\theta\right)d\theta = \left(F\left(\frac{\sin\theta}{\cos\theta}\right)\cos\theta - \sin\theta\right)dr$$

$$\frac{F\left(\frac{\sin\theta}{\cos\theta}\right)\sin\theta + \cos\theta}{F\left(\frac{\sin\theta}{\cos\theta}\right)\cos\theta - \sin\theta}\, d\theta = \frac{1}{r}\, dr.$$

53.

$f(x) = 1 - 2x$; $g(x) = x^{-2}$; General Solution: $1 - 2(xc - y) = c^{-2} \Rightarrow y = \frac{1}{2}\left[c^{-2} + 2cx - 1\right]$;

Singular Solution: $y = \frac{1}{2}\left(3x^{2/3} - 1\right)$

61. This is a homogeneous equation of degree one because

$$(tx)^{1/3}(ty)^{2/3}+(tx)=t\left(x^{1/3}y^{2/3}+x\right)$$

and

$$(tx)^{2/3}(ty)^{1/3}+(ty)=t\left(x^{2/3}y^{1/3}+y\right).$$

We proceed by letting $x=v\,y$ and substituting into the equation.

We use `capm` *and* `capn` *to represent* $M(x,y)$ *and* $N(x,y)$*, respectively, and avoid any possible ambiguity with built-in Mathematica objects, which always begin with a capital letter. Note that* `Dt[v]` *represents* dv *and* `Dt[y]` *represents* dy.

```
Clear[capm,capn]
capm[x_,y_]=x^(1/3)y^(2/3)+x;
capn[x_,y_]=x^(2/3) y^(1/3)+y;
x=v y;
step1=capm[x,y]Dt[x]+capn[x,y]Dt[y]
```

```
M:=(x,y)->x^(1/3)*y^(2/3)+x:
N:=(x,y)->x^(2/3)*y^(1/3)+y:
x:=v*y:
step1:=M(x,y)*D(x)+N(x,y)*D(y);
```

To see that the resulting equation is separable, we collect together terms involving dv and dy.

```
step2=PowerExpand[step1]
step3=Collect[step2,{Dt[y],Dt[v]}]
```

```
step2:=expand(combine(step1,radical,
  symbolic));
step3:=collect(step2,{D(v),D(y)});
```

Thus, substituting $x=v\,y$ into the equation yields the separable equation

$$y^2\left(v^{1/3}+v\right)dv+y\left(v^{4/3}+v^2+v^{2/3}+1\right)dy=0$$

$$y\left(v^{4/3}+v^2+v^{2/3}+1\right)dy=-y^2\left(v^{1/3}+v\right)dv$$

$$\frac{1}{y}dy=-\frac{v^{1/3}+v}{v^{4/3}+v^2+v^{2/3}+1}dv.$$

We find a general solution by integrating and then substituting $v=x/y$.

```
step4=step3 /. {Dt[y]->y'[v],y->y[v],
    Dt[v]->1}
solv=DSolve[step4==0,y[v],v]
imsol=solv[[1,1,2]]
Clear[x,v,y]
sol=imsol /. v->x/y
toplot=Solve[y==sol,C[1]]
ContourPlot[toplot[[1,1,2]],{x,0.01,5},
    {y,0.01,5},
    ContourShading->False,
    Contours->30,PlotPoints->120,
    Frame->False,Axes->Automatic,
    AxesOrigin->{0,0}]
```

```
step4:=subs({D(y)=diff(y(v),v),
    D(v)=1,y=y(v)},step3);
solv:=dsolve(step4=0,y(v));
v:='v':x:='x':y:='y':
toplot:=subs({y(v)=y,v=x/y},
    lhs(solv));
with(plots);
contourplot(toplot,x=0.01..5,
    y=0.01..5,grid=[60,60],
    axes=NORMAL);
```

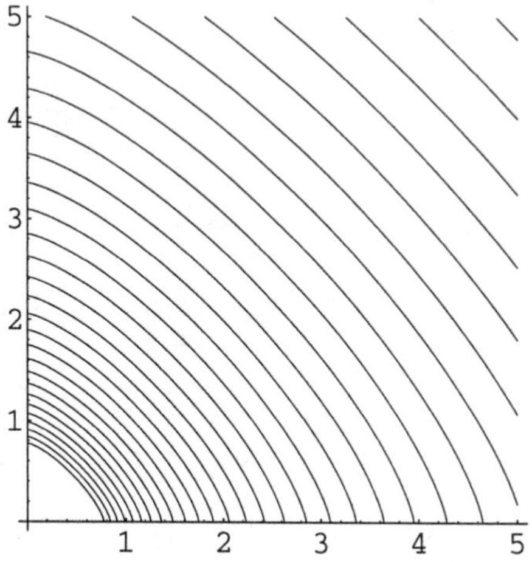

62. Unlike Problem 1 where we implemented the steps used to solve homogeneous equations, we solve the problem directly.

```
Clear[x,y]
gensol=DSolve[x y[x] y'[x]==
    (y[x]^2-x^2),y[x],x]
sol=DSolve[{x y[x] y'[x]==(y[x]^2-x^2),
    y[4]==0},y[x],x]
Plot[Evaluate[y[x] /. sol],{x,0,4},
    PlotRange->{{0,5},{-5/2,5/2}},
    AspectRatio->1]
```

```
x:='x':y:='y':
gensol:=dsolve(x*y(x)*diff(y(x),x)=
    (y(x)^2-x^2),y(x));
sol:=dsolve({x*y(x)*diff(y(x),x)=
    (y(x)^2-x^2),
    y(4)=0},y(x));
plot({rhs(sol[1]),rhs(sol[2])},
    x=0..4);
```

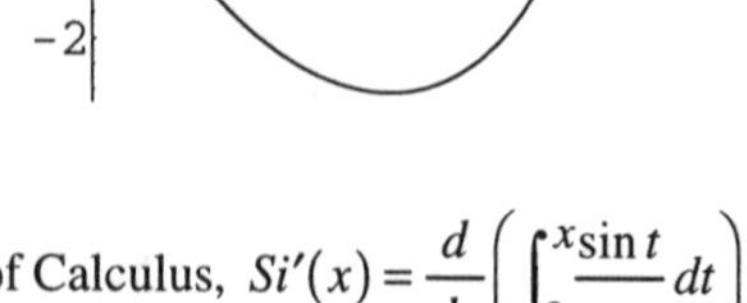

63. (a) By the Fundamental Theorem of Calculus, $Si'(x)=\frac{d}{dx}\left(\int_0^x \frac{\sin t}{t}\,dt\right)=\frac{\sin x}{x}$. Use the graph to determine the maximum value. Note that the maximum value occurs when $x=\pi$. (Why?)

```
<<NumericalMath`NLimit`
NLimit[SinIntegral[x],x->Infinity]
N[Pi/2]
Plot[SinIntegral[x],{x,0,10},
    PlotRange->All]
```

```
limit(Si(x),x=infinity);
plot(Si(x),x=0...10);
```


(b) $y\sin(x/y)\,dx - \left(x + x\sin(x/y)\right)dy = 0 \Rightarrow \frac{dy}{dx} = \frac{y\sin(x/y)}{x + x\sin(x/y)}$

```
<<Graphics`PlotField`
PlotVectorField[{1,
        y Sin[x/y]/(x+x Sin[x/y])},
    {x,-Pi,Pi},{y,-Pi,Pi},
    Axes->Automatic,HeadLength->0,
    HeadWidth->0,ScaleFunction->(0.5&),
    PlotPoints->30]
```

```
with(DEtools);
dfieldplot(y*sin(x/y)/(x+x*sin(x/y)),
    [x,y],-Pi..Pi,y=-Pi..Pi);
```

(c)

```
sol:=dsolve(diff(y(x),x)=
    y*sin(x/y)/(x+x*sin(x/y)),
    y(x));
```

By hand, we let $x = v\,y$ so $dx = ydv + vdy$. Then,

$$y\sin(x/y)dx - \left(x + x\sin(x/y)\right)dy = 0$$
$$y\sin v\,(ydv + vdy) - (vy + vy\sin v)dy = 0$$
$$y^2 \sin v dv = vydy$$
$$\frac{1}{y}dy = \frac{\sin v}{v}dv.$$

Integrating both sides yields

$$\ln|y| = Si\,v + C$$
$$\ln|y| = C + Si(x/y)$$
$$y = C\,e^{Si(x/y)}.$$

```
imsol=y==capc Exp[SinIntegral[x/y]]
eqn=imsol /. {y->2,x->1}
cval=Solve[eqn,capc]
N[cval]
toplot=y-capc Exp[SinIntegral[x/y]] /.
    cval[[1]]
ContourPlot[toplot,{x,0,2Pi},{y,0,2Pi},
    Contours->{0},ContourShading->False,
    Frame->False,Axes->Automatic]
```

```
imsol:=y=capc*exp(Si(x/y));
eqn:=subs({y=2,x=1},imsol);
cval:=solve(eqn,capc);
evalf(cval);
toplot:=subs(capc=cval,imsol);
with(plots):
implicitplot(toplot,x=0..2*Pi,
    y=0..2*Pi);
```

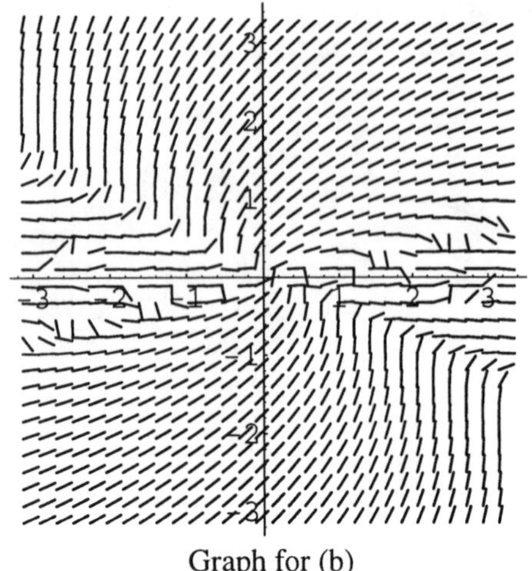

Graph for (b)

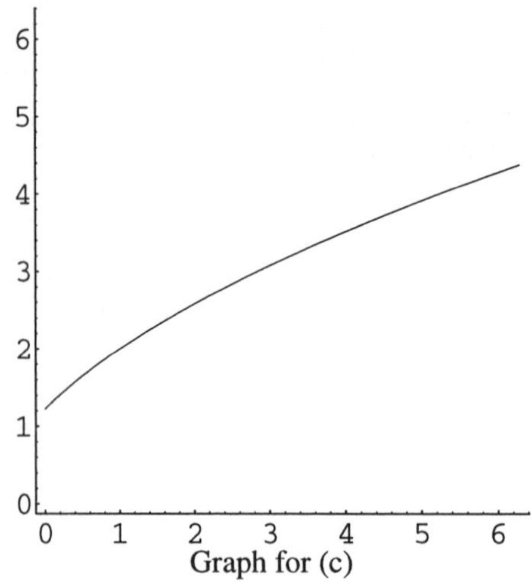

Graph for (c)

EXERCISES 2.4

3. Exact:

$$\frac{\partial}{\partial y}(y\cos(xy)) = \cos(xy) - xy\sin(xy)$$

$$= \frac{\partial}{\partial x}(x\cos(xy))$$

7. Not Exact:

$$\frac{\partial}{\partial y}(y\sin 2x) = \sin 2x$$

$$\neq 2\sin 2x = \frac{\partial}{\partial x}\left(-\left(\sqrt{y} + \cos 2x\right)\right)$$

13. Note that the equation is both separable and exact. We find a general solution by solving it as an exact equation:

$$\int y^2 dx = x\,y^2 + g(y);$$

$$\frac{\partial}{\partial y}\left(x\,y^2 + g(y)\right) = 2xy + g'(y) = 2xy$$

$$\Rightarrow g'(y) = 0 \Rightarrow g(y) = 0;$$

$$xy^2 = C.$$

If instead we solve it as a separable equation, we have:

$$y^2 dx + 2xydy = 0$$

$$ydx = -2xdy$$

$$\frac{1}{x}dx = -\frac{2}{y}dy$$

$$\ln|x| = -2\ln|y| + C$$

$$x = Cy^{-2}$$

$$xy^2 = C.$$

17. The equation is homogeneous of degree two as well as an exact equation because

$$\frac{\partial}{\partial y}(2xy) = 2x = \frac{\partial}{\partial x}\left(x^2 + y^2\right).$$

Then,

$$\int 2xy\,dx = x^2y + g(y);$$

$$\frac{\partial}{\partial y}\left(x^2y + g(y)\right) = x^2 + g'(y) = x^2 + y^2$$

$$\Rightarrow g'(y) = y^2 \Rightarrow g(y) = \tfrac{1}{3}y^3$$

so a general solution is $x^2y + \frac{1}{3}y^3 = C$.

Alternatively, solving it as a homogeneous equation, we let $x = vy \Rightarrow dx = ydv + vdy$. Then,

$$2xydx + \left(x^2 + y^2\right)dy = 0$$

$$2\cdot vy\cdot y\cdot(ydv + vdy) + \left((vy)^2 + y^2\right)dy = 0$$

$$2\cdot v(ydv + vdy) + \left(v^2 + 1\right)dy = 0$$

$$2vydv + \left(3v^2 + 1\right)dy = 0$$

so

$$\frac{1}{y}dy = -\frac{2v}{3v^2 + 1}dv$$

$$\ln|y| = -\frac{1}{3}\ln\left(3v^2 + 1\right) + C$$

$$y^3 = C\left(3(x/y)^2 + 1\right)^{-1}$$

$$y^3\left(3(x/y)^2 + 1\right) = C$$

$$3x^2y + y^3 = C.$$

```
ContourPlot[3 x^2 y + y^3, {x, -5, 5},
 {y, -5, 5}, PlotPoints -> 300,
 Contours -> 20,
 ContourShading -> False,
 Frame -> False, Axes -> Automatic,
 AxesOrigin -> {0, 0}]
```

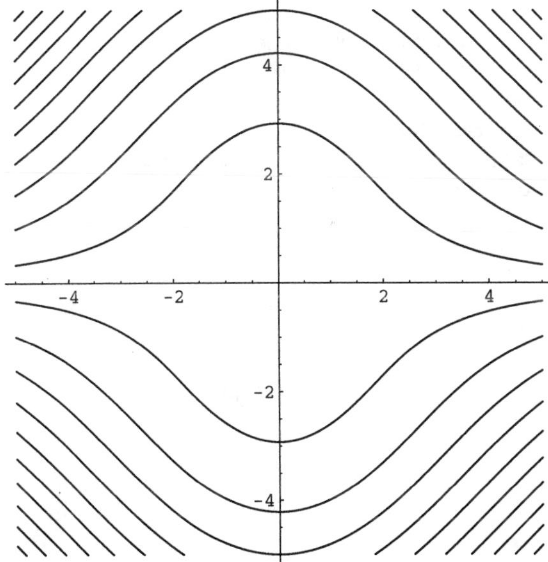

21. The equation is exact because

$$\frac{\partial}{\partial y}\left[\frac{y + y^2}{(y - x)^2}\right] = \frac{x + y + 2xy}{(x - y)^3} = \frac{\partial}{\partial x}\left[-\frac{x + x^2}{(x - y)^2}\right].$$

Then,

$$\int \frac{y+y^2}{(y-x)^2}\,dx = \frac{y+y^2}{y-x} + g(y)$$

and

$$\frac{\partial}{\partial y}\left(\frac{y+y^2}{y-x} + g(y)\right) = \frac{y^2-2xy-x}{(y-x)^2} + g'(y) = -\frac{x+x^2}{(x-y)^2}$$

so

$$g'(y) = -\frac{x+x^2}{(y-x)^2} - \frac{y^2-2xy-x}{(y-x)^2} = -1$$

$$\Rightarrow g(y) = -y.$$

Thus, a general solution is given by

$$\frac{y+y^2}{y-x} - y = C.$$

Note that if you multiply through by $(y-x)^2 = (x-y)^2$, the resulting equation is not exact but is separable:

$$\left(y+y^2\right)dx - \left(x+x^2\right)dy = 0$$

$$\frac{1}{x+x^2}\,dx = \frac{1}{y+y^2}\,dy$$

$$\left(\frac{1}{x} - \frac{1}{x+1}\right)dx = \left(\frac{1}{y} - \frac{1}{y+1}\right)dy$$

$$\ln|x| - \ln|x+1| = \ln|y| - \ln|y+1| + C$$

$$\frac{x}{x+1} = C\frac{y}{y+1}$$

$$\frac{x(y+1)}{y(x+1)} = C.$$

25. The equation is exact because

$$\frac{\partial}{\partial y}\left(1+y^2\cos(xy)\right) = 2y\cos(xy) - xy^2\sin(xy) = \frac{\partial}{\partial x}\left(xy\cos(xy)+\sin(xy)\right).$$

Note that we choose to integrate $1+y^2\cos(xy)$ with respect to x because we would have to use integration by parts to integrate $xy\cos(xy)+\sin(xy)$ with respect to y. Then, $\int\left(1+y^2\cos(xy)\right)dx = x + y\sin(xy) + g(y)$ and

$$\frac{\partial}{\partial y}\left(x+y\sin(xy)+g(y)\right) = xy\cos(xy)+\sin(xy)+g'(y) = xy\cos(xy)+\sin(xy) \Rightarrow g'(y) = 0 \Rightarrow g(y) = 0$$

so a general solution is $x + y\sin(xy) = C$.

```
ContourPlot[x + y Sin[x y],
  {x, -Pi, Pi}, {y, -Pi, Pi},
  PlotPoints -> 300, Contours -> 20,
  ContourShading -> False,
  Frame -> False, Axes -> Automatic,
  AxesOrigin -> {0, 0}]
```

29. The equation is exact because

$$\frac{\partial}{\partial y}\left(\frac{e^{y/x}(x-y)}{x}\right)=-\frac{e^{y/x}y}{x^2}=\frac{\partial}{\partial x}\left(e^{y/x}\right).$$

We choose to integrate $e^{y/x}$ with respect to y because integrating $\frac{e^{y/x}(x-y)}{x}$ with respect to x using traditional techniques would be *very* difficult (if not actually impossible). Then,

$$\int e^{y/x}dy = xe^{y/x}+g(x)\Rightarrow\frac{\partial}{\partial x}\left(xe^{y/x}+g(x)\right)=\frac{e^{y/x}(x-y)}{x}+g'(x)=\frac{e^{y/x}(x-y)}{x}\Rightarrow g'(x)=0\Rightarrow g(x)=0$$

so a general solution is $y=-x\ln|x|+Cx$ or $\left(y+x\ln|x|\right)/x=C$.

```
ContourPlot[(y + x Log[Abs[x]]) / x,
  {x, -2, 2}, {y, -2, 2},
  PlotPoints -> 300, Contours -> 20,
  ContourShading -> False,
  Frame -> False, Axes -> Automatic,
  AxesOrigin -> {0, 0}]
```

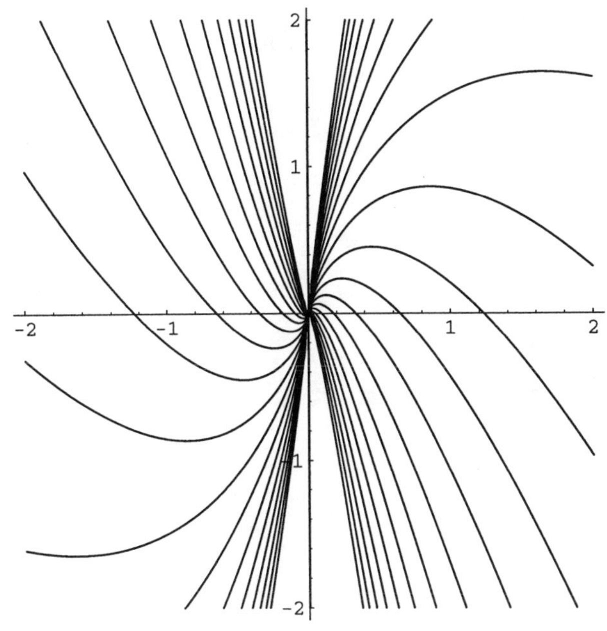


```
Plot[-(x^3 + 1) / (x^2 - 1), {:
```

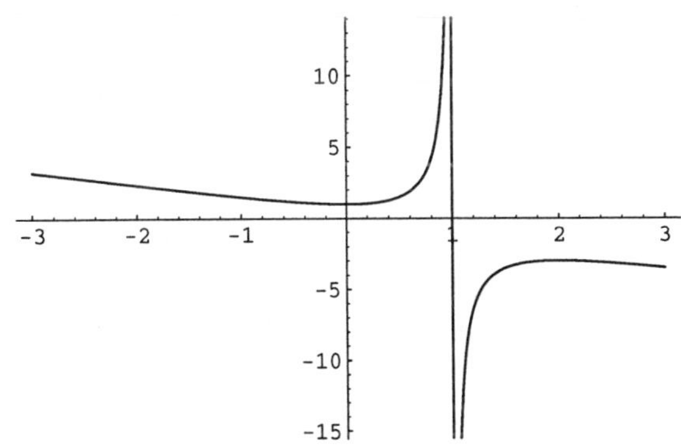

33. We find a general solution of the equation:

$$\int\left(x^2-1\right)dy = x^2y-y+g(x);$$

$$\frac{\partial}{\partial x}\left(x^2y-y+g(x)\right)=2xy+g'(x)=2xy+3x^2$$

$$\Rightarrow g'(x)=3x^2\Rightarrow g(x)=x^3;$$

$$x^2y-y+x^3=C.$$

Application of the initial condition yields $C=-1$ so the solution to the initial-value problem is

$$x^2y-y+x^3=-1 \qquad\text{or}\qquad y=-\frac{x^3+1}{x^2-1}.$$

37. We find a general solution of the equation:

$$\int (1+2xy)dy = y + xy^2 + g(x);$$

$$\frac{\partial}{\partial x}\left(y + xy^2 + g(x)\right) = y^2 + g'(x) = y^2 - 2\sin 2x \Rightarrow g'(x) = -2\sin 2x \Rightarrow g(x) = \cos 2x;$$

$$y + xy^2 + \cos 2x = C.$$

Application of the initial condition yields $C = 2$ so the solution to the initial-value problem is $xy^2 + \cos 2x + y = 2$.

```
ContourPlot[x y^2 + Cos[2 x] + y,
 {x, -Pi, 2 Pi}, {y, -3 Pi / 2, 3 Pi / 2},
 PlotPoints -> 300, Contours -> {2},
 ContourShading -> False,
 Frame -> False, Axes -> Automatic,
 AxesOrigin -> {0, 0}]
```

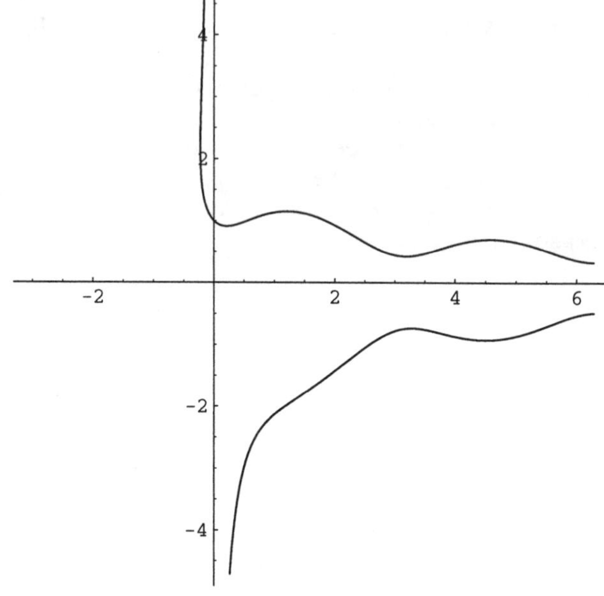

41-44. (Mathematica only) We use the direction fields to match 41 with (d), 42 with (a), 43 with (b) and 44 with (c).

```
<< Graphics`PlotField`
s =
 Map[
  PlotVectorField[#, {x, 0, 3 Pi},
    {y, 0, 3 Pi},
    Axes -> Automatic,
    AxesOrigin -> {0, 0},
    HeadLength -> 0, HeadWidth -> 0,
    ScaleFunction -> (1 &),
    AspectRatio -> 1,
    DisplayFunction -> Identity] &,
  {{1, Exp[y - x]
     (Cos[x] - Sin[x]) /
      (Cos[y] + Sin[y])},
   {1, -Exp[x - y]
     (Cos[x] + Sin[x]) /
      (Cos[y] - Sin[y])},
   {1,
    (-Exp[y] Cos[x] - Exp[x] Cos[y]) /
     (Exp[y] Sin[x] - Exp[x] Sin[y])},
   {1, (Exp[x] Sin[x] + Exp[y] Sin[y]) /
     (Exp[y] Cos[y] - Exp[x] Cos[x])}}]
```

```
Show[GraphicsArray[Pa
```

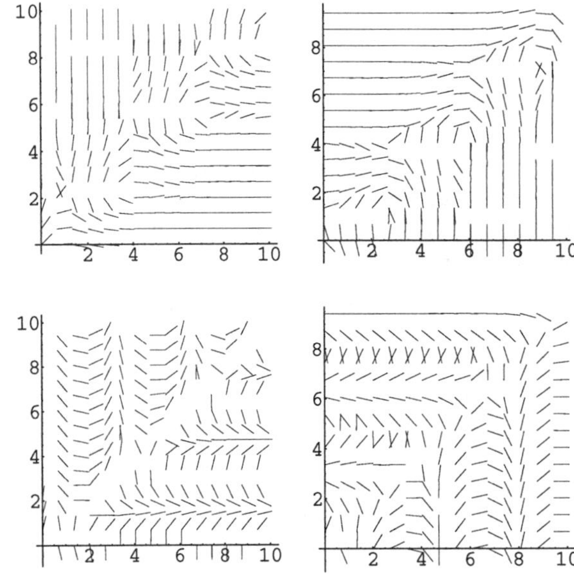

53. First, we identify $M(x,y)$ and $N(x,y)$: $\underbrace{\left(2xy^2+y\right)}_{M(x,y)}dx+\underbrace{\left(2y^3-x\right)}_{N(x,y)}=0$. Then,

$$\mu(y)=\exp\left(\int\frac{1}{M(x,y)}\left[\frac{\partial N}{\partial x}-\frac{\partial M}{\partial y}\right]dy\right)=\exp\left(\int-\frac{2}{y}\,dy\right)=e^{-2\ln|y|}=\frac{1}{y^2}$$

and multiplying the equation by $1/y^2$ yields $\frac{2xy^2+y}{y^2}dx+\frac{2y^3-x}{y^2}dy=0$, which is an exact equation because

$$\frac{\partial}{\partial y}\left(\frac{2xy^2+y}{y^2}\right)=-\frac{1}{y^2}=\frac{\partial}{\partial x}\left(\frac{2y^3-x}{y^2}\right).$$

Then, $\int\frac{2xy^2+y}{y^2}dx=\frac{x^2y^2+xy}{y^2}+g(y)$ and

$$\frac{\partial}{\partial y}\left(\frac{x^2y^2+xy}{y^2}+g(y)\right)=-\frac{x}{y^2}+g'(y)=\frac{2y^3-x}{y^2}\Rightarrow g'(y)=2y\Rightarrow g(y)=y^2.$$

Therefore, a general solution is $\frac{x^2y^2+xy}{y^2}+y^2=C$. If instead we compute $\int\frac{2y^3-x}{y^2}dy=y^2+\frac{x}{y}+g(x)$ and

$$\frac{\partial}{\partial x}\left(y^2+\frac{x}{y}+g(x)\right)=\frac{1}{y}+g'(x)=\frac{2xy^2+y}{y^2}\Rightarrow g'(x)=2x\Rightarrow g(x)=x^2,$$

we obtain the same result. We can write this solution as $\frac{1}{y} = \frac{C\sqrt{2x^2-1}-1}{x}$ or $y = \frac{x}{C\sqrt{2x^2-1}-1}$.

```
ContourPlot[(x + y)/Sqrt[-1 + 2 x^2 y],
  {x, -10, 10}, {y, -10, 10},
  PlotPoints -> 300, Contours -> 30,
  ContourShading -> False,
  Frame -> False, Axes -> Automatic,
  AxesOrigin -> {0, 0}]
```

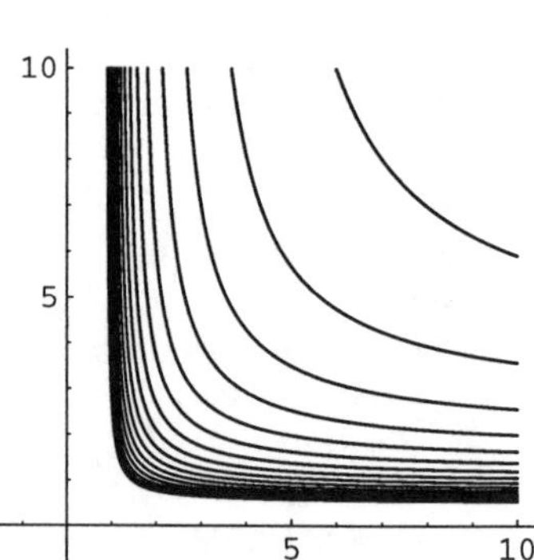
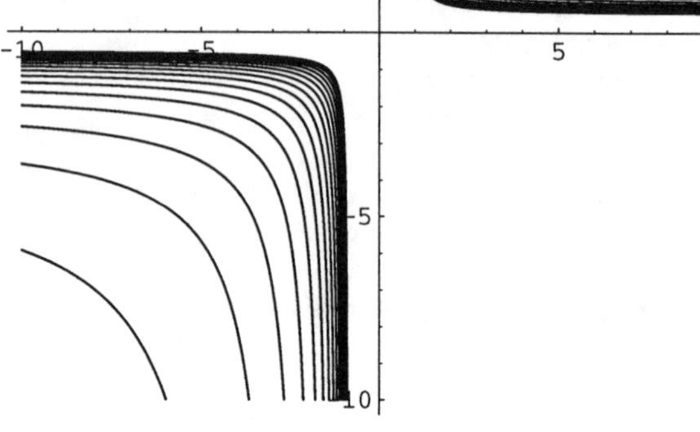

55. (a) We have

$$\begin{aligned}\frac{\partial}{\partial x}\left(N(x,y) - \frac{\partial}{\partial y}\int M(x,y)\,dx\right) &= \frac{\partial N(x,y)}{\partial x} - \frac{\partial}{\partial x}\frac{\partial}{\partial y}\int M(x,y)\,dx \\ &= \frac{\partial N(x,y)}{\partial x} - \frac{\partial}{\partial y}\frac{\partial}{\partial x}\int M(x,y)\,dx \quad \text{(The mixed partials are equal)} \\ &= \frac{\partial N(x,y)}{\partial x} - \frac{\partial M(x,y)}{\partial y} \quad \text{(Apply the Fundamental Theorem of Calculus)} \\ &= 0,\end{aligned}$$

which shows that $N(x,y) - \frac{\partial}{\partial y}\int M(x,y)\,dx$ is "x-free." Therefore, g is a function of y only.

(b)

$$\begin{aligned}\frac{\partial f}{\partial x}dx + \frac{\partial f}{\partial y}dy &= \frac{\partial}{\partial x}\left[g(y) + \int M(x,y)\,dx\right]dx + \frac{\partial}{\partial y}\left[g(y) + \int M(x,y)\,dx\right]dy \\ &= \frac{\partial}{\partial x}\int M(x,y)\,dx + g'(y)dy + \frac{\partial}{\partial y}\int M(x,y)\,dx\,dy \\ &= M(x,y)dx + N(x,y)dy - \frac{\partial}{\partial y}\int M(x,y)\,dx\,dy + \frac{\partial}{\partial y}\int M(x,y)\,dx\,dy \\ &= M(x,y)dx + N(x,y)dy\end{aligned}$$

56. We compute the implicit solution $x^2 + y\cos xy = C$ and generate its graph for $0 \le x \le 3\pi$ and $0 \le y \le 3\pi$.

```
gensol=DSolve[2x-y[x]^2Sin[x y[x]]+
    (Cos[x y[x]]-x y[x] Sin[x y[x]])y'[x]==0,
    y[x],x]
toplot=gensol[[1,1]] /. y[x]->y
ContourPlot[toplot,{x,0,3Pi},{y,0,3Pi},
    Frame->False,
    Axes->Automatic,
    AxesOrigin->{0,0},
    ContourShading->False,
    PlotPoints->150]
```

```
gensol:=dsolve(2*x-y^2*sin(x*y)+
    (cos(x*y)-x*y*sin(x*y))*
    diff(y(x),x)=0,y(x));
with(plots):
contourplot(lhs(gensol),x=0..3*Pi,
    y=0..3*Pi,grid=[40,40],
    axes=NORMAL);
```

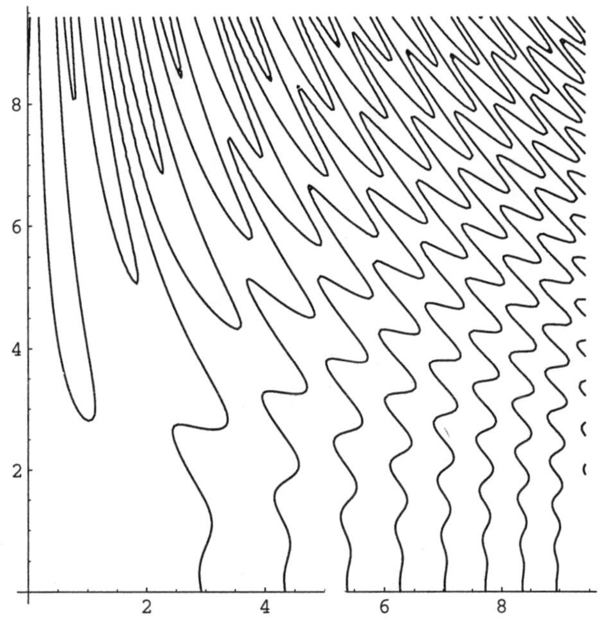

57. The general solution $e^{xy} - x + y + \sin xy = C$ is found and graphed as follows.

```
Clear[x,y]
gensol=DSolve[-1+Exp[x y[x]]y[x]+
    y[x] Cos[x y[x]]+(1+x Exp[x y[x]]+
        x Cos[x y[x]])y'[x]==0,y[x],x]
toplot=gensol[[1,1]] /. y[x]->y
ContourPlot[toplot,{x,-Pi,Pi},{y,-Pi,Pi},
    Frame->False,
    Axes->Automatic,
    AxesOrigin->{0,0},
    ContourShading->False,
    PlotPoints->150]
```

Remember that you do not need to reload the **plots** *package if you have previously loaded it during your current Maple work session.*

```
gensol:=dsolve(-1+exp(x*y)*y+
    y*cos(x*y)+(1+exp(x*y)*x+
    x*cos(x*y))*diff(y(x),x)=0,y(x));
with(plots):
contourplot(lhs(gensol),x=-Pi..Pi,
    y=-Pi..Pi,grid=[40,40],
    axes=NORMAL);
```

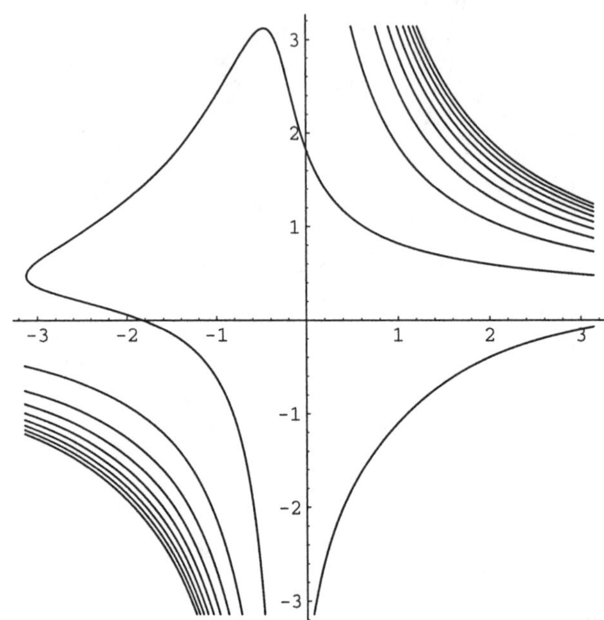

58. First, we find general solutions of the three equations. The resulting formulas indicate that the solutions to the three equations are quite different.

```
Clear[x,y]
someequations={y'[x]==
    -(2x+2y[x])/(2x+2y[x]),
    (2x+2y[x])y'[x]==-(1.8x+2y[x]),
    (1.9x+2y[x])y'[x]==-(2x+1.9y[x])}
solutions=Map[DSolve[#,y[x],x]&,
    someequations]
```

```
someequations:={diff(y(x),x)=
        -(2*x+2*y)/(2*x+2*y),
    diff(y(x),x)=
        -(1.8*x+2*y)/(2*x+2*y),
    diff(y(x),x)=
    -(2*x+1.9*y)/(1.9*x+2*y)};
somesolutions:=map(dsolve,
    someequations,y(x));
```

The direction fields for the three equations also indicate that the solutions to the three equations differ considerably.

```
<< Graphics`PlotField`
s =
 Map[
  PlotVectorField[#, {x, -1, 1},
    {y, -1, 1},
    Axes -> Automatic,
    AxesOrigin -> {0, 0},
    HeadLength -> 0, HeadWidth -> 0,
    ScaleFunction -> (1 &),
    AspectRatio -> 1,
    DisplayFunction -> Identity] &,
  {{2 x + 2 y, 2 x + 2 y},
   {1.8 x + 2 y, 2 x + 2 y},
   {2 x + 1.9 y, 1.9 x + 2 y}}]
```

```
Show[GraphicsArray[s]]
```

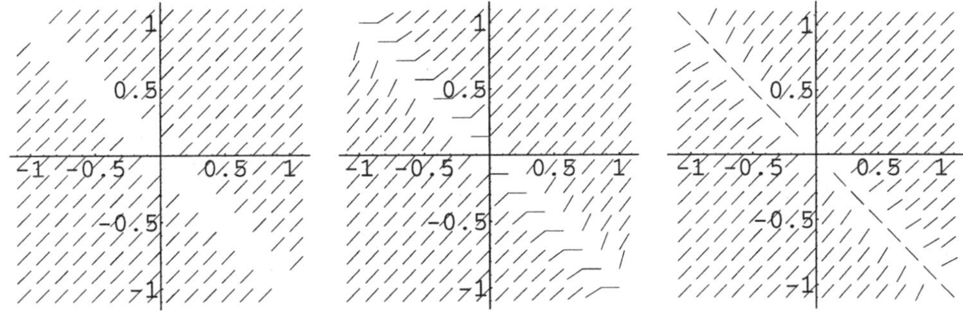

59.

```
gensol=DSolve[Sin[c y[x]]-y[x] c Sin[c x]+
    (x c Cos[c y[x]]+Cos[c x])y'[x]==
        0,y[x],x]
Solve[x Sin[c y[x]] + Cos[c x] y[x] ==
        C[1], y[x]]
eq=gensol[[1]] /. {y[x]->1,x->0}
cval=Solve[eq]
```

```
toplot = gensol[[1, 1]] /. y[x] -> y;
toshow =
 Table[ContourPlot[toplot, {x, 0, 6},
   {y, 0, 6},
    Frame -> False,
    Axes -> Automatic,
    AxesOrigin -> {0, 0},
    ContourShading -> False,
    PlotPoints -> 120,
    Contours -> {1},
   DisplayFunction -> Identity],
  {c, -2, 2, 4 / 8}]
```

```
Show[GraphicsArray[
  Partition[toshow, 3]]]
```

```
c:='c':
gensol:=dsolve(sin(c*y)-y*c*sin(c*x)+
    (x*c*cos(c*y)+cos(c*x))*
       diff(y(x),x),y(x));
eq:=subs({y(x)=1,x=0},gensol);
cval:=solve(eq);
partsol:=subs(_C1=cval,gensol);
```

```
constants:=[seq(-2+4/5*k,k=0..5)];
with(plots):
for c in constants do
   implicitplot(partsol,x=0..6,
      y=0..6)
   od;
```

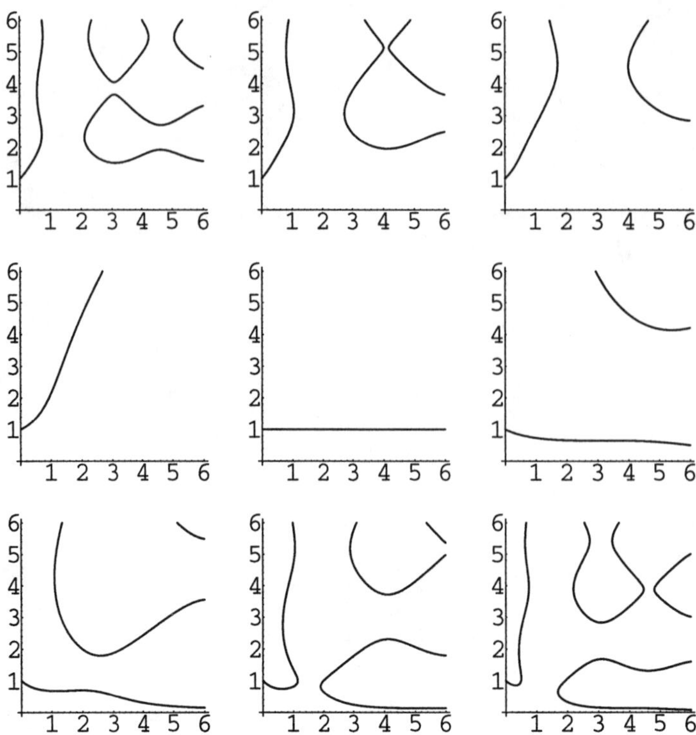

EXERCISES 2.5

3. Uniqueness is not guaranteed. In this case, $f(x,y)=y^{1/5}$, so $\frac{\partial f}{\partial y}f(x,y)=\frac{1}{5}y^{-4/5}$ is not continuous at (0,0).

Solutions: $y=\left(\frac{4}{5}x\right)^{5/4}$, $y=0$.

5. Because $f(x,y)=2\sqrt{|y|}=\begin{cases}2\sqrt{y}, y\ge 0\\ 2\sqrt{-y}, y<0\end{cases}$, $\frac{\partial f}{\partial y}(x,y)=\begin{cases}y^{-1/2}, y\ge 0\\ -(-y)^{-1/2}, y<0\end{cases}$ is not continuous at (0,0). Therefore, the hypotheses of the Existence and Uniqueness Theorem are not satisfied.

7. Yes. $\frac{dy}{dx}=y\sqrt{x}\Rightarrow\frac{1}{y}dy=x^{1/2}dx\Rightarrow\ln|y|=\frac{2}{3}x^{3/2}+C_1\Rightarrow y=Ce^{2x^{3/2}/3}$. Application of the initial condition yields $y=e^{2\left(x^{3/2}-1\right)/3}$.

```
sol =
  DSolve[{y'[x] == y[x] Sqrt[x],
    y[1] == 1}, y[x], x]
```

$\{\{y[x]\to e^{-\frac{2}{3}+\frac{2x^{3/2}}{3}}\}\}$

```
Plot[y[x] /. sol, {x, 0, 2}]
```

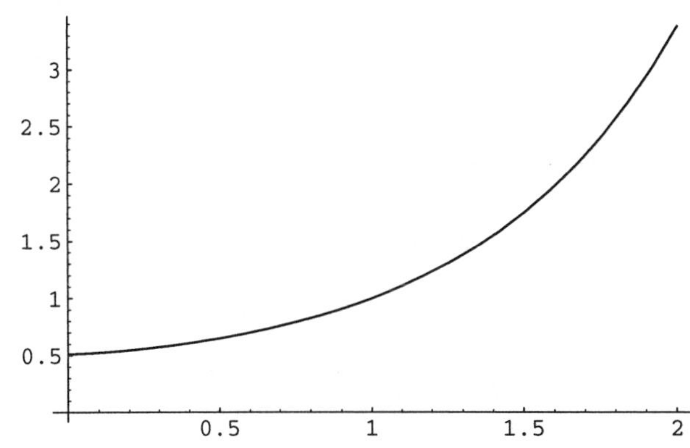

9. Yes. $f(x,y)=\sin y-\cos x$ and $\frac{\partial f}{\partial y}(x,y)=\cos y$ are continuous in a region containing $(\pi,0)$.

```
sol =
  NDSolve[{y'[x] == Sin[y[x]] - Cos[x
    y[Pi] == 0}, y[x], {x, 0, 2 Pi}]

{{y[x] → InterpolatingFunction[
    {{0., 6.28319}}, <>][x]}}

Plot[y[x] /. sol, {x, 0, 2 Pi}]
```

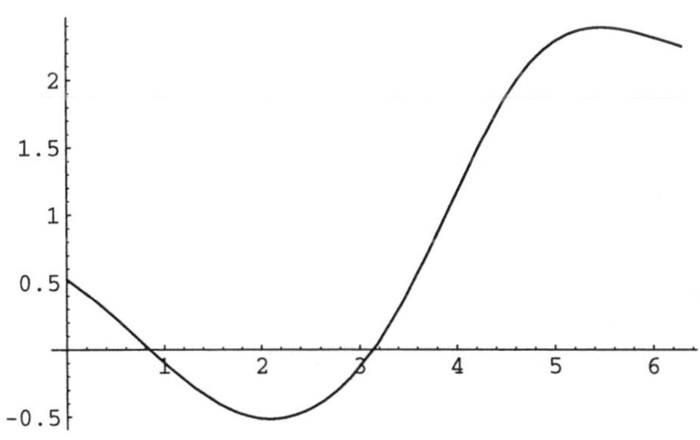

17. $(0,\infty)$ because $\ln x$ has domain $0<x<\infty$

```
sol =
 DSolve[{2 y'[x] + x y[x] == Log[x],
   y[E] == 0}, y[x], x]
```

$$\{\{y[x] \to \frac{1}{2} e^{-\frac{x^2}{4}} \left(-\sqrt{\pi}\, \text{Erfi}[\frac{e}{2}] + e\, \text{HypergeometricPFQ}[\{\frac{1}{2}, \frac{1}{2}\}, \{\frac{3}{2}, \frac{3}{2}\}, \frac{e^2}{4}] - x\, \text{HypergeometricPFQ}[\{\frac{1}{2}, \frac{1}{2}\}, \{\frac{3}{2}, \frac{3}{2}\}, \frac{x^2}{4}] + \sqrt{\pi}\, \text{Erfi}[\frac{x}{2}]\, \text{Log}[x]\right)\}\}$$

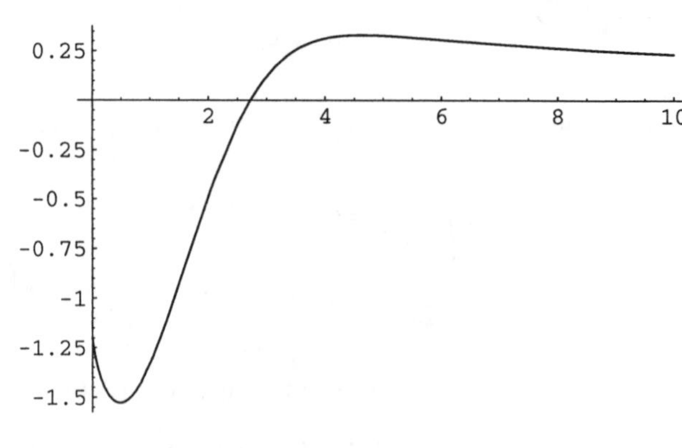

```
Plot[y[x] /. sol, {x, 0, 10}]
```

20. In standard form, the equation is $y' + \frac{x^2-4}{x-2} y = \frac{1}{x^2-4}$ so the solution is guaranteed to exist on $(-2, 2)$.

```
sol = NDSolve[
   {(x - 2) y'[x] + (x^2 - 4) y[x] ==
     1 / (x + 2), y[0] == 3}, y[x],
   {x, -2, 2}];

Plot[y[x] /. sol, {x, -2, 2}]
```

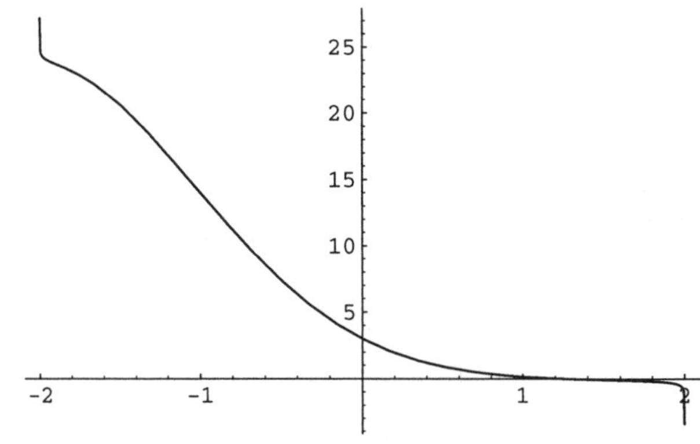

21. $(-2, 2)$

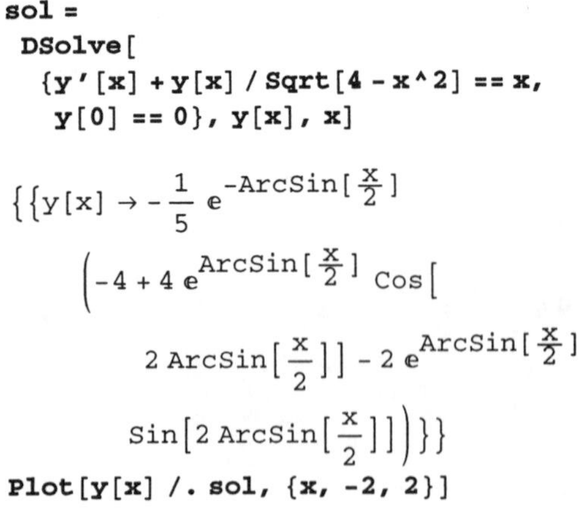

```
sol =
 DSolve[
  {y'[x] + y[x] / Sqrt[4 - x^2] == x,
   y[0] == 0}, y[x], x]
```

$$\{\{y[x] \to -\frac{1}{5} e^{-\text{ArcSin}[\frac{x}{2}]} \left(-4 + 4 e^{\text{ArcSin}[\frac{x}{2}]} \text{Cos}[2\, \text{ArcSin}[\frac{x}{2}]] - 2 e^{\text{ArcSin}[\frac{x}{2}]} \text{Sin}[2\, \text{ArcSin}[\frac{x}{2}]]\right)\}\}$$

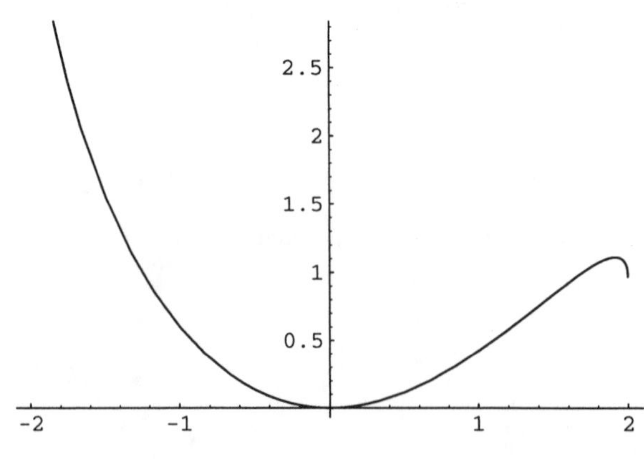

```
Plot[y[x] /. sol, {x, -2, 2}]
```

23. $x>0$, $y=\dfrac{\sin x}{x}-\cos x$, $-\infty<x<+\infty$

```
sol =
  DSolve[{y'[x] + 1 / x y[x] == Sin[x],
    y[Pi] == 1}, y[x], x]
```

$$\left\{\left\{y[x] \to \frac{-x\,\mathrm{Cos}[x] + \mathrm{Sin}[x]}{x}\right\}\right\}$$

```
Plot[y[x] /. sol, {x, -3 Pi, 3 Pi}]
```

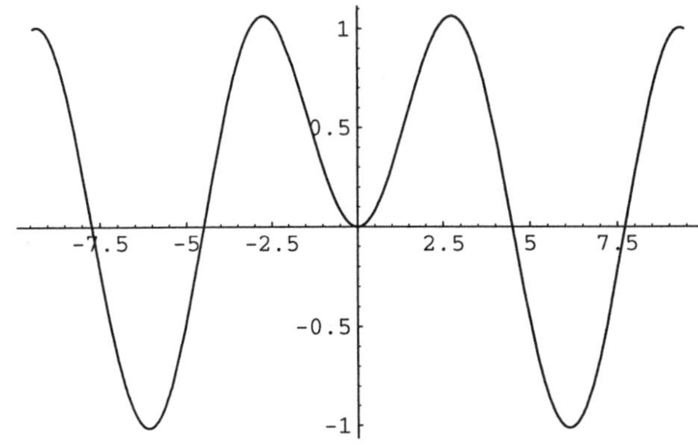

27. Separating variables gives us $y\,dy=-x\,dx \Rightarrow \frac{1}{2}y^2=\frac{1}{2}x^2+C \Rightarrow y=\pm\sqrt{C-x^2}$. Applying the initial condition indicates that $C=a^2$ so $y=\sqrt{a^2-x^2}$ (because $y(0)=a$ is positive). Thus, the interval of definition of the solution is $|x|<a$.

```
Clear[a, y, x, c]
sol =
  DSolve[{y'[x] == -x / y[x],
    y[0] == a}, y[x], x]
```

$$\left\{\left\{y[x] \to -\sqrt{a^2 - x^2}\right\},\right.$$

```
toplot = Table[sol[[2, 1, 2]],
    {a, 1, 5}];
Plot[Evaluate[toplot], {x, -5, 5},
  PlotRange -> {-3, 7}, AspectRatio -> 1]
```

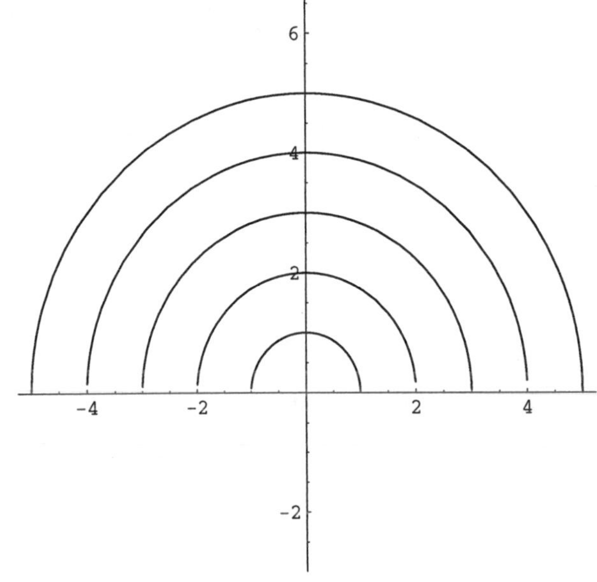

29. By the Existence and Uniqueness theorem, the solution *may not* be unique if $(\cos x-\sin y)\cos y=0$.

```
ContourPlot[Cos[y] (Cos[x] - Sin[y]),
  {x, 0, 2 π}, {y, -π, π}, Contours → {0},
  ContourShading → False, Frame → False,
  Axes → Automatic, AxesOrigin → {0, 0},
  PlotPoints → 160]
```

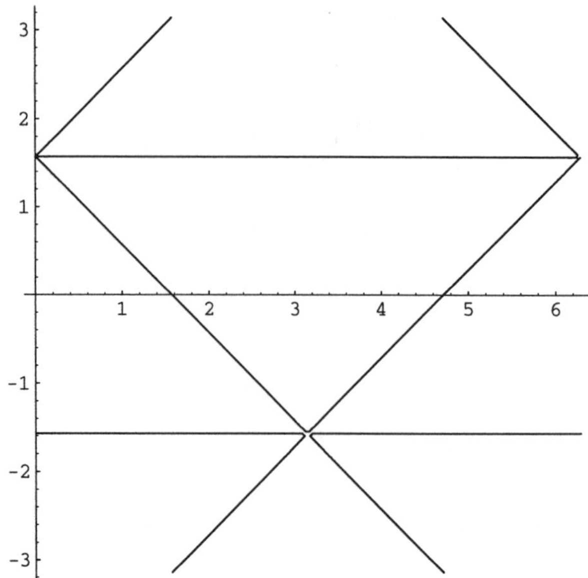

Thus, the solution *may not* be unique if $x_0 = \pi/2$ or $x_0 = 3\pi/2$.

In differential form the equation is $(\cos x \sin x + \sin x \sin y)dx + (\cos y \sin y - \cos x \cos y)dy = 0$ and is exact because

$$\frac{\partial}{\partial y}(\cos x \sin x + \sin x \sin y) = \cos y \sin x = \frac{\partial}{\partial x}(\cos y \sin y - \cos x \cos y).$$

Then,

$$\int (\cos x \sin x + \sin x \sin y)\, dx = -\frac{1}{2}(\cos x + 2\sin y)\cos x + g(y)$$
$$= \frac{1}{2}\sin^2 x - \sin y \cos x + g(y)$$

(because $\frac{1}{2}\sin^2 x = -\frac{1}{2}\cos^2 x + 1$) and

$$\frac{\partial}{\partial y}\left(-\frac{1}{2}(\cos x + 2\sin y)\cos x + g(y)\right) = -\cos x \cos y + g'(y) = \cos y \sin y - \cos x \cos y \Rightarrow g'(y) = \cos y \sin y$$

so $g(y) = \frac{1}{2}\sin^2 y$ and thus a general solution is

$$\frac{1}{2}\sin^2 x - \sin y \cos x + \frac{1}{2}\sin^2 y = C,$$

which is not unique if $x_0 = \pi/2$ or $3\pi/2$.

```
ContourPlot[
  Sin[x]^2/2 - Cos[x] Sin[y] + Sin[y]^2/2,
  {x, 0, 2 Pi}, {y, -Pi, Pi},
  Contours -> {1 / 2},
  ContourShading -> False,
  PlotPoints -> 120, Axes -> Automatic,
  AxesOrigin -> {0, 0}, Frame -> False]
```

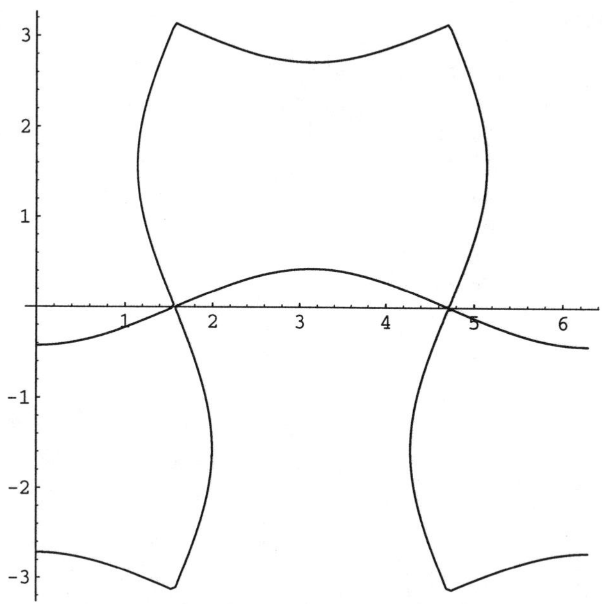

30. The graphs of the direction field and various solutions indicate the points at which there is no solution,

```
<< Graphics`PlotField`
PlotVectorField[
 {1, (Cos[y] - y Cos[x]) /
    (x Sin[y] + Sin[x] - 1)}, {x, 0, 4 Pi},
      {y, 0, 4 Pi},
      PlotPoints -> 30,
      Axes -> Automatic,
      ScaleFunction -> (0.5 &),
      HeadLength -> 0,
      HeadWidth -> 0]
```

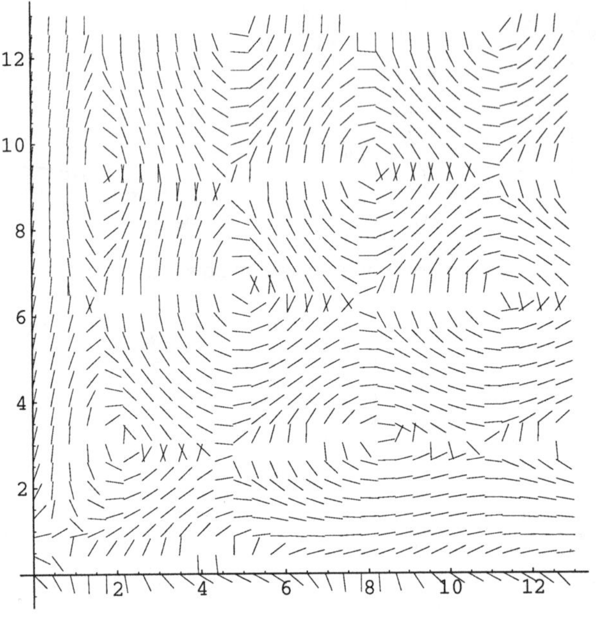


```
gensol = DSolve[(x Sin[y[x]] +
    Sin[x] - 1) y'[x] ==
       Cos[y[x]] - y[x] Cos[x],
  y[x], x];
toplot = gensol[[1, 1]] /. {y[x] -> y}
```

```
-y - x Cos[y] + y Sin[x]
```

```
ContourPlot[toplot, {x, 0, 4 Pi},
    {y, 0, 4 Pi},
    Axes -> Automatic,
  AxesOrigin -> {0, 0},
    Frame -> False,
    ContourShading -> False,
    Contours -> 30, PlotPoints -> 160]
```

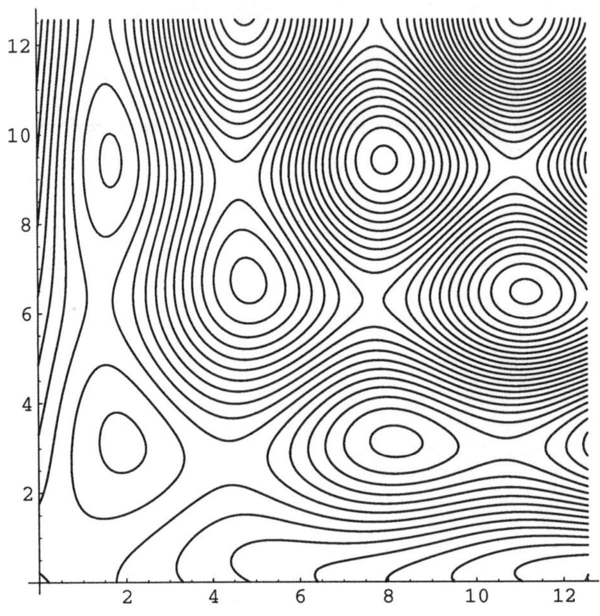

which are approximated as follows.

```
eq1 = Cos[y] - y Cos[x] == 0;
eq2 = 1 - Sin[x] - x Sin[y] == 0;
Map[
 FindRoot[{eq1, eq2}, {x, #[[1]]},
    {y, #[[2]]}] &,
 {{1.89, 3.12}, {1.64, 9.47},
  {8.24, 3.17}, {7.95, 9.47},
  {11.1, 6.49}}]
```

```
{{x → 1.89762, y → 3.11369},
 {x → 1.67714, y → 9.42141},
 {x → 8.1786, y → 3.13521},
 {x → 7.96029, y → 9.42407},
 {x → 11.1484, y → 6.4625}}
```

EXERCISES 2.6

We state the following algorithms that can be used to generate numerical solutions of some initial-value problems of the form

$$\begin{cases} y' = f(x,y) \\ y(x_0) = y_0 \end{cases}.$$

The following three algorithms implement Euler's method., the improved Euler's method, and the Runge-Kutta method of order four.
For all three algorithms, first define $f(x,y)$, x_0, y_0, and the desired step size, *h*.

Mathematica

```
Clear[f,h,xe,ye,x0,y0,xr,yr,
    xi,yi,k1,k2,k3,k4]
f[x_,y_]=?;
h=?;
x0=?;
y0=?;
```

Maple V

```
f:='f':h:='h':xe:='xe':ye:='ye':
    x0:='x0':xr:='xr':yr:='yr':
    xi:='xi':yi:='yi':
f:=(x,y)->?:
h:=?:
x0:=?:
y0:=?:
```

The following implements Euler's method.

```
xe[n_]=x0+n h;
ye[n_]:=ye[n]=h*f[xe[n-1],ye[n-1]]+ye[n-1];
ye[0]=y0;
```

```
xe:=n->x0+n*h:
ye:=proc(n) option remember;
    ye(n-1)+h*f(xe(n-1),ye(n-1))
    end:
ye(0):=y0:
```

The following implements the improved Euler's method.

```
xi[n_]=x0+n h;
yi[n_]:=yi[n]=h(f[xi[n-1],yi[n-1]]+
    f[xi[n],h*f[xi[n-1],yi[n-1]]+yi[n-1]])/2+
    yi[n-1];
yi[0]=y0;
```

```
xi:=n->x0+n*h:
yi:=proc(n) option remember;
    yi(n-1)+h/2*
    (f(xi(n-1),yi(n-1))+f(xi(n-1),
        yi(n-1)+h*f(xi(n-1),yi(n-1))))
    end:
yi(0):=y0:
```

The following implements the fourth-order Runge-Kutta method.

```
xr[n_]=x0+n h;
yr[n_]:=yr[n]=yr[n-1]+h/6(k1[n-1]+
    2k2[n-1]+2k3[n-1]+k4[n-1]);
yr[0]=y0;
k1[n_]:=k1[n]=f[xr[n],yr[n]];
k2[n_]:=k2[n]=f[xr[n]+h/2,yr[n]+h k1[n]/2];
k3[n_]:=k3[n]=f[xr[n]+h/2,yr[n]+h k2[n]/2];
k4[n_]:=k4[n]=f[xr[n+1],yr[n]+h k3[n]]
```

```
xr:=n->x0+n*h:
yr:=proc(n)
    local k1,k2,k3,k4;
    option remember;
    k1:=f(xr(n-1),yr(n-1));
    k2:=f(xr(n-1)+h/2,
        yr(n-1)+h*k1/2);
    k3:=f(xr(n-1)+h/2,
        yr(n-1)+h*k2/2);
    k4:=f(xr(n),yr(n-1)+h*k3);
    yr(n-1)+h/6*(k1+2*k2+2*k3+k4)
    end:
yr(0):=y0:
```

3, 11, and 19.
For $h = 0.1$:

n	x_n	y_n (Euler's method)	y_n (improved Euler's method)	y_n (Runge-Kutta method of order 4)
0	0	1	1	1
1	0.1	1	1.005	1.00535
2	0.2	1.01	1.02211	1.02298
3	0.3	1.03201	1.05412	1.05572
4	0.4	1.06851	1.10465	1.10733
5	0.5	1.12269	1.17872	1.18299
6	0.6	1.19873	1.28354	1.29026
7	0.7	1.30242	1.43008	1.44075
8	0.8	1.44206	1.63607	1.65355
9	0.9	1.63001	1.93233	1.9626
10	1.	1.8857	2.37754	2.43501

For $h = 0.05$:

n	x_n	y_n (Euler's method)	y_n (improved Euler's method)	y_n (Runge-Kutta method of order 4)
0	0	1	1	1
1	0.05	1	1.00125	1.00129
2	0.1	1.0025	1.00526	1.00535
3	0.15	1.00775	1.01231	1.01247
4	0.2	1.01603	1.02275	1.02298
5	0.25	1.02764	1.03693	1.03725
6	0.3	1.04295	1.05529	1.05572
7	0.35	1.06233	1.07832	1.07889
8	0.4	1.08626	1.10661	1.10733
9	0.45	1.11526	1.14084	1.14176
10	0.5	1.14995	1.18185	1.183
11	0.55	1.19107	1.23064	1.23208
12	0.6	1.2395	1.28845	1.29027
13	0.65	1.29632	1.35687	1.35916
14	0.7	1.36284	1.43787	1.44077
15	0.75	1.44071	1.53404	1.53774
16	0.8	1.53199	1.6488	1.65357
17	0.85	1.63934	1.78675	1.79301
18	0.9	1.76621	1.9543	1.96266
19	0.95	1.91719	2.16055	2.17199
20	1.	2.09847	2.41897	2.43514

7, 15, and 23.
For $h = 0.1$:

n	x_n	y_n (Euler's method)	y_n (improved Euler's method)	y_n (Runge-Kutta method of order 4)
0	0.	1.	1.	1.
1	0.1	1.08415	1.08627	1.08636
2	0.2	1.17254	1.17664	1.17682
3	0.3	1.26471	1.27055	1.27082
4	0.4	1.36006	1.36727	1.36763
5	0.5	1.45785	1.46596	1.4664
6	0.6	1.55721	1.56568	1.56621
7	0.7	1.6572	1.66546	1.66607
8	0.8	1.75683	1.76429	1.76498
9	0.9	1.85511	1.86125	1.86201
10	1.	1.95109	1.95547	1.95629

For $h = 0.05$:

n	x_n	y_n (Euler's method)	y_n (improved Euler's method)	y_n (Runge-Kutta method of order 4)
0	0.	1.	1.	1.
1	0.05	1.04207	1.04262	1.04263
2	0.1	1.08525	1.08633	1.08636
3	0.15	1.12947	1.13107	1.13111
4	0.2	1.17468	1.17677	1.17682
5	0.25	1.2208	1.22336	1.22342
6	0.3	1.26777	1.27075	1.27082
7	0.35	1.3155	1.31884	1.31892
8	0.4	1.36387	1.36753	1.36763
9	0.45	1.41281	1.41672	1.41683
10	0.5	1.46219	1.46629	1.4664
11	0.55	1.51189	1.51612	1.51624
12	0.6	1.5618	1.56608	1.56621
13	0.65	1.6118	1.61605	1.6162
14	0.7	1.66176	1.66591	1.66607
15	0.75	1.71155	1.71554	1.7157
16	0.8	1.76106	1.76481	1.76498
17	0.85	1.81016	1.8136	1.81379
18	0.9	1.85873	1.86182	1.86201
19	0.95	1.90667	1.90934	1.90954
20	1.	1.95388	1.95609	1.95629

25-27. (a) $y(t) = e^{-t}$, $y(1) = 1/e \approx 0.367879$

	$h = 0.1$	$h = 0.05$	$h = 0.025$
25. (Euler's)	$y(1) \approx 0.348678$	$y(1) \approx 0.358486$	$y(1) \approx 0.363232$
26. (Improved Euler's)	$y(1) \approx 0.368541$	$y(1) \approx 0.368039$	$y(1) \approx 0.367918$
27. (4th Order Runge Kutta)	$y(1) \approx 0.36788$	$y(1) \approx 0.414831$	$y(1) \approx 0.429069$

25.

```
Exp[-1] // N

0.367879
Remove[f, x, y]
f[x_, y_] = -y;
h = .1;
x0 = 0;
y0 = 1;
xe[n_] = x0 + n h;
ye[n_] :=
 ye[n] = h f[xe[n - 1], ye[n - 1]] +
   ye[n - 1];
ye[0] = y0;

ye[10]

0.348678
Remove[f, x, y]
f[x_, y_] = -y;
h = .05;
x0 = 0;
y0 = 1;
xe[n_] = x0 + n h;
ye[n_] :=
  ye[n] = h f[xe[n - 1], ye[n - 1]] +
    ye[n - 1];
ye[0] = y0;

ye[20]

0.358486
```

```
Remove[f, x, y]
f[x_, y_] = -y;
h = .025;
x0 = 0;
y0 = 1;
xe[n_] = x0 + n h;
ye[n_] :=
  ye[n] = h f[xe[n - 1], ye[n - 1]] +
    ye[n - 1];
ye[0] = y0;

ye[40]

0.363232
```

26.

```
Remove[f, x, y]
f[x_, y_] = -y;
h = .1;
x0 = 0;
y0 = 1;
xi[n_] = x0 + n h;
yi[n_] :=
  yi[n] =
   1
   — h (f[xi[n - 1], yi[n - 1]] +
   2
       f[xi[n],
        h f[xi[n - 1], yi[n - 1]] +
         yi[n - 1]]) + yi[n - 1];
yi[0] = y0;

yi[10]

0.368541
```

```
Remove[f, x, y]
f[x_, y_] = -y;
h = .05;
x0 = 0;
y0 = 1;
xi[n_] = x0 + n h;
yi[n_] :=
  yi[n] =
   1
   — h (f[xi[n - 1], yi[n - 1]] +
   2
       f[xi[n],
        h f[xi[n - 1], yi[n - 1]] +
         yi[n - 1]]) + yi[n - 1];
yi[0] = y0;

yi[20]

0.368039
```

```
Remove[f, x, y]
f[x_, y_] = -y;
h = .025;
x0 = 0;
y0 = 1;
xi[n_] = x0 + n h;
yi[n_] :=
  yi[n] =
   1
   — h (f[xi[n - 1], yi[n - 1]] +
   2
       f[xi[n],
        h f[xi[n - 1], yi[n - 1]] +
         yi[n - 1]]) + yi[n - 1];
yi[0] = y0;

yi[40]

0.367918
```

28.

```
Clear[x, y]
partsol =
  NDSolve[{y'[x] == Sin[2 x - y[x]],
    y[0] == 0.5}, y[x], {x, 0, 15}];
Plot[y[x] /. partsol, {x, 0, 15}]
```

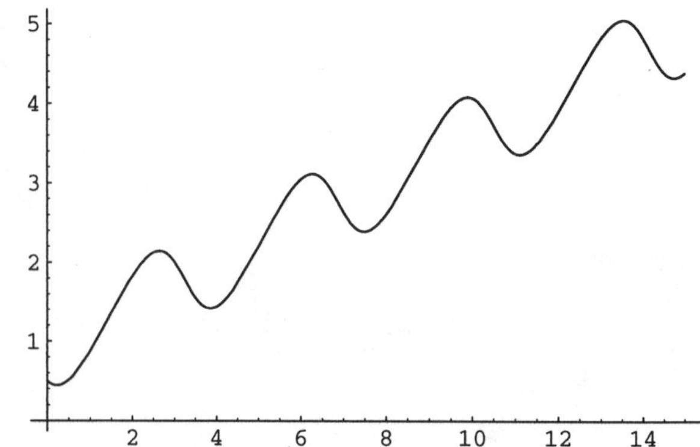

29.

```
numericalsols = Map[NDSolve[
      {y'[x] == Sin[x y[x]],
     y[0] == #},
      y[x], {x, 0, 7}][[1, 1, 2]]
   {0.5, 1.0, 1.5, 2.0, 2.5}];
Plot[Evaluate[numericalsols],
 {x, 0, 7}]
```

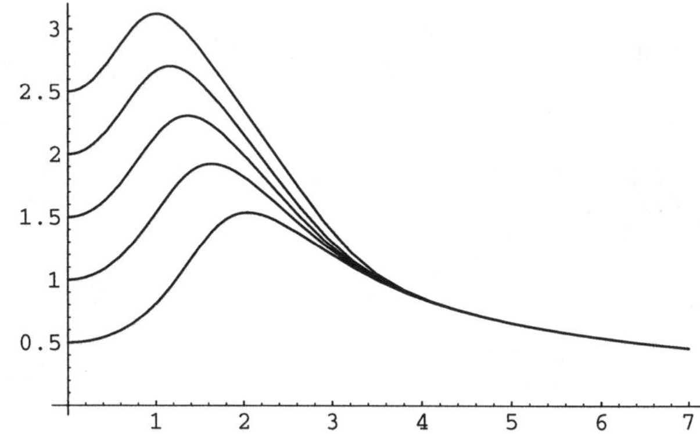

```
numericalsols /. x -> 0.5
```

```
{0.566144, 1.12971,
 1.68832, 2.23992, 2.78297}
```

30.

```
<< "Graphics`PlotField`"
dirfield = PlotVectorField[
  {1, x^2 + y^2}, {x, -1, 1}, {y, -1, 1
  Axes → Automatic,
  AxesOrigin → {0, 0}, HeadWidth → 0
  HeadLength → 0,
  ScaleFunction → (1 &),
  PlotPoints → 30]
```

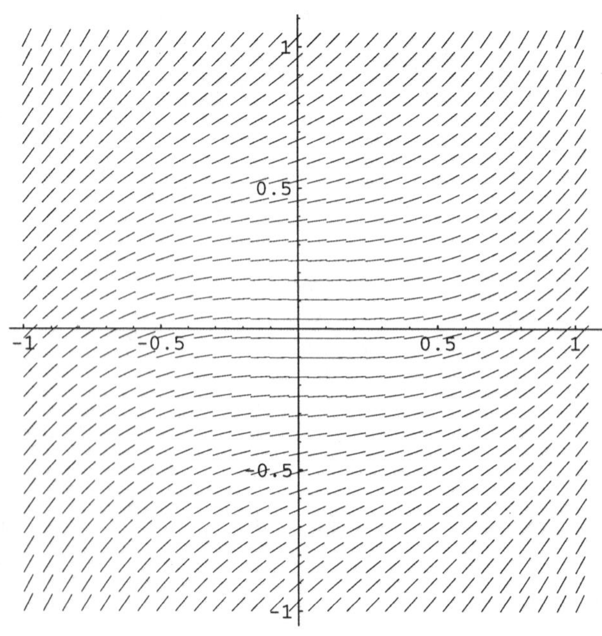

```
numsol =
  NDSolve[{y'[x] == x^2 + y[x]^2,
    y[0] == 0}, y[x], {x, -1, 1}];
Plot[y[x] /. numsol, {x, -1, 1},
  PlotRange → {-1, 1}, AspectRatio -
```

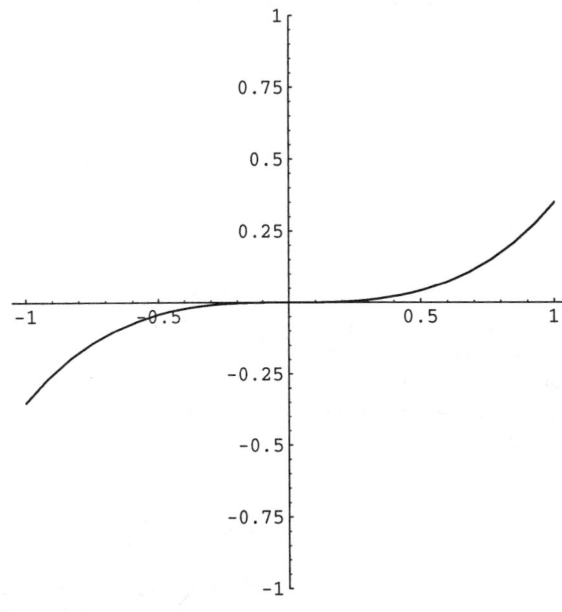


```
numsol /. x -> 1

{{y[1] → 0.350247}}
```

CHAPTER 2 REVIEW EXERCISES

3. Use separation of variables:

$$\frac{dy}{dx} = \frac{\sinh x}{2\cosh y} \Rightarrow \cosh y\, dy = \frac{1}{2}\sinh x\, dx$$

$$\Rightarrow \sinh y - \frac{1}{2}\cosh x = C$$

```
ContourPlot[Sinh[y] - 1 / 2 Cosh[x],
 {x, -2 Pi, 2 Pi}, {y, -2 Pi, 2 Pi},
 PlotPoints -> 300, Contours -> 20,
 ContourShading -> False,
 Frame -> False, Axes -> Automatic,
 AxesOrigin -> {0, 0}]
```

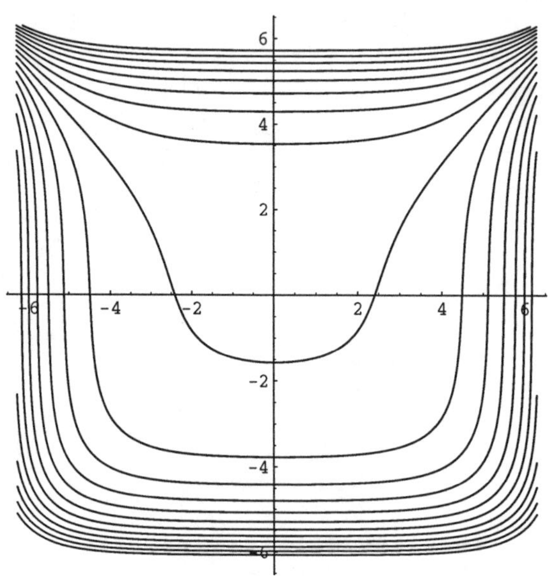

7. Use separation of variables:

$$\frac{dy}{dx} = \frac{y}{e^{2x}\ln y} \Rightarrow \frac{\ln y}{y} dy = e^{-2x} dx$$

$$\Rightarrow \frac{1}{2}(\ln y)^2 + \frac{1}{2}e^{-2x} = C$$

11. The equation is homogeneous of degree one. We let $y = u\,x$ so $dy = x\,du + u\,dx$. Then,

$$(y-x)dx + (x+y)dy = 0$$

$$x(u-1)dx + x(1+u)(x\,du + u\,dx) = 0$$

$$x\left(u^2+2u-1\right)dx + x^2(1+u)du = 0$$

$$x\left(u^2+2u-1\right)dx = -x^2(1+u)du$$

$$\frac{1}{x}dx = -\frac{u+1}{u^2+2u-1}du$$

$$\ln|x| = -\frac{1}{2}\ln\left|u^2+2u-1\right| + C_1$$

$$\ln|x| = -\frac{1}{2}\ln\left|\frac{y^2}{x^2}+2\frac{y}{x}-1\right| + C_1 = \ln\left|\frac{x}{\sqrt{y^2+2xy-x^2}}\right| + C_1$$

$$x = C_2\frac{x}{\sqrt{y^2+2xy-x^2}}$$

$$-\frac{1}{2}x^2 + xy + \frac{1}{2}y^2 = C.$$

```
ContourPlot[-x^2/2 + x y + y^2/2,
  {x, -5, 5}, {y, -5, 5},
  PlotPoints → 300, Contours → 20,
  ContourShading → False, Frame → Fals
  Axes → Automatic, AxesOrigin → {0, 0
```

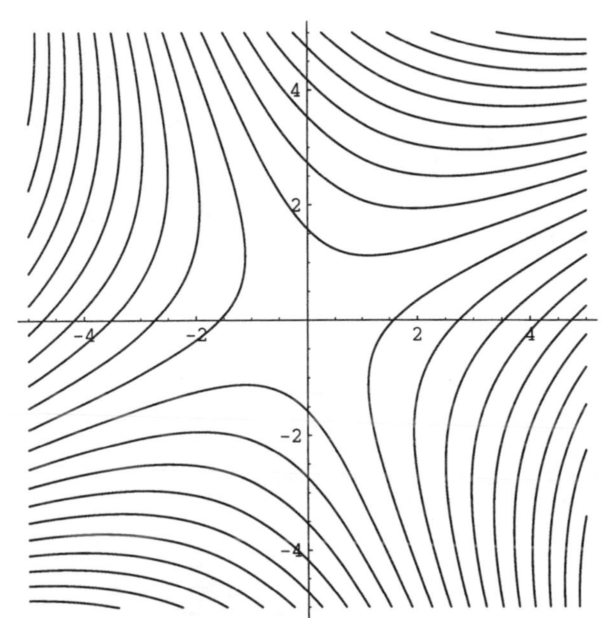

15. The equation is homogeneous of degree two:

$$\frac{dy}{dx} = \frac{5xy}{y^2+x^2}$$

$$5xy\,dx - \left(y^2+x^2\right)dy = 0.$$

Letting $x = v\,y$ yields $dx = y\,dv + v\,dy$ and substitution into the equation gives us

$$5vy^2(y\,dv + v\,dy) - y^2\left(1+v^2\right)dy = 0$$
$$5vy^3dv + y^2\left(4v^2-1\right)dy = 0$$
$$5vy^3dv = y^2\left(1-4v^2\right)dy$$
$$\frac{1}{y}dy = \frac{5v}{1-4v^2}dv = \left(\frac{5}{4}\frac{1}{1-2v} - \frac{5}{4}\frac{1}{1+2v}\right)dv$$
$$\ln|y| = -\frac{5}{8}\ln|1-2v| - \frac{5}{8}\ln|1+2v| + C.$$

Because $v = x/y$, we can rewrite the solution as

$$\ln|y| = -\frac{5}{8}\ln|1-2\,x/y| - \frac{5}{8}\ln|1+2\,x/y| + C$$
$$-\frac{8}{5}\ln|y| = \ln|1-2\,x/y| + \ln|1+2\,x/y| + C$$
$$y^{-8/5} = C\,(1-2\,x/y)(1+2\,x/y)$$
$$y^{-8/5} = C\frac{y^2-4x^2}{y^2}$$
$$y^{2/5} = C\left(y^2-4x^2\right)$$
$$y = C\left(y^2-4x^2\right)^{5/2}$$

19. This equation is exact because $\frac{\partial}{\partial y}(x\ln y) = \frac{x}{y} = \frac{\partial}{\partial x}\left(\frac{x^2}{2y}+1\right)$. Then, $\int x\ln y\,dx = \frac{1}{2}x^2\ln y + g(y)$ and

$$\frac{\partial}{\partial y}\left(\frac{1}{2}x^2\ln y + g(y)\right) = \frac{x^2}{2y} + g'(y) = \frac{x^2}{2y} + 1 \Rightarrow g'(y) = 1 \Rightarrow g(y) = y$$

so a general solution is $x^2\ln y + 2y = C$.

21. The equation is first-order linear:

$$y' + xy = x \qquad \left(\mu(x) = e^{\int x\,dx} = e^{x^2/2}\right)$$
$$e^{x^2/2}y' + xe^{x^2/2}y = xe^{x^2/2}$$
$$\frac{d}{dx}\left(e^{x^2/2}y\right) = xe^{x^2/2}$$
$$e^{x^2/2}y = e^{x^2/2} + C$$
$$y = 1 + Ce^{-x^2/2}.$$

25. We let $w = y^{1-(-2)} = y^3$. Then, $dw/dx = 3y^2\,dy/dx$ so $dy/dx = \dfrac{1}{3y^2}dw/dx$. Substitution into the equation yields

$$\begin{aligned}\frac{1}{3y^2}\frac{dw}{dx} + y &= e^x y^{-2}\\ \frac{dw}{dx} + 3y^3 &= 3e^x\\ \frac{dw}{dx} + 3w &= 3e^x.\end{aligned}$$

We solve this linear equation for w:

$$\begin{aligned}e^{3x}\frac{dw}{dx} + 3e^{3x}w &= 3e^{4x} \qquad \left(\mu(x) = e^{\int 3dx} = e^{3x}\right)\\ \frac{d}{dx}\left(e^{3x}w\right) &= 3e^{4x}\\ e^{3x}w &= \frac{3}{4}e^{4x} + C\\ w &= \frac{3}{4}e^x + Ce^{-3x}\\ y^3 &= \frac{3}{4}e^x + Ce^{-3x} \qquad \left(w = y^3\right).\end{aligned}$$

27. The equation $y - xy' = 2\ln(y')$ is a Clairaut equation with $f(t) = -t$ and $g(t) = 2\ln t$

$y = cx + 2\ln c$; *Sing*: $y = 2\ln\left(\dfrac{-2}{x}\right) - 2,\ x < 0$

29. $x = \dfrac{8}{3}p + Cp^{-2}$; $y = 2xy' - 4(y')^2 = 2\left(\dfrac{8}{3}p + Cp^{-2}\right)p - 4p^2 = \dfrac{4}{3}p^2 + 2Cp^{-1}$

32. This is an exact equation because $\dfrac{\partial}{\partial y}\left(y\,e^{xy} - 2x\right) = e^{xy} + x\,y\,e^{xy} = \dfrac{\partial}{\partial x}\left(x\,e^{xy}\right)$. Then,

$$\int\left(y\,e^{xy} - 2x\right)dx = e^{xy} - x^2 + g(y)$$

and

$$\frac{\partial}{\partial y}\left(e^{xy} - x^2 + g(y)\right) = x\,e^{xy} + g'(y) = x\,e^{xy}$$

so $g'(y) = 0$ and $g(y) = C$. Therefore, a general solution of the equation is

$$e^{xy} - x^2 = C.$$

Applying the condition $y(0) = 0$ yields

$$\begin{aligned}e^0 - 0 &= C\\ C &= 1\end{aligned}$$

so $e^{xy} - x^2 = 1 \Rightarrow y = \begin{cases}\ln\left(1 + x^2\right)/x, & x \neq 0\\ 0, & x = 0\end{cases}$.

```
Plot[Log[1 + x^2]/x, {x, -5, 5}]
```

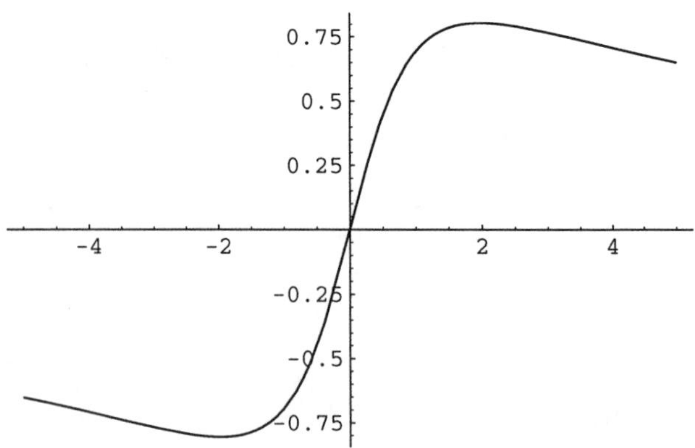

33. This equation is exact because $\frac{\partial}{\partial y}(\sin y - y\cos x) = \cos y - \cos x = \frac{\partial}{\partial x}(x\cos y - \sin x)$. Then,

$$\int (\sin y - y\cos x)\, dx = x\sin y - y\sin x + g(y)$$

and

$$\frac{\partial}{\partial y}(x\sin y - y\sin x + g(y)) = x\cos y - \sin x + g'(y) = x\cos y - \sin x \Rightarrow g'(y) = 0 \Rightarrow g(y) = 0$$

so a general solution of the equation is $x\sin y - y\sin x = C$. Application of the initial condition yields $C = 0$ so the solution to the initial-value problem is $x\sin y - y\sin x = 0$.

```
ContourPlot[x Sin[y] - y Sin[x],
  {x, -8 π, 8 π}, {y, -8 π, 8 π},
  PlotPoints → 600, Contours → {0},
  ContourShading → False, Frame → Fals
  Axes → Automatic, AxesOrigin → {0, 0}
```

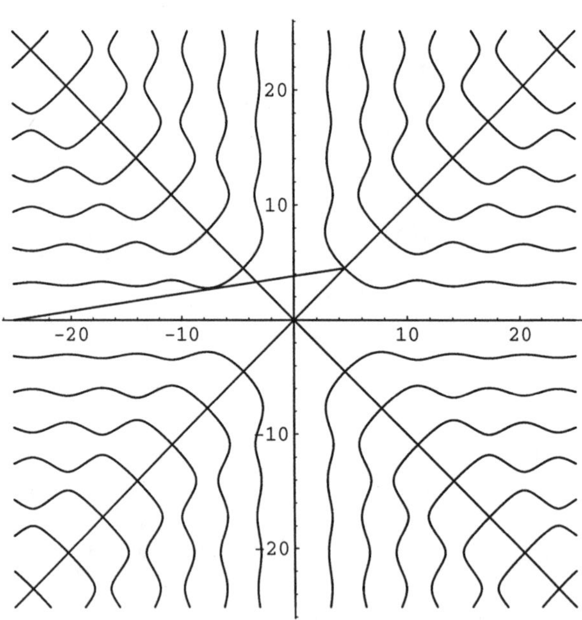

39.

n	x_n	y_n (Euler's method)	y_n (improved Euler's method)	y_n (Runge-Kutta method of order 4)
0	0	1	1	1
1	0.05	1	1.00125	1.00125
2	0.1	1.0025	1.00501	1.00501
3	0.15	1.0075	1.01129	1.01129
4	0.2	1.01503	1.02012	1.02013
5	0.25	1.02511	1.03156	1.03157
6	0.3	1.03779	1.04564	1.04565
7	0.35	1.0531	1.06242	1.06245
8	0.4	1.07112	1.08198	1.08201
9	0.45	1.09189	1.10437	1.10441
10	0.5	1.11548	1.12966	1.12971
11	0.55	1.14194	1.15789	1.15795
12	0.6	1.17132	1.1891	1.18918
13	0.65	1.20364	1.22329	1.2234
14	0.7	1.23889	1.26043	1.26056
15	0.75	1.27702	1.30042	1.30058
16	0.8	1.31791	1.34309	1.34329
17	0.85	1.36139	1.38818	1.38842
18	0.9	1.40717	1.43532	1.43561
19	0.95	1.45488	1.48403	1.48438
20	1.	1.50399	1.53369	1.5341

Differential Equations at Work: A. Modeling the Spread of a Disease

1.

Mathematica

```
gensol=DSolve[y'[t]-((gamma+mu)-lambda)y[t]
    ==lambda,y[t],t]
```

Note that we use capi *to represent* $I(t)$, *thus avoiding any ambiguity with the built-in symbol* I, *which represents the imaginary unit* $i = \sqrt{-1}$.

```
step2=Solve[gensol[[1,1,2]]==1/capi[t],
    capi[t]]
```

Maple V

```
gensol:=dsolve(diff(y(t),t)-
    ((gamma+mu)-lambda)*y=lambda,
    y(t));
```

Note that we use capi *to represent* $I(t)$, *thus avoiding any ambiguity with the built-in symbol* I, *which represents the imaginary unit* $i = \sqrt{-1}$.

```
step2:=subs(y(t)=1/capi(t),gensol);
step3:=solve(step2,capi(t));
```

2.

```
step3=step2[[1,1,2]] /. t->0
step4=Solve[step3==capi0,C[1])
```

```
step2[[1,1,2]] /. step4[[1]] // Simplify
```

```
step4:=eval(subs(t=0,step3));
step5:=solve(step4=capi0,_C1);
solve(i0=1/c,c);
```

```
step6:=subs(_C1=step5,step3);
simplify(step6);
```

(b)

```
gensol=DSolve[i'[t]==
    (lambda-(gamma+mu))i[t]-lambda i[t]^2,
    i[t],t]
gensol[[1,1,2]]
s1=gensol[[1,1,2]] /. t->0
cval=Solve[s1==i0,C[1]]
sol=gensol[[1,1,2]] /. cval[[1]]
lambda=3.6;
gamma=2;
mu=1;
toplot=Table[sol,{i0,.1,.9,.1}]
Plot[Evaluate[toplot],{t,0,2},
    PlotRange->{0,1},AspectRatio->1/2]
```

```
i:='i':
Eq:=diff(i(t),t)+
    (Gamma+Mu-Lambda)*i(t)=
    -Lambda*i(t)^2;
Sol:=dsolve({Eq,i(0)=i0},i(t));
assign(Sol):
Lambda:=3.6:
Gamma:=2:
Mu:=1:
Lambda/(Gamma+Mu);
eval(i(t));
to_plot1:={seq(
    subs(i0=.1*j,eval(i(t))),
      j=1..9)}:
plot(to_plot1,t=0..5);
```

4.

```
initconds=Table[i,{i,.1,.9,.1}]
Clear[lambda,gamma,mu]
lambda[t_]=5-2Sin[6t];
gamma=1;
mu=4;
numericalsols=Map[
    NDSolve[{i'[t]==
        (lambda[t]-(gamma+mu))i[t]-
            lambda[t]i[t]^2,
        i[0]==#},i[t],{t,0,2}][[1,1,2]]&,
    initconds]
Plot[Evaluate[numericalsols],{t,0,2},
    PlotRange->{0,1},AspectRatio->1/2]
```

```
with(DEtools):
Lambda:=t->5-2*sin(6*t):
Gamma:=1:
Mu:=4:
i:='i':
Eq:=diff(i(t),t)=
    (Lambda(t)-(Gamma+Mu))*i-
    Lambda(t)*i^2:
inits:={seq([0,i/10],i=1..9)};
DEplot1(Eq,i(t),0..10,inits,
    stepsize=0.05,arrows=NONE);
```

1
0.8
0.6
0.4
0.2
0 0.5 1 1.5 2

Differential Equations at Work: B. Population Model with Harvesting

3.

```
sola =
 DSolve[{y'[t] == y[t] - 1,
   y[0] == 1 / 2}, y[t], t]
```

$\{\{y[t] \to 1 - \frac{e^t}{2}\}\}$

```
solb =
 DSolve[{y'[t] == 1 / 2 y[t] - 1,
   y[0] == 1 / 2}, y[t], t]
```

$\{\{y[t] \to 2 - \frac{3\, e^{t/2}}{2}\}\}$

```
p1 = Plot[y[t] /. sola, {t, 0, 2},
    DisplayFunction -> Identity];
p2 = Plot[y[t] /. solb, {t, 0, 2},
    PlotStyle -> GrayLevel[.5],
    DisplayFunction -> Identity];
```

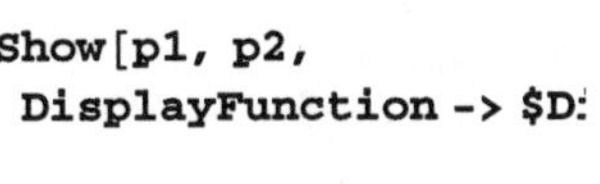

```
Show[p1, p2,
 DisplayFunction -> $D:
```

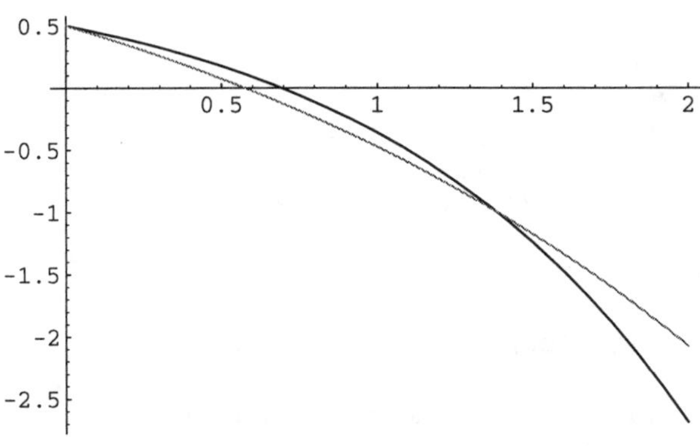

Differential Equations at Work: C. Logistic Model with Harvesting

1.

```
sol =
 DSolve[{y'[t] == y[t] - y[t]^2 - 1,
   y[0] == 1}, y[t], t]
```

$\{\{y[t] \to \frac{1}{2}\left(1 + \sqrt{3}\,\mathrm{Tan}\left[\frac{1}{6}(\pi - 3\sqrt{3}\,t)\right]\right)\}\}$

```
Plot[y[t] /. sol, {t, 0, 2}]
```

2.

```
sols =
  Map[
   NDSolve[
      {y'[t] == 7 / 10 y[t] -
         1 / 10 y[t]^2 - 1, y[0] == #},
      y[t], {t, 0, 4}][[1, 1, 2]] &,
   {3, 6, 1}];

Plot[Evaluate[sols], {t, 0, 4},
 PlotStyle ->
  {GrayLevel[0], GrayLevel[.5],
   Dashing[{0.01}]}]
```

Differential Equations at Work: D. Logistic Model with Predation

4.

```
eq = w'[t] == r w[t] (1 - 1 / k w[t]) -
     a w[t] ^ 2 / (b ^ 2 + w[t] ^ 2);
eqa = eq /. {r -> 1, k -> 15, a -> 5,
     b -> 2};

solsa =
  Map[
   NDSolve[{eqa, w[0] == #}, w[t],
     {t, 0, 15}] &,
   Table[i / 2, {i, 1, 30}]];

grays = Table[GrayLevel[i],
   {i, 0, 0.7, .7 / 29}];
p1 = Plot[Evaluate[w[t] /. solsa],
   {t, 0, 15},
   PlotRange -> {{0, 15}, {0, 15}},
   PlotStyle -> grays,
   AspectRatio -> 1];
```

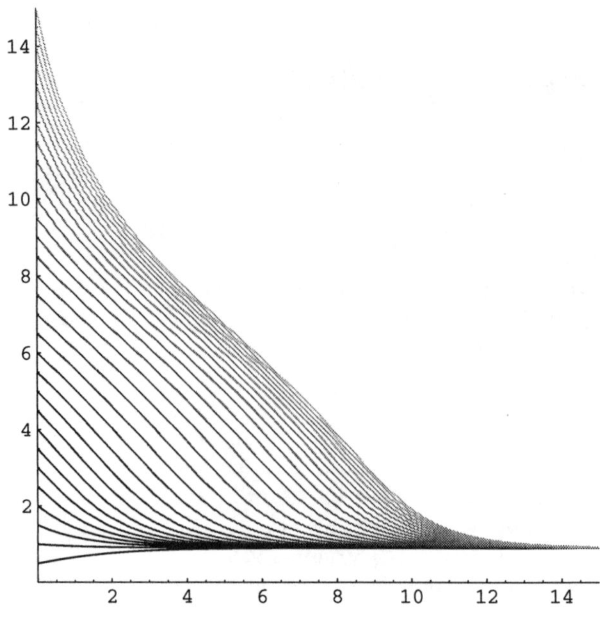

Applications of First-Order Ordinary Differential Equations

3

EXERCISES 3.1

3. We assume that $y(0)=y_0$. Then, $y(1)=y_0e^k=\frac{2}{3}y_0 \Rightarrow e^k=\frac{2}{3} \Rightarrow y(t)=y_0e^{kt}=y_0\left(\frac{2}{3}\right)^t$. To find when $\frac{1}{3}$ of the population remains, we solve

$$y(t)=y_0\left(\tfrac{2}{3}\right)^t=\tfrac{1}{3}y_0 \Rightarrow \left(\tfrac{2}{3}\right)^t=\tfrac{1}{3} \Rightarrow t\ln\tfrac{2}{3}=\ln\tfrac{1}{3} \Rightarrow t=\frac{\ln\frac{1}{3}}{\ln\frac{2}{3}} \approx 2.709511291$$

so $t \approx 2.71$ *days*.

7. Let $y(t)$ represent the amount ofthe mold at time t, where t is the number of hours after there are initially 500 grams of mold. Then, $y(0)=500$ and $dy/dt=k\,y \Rightarrow y(t)=500e^{kt}$. To find e^k we use $y(6)=600$: $y(6)=500e^{6k}=600 \Rightarrow e^{6k}=\frac{6}{5} \Rightarrow e^k=\left(\frac{6}{5}\right)^{1/6} \Rightarrow y(t)=500\left(\frac{6}{5}\right)^{t/6}$. After one day, which is twenty-four hours, we have $y(24)=500\left(\frac{6}{5}\right)^4 \approx 1036.8$. To find when there are 1000 g of mold, we solve

$$500\left(\tfrac{6}{5}\right)^{t/6}=1000 \Rightarrow \left(\tfrac{6}{5}\right)^{t/6}=2 \Rightarrow t=\frac{6\ln 2}{\ln\frac{6}{5}} \approx 22.8107041.$$

11. Let H be the half-life of the radioactive substance. Then,

$$y(t)=e^{H\,k}=\tfrac{1}{2} \Rightarrow e^k=\left(\tfrac{1}{2}\right)^{1/H} \Rightarrow y(t)=\left(\tfrac{1}{2}\right)^{t/H}$$

Thus,

$$y(t_1)=\left(\tfrac{1}{2}\right)^{t_1/H} \Rightarrow \ln[y(t_1)]=-\frac{t_1}{H}\ln 2 \qquad \text{and} \qquad y(t_2)=\left(\tfrac{1}{2}\right)^{t_2/H} \Rightarrow \ln[y(t_2)]=-\frac{t_2}{H}\ln 2$$

so

$$\ln[y(t_1)]-\ln[y(t_2)]=-\frac{t_1}{H}\ln 2+\frac{t_2}{H}\ln 2$$

$$\ln[y(t_1)]-\ln[y(t_2)]=\frac{1}{H}(t_2-t_1)\ln 2$$

$$H=\frac{(t_2-t_1)\ln 2}{\ln[y(t_1)]-\ln[y(t_2)]}.$$

15. The half-life of ^{226}Ra is approximately 1700 years. Then, $y(1700)=y_0e^{1700k}=\frac{1}{2}y_0 \Rightarrow e^k=\left(\frac{1}{2}\right)^{1/1700}$ so $y(t)=y_0\left(\frac{1}{2}\right)^{t/1700}$, $y(100) \approx 0.96y_0$ and approximately 96% of the original amount remains.

19. First, we write the equation as $\dfrac{dy}{dt}-r\,y=-ay^2$. Now, we let

$$w=y^{1-2}=y^{-1} \Rightarrow \frac{dw}{dt}=-y^{-2}\frac{dy}{dt} \Rightarrow -y^2\frac{dw}{dt}=\frac{dy}{dt}.$$

Then,

$$-y^2\frac{dw}{dt}-r\,y=-ay^2$$
$$\frac{dw}{dt}+r\,w=a$$
$$e^{rt}\frac{dw}{dt}+r\,e^{rt}w=ae^{rt}$$
$$\frac{d}{dt}\left(e^{rt}w\right)=ae^{rt}$$
$$e^{rt}w=\frac{a}{r}e^{rt}+C$$
$$w=\frac{a}{r}+Ce^{-rt}\Rightarrow\frac{1}{y}=\frac{a}{r}+Ce^{-rt}\Rightarrow y=\frac{1}{\frac{a}{r}+Ce^{-rt}}.$$

Application of the initial condition $y(0)=y_0$ gives us

$$\frac{1}{\frac{a}{r}+C}=y_0\Rightarrow C+\frac{a}{r}=\frac{1}{y_0}\Rightarrow C=\frac{a\,y_0+r}{r\,y_0}$$

so

$$y=\frac{1}{\frac{a}{r}+Ce^{-rt}}=\frac{1}{\frac{a}{r}+\frac{a\,y_0+r}{r\,y_0}e^{-rt}}=\frac{r\,y_0}{ay_0+\left(ay_0+r\right)e^{-rt}}.$$

23. We solve the initial-value problem $\left\{\frac{dy}{dt}=ky(200-y),y(0)=1\right\}$ subject to the additional condition $y(1)=50$. Separating variables, we find that

$$\frac{dy}{y(200-y)}=kdt$$
$$\frac{1}{200}\left(\frac{1}{y}+\frac{1}{200-y}\right)dy=kdt$$
$$\ln y-\ln(200-y)=200kt+C$$
$$\ln\frac{y}{200-y}=200kt+C$$
$$\frac{y}{200-y}=K_1e^{200kt}$$

Applying the condition $y(0)=1$ indicates that $K_1=\frac{1}{200-1}=\frac{1}{199}$. Then, the condition $y(1)=50$ implies that $\frac{50}{150}=\frac{1}{199}e^{200k}$, so $e^{200k}=\frac{199}{3}$. Therefore, $\frac{y}{200-y}=\frac{1}{199}\left(\frac{199}{3}\right)^t$. Solving for y, we find $\frac{200}{199}\left(\frac{199}{3}\right)^t-\frac{y}{199}\left(\frac{199}{3}\right)^t=y$ so that $y=\frac{\frac{200}{199}\left(\frac{199}{3}\right)^t}{1+\frac{1}{199}\left(\frac{199}{3}\right)^t}$. Multiplying both numerator and denominator by $\left(\frac{3}{199}\right)^t$ yields

$y = \dfrac{200}{1+199\left(\dfrac{3}{199}\right)^t}$; $y(2) \approx 191$; Because there is no t so that $y = 200$, all students do not learn of the rumor.

29. (a) $-r\left(1-\dfrac{1}{A}y\right)y = 0 \Rightarrow y = 0, A$; (c) $y(t) = \dfrac{2}{1+e^{rt}}$ and $\lim_{t\to\infty} y(t) = 0$; (d) $y(t) = \dfrac{6}{3-e^{rt}}$ and $\lim_{t\to\infty} y(t) = 0$

31. (a) $y = 0$, unstable; $y = 2$, semistable (b) $y = 0$, semistable; $y = 1$ unstable (c) $y = 0$, stable; $y = 1$ unstable (d) $y = -3$, unstable; $y = 0$, semistable; $y = 3$, stable.

```
<< Graphics`PlotField`
p1 = PlotVectorField[{1, y (y - 2) ^2},
    {x, 0, 4}, {y, -1, 3}, Axes → Automati‹
    ScaleFunction → (1 &),
    AxesOrigin → {0, 0}, HeadLength -> 0,
    PlotPoints -> 30,
    DisplayFunction -> Identity];

p2 = PlotVectorField[{1, y^2 (y - 1)},
    {x, 0, 4}, {y, -1, 3}, Axes → Automati‹
    ScaleFunction → (1 &),
    AxesOrigin → {0, 0}, HeadLength -> 0,
    PlotPoints -> 30,
    DisplayFunction -> Identity];

p3 = PlotVectorField[{1, y - Sqrt[y]},
    {x, 0, 4}, {y, 0, 2}, Axes → Automatic,
    ScaleFunction → (1 &),
    AxesOrigin → {0, 0}, HeadLength -> 0,
    PlotPoints -> 30,
    DisplayFunction -> Identity];

p4 = PlotVectorField[{1, y^2 (9 - y^2)},
    {x, 0, 8}, {y, -4, 4}, Axes → Automati‹
    ScaleFunction → (1 &),
    AxesOrigin → {0, 0}, HeadLength -> 0,
    PlotPoints -> 30,
    DisplayFunction -> Identity];

Show[GraphicsArray[{{p1,
```

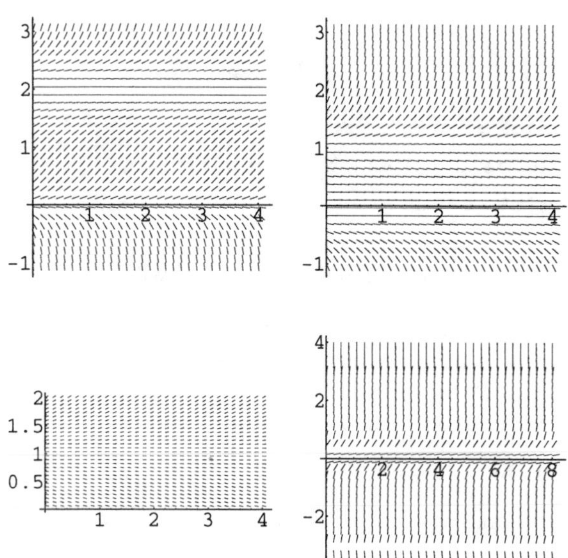

34.

```
Clear[f]
f[{r_,a_}]=r/(a+(r-a) Exp[-r t])
pairs={{0.5,0.5},{0.5,1.5},{1,0.5},
    {1,1.5},{1.5,.5},{1.5,1.5}};
grays=Table[GrayLevel[i],
    {i,0,.7,.7/5}];
toplot=Map[f,pairs];
Plot[Evaluate[toplot],{t,0,5},
    PlotStyle->grays,
    PlotRange->{0,5},AspectRatio->1]
```

```
f:='f':
f:=(r,a)->r/(a+(r-a) *exp(-r*t));
toplot:={f(0.5,0.5),f(0.5,1.5),
   f(1,0.5),f(1,1.5),f(1.5,.5),
   f(1.5,1.5)};
plot(toplot,t=0..5);
```

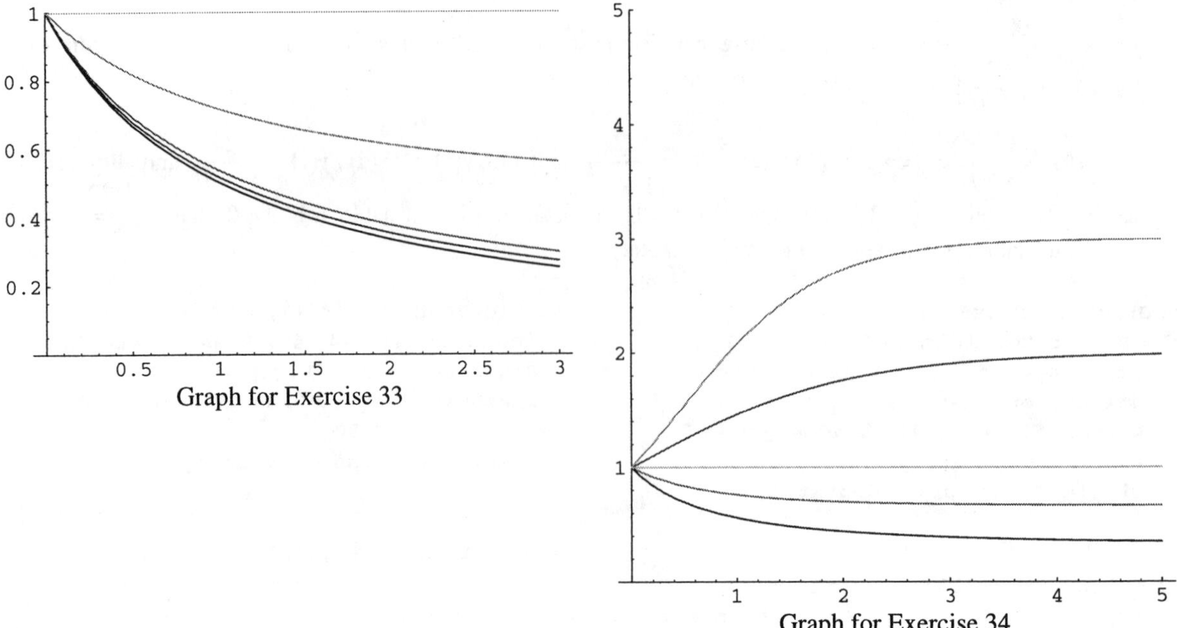

Graph for Exercise 33

Graph for Exercise 34

35.

```
gensol =
 DSolve[p'[t] == r p[t] - a p[t]^2 - h,
  p[t], t]
```

```
gensol:=dsolve(diff(p(t),t)=
   r*p(t)-a*p(t)^2-h,p(t));
gensol:=solve(gensol,p(t));
```

(b) The graph indicates that the species will become extinct after approximately 2958.7 years.

```
toplot = gensol[[1, 1, 2]] /.
    {r -> 0.03, a -> 0.0001, h -> 2.26,
     C[1] -> -1501.85};
Plot[toplot, {t, 0, 3000}]

FindRoot[toplot == 0, {t, 3000}]

{t → 2958.7}
```

(c),(d) The graphs indicate that h must be slightly smaller than $h \approx 0.1667$.

```
Clear[p]
partsol =
 DSolve[{p'[t] == r p[t] - a p[t]^2 - h,
    p[0] == 53/10}, p[t], t];

p[h_] = partsol[[1, 1, 2]] /.
  {r -> 0.03, a -> 0.0001}
```

$$5000.\left(0.03+\sqrt{-0.0009+0.0004\,h}\ \mathrm{Tan}\left[\frac{1}{2}\left(-\sqrt{-0.0009+0.0004\,h}\ t-2\,\mathrm{ArcTan}\left[\frac{0.02894}{\sqrt{-0.0009+0.0004\,h}}\right]\right)\right]\right)$$

```
hvals = {0, 0.5, 1, 1.5, 2, 2.25, 2.5};
Clear[toplot]
toplot = Map[p, hvals] // Chop;
grays = Table[GrayLevel[i],
    {i, 0, .7, .7 / 6}];
p1 = Plot[Evaluate[toplot], {t, 0, 10},
    PlotRange -> {-2, 8},
     PlotStyle -> grays, AspectRatio -> 1,
    DisplayFunction -> Identity];
```

```
Table[h, {h, 0, .5, .5 / 9}];
toplot2 =
  Table[p[h], {h, 0, .5, .5 / 9}];
grays = Table[GrayLevel[i],
    {i, 0, 7, .7 / 9}];
p2 = Plot[Evaluate[toplot2], {t, 0, 10}
    PlotRange -> {-2, 8},
    AspectRatio -> 1,
     PlotStyle -> grays,
    DisplayFunction -> Identity];
```

```
Show[GraphicsArray[{p1, p2}]]
```

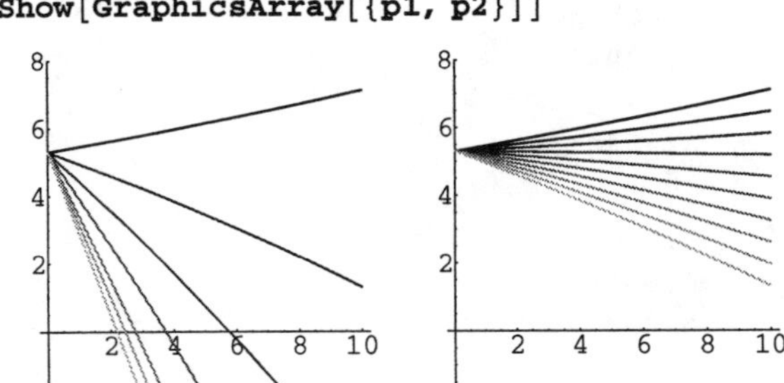

36. From the table we see that it took 13 days for milk to become safe for human consumption.

```
y[t_] = 3 Exp[k t];
kval = Solve[y[8.05] == 1.5, k];
y[t_] = y[t] /. kval[[1]]
Table[{t, y[t]}, {t, 1, 21}] //
 TableForm
```

$3\,e^{-0.0861052\,t}$

```
k:='k':y:='y':
y:=t->3*exp(k*t):
kval:=solve(y(8.05)=1.5,k);
y:=subs(k=kval,y(t));
array([seq([t0,evalf(subs(t=t0,y))],
   t0=1..21)]);
```

1	2.75249
2	2.52541
3	2.31705
4	2.12589
5	1.9505
6	1.78958
7	1.64194
8	1.50647
9	1.38218
10	1.26815
11	1.16353
12	1.06753
13	0.97945
14	0.89865
15	0.82451
16	0.75648
17	0.69407
18	0.63681
19	0.58427
20	0.53606
21	0.49184

EXERCISES 3.2

3. First, we have that $T(t) = -35e^{5k} + 75$. Now we find e^k:

$$T(5) = -35e^{5k} + 75 = 50$$
$$-35e^{5k} = -25$$
$$e^{5k} = \tfrac{5}{7} \Rightarrow e^k = \left(\tfrac{5}{7}\right)^{1/5}$$

so $T(t) = -35\left(\frac{5}{7}\right)^{t/5} + 75$. Next,

$$-35\left(\tfrac{5}{7}\right)^{t/5} + 75 = 60$$
$$-35\left(\tfrac{5}{7}\right)^{t/5} = -15$$
$$\left(\tfrac{5}{7}\right)^{t/5} = \tfrac{3}{7}$$
$$t = \frac{5\ln\frac{3}{7}}{\ln\frac{5}{7}} \approx 12.59 \text{ min.}$$

7. $T(t) = 20e^{kt} + 70$ and $T(3) = 20e^{3k} + 70 = 80 \Rightarrow 20e^{3k} = 10 \Rightarrow e^k = \left(\frac{1}{2}\right)^{1/3}$ so $T(t) = 20\left(\frac{1}{2}\right)^{t/3} + 70$. Then, $T(5) = 20\left(\frac{1}{2}\right)^{5/3} + 70 \approx 76.3°\text{F}$.

11. We have $T(t) = 132e^{kt} + 68$. To find e^k we use $T(2) = 170$:

$$T(2) = 132e^{2k} + 68 = 170 \Rightarrow 132e^{2k} = 102 \Rightarrow e^k = \left(\tfrac{17}{22}\right)^{1/2}$$

so $T(t) = 132\left(\frac{17}{22}\right)^{t/2} + 68$. Then, $T(t) = 132\left(\frac{17}{22}\right)^{t/2} + 68 = 140 \Rightarrow 132\left(\frac{17}{22}\right)^{t/2} = 72 \Rightarrow t = \dfrac{2\ln\frac{6}{11}}{\ln\frac{17}{22}} \approx 4.70$ min.

15. We solve the initial-value problem $\left\{\dfrac{du}{dt} = \dfrac{1}{4}\left[70 - 5\cos\dfrac{\pi t}{12} - u\right], u(0) = 65\right\}$. Writing the differential equation as $\dfrac{du}{dt} + \dfrac{u}{4} = \dfrac{1}{4}\left(70 - 5\cos\dfrac{\pi t}{12}\right)$, we find the integrating factor $\mu(t) = e^{t/4}$. Therefore, $\dfrac{d}{dt}\left[e^{t/4}u\right] = \dfrac{1}{4}\left(70 - 5\cos\dfrac{\pi t}{12}\right)e^{t/4}$ so that $e^{t/4}u = \dfrac{5e^{t/4}}{9+\pi^2}\left(126 + 14\pi^2 - 9\cos\dfrac{\pi t}{12} - 3\pi\sin\dfrac{\pi t}{12}\right) + C$ or $u = \dfrac{5}{9+\pi^2}\left(126 + 14\pi^2 - 9\cos\dfrac{\pi t}{12} - 3\pi\sin\dfrac{\pi t}{12}\right) + Ce^{-t/4}$. Applying $u(0) = 65$ indicates that $u(0) = \dfrac{5\left(126 + 14\pi^2 - 9\right)}{9+\pi^2} + C = 65$, so $C = \dfrac{-5\pi^2}{9+\pi^2}$ and

$$u = \frac{5}{9+\pi^2}\left(126 + 14\pi^2 - 9\cos\frac{\pi t}{12} - 3\pi\sin\frac{\pi t}{12} - \pi^2 e^{-t/4}\right).$$

or

$$u(t) = \frac{-5}{9+\pi^2}\left(-14\pi^2 + \pi^2 e^{-t/4} + 3\pi\sin\frac{\pi t}{12} + 9\cos\frac{\pi t}{12} - 126\right).$$

19.

$$R_1 = 4\frac{\text{gal}}{\text{min}},\ R_2 = 3\frac{\text{gal}}{\text{min}}$$

$$\frac{dV}{dt} = 4-3 = 1,\ V(0) = 200 \Rightarrow V(t) = t+200$$

$$\frac{dy}{dt} = \left(2\frac{\text{lb}}{\text{gal}}\right)\left(4\frac{\text{gal}}{\text{min}}\right) - \left(\frac{y}{t+200}\right)\left(3\frac{\text{gal}}{\text{min}}\right) = 8 - \frac{3y}{t+200};\ y(0) = 10$$

$$\frac{dy}{dt} + \frac{3y}{t+200} = 8 \Rightarrow \mu = e^{\int 3/(t+200)dt} = e^{3\ln(t+200)} = (t+200)^3$$

$$\frac{d}{dt}\left[(t+200)^3 y\right] = 8(t+200)^3 \Rightarrow (t+200)^3 y = 2(t+200)^4 + C \Rightarrow y = 2(t+200) + C(t+200)^{-3}$$

$$10 = 400 + C200^{-3} \Rightarrow C = -390\cdot 200^3 \Rightarrow y = 2t + 400 - 390\cdot 200^3(t+200)^{-3}$$

Then, $V(t) = t+200 = 400 \Rightarrow t = 200$; $y(200) = 2(200) + 400 - 390(200)^3(200+200)^{-3} = 605$ lb;

Concentration at t=200 is $\dfrac{y(200)}{V(200)} = \dfrac{605\text{ lb}}{400\text{ gal}} = 1.5125\text{ lb / gal}$.

21.

```
Clear[a, c, b]
sol[a_, c_, b_] := DSolve[{
    u'[t] == 1 / 4 (c - u[t]) + a + b,
    u[0] == 70}, u[t], t]
sola = sol[1 / 4, 75, 0] // Simplify
p1 = Plot[u[t] /. sola, {t, 0, 24},
   DisplayFunction -> Identity];
```

$$\{\{u[t] \to 76 - 6\, e^{-t/4}\}\}$$

```
solb = sol[1 / 4, 70 - 10 Cos[Pi t / 12], 0]
  Simplify
p2 = Plot[u[t] /. solb, {t, 0, 24},
   DisplayFunction -> Identity];
```

$$\left\{\left\{u[t] \to \frac{1}{9+\pi^2}\left(639 + 81\, e^{-t/4} + 71\,\pi^2 - e^{-t/4}\,\pi^2 - 90\,\text{Cos}\left[\frac{\pi t}{12}\right] - 30\,\pi\,\text{Sin}\left[\frac{\pi t}{12}\right]\right)\right\}\right\}$$

```
sol:=proc(a,c,b)
   dsolve({diff(u(t),t)=
      1/4*(c-u(t))+a(t)+b(t),
   u(0)=70},u(t))
      end:
sola:=sol(1/4,75,0);
plot(rhs(sola),t=0..24);
solb:=sol(1,70-10*cos(Pi*t/12),0);
plot(rhs(solb),t=0..24);
solc:=sol(1/4,70-10*cos(Pi*t/12),0);
plot(rhs(solc),t=0..24);
```

```
solc = sol[1, 70 - 10 Cos[Pi t / 12], 0] //
  Simplify
p3 = Plot[u[t] /. solc, {t, 0, 24},
   DisplayFunction -> Identity];
```

$$\left\{\left\{u[t] \to \frac{1}{9+\pi^2}\left(2\left(333+27\,e^{-t/4}+37\,\pi^2-2\,e^{-t/4}\,\pi^2-45\,\mathrm{Cos}\left[\frac{\pi t}{12}\right]-15\,\pi\,\mathrm{Sin}\left[\frac{\pi t}{12}\right]\right)\right)\right\}\right\}$$

```
Show[GraphicsArray[{p1, p2, p3}]]
```

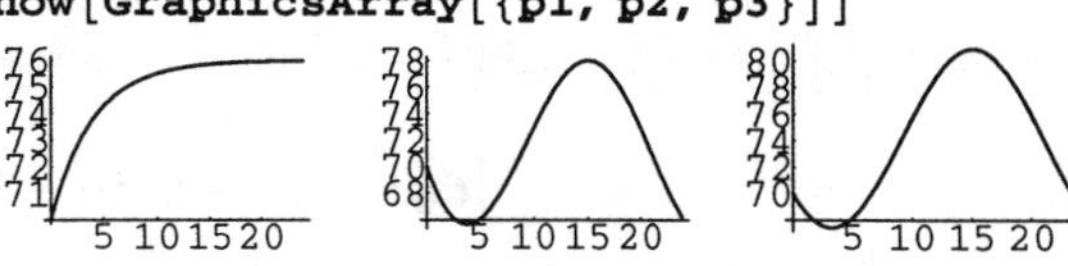

23.

```
temp = sol[0.25, 70 - 10 Cos[Pi t / 12],
  1.75 (ud - u[t])]
```

$$\{\{u[t] \to (0.000213358 - 1.05929\times 10^{-21}\,i)\,e^{-2.\,t}\,(41596.7\,e^{2.\,t} + 4101.09\,e^{2.\,t}\,ud + 1.\,(9.8696\,(489.-7.\,ud) - 576.\,(-499.+7.\,ud)) - 5760.\,e^{2.\,t}\,\mathrm{Cos}[0.261799\,t] - 753.982\,e^{2.\,t}\,\mathrm{Sin}[0.261799\,t])\}\}$$

```
avgtemp =
 1 / 24 Integrate[temp[[1, 1, 2]],
   {t, 0, 24}]
```

$$(8.88992\times 10^{-6} - 4.4137\times 10^{-23}\,i)\,(-4101.09\,(-34.9286+0.5\,ud) + 5.84472\times 10^{-18}\,(1.713\times 10^{23} + 1.68402\times 10^{22}\,ud)$$

```
avgtemp // N
```

$$(8.88992\times 10^{-6} - 4.4137\times 10^{-23}\,i)\,(-4101.09\,(-34.9286+0.5\,ud) + 5.84472\times 10^{-18}\,(1.713\times 10^{23} + 1.68402\times 10^{22}\,ud$$

```
exsol = Solve[avgtemp == 70, ud] // Ch
```

$\{\{ud \to 69.8273\}\}$

EXERCISES 3.3

3. First, we find the mass $m = W/g \Rightarrow m = 1/32$. Now we solve $\left\{\frac{1}{32}\frac{dv}{dt} = 1 - 2v,\ v(0) = 8\right\}$:

$$\frac{dv}{dt} + 64v = 32$$

$$e^{64t}\frac{dv}{dt} + 64e^{64t}v = 32e^{64t}$$

$$\frac{d}{dt}\left(e^{64t}v\right) = 32e^{64t}$$

$$e^{64t}v = \frac{1}{2}e^{64t} + C$$

$$v = \frac{1}{2} + Ce^{-64t}$$

$$v(0) = 8 \Rightarrow v = \frac{1}{2} + \frac{15}{2}e^{-64t}.$$

$v(1) \approx 0.5$ ft / s

7. The weight of the rock is 0.5 lb, so the mass is $m = \frac{1/2}{32} = \frac{1}{64}$, and we solve $\left\{\frac{1}{64}\frac{dv}{dt} = \frac{1}{2} - \frac{v}{64}, v(0) = 0\right\}$ or

$$\{dv/dt = 32 - v, v(0) = 0\} \Rightarrow v(t) = 32 - 32e^{-t}$$

$$\{ds/dt = v(t), s(0) = 0\} \Rightarrow s(t) = 32t + 32e^{-t} - 32;$$

$$s(4) \approx 96.59 \text{ ft} < 300 \Rightarrow 203.41 \text{ ft above the ground}$$

11. Because the mass of the object is 100 kg, the weight of it is $mg = (100)(9.8)$. Therefore, we solve $\{100\,dv/dt = -(100)(9.8) - v/10, v(0) = 100\}$ or $\{dv/dt = -9.8 - v/1000, v(0) = 100\}$. Using separation of variables gives us

$$\frac{1}{\frac{49}{5} + \frac{1}{1000}v}\,dv = -dt$$

$$1000 \ln\left|\frac{49}{5} + \frac{1}{1000}v\right| = -t + C$$

$$\frac{49}{5} + \frac{1}{1000}v = Ce^{-t/1000}$$

$$v = -9800 + Ce^{-t/1000}$$

and applying the initial condition results in

$$v(t) = -9800 + 9900e^{-t/1000};\ v(t) = 0 \Rightarrow t \approx 10.152 \text{ s};\ s(t) = -9800t + 9900000\left(-e^{-t/1000} + 1\right)$$

$$s(10.152) \approx 506.76 \text{ m}$$

15. Because the object reaches its max. height when $v = -gt + v_0 = 0$ or $t = v_0/g$ and the air resistance is ignored, the object hits the ground when $t = 2v_0/g$. Therefore, the velocity at this time is $v(2v_0/g) = -g \cdot 2v_0/g + v_0 = -v_0$.

19. The parachutist's mass is

$$192 \text{ lb} = m \cdot 32 \text{ ft/s}^2$$

$$m = 6 \text{ slugs.}$$

so we solve the initial-value problem

$$\begin{cases} 6\dfrac{dv}{dt} = 192 - 3v^2 \\ v(0) = 60 \end{cases} \quad \text{or} \quad \begin{cases} \dfrac{dv}{dt} = 32 - \dfrac{1}{2}v^2 \\ v(0) = 60 \end{cases}$$

Here, we use separation of variables:

$$\frac{dv}{32 - \frac{1}{2}v^2} = dt$$

$$\frac{1}{8}\left(\frac{1}{8+v} + \frac{1}{8-v}\right) = dt$$

$$(\ln|8+v| - \ln|8-v|) = 8t + C$$

$$\frac{8+v}{8-v} = C\,e^{8t}$$

Now, we find v:

$$\frac{8+v}{8-v} = C e^{8t}$$

$$8 + v = 8Ce^{8t} - C ve^{8t}$$

$$v + C ve^{8t} = 8Ce^{8t} - 8$$

$$v = \frac{8Ce^{8t} - 8}{1 + Ce^{8t}}$$

and apply the initial condition

$$v(0) = 60 \Rightarrow \frac{8C-8}{C+1} = 60 \Rightarrow C = -\frac{17}{13}$$

to see that $v(t) = 8\frac{17e^{8t}+13}{17e^{8t}-13}$. Last, we compute the limiting velocity:

$$\lim_{t\to\infty} 8\frac{17e^{8t}+13}{17e^{8t}-13} = 8 \lim_{t\to\infty} \frac{17+13e^{-8t}}{17-13e^{-8t}} \qquad \left(\text{multiply by } \frac{e^{-8t}}{e^{-8t}}\right)$$

$$= 8. \qquad \left(\lim_{t\to\infty} e^{-8t} = 0\right)$$

23. $g \approx 32\,\frac{\text{ft}}{\text{s}^2} \approx 0.006\,\frac{\text{mi}}{\text{s}^2}$; $v_0 = \sqrt{2gR} = \sqrt{2(.165)(0.006)(1080)} \approx 1.46 \text{ mi / s}$

27. We use an integrating factor:

$$\frac{dv}{dt} + \frac{1}{3}v = 16 - 4\sqrt{3}$$

$$e^{t/3}\frac{dv}{dt} + \frac{1}{3}e^{t/3}v = \left(16 - 4\sqrt{3}\right)e^{t/3} \qquad \left(\mu(t) = e^{\int 1/3\,dt} = e^{t/3}\right)$$

$$\frac{d}{dt}\left(e^{t/3}v\right) = \left(16 - 4\sqrt{3}\right)e^{t/3}$$

$$e^{t/3}v = 3\left(16 - 4\sqrt{3}\right)e^{t/3} + C$$

$$v = 3\left(16 - 4\sqrt{3}\right) + C e^{-t/3}.$$

Application of the initial condition yields

$$v(0) = 3\left(16 - 4\sqrt{3}\right) + C = 0 \Rightarrow C = -12\left(4 - \sqrt{3}\right)$$

so

$$v(t) = 12\left(4 - \sqrt{3}\right) - 12\left(4 - \sqrt{3}\right)e^{-t/3}.$$

Integrating we obtain

$$x(t) = 12\left(4 - \sqrt{3}\right)t + 36\left(4 - \sqrt{3}\right)e^{-t/3} + C$$

$$= 12\left(4 - \sqrt{3}\right)t + 36\left(4 - \sqrt{3}\right)e^{-t/3} - 36\left(4 - \sqrt{3}\right) \qquad \left(x(0) = 0 \Rightarrow C = -36\left(4 - \sqrt{3}\right)\right).$$

31. $v(t) = -gm/c$, $\lim_{t\to\infty}\left(-gm/c + Ce^{-ct/m}\right) = -gm/c$

33. $c = 1/2$: $v(t) = 64e^{-t/2}\left(-1 + e^{t/2}\right)$, $c = 1$: $v(t) = 32e^{-t}\left(-1 + e^{t}\right)$, $c = 2$: $v(t) = 16e^{-2t}\left(-1 + e^{2t}\right)$

```
sol1 =
 Map[
  DSolve[{v'[t] == 32 - # v[t],
       v[0] == 0}, v[t], t][[1, 1, 2]]
  {1/2, 1, 2}]
```

$\{e^{-t/2}(-64+64e^{t/2}),$
$e^{-t}(-32+32e^{t}),\ e^{-2t}(-16+16e^{2t}$

```
Plot[Evaluate[sol1], {t, 0, 3},
 PlotStyle ->
  {GrayLevel[0], GrayLevel[0.25],
   GrayLevel[0.5]}]
```

35. (Numerical solutions)

```
sol1 =
 NDSolve[{1/2 v'[t] == 16 - 16 v[
   v[0] == 0}, v[t], {t, 0, 5}]
```

```
{{v[t] → InterpolatingFunction[
    {{0., 5.}}, <>][t]}}
```

```
Plot[v[t] /. sol1, {t, 0, 1},
 PlotRange -> All]
```

$F_R = 16v^3$

```
sol2 =
 NDSolve[
  {1/2 v'[t] == 16 - 16 Sqrt[v[t]],
   v[0] == 0}, v[t], {t, 0, 5}]
```

```
{{v[t] → InterpolatingFunction[
    {{0., 5.}}, <>][t]}}
```

```
Plot[v[t] /. sol2, {t, 0, 1},
 PlotRange -> All]
```

$F_R = 16\sqrt{v}$

36. We use $v(t) = -\frac{gm}{c} + \frac{cv_0 + gm}{c}e^{-ct/m}$ and $s(t) = -\frac{gm}{c}t - \frac{cmv_0 + gm^2}{c^2}e^{-ct/m} + \frac{gm^2 + c^2 s_0 + cmv_0}{c^2}$ with $m = \frac{1}{128}$, $c = \frac{1}{160}$, and $g = 32$.

```
eq=v'[t]==-g-c/m v[t];
sol=DSolve[{eq,v[0]==v0},v[t],t]
sol2=DSolve[{s'[t]==sol[[1,1,2]],
    s[0]==s0},s[t],t]
height[t_,g_,c_,m_,v0_,s0_]=sol2[[1,1,2]]
```

```
eq:=diff(v(t),t)=-g-c/m*v(t);
sol:=dsolve({eq,v(0)=v0},v(t));
sol2:=dsolve({diff(s(t),t)=rhs(sol),
   s(0)=s0},s(t));
assign(sol2):
```

(a) The object reaches a greater maximum height on the first toss.

```
Plot[{height[t,32,1/160,1/128,48,0],
    height[t,32,1/160,1/128,36,6]},
    {t,0,3},
    PlotStyle->{GrayLevel[0],
        GrayLevel[0.5]}]
```

```
plot({subs({g=32,c=1/160,m=1/128,
        v0=48,s0=0},s(t)),
    subs({g=32,c=1/160,m=1/128,
        v0=36,s0=6},s(t))},
    t=0..3);
```

(b) Increasing v_0 increases the maximum height and the length of time the object is in the air.

```
Plot[{height[t,32,1/160,1/128,48,0],
    height[t,32,1/160,1/128,64,0],
    height[t,32,1/160,1/128,80,0]},
    {t,0,3},
    PlotStyle->{GrayLevel[0],
        GrayLevel[0.3],GrayLevel[0.6]}]
```

```
plot({subs({g=32,c=1/160,m=1/128,
        v0=48,s0=0},s(t)),
    subs({g=32,c=1/160,m=1/128,
        v0=64,s0=0},s(t)),
    subs({g=32,c=1/160,m=1/128,
        v0=80,s0=0},s(t))},
    t=0..3);
```

(c) The graph of $s(t)$ is translated vertically by the value of s_0.

```
Plot[{height[t,32,1/160,1/128,48,0],
    height[t,32,1/160,1/128,48,10],
    height[t,32,1/160,1/128,48,20]},
    {t,0,3},
    PlotStyle->{GrayLevel[0],
    GrayLevel[0.3],GrayLevel[0.6]}]
```

```
plot({subs({g=32,c=1/160,m=1/128,
        v0=48,s0=0},s(t)),
    subs({g=32,c=1/160,m=1/128,
        v0=48,s0=10},s(t)),
    subs({g=32,c=1/160,m=1/128,
        v0=48,s0=20},s(t))},
    t=0..3);
```

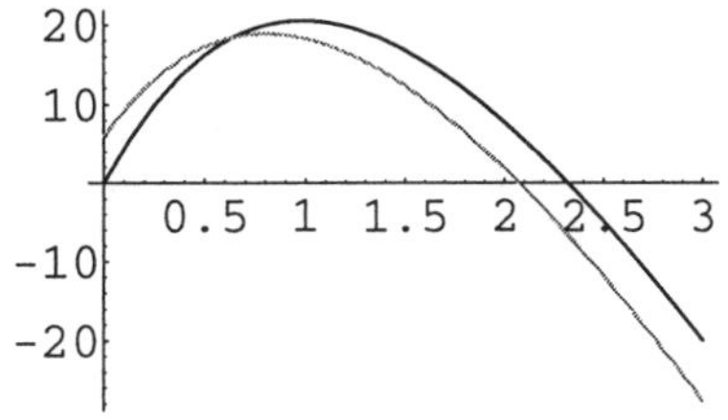

Graph for 36 (a)

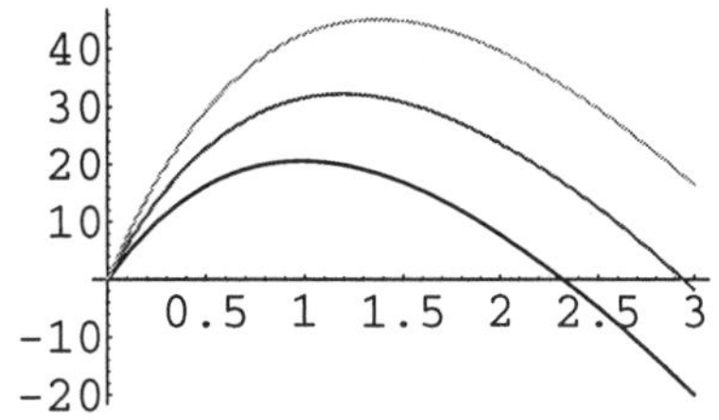

Graph for 36 (b)

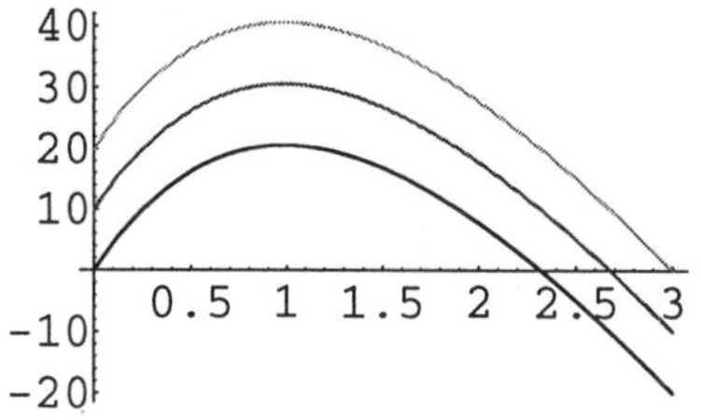

Graph for 36 (c)

37. The woman's mass is $m = 125/32$ slugs. At $t = 5$ s, she falls $s_1 \approx 272.479$ ft. After the parachute opens, she falls $4000 - 272.479 \approx 3727.52$ ft in approximately 295.772 s. The total time of the fall is $295.772 + 5 \approx 300.772$ s.

```
sol1=DSolve[{v1'[t]==32-32/125
v1[t],v1[0]==0},v1[t],t]
vopen=sol1[[1,1,2]]/.t->5
sol2=DSolve[{v2'[t]==32-10 32/125 v2[t],
    v2[0]==vopen},v2[t],t]
Clear[s,sol3]
sol3=DSolve[{s'[t]==sol1[[1,1,2]],
    s[0]==0},s[t],t]
s1=sol3[[1,1,2]]/.t->5//N
dist=4000-s1
Clear[s,sol4]
sol4=DSolve[{s'[t]==sol2[[1,1,2]],
    s[0]==0},s[t],t]
Plot[{sol4[[1,1,2]],dist},{t,0,400}]
t2=FindRoot[sol4[[1,1,2]]==dist,{t,280}]
5+t2[[1,2]]
```

```
sol:=dsolve({diff(v(t),t)=
    32-32/125*v(t),v(0)=0},v(t));
assign(sol):
v0:=subs(t=5,v(t)):
sol2:=dsolve({diff(v1(t),t)=
    32-10*32/125*v1(t),v1(0)=v0},
    v1(t));
assign(sol2):
sol3:=dsolve({diff(s(t),t)=v(t),
    s(0)=0},s(t));
assign(sol3):
fall:=evalf(subs(t=5,s(t)));
dist:=4000-fall;
s1:='s1':
sol4:=dsolve({diff(s1(t),t)=v1(t),
    s1(0)=0},s1(t));
t2:=fsolve(s1(t)=dist,t);
t2+5;
```

38. The time required for the object to reach the pond's surface and its velocity when it reaches the surface are the same as those calculated in Example 4. Therefore, under the pond's surface we must solve $\{dv/dt = 32 - 6v^2, v(0) = v_0\}$ where $v_0 \approx 29.3166$ ft/s. Following steps similar to those in Example 2, we find that $v(t) = \dfrac{4\left(Ke^{16\sqrt{3}t} - 1\right)}{\sqrt{3}\left(1 + Ke^{16\sqrt{3}t}\right)}$ where $K = \left|\dfrac{4+\sqrt{3}v_0}{4-\sqrt{3}v_0}\right|$. The distance traveled while in the pond is found by solving $\{ds/dt = v(t), s(0) = 0\}$. The object travels 25 ft in approximately $t \approx 10.8699$ s. The total time is $10.8699 + 2.47864 \approx 13.3485$ s (which is about twice the time required in Example 4).

```
v0=29.3166;
k=Abs[(4+Sqrt[3] v0)/(4-Sqrt[3] v0)]
v[t_]=4(k Exp[16 Sqrt[3] t]-1)/
    (Sqrt[3] (1+k Exp[16 Sqrt[3] t]))
Clear[s]
sol=DSolve[{s'[t]==v[t],s[0]==0},s[t],t]
Plot[{sol[[1,1,2]],25},{t,0,15}]
t2=FindRoot[sol[[1,1,2]]==25,{t,10}]
t2[[1,2]]+2.47864
```

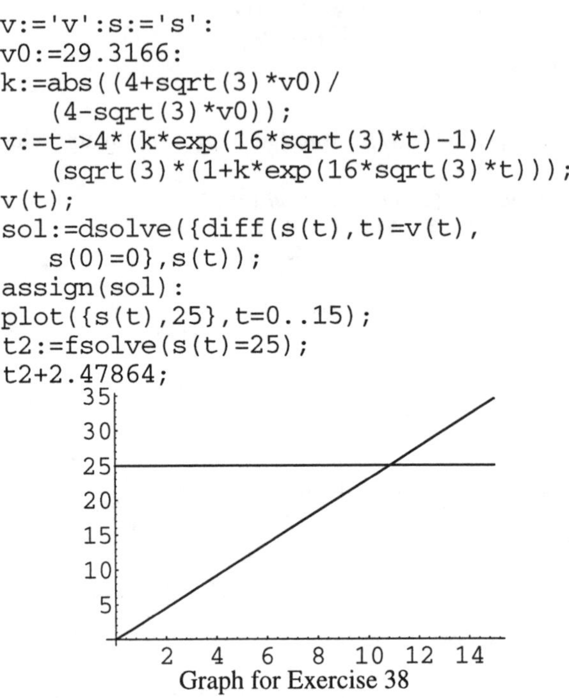

```
v:='v':s:='s':
v0:=29.3166:
k:=abs((4+sqrt(3)*v0)/
    (4-sqrt(3)*v0));
v:=t->4*(k*exp(16*sqrt(3)*t)-1)/
    (sqrt(3)*(1+k*exp(16*sqrt(3)*t)));
v(t);
sol:=dsolve({diff(s(t),t)=v(t),
    s(0)=0},s(t));
assign(sol):
plot({s(t),25},t=0..15);
t2:=fsolve(s(t)=25);
t2+2.47864;
```


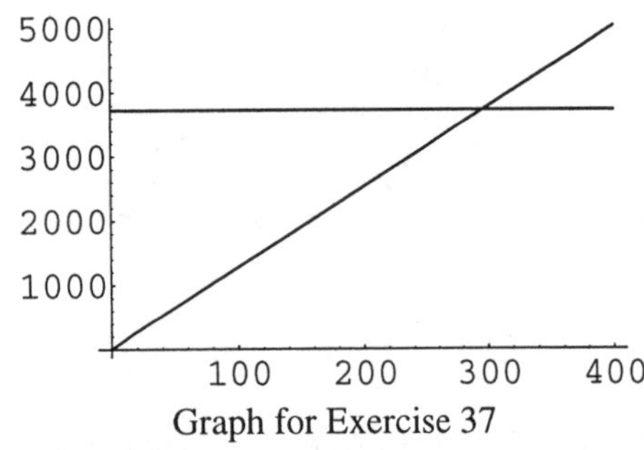

Graph for Exercise 37

Graph for Exercise 38

CHAPTER 3 REVIEW EXERCISES

3. $y<0$: $y'<0$; $0<y<4$, $y'>0$; $y>4$, $y'<0$; $y=0$: Unstable; $y=4$: Stable

```
Plot[-1/4 y (y-4), {y, -1, 5}]
```

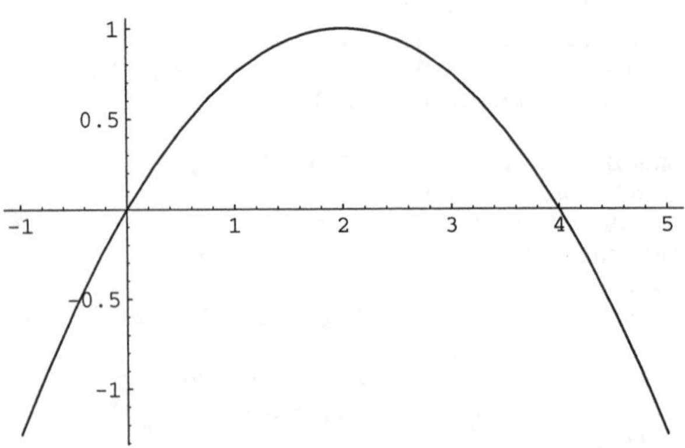

7. $y(t) = y_0 e^{kt}$; $y(1700) = y_0 e^{1700k} = \frac{1}{2} y_0 \Rightarrow e^k = \left(\frac{1}{2}\right)^{1/1700}$; Then, $y(t) = y_0 \left(\frac{1}{2}\right)^{t/1700}$ so that the amount that remains after 50 years is $y(50) = y_0 \left(\frac{1}{2}\right)^{50/1700} \approx 0.9798 y_0$. This indicates that approximately 97.98% of the original amount remains.

11. Using Newton's law of cooling, we have $T_0 = 40$ and $T_s = 90$ so $T(t) = 90 - 50e^{kt}$. Next,

$$T(20) = 90 - 50e^{20k} = 65 \Rightarrow -50e^{20k} = -25 \Rightarrow e^{20k} = \tfrac{1}{2} \Rightarrow e^k = \left(\tfrac{1}{2}\right)^{1/20}$$

so $T(t) = 90 - 50\left(\frac{1}{2}\right)^{t/20}$. After 30 minutes, the water's temperature is $T(30) = 90 - 50\left(\frac{1}{2}\right)^{3/2} \approx 72.32^o F$.

14. Solve $\left\{du/dt = \frac{1}{4}(85 - 10\cos(\pi t/12) - u), u(0) = 70\right\}$:

$$\frac{du}{dt} + \frac{1}{4}u = \frac{1}{4}(85 - 10\cos(\pi t/12) - u)$$

$$e^{t/4}\frac{du}{dt} + \frac{1}{4}e^{t/4}u = \frac{1}{4}e^{t/4}(85 - 10\cos(\pi t/12)) \qquad \left(\mu(t) = e^{\int 1/4\,dt} = e^{t/4}\right)$$

$$\frac{d}{dt}\left(e^{t/4}u\right) = \frac{1}{4}e^{t/4}(85 - 10\cos(\pi t/12))$$

$$e^{t/4}u = \frac{5}{9+\pi^2}e^{t/4}\left(17\left(9+\pi^2\right) - 18\cos(\pi t/12) - 6\pi\sin(\pi t/12)\right) + C$$

$$u = \frac{5}{9+\pi^2}\left(17\left(9+\pi^2\right) - 18\cos(\pi t/12) - 6\pi\sin(\pi t/12)\right) + Ce^{-t/4}.$$

Applying the initial condition yields

$$u(0) = \frac{5}{9+\pi^2}\left(17\left(9+\pi^2\right) - 18\right) + C = 70 \Rightarrow C = -\frac{15}{9+\pi^2}\left(3+\pi^2\right)$$

so

$$u = \frac{5}{9+\pi^2}\left(17\left(9+\pi^2\right) - 18\cos(\pi t/12) - 6\pi\sin(\pi t/12)\right) - \frac{15}{9+\pi^2}\left(3+\pi^2\right)e^{-t/4}.$$

```
Clear[u]
sol =
 DSolve[
  {u'[t] ==
    1 / 4 (85 - 10 Cos[Pi t / 12] - u[t]),
   u[0] == 70}, u[t], t] // Simplify
```

$$\left\{\left\{u[t] \to \frac{1}{9+\pi^2}\left(5\left(153 - 9\,e^{-t/4} + 17\,\pi^2 - 3\,e^{-t/4}\,\pi^2 - 18\,\mathrm{Cos}\left[\frac{\pi t}{12}\right] - 6\,\pi\,\mathrm{Sin}\left[\frac{\pi t}{12}\right]\right)\right)\right\}\right\}$$

```
Plot[u[t] /. sol, {t, 0, 24}]
```

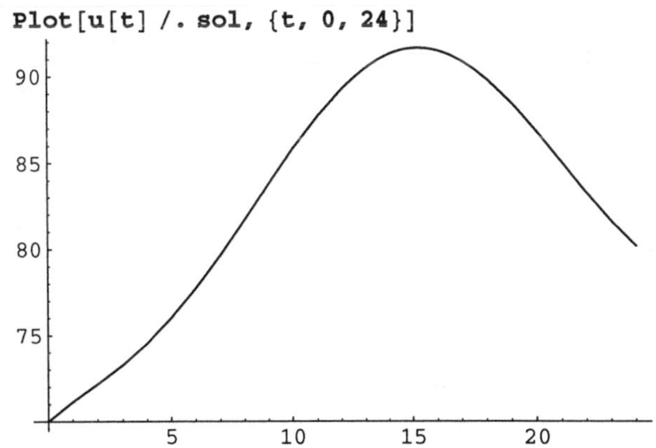

15. If the object weighs 4 lb, then its mass is found with $F = mg$ or $4 = m \cdot 32$ to be $m = 1/8$ slug. Therefore, we solve the initial-value problem $\left\{\frac{1}{8}dv/dt = 4 - v, v(0) = 0\right\}$ to find that $v(t) = 4 - 4e^{-8t}$. Then, $v(3) = 4 - 4e^{-8(3)} \approx 4$ ft / s. The distance traveled by the rock after t seconds is found by solving

$\left\{ds/dt = 4 - 4e^{-8t}, s(0) = 0\right\}$ which has solution $s(t) = 4t + \frac{1}{2}e^{-8t} - \frac{1}{2}$. After 3 seconds, the rock has fallen $s(3) = 4(3) + \frac{1}{2}e^{-8(3)} - \frac{1}{2} \approx 11.5$ ft

19. We solve the initial value-problem

$$\begin{cases} 4\dfrac{dv}{dt} = 128 - 2v^2 \\ v(0) = 30 \end{cases} \quad \text{or} \quad \begin{cases} \dfrac{dv}{dt} = 32 - \dfrac{1}{2}v^2 \\ v(0) = 30 \end{cases}$$

using separation of variables:

$$\frac{1}{32 - \frac{1}{2}v^2}\,dv = dt$$

$$\left(\frac{1}{8}\frac{1}{8+v} + \frac{1}{8}\frac{1}{8-v}\right)dv = dt$$

$$\frac{1}{8}\ln|8+v| - \frac{1}{8}\ln|8-v| = t + C$$

$$\frac{8+v}{8-v} = C\,e^{8t}$$

$$v = \frac{-8\left(1 - C\,e^{8t}\right)}{1 + C\,e^{8t}}$$

$$v = \frac{-8\left(1 + \frac{19}{11}e^{8t}\right)}{1 + \frac{19}{11}e^{8t}} \qquad \left(v(0) = 30 \Rightarrow \frac{-8(1-C)}{1+C} = 30 \Rightarrow C = -\frac{19}{11}\right)$$

$$v = \frac{8\left(19e^{8t} + 11\right)}{19e^{8t} - 11}.$$

The limiting velocity of the parachutist is

$$\lim_{t\to\infty} \frac{8\left(19e^{8t} + 11\right)}{19e^{8t} - 11} = 8\lim_{t\to\infty} \frac{19 + 11e^{-8t}}{19 - 11e^{-8t}} = 8.$$

23. (a)

$$r\frac{dr}{d\theta} + 4\sin 2\theta = 0$$

$$r\,dr = -4\sin 2\theta$$

$$\frac{1}{2}r^2 = 2\cos 2\theta + C$$

$$\frac{1}{2}r^2 = 2\cos 2\theta \quad \left(r(0) = 2 \Rightarrow C = 0\right).$$

Converting to rectangular coordinates results in

$$r^2 = 4\cos 2\theta = 4\left(2\cos^2\theta - 1\right).$$

$$r^4 = 4\cos 2\theta = 4\left(2r^2\cos^2\theta - r^2\right)$$

$$\left(x^2 + y^2\right)^2 = 4\left(2x^2 - x^2 - y^2\right)$$

$$\left(x^2 + y^2\right)^2 = 4\left(x^2 - y^2\right).$$

```
ContourPlot[
  (x^2 + y^2)^2 - 4 (x^2 - y^2),
  {x, -5/2, 5/2}, {y, -5/2, 5/2},
  ContourShading -> False,
  Contours -> {0}, Frame -> False,
  Axes -> Automatic, AxesOrigin -> {0,
  PlotPoints -> 120]
```

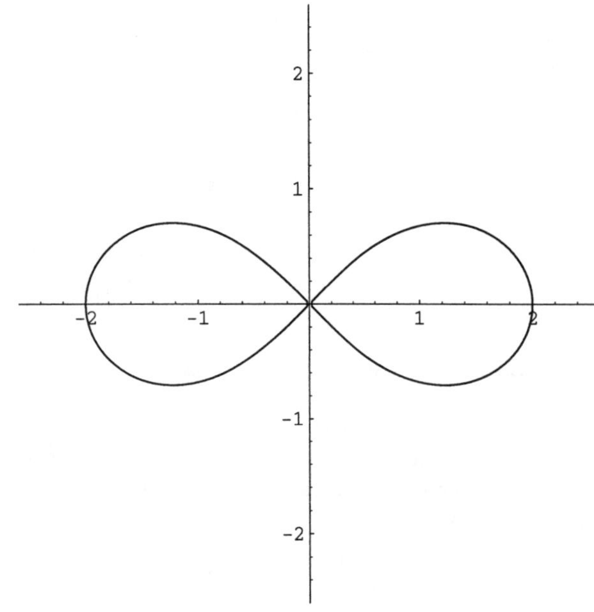

(b) Separate variables to solve:

$$\frac{dr}{d\theta} - 2\sec\theta\tan\theta$$

$$dr = 2\sec\theta\tan\theta\,d\theta$$

$$r = 2\sec\theta + C$$

$$r = 2\sec\theta + 2 \qquad \left(r(0) = 4 \Rightarrow C = 2\right).$$

```
PolarPlot[2 Sec[θ] + 2, {θ, 0, 2 Pi},
  PlotRange -> {{-5, 5}, {-5, 5}}]
```

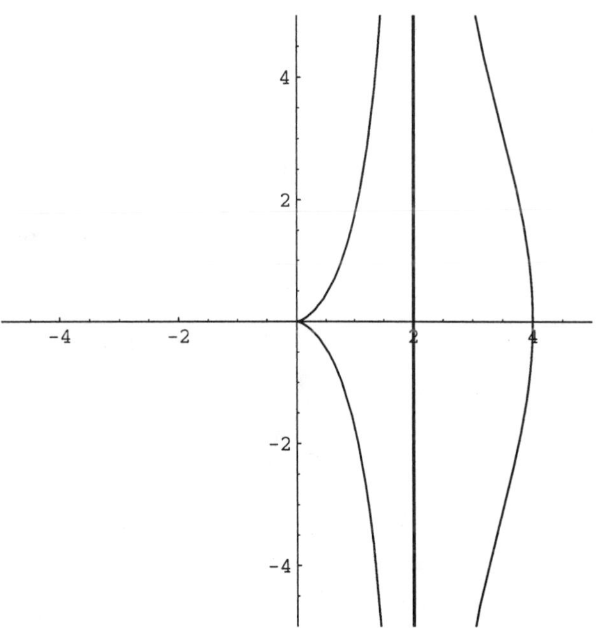

Alternatively, we convert to rectangular coordinates as follows:

$$\begin{aligned}
r &= \frac{2}{\cos\theta} + 2 \\
r\cos\theta &= 2 + 2\cos\theta \\
x &= 2 + 2\cos\theta \\
(x-2)^2 &= 4\cos^2\theta \\
(x-2)^2 r^2 &= 4r^2\cos^2\theta \\
(x-2)^2\left(x^2+y^2\right) &= 4x^2.
\end{aligned}$$

```
ContourPlot[(x - 2)^2 (x^2 + y^2) - 4
 {x, -5, 5}, {y, -5, 5},
 ContourShading -> False,
 Contours -> {0}, Frame -> False,
 Axes -> Automatic, AxesOrigin -> {0
 PlotPoints -> 120]
```

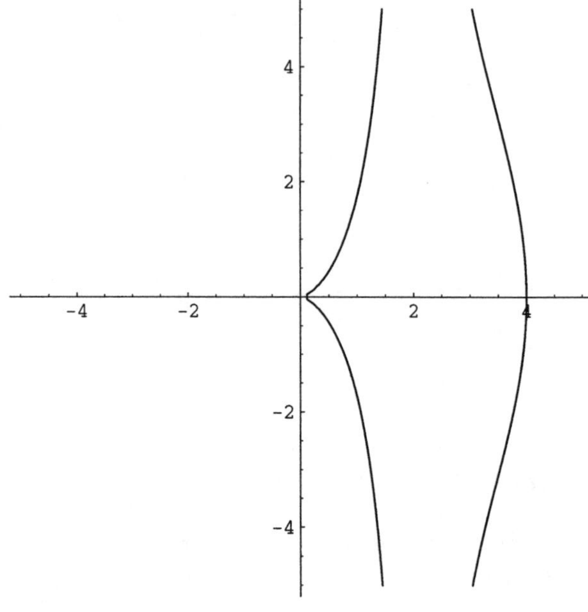

(c)

$$\frac{dr}{d\theta} - 6\sin(\theta/2)\cos(\theta/2) = 0$$

$$\begin{aligned}
dr &= 6\sin(\theta/2)\cos(\theta/2)d\theta \\
r &= 12\sin^2(\theta/2) + C \qquad \left(u = \sin(\theta/2) \Rightarrow 2du = \cos(\theta/2)d\theta\right) \\
r &= 12\sin^2(\theta/2) \quad \left(r(0) = 0 \Rightarrow C = 0\right).
\end{aligned}$$

```
PolarPlot[12 Sin[θ / 2] ^2, {6
```

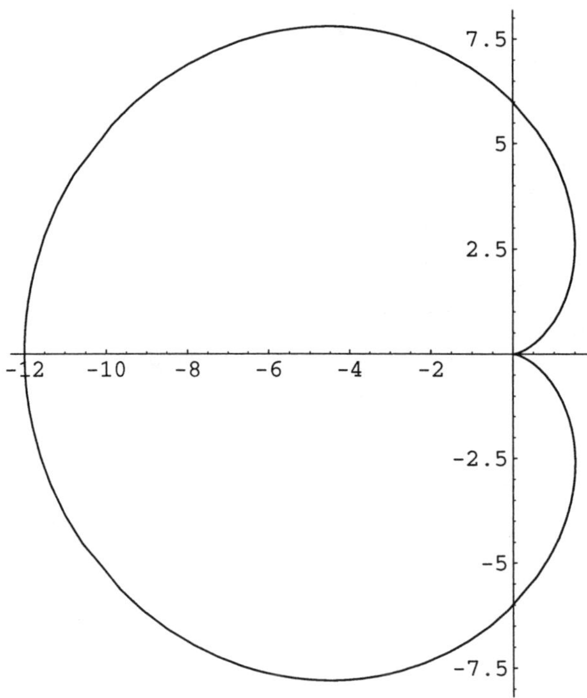

Alternatively, we can convert to rectangular coordinates after using the identity $\cos 2x = 1 - 2\sin^2 x \Rightarrow 12\sin^2(\theta/2) = 6 - 6\cos\theta$:

$$r = 6 - 6\cos\theta$$

$$r^2 = 6r - 6r\cos\theta$$

$$x^2 + y^2 = 6r - 6x$$

$$\left(x^2 + 6x + y^2\right)^2 = 36r^2$$

$$\left(x^2 + 6x + y^2\right)^2 = 36\left(x^2 + y^2\right).$$

```
ContourPlot[
 (x^2 + 6 x + y^2) ^2 - 36 (x^2 + y^2),
 {x, -12, 12}, {y, -12, 12},
 ContourShading -> False,
 Contours -> {0}, Frame -> False,
 Axes -> Automatic, AxesOrigin -> {0, 0]
 PlotPoints -> 120]
```

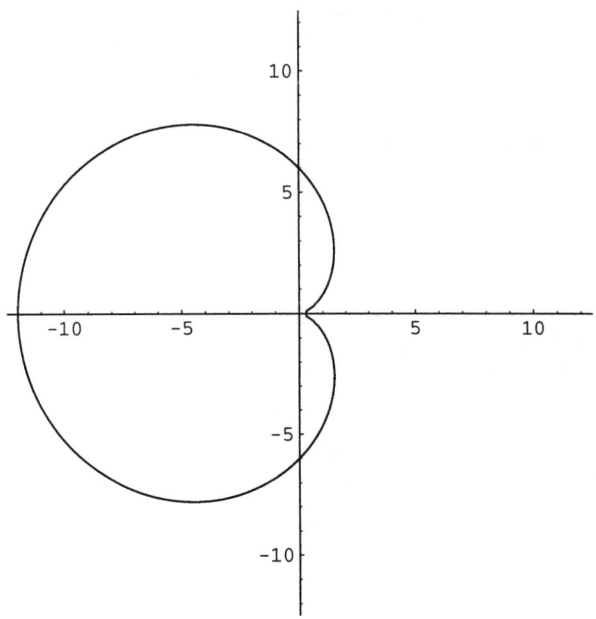

26.

```
f[x_] = Sqrt[1 - x^2];
g[x_] = x;
Plot[
 Evaluate[
  {f[x], g[x],
   f'[1 / Sqrt[2]] (x - 1 / Sqrt[2]) +
    f[1 / Sqrt[2]]}], {x, -1, 1},
 PlotRange -> {-1 / 2, 3 / 2},
 AspectRatio -> Automatic]
```

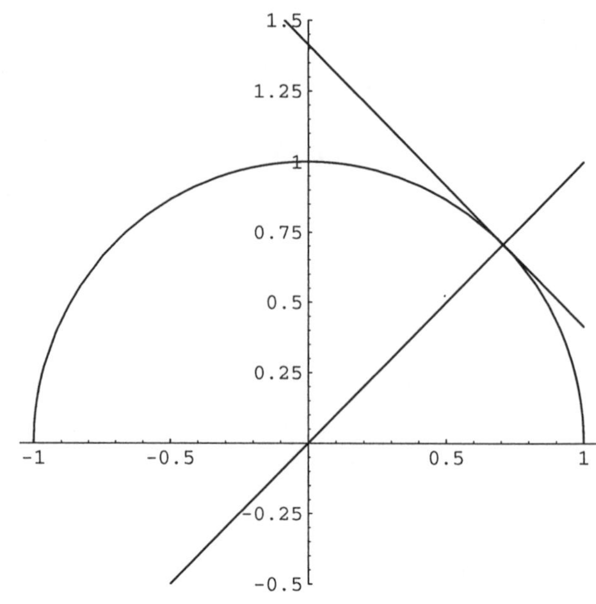

27.

```
cp1 = ContourPlot[y / x^2, {x, -10, 10},
    {y, -10, 10},
    ContourShading -> False,
    Frame -> False, Axes -> Automatic,
    AxesOrigin -> {0, 0}, Contours -> 30,
    PlotPoints -> 200,
    DisplayFunction -> Identity];

cp2 = ContourPlot[x^2 / 2 + y^2,
    {x, -10, 10}, {y, -10, 10},
    ContourShading -> False,
    ContourStyle -> GrayLevel[0.5],
    Frame -> False, Axes -> Automatic,
    AxesOrigin -> {0, 0}, Contours -> 30,
    PlotPoints -> 200,
    DisplayFunction -> Identity];

Show[cp1, cp2,
 DisplayFunction -> $D:
```

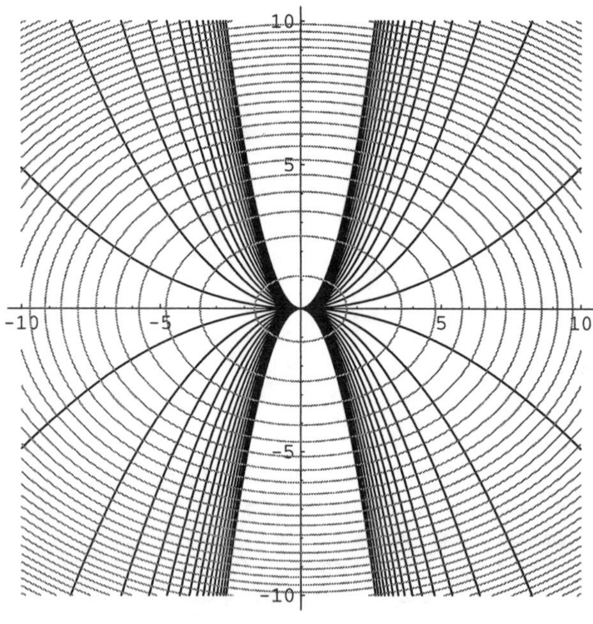

31. Differentiating the equation gives us $2y\,dy/dx = 2x$ so we must solve the separable (and linear) equation $dy/dx = -y/x$:

$$\begin{aligned}\frac{dy}{dx} &= -\frac{y}{x}\\ \frac{1}{y}dy &= -\frac{1}{x}dx\\ \ln|y| &= -\ln|x| + k\\ \ln|x\,y| &= k\\ xy &= k.\end{aligned}$$

```
cp1 = ContourPlot[y^2 - x^2,
   {x, -10, 10}, {y, -10, 10},
   ContourShading -> False,
   Frame -> False, Axes -> Automatic,
   AxesOrigin -> {0, 0}, Contours -> 30,
   PlotPoints -> 200,
   DisplayFunction -> Identity];

cp2 = ContourPlot[x y, {x, -10, 10},
   {y, -10, 10},
   ContourShading -> False,
   ContourStyle -> GrayLevel[0.5],
   Frame -> False, Axes -> Automatic,
   AxesOrigin -> {0, 0}, Contours -> 30,
   PlotPoints -> 200,
   DisplayFunction -> Identity];

Show[cp1, cp2,
  DisplayFunction -> $D:
```

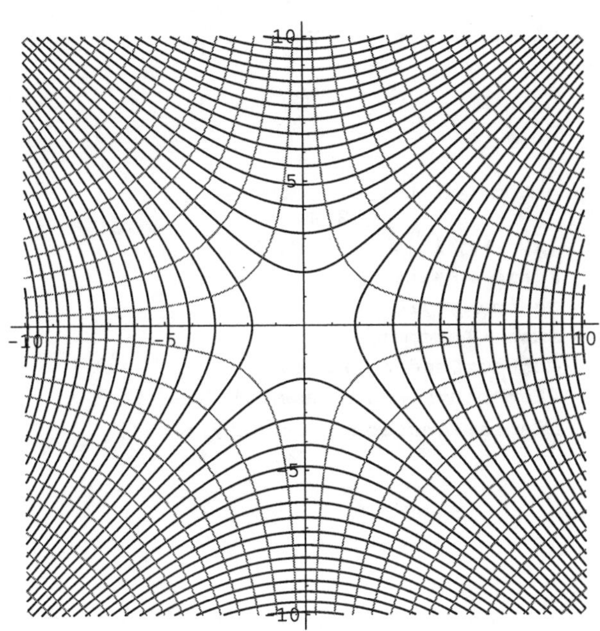

37. (a) We solve $\tan\theta = \dfrac{m_2 - m_1}{1 + m_2 m_1}$ for m_2:

$$\begin{aligned}\tan\theta &= \frac{m_2 - m_1}{1 + m_2 m_1}\\ (1 + m_2 m_1)\tan\theta &= m_2 - m_1\\ \tan\theta + m_2 m_1 \tan\theta &= m_2 - m_1\\ m_2 m_1 \tan\theta - m_2 &= -\tan\theta - m_1\\ m_2(m_1 \tan\theta - 1) &= -\tan\theta - m_1\\ m_2 &= \frac{-\tan\theta - m_1}{m_1\tan\theta - 1} = \frac{m_1 + \tan\theta}{1 - m_1\tan\theta}.\end{aligned}$$

(c) Implicit differentiation of $x^2 - y^2 = c$ yields $2x - 2yy' = 0 \Rightarrow y' = x/y = f(x,y)$. With $\tan\theta = \tan(\pi/4) = 1$, we solve $\dfrac{dy}{dx} = \dfrac{x/y + 1}{1 - x/y} = \dfrac{x + y}{y - x}$ to find one of the families of curves. Notice that when written in differential form $(x + y)dx - (y - x)dy = 0$, the equation is exact because $(x + y)_y = 1 = -(y - x)_x$. Integrating $(x + y)$ with respect to x yields $f(x,y) = \frac{1}{2}x^2 + xy + g(y)$ so that $f_y(x,y) = x + g'(y) = -(y - x)$ implies that $g'(y) = -y$. Therefore, $g(y) = -\frac{1}{2}y^2$ and part of the family of oblique trajectories is $\frac{1}{2}x^2 + xy - \frac{1}{2}y^2 = k_1$. Similarly, for

$\dfrac{dy}{dx}=\dfrac{x/y-1}{1+x/y}=\dfrac{x-y}{y+x}$, we find that $\frac{1}{2}x^2-xy-\frac{1}{2}y^2=k_2$. To confirm these results graphically, we graph several members of both families and then show the graphs together. Note that the curves appear to intersect at an angle of $\frac{\pi}{4}$

```
cp1 = ContourPlot[x^2 - y^2, {x, -10, 1(
   {y, -10, 10}, Frame → False,
   ContourStyle → GrayLevel[.5],
   Axes → Automatic,
   ContourShading → False,
   PlotPoints → 120, AxesOrigin → {0,
   DisplayFunction → Identity];
```

```
cp2 = ContourPlot[x^2/2 + x y - y^2/2,
   {x, -10, 10}, {y, -10, 10},
   Frame → False,
   ContourStyle → GrayLevel[0],
   Axes → Automatic,
   ContourShading → False,
   PlotPoints → 120, AxesOrigin → {C
   DisplayFunction → Identity];
```

```
cp3 = ContourPlot[x^2/2 - x y - y^2/2,
   {x, -10, 10}, {y, -10, 10},
   Frame → False,
   ContourStyle → GrayLevel[0],
   Axes → Automatic,
   ContourShading → False,
   PlotPoints → 120, AxesOrigin → {C
   DisplayFunction → Identity];
```

```
Show[GraphicsArray[
  {Show[cp1, cp2], Show[cp1, cp3]}]]
```

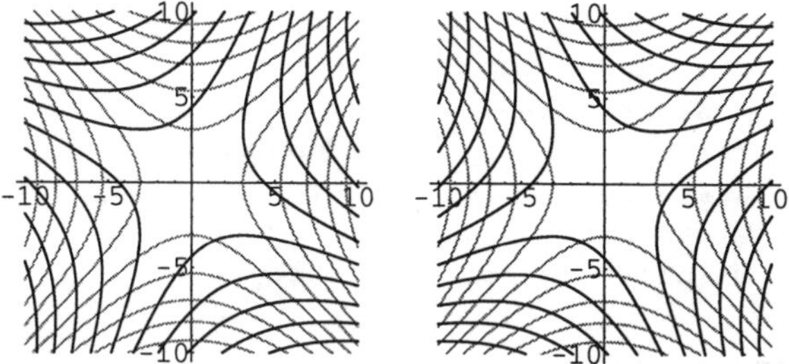

(d) Implicit differentiation of $x^2+y^2=c^2$ yields $2x+2yy'=0 \Rightarrow y'=-x/y=f(x,y)$. Because $\tan\theta=\tan(\pi/6)=1/\sqrt{3}$, we solve $\dfrac{dy}{dx}=\dfrac{-x/y+1/\sqrt{3}}{1-(-x/y)(1/\sqrt{3})}=\dfrac{-x\sqrt{3}+y}{y\sqrt{3}+x}$ which is a first-order homogeneous equation. With the substitution $x=vy$, we obtain the separable equation $\dfrac{1-v\sqrt{3}}{1+v^2}dv=\dfrac{\sqrt{3}}{y}dy$. Integrating yields $-\dfrac{\sqrt{3}}{2}\ln\left(1+v^2\right)+\tan^{-1}v=\sqrt{3}\ln|y|+k_1$, so $-\dfrac{\sqrt{3}}{2}\ln\left(1+\dfrac{x^2}{y^2}\right)+\tan^{-1}\dfrac{x}{y}=\sqrt{3}\ln|y|+k_1$. Similarly, for $\dfrac{dy}{dx}=\dfrac{-x/y-1/\sqrt{3}}{1+(-x/y)(1/\sqrt{3})}=\dfrac{-x\sqrt{3}-y}{y\sqrt{3}-x}$, we obtain $\dfrac{1+v\sqrt{3}}{1+v^2}dv=-\dfrac{\sqrt{3}}{y}dy$ so that the trajectories are

$$\frac{\sqrt{3}}{2}\ln\left(1+\frac{x^2}{y^2}\right)+\tan^{-1}\frac{x}{y}=-\sqrt{3}\ln|y|+k_2.$$

To confirm the result graphically, we proceed in the same manner as in (c).

```
cp1 = ContourPlot[x^2 + y^2, {x, -10, 1(
    {y, -10, 10}, Frame → False,
    ContourStyle → GrayLevel[.5],
    Axes → Automatic,
    ContourShading → False,
    PlotPoints → 120, AxesOrigin → {0,
    DisplayFunction → Identity];
```

```
cp2 = ContourPlot[
    1/2 (-√3) Log[1 + x^2/y^2] + ArcTan[x/y] -
     √3 Log[Abs[y]], {x, -10, 10},
    {y, -10, 10}, Frame → False,
    ContourStyle → GrayLevel[0],
    Axes → Automatic,
    ContourShading → False,
    PlotPoints → 120, AxesOrigin → {0, 0},
    DisplayFunction → Identity];
```

```
cp3 = ContourPlot[
    1/2 √3 Log[1 + x^2/y^2] + ArcTan[x/y] +
     √3 Log[Abs[y]], {x, -10, 10},
    {y, -10, 10}, Frame → False,
    ContourStyle → GrayLevel[0],
    Axes → Automatic,
    ContourShading → False,
    PlotPoints → 120, AxesOrigin → {0, 0
    DisplayFunction → Identity];
```

```
Show[GraphicsArray[
  {Show[cp1, cp2], Show[cp1, cp3]}]]
```

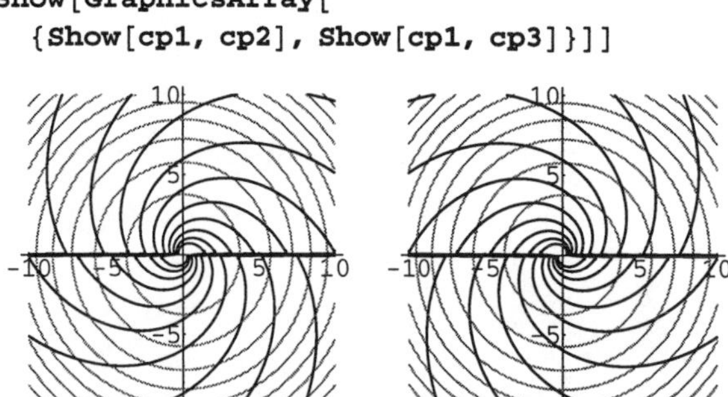

Differential Equations at Work A: Mathematics of Finance

1. This equation is both separable and linear. We quickly calculate the solution to the initial-value problem.

```
Clear[r,x,t,capp]
partsol=DSolve[{x'[t]==r x[t],
    x[0]==capp},x[t],t]
```

```
x:='x':r:='r':t:='t':capp:='capp':
partsol:=dsolve(
    {diff(x(t),t)=r*x,x(0)=capp},
    x(t));
```

2. Next, we compute the value of the account at the end of each five year period.

```
Table[{t,partsol[[1,1,2]] /.
    {r->0.08,capp->1000}},
    {t,5,20,5}]//TableForm
```

```
evalf(seq(
    {t,subs({r=0.08,capp=1000},
       rhs(partsol))},
    t={5,10,15,20}));
```

4

Higher-Order Differential Equations

EXERCISES 4.1

3. $W(S)=\begin{vmatrix} e^{-6t} & e^{-4t} \\ -6e^{-6t} & -4e^{-4t} \end{vmatrix} = -4e^{-10t}-\left(-6e^{-10t}\right)=2e^{-10t}\neq 0$ for any t; Lin. Indep.

5.

$$W(S)=\begin{vmatrix} e^{-3t}\cos 3t & e^{-3t}\sin 3t \\ e^{-3t}(-3\cos 3t-3\sin 3t) & e^{-3t}(3\cos 3t-3\sin 3t) \end{vmatrix}$$
$$=e^{-6t}\left[\left(3\cos^2 3t-3\cos 3t\sin 3t\right)-\left(-3\cos 3t\sin 3t-3\sin^2 3t\right)\right]$$
$$=e^{-6t}\left(3\cos^2 3t+3\sin^2 3t\right)=3e^{-6t}\neq 0 \text{ for any } t;\text{ Lin. Indep.}$$

9. $y''+2y'+y=\left(c_1e^{-t}-2c_2e^{-t}+c_2te^{-t}\right)+2\left(-c_1e^{-t}+c_2e^{-t}-c_2te^{t}\right)+\left(c_1e^{-t}+c_2te^{-t}\right)=0$; the functions are linearly independent because $W\left(\left\{e^{-t},te^{-t}\right\}\right)=\begin{vmatrix} e^{-t} & te^{-t} \\ -e^{-t} & e^{-t}-te^{-t} \end{vmatrix}=e^{-2t}[(1-t)-(-t)]=e^{-2t}\neq 0$.

15. $y=c_1e^{4t}+c_2te^{4t}$ so

$$y'=(4c_1+c_2)e^{4t}+4c_2te^{4t}$$

and application of the initial conditions yields the system of equations $\begin{cases} c_1e^4+c_2e^4=0 \\ -4c_1e^4+5c_2e^4=-e^4 \end{cases}$, which has solution $c_1=1$ and $c_2=-1$. Therefore, $y=e^{4t}(1-t)$.

```
sol = DSolve[{y''[t] - 8 y'[t] + 16 y[t] == 0,
   y[1] == 0, y'[1] == -Exp[4]}, y[t], t]
```

$\{\{y[t]\to e^{4t}(1-t)\}\}$

```
Plot[y[t] /. sol, {t, 0, 2}, Pl
```

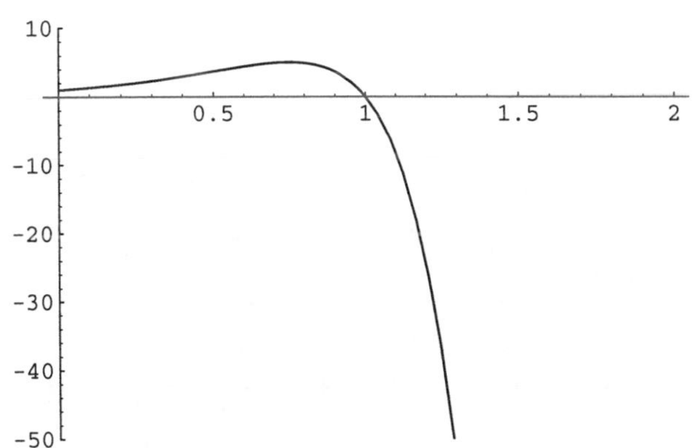

17. First, we compute

$$y'=c_2\cos t+t\cos t+\sin t-c_1\sin t.$$

Applying the initial conditions results in the system of equations

$$\begin{cases} c_1=1 \\ c_2=1 \end{cases}$$

so $y=\cos t+\sin t+t\sin t$.

18. Because y_1 and y_2 are both solutions, we have

$$\begin{cases} y_1{}' + p(t)y_1 = 0 \\ y_2{}' + p(t)y_2 = 0. \end{cases}$$

Now, multiply the first equation by y_2, the second by y_1,

$$\begin{cases} y_1{}'y_2 + p(t)y_1y_2 = 0 \\ y_1y_2{}' + p(t)y_1y_2 = 0 \end{cases}$$

and subtract the first from the second:

$$y_1y_2{}' - y_1{}'y_2 = 0.$$

This is the Wronskian of y_1 and y_2:

$$W(\{y_1, y_2\}) = \begin{vmatrix} y_1 & y_2 \\ y_1{}' & y_2{}' \end{vmatrix} = y_1y_2{}' - y_1{}'y_2.$$

Because the Wronskian is zero, the functions are linearly dependent.

19. (a) $\frac{d}{dt}(\cosh t) = \frac{d}{dt}\left(\frac{1}{2}\left(e^t + e^{-t}\right)\right) = \frac{1}{2}\left(e^t - e^{-t}\right) = \sinh t$

(b) $\frac{d}{dt}(\sinh t) = \frac{d}{dt}\left(\frac{1}{2}\left(e^t - e^{-t}\right)\right) = \frac{1}{2}\left(e^t + e^{-t}\right) = \cosh t$

(c) $\cosh^2 t - \sinh^2 t = \frac{1}{4}\left(e^{2t} + 2 + e^{-2t}\right) - \frac{1}{4}\left(e^{2t} - 2 + 2^{-2t}\right) = 1$

(d) $W(\{\cosh t, \sinh t\}) = \cosh^2 t - \sinh^2 t = 1$

23.

$$\frac{d^2}{dt^2}\left(e^{5t}\cos 4t\right) - 10\frac{d}{dt}\left(e^{5t}\cos 4t\right) + 41\left(e^{5t}\cos 4t\right)$$

$$= \left(9e^{5t}\cos 4t - 40e^{5t}\sin 4t\right) - 10\left(5e^{5t}\cos 4t - 4e^{5t}\sin 4t\right) + 41e^{5t}\cos 4t = 0$$

$$\frac{d^2}{dt^2}\left(e^{5t}\sin 4t\right) - 10\frac{d}{dt}\left(e^{5t}\sin 4t\right) + 41\left(e^{5t}\sin 4t\right)$$

$$= \left(9e^{5t}\sin 4t + 40e^{5t}\cos 4t\right) - 10\left(5e^{5t}\sin 4t + 4e^{5t}\cos 4t\right) + 41e^{5t}\sin 4t = 0$$

$$W\left(e^{5t}\cos 4t,\ e^{5t}\sin 4t\right) = 4e^{10t}$$

27. $y_2(t) = e^{2t}\int \frac{e^{-\int -4\,dt}}{\left(e^{2t}\right)^2}dt = e^{2t}\int \frac{e^{4t}}{e^{4t}}dt = te^{2t}$

31. Dividing by t^2 yields $y'' + \frac{4}{t}y' - \frac{4}{t^2}y = 0$ so $p(t) = \frac{4}{t}$. Therefore,

$$y_2(t) = t^{-4}\int \frac{e^{-\int \frac{4}{t}dt}}{\left(t^{-4}\right)^2}dt = t^{-4}\int \frac{e^{-4\ln t}}{t^{-8}}dt = t^{-4}\int \frac{t^{-4}}{t^{-8}}dt$$

$$= t^{-4}\int t^4 dt = \tfrac{1}{5}t^{-4}\cdot t^5 = \tfrac{1}{5}t.$$

Because $t^2y'' + 4ty' - 4y = 0$ is a linear homogeneous equation, any constant multiple of $\frac{1}{5}t$ is also a solution. This means that the constant $\frac{1}{5}$ in $\frac{1}{5}t$ is not important. A second solution is t.

37. By the Fundamental Theorem of Calculus

$$\frac{d}{dx}\left(f(t)\int\frac{e^{-\int p(t)dt}}{\left[f(t)^2\right]}dt\right)=f'(t)\int\frac{e^{-\int p(t)dt}}{\left[f(t)^2\right]}dt+f(t)\frac{e^{-\int p(t)dt}}{\left[f(t)^2\right]}$$

so

$$W\left(\left\{f(t),f(t)\int\frac{e^{-\int p(t)dt}}{\left[f(t)^2\right]}dt\right\}\right)=\begin{vmatrix} f(t) & f(t)\int\frac{e^{-\int p(t)dt}}{\left[f(t)^2\right]}dt \\ f'(t) & f'(t)\int\frac{e^{-\int p(t)dt}}{\left[f(t)^2\right]}dt+f(t)\frac{e^{-\int p(t)dt}}{\left[f(t)^2\right]}\end{vmatrix}$$

$$=f'(t)f(t)\int\frac{e^{-\int p(t)dt}}{\left[f(t)^2\right]}dt+e^{-\int p(t)dt}-f'(t)f(t)\int\frac{e^{-\int p(t)dt}}{\left[f(t)^2\right]}dt$$

$$=e^{-\int p(t)dt}\neq 0.$$

41. $y=c_1+c_2\tan(c_3+c_4\ln t)$ is a solution of the equation if

(a) $y=c_1$ (note that if $c_2=0$ or $c_4=0$, y is a constant function); or

(b) $y=-\frac{1}{2}+c_2\tan(c_3+c_2\ln t)$.

The Principle of Superposition does not hold because if $c_1\neq 0$ and $y=c_1-\frac{1}{2}+c_2\tan(c_3+c_2\ln t)$,

$$ty''-2yy'=-\frac{2}{t}c_1c_2^2\sec^2(c_3+c_2\ln t)\neq 0.$$

43. We use reduction of order to find a second linearly independent solution. Dividing by t yields

$$y''+\frac{2}{t}y'+\frac{16}{t}y=0\Rightarrow p(t)=\frac{2}{t}.$$

Then a second linearly indepent solution is

$$y_2(t)=\frac{\sin 4t}{t}\int\frac{e^{-\int 2/t\,dt}}{\left(\frac{\sin 4t}{t}\right)^2}dt=\frac{\sin 4t}{t}\int\frac{t^2\cdot\frac{1}{t^2}}{\sin^2 4t}dt=\frac{\sin 4t}{t}\cdot-\frac{1}{4}\cot 4t=-\frac{1}{4}\frac{\cos 4t}{t}$$

and a general solution is

$$y=\frac{1}{t}(c_1\cos 4t+c_2\sin 4t)\Rightarrow y'=\frac{1}{t^2}(-c_1\cos 4t+4c_2t\cos 4t-c_2\sin 4t-4c_1t\sin 4t).$$

Application of the initia conditions yields

$$\begin{cases} y(\pi/8)=8c_2/\pi=0 \\ y'(\pi/8)=-64c_2/\pi^2-32c_1/\pi=-32/\pi \end{cases}\Rightarrow c_1=1,\ c_2=0$$

so the solution to the initial-value problem is

$$y=\frac{\cos 4t}{t}.$$

```
y[t_] = (c1 Cos[4 t] + c2 Sin[4 t])/t;
vals1 = Solve[{y[π/8] == -32/π,
   y'[π/8] == -32/π}]
```

$\{\{c1 \to -\frac{-8-\pi}{\pi},\ c2 \to -4\}\}$

```
Plot[y[t] /. vals1, {t, 0, 2 Pi}]
```

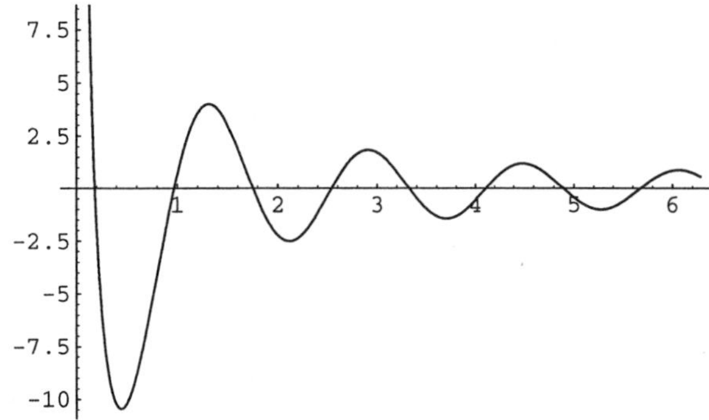

```
y[t_] = (c1 Cos[4 t] + c2 Sin[4 t])/t;
vals1 = Solve[{y[π/8] == 0, y'[π/8] == -32/π}]
```

{{c1 → 1, c2 → 0}}

```
Plot[y[t] /. vals1, {t, 0, 2 Pi}]
```

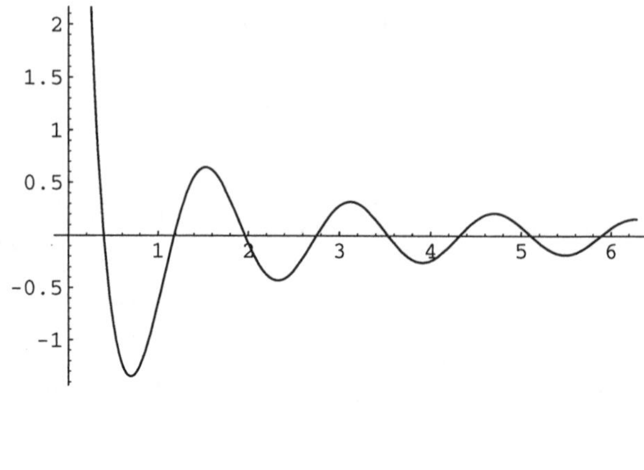

45. $y=\dfrac{\cos t}{t^3}\Rightarrow y'=\dfrac{-3\cos t-t\sin t}{t^4},\ y''=\dfrac{12\cos t-t^2\cos t+6t\sin t}{t^5}$ and substitution into the equation gives us

$$\left(\frac{12}{t^5}-\frac{1}{t^3}-\frac{3b(t)}{t^4}+\frac{c(t)}{t^3}\right)\cos t+\left(\frac{6}{t^4}-\frac{b(t)}{t^3}\right)\sin t=0.$$

Assuming that $\dfrac{12}{t^5}-\dfrac{1}{t^3}-\dfrac{3b(t)}{t^4}+\dfrac{c(t)}{t^3}=0$ and $\dfrac{6}{t^4}-\dfrac{b(t)}{t^3}=0$ gives us $b(t)=\dfrac{6}{t}$ and $c(t)=-\dfrac{t^2+6}{t^2}$. By reduction of order, a second linearly independent solution is

$$y_2(t)=\frac{\cos t}{t^3}\int\frac{e^{-\int 6/t\,dt}}{\left(\dfrac{\cos t}{t^3}\right)^2}dt=\frac{\cos t}{t^3}\int\frac{t^6\cdot\dfrac{1}{t^6}}{\cos^2 t}dt=\frac{\cos t}{t^3}\cdot\tan t=\frac{\sin t}{t^3}$$

and a general solution is

$$y=\frac{1}{t^3}\left(c_1\cos t+c_2\sin t\right)\Rightarrow y'=\frac{1}{t^4}\left(-3c_1\cos t+c_2t\cos t-3c_2\sin t-c_1t\sin t\right).$$

Application of the boundary conditions gives us

$$\begin{cases} y(\pi)=-c_1/\pi^3=0 \\ y(2\pi)=c_1/\left(8\pi^3\right)=0 \end{cases}\Rightarrow c_1=0.$$

Thus, there are infinitely many solutions of the form $y=C\dfrac{\sin t}{t^3}$, C arbitrary, that satisfy the boundary conditions.

```
toplot = Table[c Sin[t]/t^3, {c, -5, 5}];

grays = Table[GrayLevel[i], {i, 0, 0.7, -

Plot[Evaluate[toplot], {t, 0, 3 π},
  PlotStyle → grays,
  PlotRange → {-.2, .2}]
```

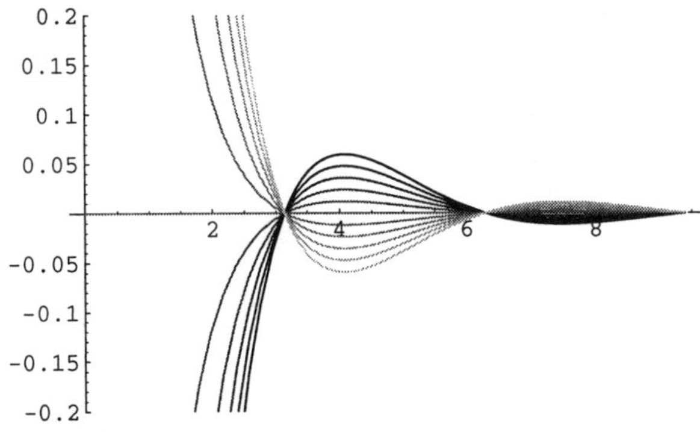

EXERCISES 4.2

3. The characteristic equation is

$$r^2 + 3r - 4 = 0$$
$$(r+4)(r-1) = 0$$

so $r = -4$ or $r = 1$ anda general solution of the equation is $y = c_1 e^{-4t} + c_2 e^t$.

7. $r^2 + 7 = 0 \Rightarrow r = \pm\sqrt{7}\,i$ so

$$y = c_1 \cos\sqrt{7}t + c_2 \sin\sqrt{7}t$$

11.

$$r^2 + 6r + 18 = 0 \Rightarrow$$
$$r = \frac{-6 \pm \sqrt{6^2 - 4 \cdot 1 \cdot 18}}{2 \cdot 1}$$
$$= -3 \pm 3i$$

so $y = c_1 e^{-3t} \cos 3t + c_2 e^{-3t} \sin 3t$

15. The characteritic equation

$$r^2 - 6r + 9 = (r-3)^2 = 0$$

has one solution $r = 3$ of multiplicity two so $y = c_1 e^{3t} + c_2 t e^{3t}$.

19. A general solution of the differential equation is

$$y = c_1 e^{3t} + c_2 e^{4t}.$$

Differentiating yields $y' = 3c_1 e^{3t} + 4c_2 e^{4t}$ and applying the initial conditions results in the system of equations

$$\begin{cases} c_1 + c_2 = 3 \\ 3c_1 + 4c_2 = -2 \end{cases},$$

which has solution $c_1 = 14$ and $c_2 = -11$ so

$$y = 14e^{3t} - 11e^{4t}.$$

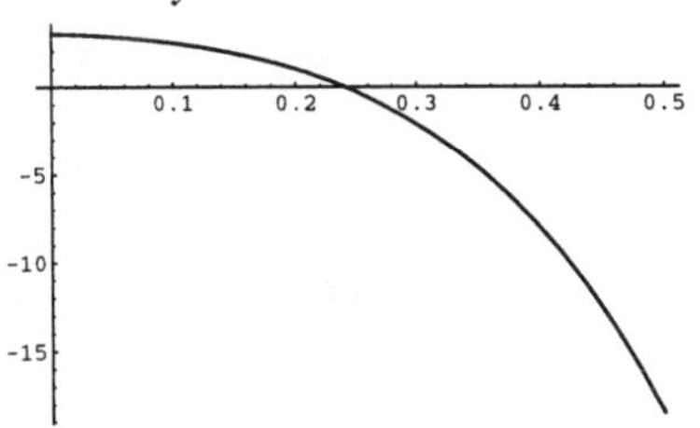

23. A general solution of the equation is $y = c_1 \cos 10t + c_2 \sin 10t$. Application of the initial conditions yields

$$\begin{cases} c_1 = 1 \\ 10c_2 = 10 \end{cases} \Rightarrow (c_1, c_2) = (1,1)$$

so $y = \cos 10t + \sin 10t$

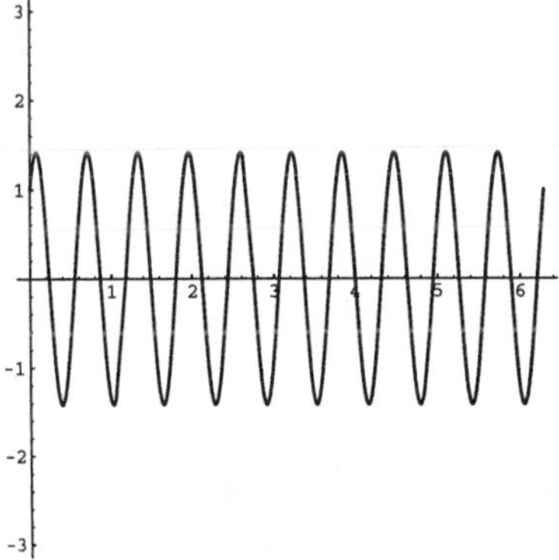

27. A general solution of the differential equation is $y = e^{-2t}(c_1 \cos 4t + c_2 \sin 4t)$ with $y' = 2e^{-2t}(-c_1 \cos 4t + 2c_2 \cos 4t - 2c_1 \sin 4t - c_2 \sin 4t)$. Application of the initial conditions results in the system of equations $\begin{cases} c_1 = 2 \\ -2c_1 + 4c_2 = 0 \end{cases}$ which has solutions $c_1 = 2$ and $c_2 = 1$ so $y = e^{-2t}(2\cos 4t + \sin 4t)$.

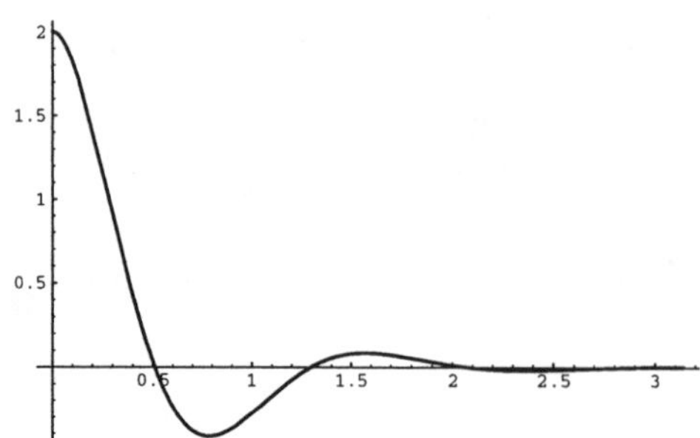

30.

$$a(t)=\frac{1}{t^2+1};\ b(t)=-\frac{2\left(-6+t-6t^2\right)}{t^2+1};\ c(t)=45\left(t^2+1\right)$$

$$w''-\left[\frac{-2t/\left(t^2+1\right)^2}{1/\left(t^2+1\right)}+\frac{2\left(-6+t-6t^2\right)}{t^2+1}\right]w'+45w=0$$

$$w''-\left[\frac{-2t}{t^2+1}+\frac{2\left(-6+t-6t^2\right)}{t^2+1}\right]w'+45w=0$$

$$w''+12w'+45w=0 \Rightarrow w(t)=c_1e^{-6t}\cos 3t+c_2e^{-6t}\sin 3t$$

$$y(t)=\frac{-6\left(c_1e^{-6t}\cos 3t+c_2e^{-6t}\sin 3t\right)+3\left(-c_1e^{-6t}\sin 3t+c_2e^{-6t}\cos 3t\right)}{c_1e^{-6t}\cos 3t+c_2e^{-6t}\sin 3t}\left(t^2+1\right)$$

$$=\frac{\left(-6c_1+3c_2\right)\cos 3t+\left(-6c_2-3c_1\right)\sin 3t}{c_1\cos 3t+c_2\sin 3t}\left(t^2+1\right)$$

42. (a) $(y'')^2-5y''y+4y^2=(y''-y)(y''-4y)=0$. Therefore, $y''-y=0$ or $y''-4y=0$, so $y=c_1e^{-t}+c_2e^{t}$ or $y=c_3e^{-2t}+c_4e^{2t}$; the Principle of Superposition is not valid: $c_1e^{-t}+c_2e^{t}+c_3e^{-2t}+c_4e^{2t}$ is not a solution of the equation unless $c_1=c_2=0$ or $c_3=c_4=0$.

(b) $(y'')^2-2y''y+y^2=(y''-y)(y''-y)=0$, so $y''-y=0$. $y=c_1e^{-t}+c_2e^{t}$; the Principle of Superposition is valid.

43.

```
Clear[x,y,a,b]
partsol=DSolve[{y''[x]+4y'[x]+3y[x]==0,
    y[0]==a,y'[0]==b},y[x],x]
dsol=D[partsol[[1,1,2]],x]//Simplify
Solve[dsol==0,x]
```

```
x:='x':y:='y':a:='a':b:='b':
partsol:=dsolve({diff(y(x),x$2)+
    4*diff(y(x),x)+3*y(x)=0,
    y(0)=a,D(y)(0)=b},y(x));
assign(partsol):
dsol:=diff(y(x),x);
solve(dsol=0,x);
```

The local extrema occur at the values of x where the derivative of the solution is zero:

$$\begin{aligned} 3a+3b-3ae^{2x}-be^{2x} &= 0 \\ -(3a+b)e^{2x} &= -(3a+3b) \\ e^{2x} &= \frac{3a+3b}{3a+b} \\ x &= \ln\sqrt{\frac{3a+3b}{3a+b}}. \end{aligned}$$

Because the domain of the natural logarithm function is $(0,\infty)$, there will be no local extrema if

$$\frac{3a+3b}{3a+b} \le 0 \Rightarrow a \le -b \text{ and } a > -\tfrac{1}{3}b \text{ OR } a \ge -b \text{ and } a < -\tfrac{1}{3}b.$$

44. We illustrate typical results using one point, instead of three, in each region. (a), (b) The characteristic equation of $\dfrac{d^2x}{dt^2}+a_1\dfrac{dx}{dt}+a_0x=0$ is $r^2+a_1r+a_0=0$.

```
Clear[x,y,t,a1,a0]
Solve[m^2+a1 m+a0==0,m]
Plot[1/4 a1^2,{a1,-2,2},PlotRange->{-2,2},
    AspectRatio->1]
```

```
x:='x':y:='y':t:='t':a1:='a1':
    a0:='a0':
solve(m^2+a1*m+a0=0,m);
plot(1/4*a1^2,a1=-2..2,-2..2);
```

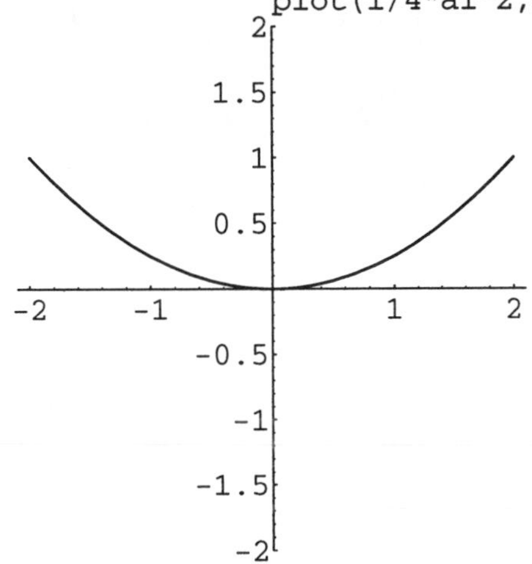

(c) $y=x'=\dfrac{dx}{dt} \Rightarrow y'=x''=-a_1\dfrac{dx}{dt}-a_0x=-a_0x-a_1y$ so we obtain the system $\begin{cases} x'=y \\ y'=-a_0x-a_1y \end{cases}$.

(d)

```
<<Graphics`PlotField`
a0=1;
a1=1;
PlotVectorField[{y,-a0 x-a1 y},{x,-1,1},
    {y,-1,1},
    ScaleFunction->(1&),Axes->Automatic,
    AxesOrigin->{0,0}]
a0=1;
a1=2;
PlotVectorField[{y,-a0 x-a1 y},{x,-1,1},
    {y,-1,1},
    ScaleFunction->(1&),Axes->Automatic,
    AxesOrigin->{0,0}]
a0=-1;
a1=-1;
PlotVectorField[{y,-a0 x-a1 y},{x,-1,1},
    {y,-1,1},
    ScaleFunction->(1&),Axes->Automatic,
    AxesOrigin->{0,0}]
```

```
with(DEtools);
a0:=1:
a1:=1:
dfieldplot([y,-a0*x-a1*y],[x,y],
    -1..1,y=-1..1);
a0:=1:
a1:=2:
dfieldplot([y,-a0*x-a1*y],[x,y],
    -1..1,y=-1..1);
a0:=-1:
a1:=-1:
dfieldplot([y,-a0*x-a1*y],[x,y],
    -1..1,y=-1..1);
```

-1 -0.5 0.5 1 0.5 -0.5 1 -1

-1 -0.5 0.5 1 0.5 -0.5 1 -1

-1 -0.5 0.5 1 0.5 -0.5 1 -1

EXERCISES 4.3

1.

$$W(S)=\begin{vmatrix}3t^2 & t & 2t-t^2\\ 6t & 1 & 2-2t\\ 6 & 0 & -2\end{vmatrix}=3t^2\begin{vmatrix}1 & 2-2t\\ 0 & -2\end{vmatrix}-t\begin{vmatrix}6t & 2-2t\\ 6 & -2\end{vmatrix}+\left(2t-t^2\right)\begin{vmatrix}6t & 1\\ 6 & 0\end{vmatrix}$$

$$=-6t^2+12t-12t+6t^2=0 \text{ for all } t; \text{ Lin. Dep.}$$

5.

$$W(S)=\begin{vmatrix}e^t & e^{-5t}\sin t & e^{-5t}\cos t\\ e^t & e^{-5t}(\cos t-5\sin t) & e^{-5t}(-5\cos t-\sin t)\\ e^t & e^{-5t}(-10\cos t+24\sin t) & e^{-5t}(24\cos t+10\sin t)\end{vmatrix}$$

$$=e^t\begin{vmatrix}e^{-5t}(\cos t-5\sin t) & e^{-5t}(-5\cos t-\sin t)\\ e^{-5t}(-10\cos t+24\sin t) & e^{-5t}(24\cos t+10\sin x)\end{vmatrix}-e^{-5t}\sin t\begin{vmatrix}e^t & e^{-5t}(-5\cos t-\sin t)\\ e^t & e^{-5t}(24\cos t+10\sin t)\end{vmatrix}$$

$$+e^{-5t}\cos t\begin{vmatrix}e^t & e^{-5t}(\cos t-5\sin t)\\ e^t & e^{-5t}(-10\cos t+24\sin t)\end{vmatrix}$$

$$=-37e^{-9t}\neq 0 \text{ for any } t; \text{ Lin. Indep.}$$

11. $\dfrac{d}{dt^3}\left(e^t\right)+9\dfrac{d}{dt^2}\left(e^t\right)+16\dfrac{d}{dt}\left(e^t\right)-26e^t=e^t+9e^t+16e^2-26e^t=0,$

$$\frac{d}{dt^3}\left(e^{-5t}\cos t\right)+9\frac{d}{dt^2}\left(e^{-5t}\cos t\right)+16\frac{d}{dt}\left(e^{-5t}\cos t\right)-26\left(e^{-5t}\cos t\right)=-136e^{-5t}\cos t-74e^{-5t}\sin t+$$
$$16\left(-5e^{-5t}\cos t-e^{-5t}\sin t\right)+9\left(24e^{-5t}\cos t+10e^{-5t}\sin t\right)=0$$

and

$$\frac{d}{dt^3}\left(e^{-5t}\sin t\right)+9\frac{d}{dt^2}\left(e^{-5t}\sin t\right)+16\frac{d}{dt}\left(e^{-5t}\sin t\right)-26\left(e^{-5t}\sin t\right)=74e^{-5t}\cos t-136e^{-5t}\sin t+$$
$$16\left(e^{-5t}\cos t-5e^{-5t}\sin t\right)+9\left(-10e^{-5t}\cos t+24e^{-5t}\sin t\right)=0.$$

The Wronskian is

$$W\left(\left\{e^t,e^{-5t}\cos t,e^{-5t}\sin t\right\}\right)=\begin{vmatrix} e^t & e^{-5t}\cos t & e^{-t}\sin t \\ e^t & -5e^{-5t}\cos t-e^{-5t}\sin t & e^{-5t}\cos t-5e^{-5t}\sin t \\ e^t & 24e^{-5t}\cos t+10e^{-5t}\sin t & -10e^{-5t}\cos t+24e^{-5t}\sin t \end{vmatrix}$$

$$=e^t\left(\begin{vmatrix} -5e^{-5t}\cos t-e^{-5t}\sin t & e^{-5t}\cos t-5e^{-5t}\sin t \\ 24e^{-5t}\cos t+10e^{-5t}\sin t & -10e^{-5t}\cos t+24e^{-5t}\sin t \end{vmatrix}-\right.$$
$$\begin{vmatrix} e^{-5t}\cos t & e^{-t}\sin t \\ 24e^{-5t}\cos t+10e^{-5t}\sin t & -10e^{-5t}\cos t+24e^{-5t}\sin t \end{vmatrix}+$$
$$\left.\begin{vmatrix} e^{-5t}\cos t & e^{-t}\sin t \\ -5e^{-5t}\cos t-e^{-5t}\sin t & e^{-5t}\cos t-5e^{-5t}\sin t \end{vmatrix}\right)$$

$$=37e^{-9t}.$$

16. The functions are linearly dependent.

Mathematica

```
caps={1-2Sin[x]^2,Cos[2x]};
wrmat={caps,D[caps,x]};
MatrixForm[wrmat]
Det[wrmat]//Simplify
```

Maple V

```
with(linalg);
caps:=[1-2*sin(x)^2,cos(2*x)];
wr:=Wronskian(caps,x);
simplify(det(wr));
```

20. The set of functions $S=\left\{1,t^{-5/2},t^{-3/2},t^{-1/2},t^{1/2},t^{3/2},t^{5/2}\right\}$ is linearly independent so the answer is "no."

```
caps={1,x^(-5/2),x^(-3/2),x^(-1/2),x^(1/2),
    x^(3/2),x^(5/2)};
wrmat={caps,D[caps,x],D[caps,{x,2}],
    D[caps,{x,3}],D[caps,{x,4}],
    D[caps,{x,5}],D[caps,{x,6}]};
MatrixForm[wrmat]
Det[wrmat]//Simplify
```

```
caps:=[1,x^(-5/2),x^(-3/2),
    x^(-1/2),x^(1/2),
    x^(3/2),x^(5/2)];
    wr:=Wronskian(caps,x);
simplify(det(wr));
```

21.

```
caps={f[x],x f[x]};
wrmat={caps,D[caps,x]};
MatrixForm[wrmat]
Det[wrmat]//Simplify
caps={f[x],x f[x],x^2 f[x]};
wrmat={caps,D[caps,x],D[caps,{x,2}]};
MatrixForm[wrmat]
Det[wrmat]//Simplify
caps={f[x],x f[x],x^2 f[x],x^3 f[x]};
wrmat={caps,D[caps,x],D[caps,{x,2}],
    D[caps,{x,3}]};
MatrixForm[wrmat]
Det[wrmat]//Simplify
```

```
caps:=[f(x),x*f(x)];
wr:=Wronskian(caps,x);
simplify(det(wr));
caps:=[f(x),x*f(x),x^2*f(x)];
wr:=Wronskian(caps,x);
simplify(det(wr));
caps:=[f(x),x*f(x),x^2*f(x),
    x^3*f(x)];
wr:=Wronskian(caps,x);
simplify(det(wr));
```

EXERCISES 4.4

3. $y(t) = c_1 + c_2 t + c_3 \cos 3t + c_4 \sin 3t$, 4. The characteristic equation is

$$r^2(r+3i)(r-3i) = r^2\left(r^2+9\right) = r^4 + 9r^2 = 0.$$

A fourth-order homogeneous equation with this charactertistic equation is

$$y^{(4)} + 9y'' = 0.$$

9. These functions are linearly independent:

$$W\left(\left\{e^{2t}, e^{t/4}, e^{-t}\cos t, e^{-t}\sin t\right\}\right) = \begin{vmatrix} e^{2t} & e^{t/4} & e^{-t}\cos t & e^{-t}\sin t \\ 2e^{2t} & \frac{1}{4}e^{t/4} & -e^{-t}\cos t - e^{-t}\sin t & e^{-t}\cos t - e^{-t}\sin t \\ 4e^{2t} & \frac{1}{16}e^{t/4} & 2e^{-t}\sin t & -2e^{-t}\cos t \\ 8e^{2t} & \frac{1}{64}e^{t/4} & 2e^{-t}\cos t - 2e^{-t}\sin t & 2e^{-t}\cos t + 2e^{-t}\sin t \end{vmatrix}$$

$$= -\frac{1435}{32}e^{t/4}\cos^2 t - \frac{1435}{32}e^{t/4}\sin^2 t = -\frac{1435}{32}e^{t/4}.$$

To have $\left\{e^{2t}, e^{t/4}, e^{-t}\cos t, e^{-t}\sin t\right\}$ be a fundamental set of solutions to a homogeneous differential equation with constant coefficients, the solutions of the characterstic equation would be $r_1 = 2, r_2 = 1/4, r_3 = -1+i, r_4 = -1-i$ so the characteristic equation would be

$$(r-2)(r-1/4)(r-(-1+i))(r-(-1-i)) = r^4 - \frac{1}{4}r^3 - 2r^2 - \frac{7}{2}r + 1 = 0$$

and a homogeneous differential equation with this characteristic equation is

$$y^{(4)} - \frac{1}{4}y''' - 2y'' - \frac{7}{2}y' + y = 0$$

or

$$4y^{(4)} - y''' - 8y'' - 14y' + 4y = 0.$$

13. The characteristic equation is

$$r^3 + 7r^2 + 17r + 15 = 0$$

$$(r+3)\left(r^2 + 4r + 5\right) = 0$$

so $r = -3$ or $r = \dfrac{-4 \pm \sqrt{4^2 - 4 \cdot 1 \cdot 5}}{2} = -2 \pm i$ and

$$y = c_1 e^{-3t} + e^{-2t}\left(c_2 \cos t + c_3 \sin t\right).$$

17. $r^3 + r^2 - 16r + 20 = (r-2)^2(r+5) = 0$ so $y = c_1 e^{-5t} + c_2 e^{2t} + c_3 t e^{2t}$

21. $r^4 - 9r^2 = r^2(r+3)(r-3) = 0$ so $y = c_1 e^{-3t} + c_2 + c_3 t + c_4 e^{3t}$

25. The characteristic equation is

$$r^4 - 6r^3 - r^2 + 54r - 72 = (r-4)(r-3)(r-2)(r+3) = 0$$

so $r=4$, $r=3$, $r=2$, or $r=-3$ and $y=c_1e^{-3t}+c_2e^{2t}+c_3e^{3t}+c_4e^{4t}$.

29. The characteristic equation is $r^4+2r^3-2r^2+8=(r+2)^2\left(r^2-2r+2\right)=0$ so $r=-2$ (multiplicity two) or $r=1\pm i$ and $y=c_1e^{-2t}+c_2te^{-2t}+c_3e^t\cos t+c_4e^t\sin t$.

33. The characteristic equation is

$$r^3-1=(r-1)\left(r^2+r+1\right)=0\Rightarrow r=1,\, r=\frac{-1\pm\sqrt{1^2-4\cdot1\cdot1}}{2}=-\frac{1}{2}\pm\frac{\sqrt{3}}{2}i$$

so a general solution is

$$y=c_1e^t+e^{-t/2}\left(c_2\cos\left(\frac{\sqrt{3}}{2}t\right)+c_3\sin\left(\frac{\sqrt{3}}{2}t\right)\right)\Rightarrow$$

$$y'=c_1e^t+e^{-t/2}\left(\left(-\frac{1}{2}c_2+\frac{\sqrt{3}}{2}c_3\right)\cos\left(\frac{\sqrt{3}}{2}t\right)+\left(-\frac{\sqrt{3}}{2}c_2-\frac{1}{2}c_3\right)\sin\left(\frac{\sqrt{3}}{2}t\right)\right)$$

and

$$y''=c_1e^t+e^{-t/2}\left(\left(-\frac{1}{2}c_2-\frac{\sqrt{3}}{2}c_3\right)\cos\left(\frac{\sqrt{3}}{2}t\right)+\left(\frac{\sqrt{3}}{2}c_2-\frac{1}{2}c_3\right)\sin\left(\frac{\sqrt{3}}{2}t\right)\right).$$

Application of the initial conditions yields

$$\begin{cases} c_1+c_2=0 \\ c_1-\frac{1}{2}c_2+\frac{\sqrt{3}}{2}c_3=0\Rightarrow c_1=1,\, c_2=-1,\, c_3=-\sqrt{3} \\ c_1-\frac{1}{2}c_2-\frac{\sqrt{3}}{2}c_3=3 \end{cases}$$

so

$$y=e^t+e^{-t/2}\left(-\cos\left(\frac{\sqrt{3}}{2}t\right)+-\sqrt{3}\sin\left(\frac{\sqrt{3}}{2}t\right)\right).$$

34. $r^4-16=(r+2)(r-2)\left(r^2+4\right)=0\Rightarrow r=-2,2,\pm 2i\Rightarrow y=c_1e^{-2t}+c_2e^{2t}+c_3\cos 2t+c_4\sin 2t$.

Application of the initial conditions yields

$$\begin{cases} c_1+c_2+c_3=0 \\ -2c_1+2c_2+2c_4=-8 \\ 4c_1+4c_2-4c_3=0 \\ -8c_1+8c_2-8c_4=0 \end{cases}\Rightarrow c_1=1,\, c_2=-1,\, c_3=0,\, c_4=-2$$

so $y=e^{-2t}-e^{2t}-2\sin 2t$.

```
Factor[r^4 - 16]

(-2 + r) (2 + r) (4 + r^2
y[t_] = c1 Exp[-2 t] + c2 Exp[2 t] + c3 Co
  c4 Sin[2 t]

c1 e^-2 t + c2 e^2 t + c3 Cos[2 t] + c4 Sin[2
vals = Solve[{y[0] == 0, y'[0] == ·
  y''[0] == 0, y'''[0] == 0}]

{{c1 → 1, c2 → -1, c3 → 0, c4 → -2}
```

```
Plot[y[t] /. vals, {t, 0, 2}]
```

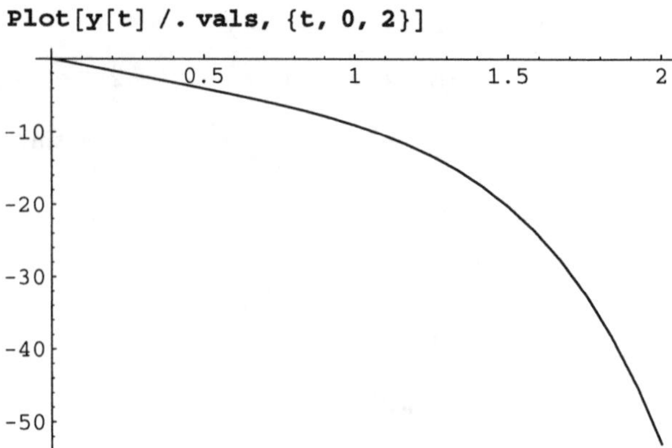

35. The characteristic equation of $y^{(4)} - 8y'' + 16y = 0$ is

$$r^4 - 8r^2 + 16 = \left(r^2 - 4\right)^2 = (r+2)^2(r-2)^2 = 0 \Rightarrow r_{1,2} = -2, r_{3,4} = 2$$

so a general solution is

$$y = c_1e^{-2t} + c_2te^{-2t} + c_3e^{2t} + c_4te^{2t} \Rightarrow y' = \left(-2c_1 + c_2\right)e^{-2t} - 2c_2te^{-2t} + \left(2c_3 + c_4\right)e^{2t} + 2c_4te^{2t},$$

$$y'' = \left(4c_1 - 4c_2\right)e^{-2t} + 4c_2te^{-2t} + \left(4c_3 + 4c_4\right)e^{2t} + 4c_4te^{2t}, \text{ and}$$

$$y''' = \left(-8c_1 + 12c_2\right)e^{-2t} - 8c_2te^{-2t} + \left(8c_3 + 12c_4\right)e^{2t} + 8c_4te^{2t}.$$

Application of the initial conditions yields

$$\begin{cases} c_1 + c_3 = 0 \\ -2c_1 + c_2 + 2c_3 + c_4 = 0 \\ 4c_1 - 4c_2 + 4c_3 + 4c_4 = 8 \\ -8c_1 + 12c_2 + 8c_3 + 12c_4 = 0 \end{cases} \Rightarrow c_1 = 0, c_2 = -1, c_3 = 0, c_4 = 1$$

so $y = -te^{-2t} + te^{2t}$.

```
Clear[y]
sol =
 DSolve[{y''''[t] - 8 y''[t] + 16 y[t] == 0,
   y[0] == 0, y'[0] == 0, y''[0] == 8,
   y'''[0] == 0}, y[t], t]

{{y[t] → e^-2 t (-t + e^4 t t)}}

Plot[y[t] /. sol, {t, 0, 1}]
```

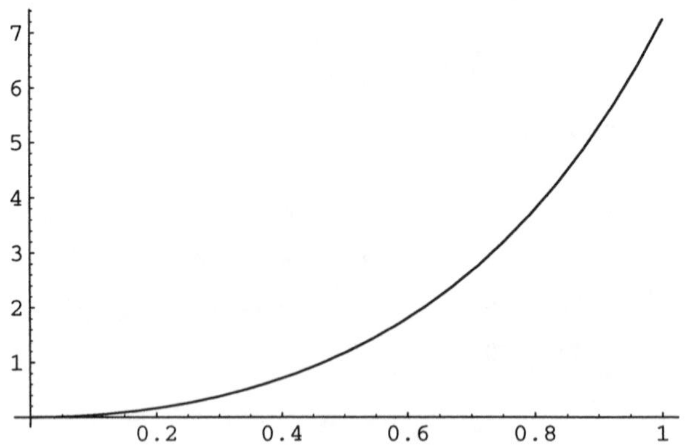

37. The characteristic equation is $r^5 + 8r^4 = r^4(r+8) = 0 \Rightarrow r_{1,2,3,4} = 0, r_5 = -8$ so a general solution of the equation is $y = c_1 + c_2t + c_3t^2 + c_4t^3 + c_5e^{-8t}$. Then,

$$y' = c_2 + 2c_3t + 3c_4t^2 - 8c_5e^{-8t},$$

$$y'' = 2c_3 + 6c_4t + 64c_5e^{-8t},$$

$$y''' = 6c_4 - 512c_5e^{-8t}, \text{ and}$$

$$y^{(4)} = 4096c_5e^{-8t}.$$

Application of the intial conditions yields

$$\begin{cases} c_1 + c_5 = 8 \\ c_2 - 8c_5 = 4 \\ 2c_3 + 64c_5 = 0 \\ 6c_4 - 512c_5 = 48 \\ 4096c_5 = 0 \end{cases} \Rightarrow c_1 = 8, c_2 = 4, c_3 = 0, c_4 = 8, c_5 = 0$$

so the solution to the initial value problem is $y = 8 + 4t + 8t^3$.

```
Clear[y]
sol =
 DSolve[{y^(5)[t] + 8 y^(4)[t] == 0,
   y[0] == 8, y'[0] == 4, y''[0] == 0,
   y^(3)[0] == 48, y^(4)[0] == 0}, y[t], t]

{{y[t] → 8 + 4 t + 8 t^3}}

Plot[y[t] /. sol, {t, 0, 2}]
```

80
60
40
20
0.5 1 1.5 2

44. (a)

```
sola1 =
 DSolve[
  {y^(3)[t] + 2 y''[t] + 5 y'[t] -
     26 y[t] == 0}, y[t], t]

{{y[t] →
   e^(2 t) C[3] + e^(-2 t) C[2] Cos[3 t]
    e^(-2 t) C[1] Sin[3 t]}}

sola2 =
 DSolve[
  {y^(3)[t] + 2 y''[t] + 5 y'[t] -
     26 y[t] == 0, y[0] == y0,
   y'[0] == v0, y''[0] == a0}, y[t],
  t]

{{y[t] →
   e^(-2 t) (1/25 e^(4 t) (a0 + 4 v0 + 13 y0) +
     1/25 (-a0 - 4 v0 + 12 y0) Cos[3 t]
     1/75 (4 a0 - 9 v0 + 2 y0)
      Sin[3 t])}}

grays = Table[GrayLevel[i],
   {i, 0, 0.7, 0.7 / 10}];
```

```
p1 =
  Plot[Evaluate[
    Table[sola2[[1, 1, 2]] /.
      {y0 -> i, v0 -> 0, a0 -> 0},
     {i, -5, 5}]], {t, 0, 2},
   PlotStyle -> grays,
   DisplayFunction -> Identity];

p2 =
  Plot[Evaluate[
    Table[sola2[[1, 1, 2]] /.
      {y0 -> 0, v0 -> i, a0 -> 0},
     {i, -5, 5}]], {t, 0, 2},
   PlotStyle -> grays,
   DisplayFunction -> Identity];

p3 =
  Plot[Evaluate[
    Table[sola2[[1, 1, 2]] /.
      {y0 -> 0, v0 -> 0, a0 -> i},
     {i, -5, 5}]], {t, 0, 2},
   PlotStyle -> grays,
   DisplayFunction -> Identity];

p4 =
  Plot[Evaluate[
    Table[sola2[[1, 1, 2]] /.
      {y0 -> 0, v0 -> -i, a0 -> i},
     {i, -5, 5}]], {t, 0, 2},
   PlotStyle -> grays,
   DisplayFunction -> Identity];
```

```
Show[GraphicsArray[
  {{p1, p2}, {p3, p4}}]]
```

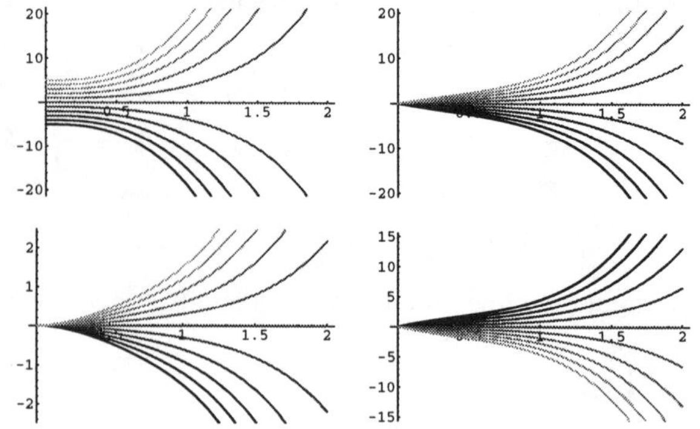

(b)

```
sola1 =
 DSolve[
  {0.9 y'''[t] + 18.78 y'[t] - 0.2987
    0}, y[t], t]
```

```
{{y[t] → e^(0.015905 t) C[1] +
    e^((-0.00795251-4.56803 i) t) C[2]
    e^((-0.00795251+4.56803 i) t) C[3]
```

```
sola2 =
 DSolve[
  {0.9 y'''[t] + 18.78 y'[t] - 0.2987 y[t
    0, y[0] == y0, y'[0] == v0,
   y''[0] == a0}, y[t], t]
```

```
{{y[t] → (-0.0239608 + 0.000125141 i)
     e^((-0.00795251+4.56803 i) t)
     (a0 - (0.00795251 - 4.56803 i) v0 -
       (0.000126485 + 0.0726546 i) y0) -
    (0.0239608 + 0.000125141 i)
     e^((-0.00795251-4.56803 i) t)
     (a0 - (0.00795251 + 4.56803 i) v0 -
       (0.000126485 - 0.0726546 i) y0) +
    (0.0479216 - 2.56563 × 10^-21 i)
     e^(0.015905 t) (a0 + 0.015905 v0 +
       (20.8669 - 3.0564 × 10^-18 i) y0)}}
```

```
grays = Table[GrayLevel[i],
   {i, 0, 0.7, 0.7 / 10}];
```

```
p1 =
  Plot[
   Evaluate[
    Table[sola2[[1, 1, 2]] /.
      {y0 -> i, v0 -> 0, a0 -> 0},
     {i, -5, 5}]], {t, 0, 2},
   PlotStyle -> grays,
   DisplayFunction -> Identity];
p2 =
  Plot[
   Evaluate[
    Table[sola2[[1, 1, 2]] /.
      {y0 -> 0, v0 -> i, a0 -> 0},
     {i, -5, 5}]], {t, 0, 2},
   PlotStyle -> grays,
   DisplayFunction -> Identity];
p3 =
  Plot[
   Evaluate[
    Table[sola2[[1, 1, 2]] /.
      {y0 -> 0, v0 -> 0, a0 -> i},
     {i, -5, 5}]], {t, 0, 2},
   PlotStyle -> grays,
   DisplayFunction -> Identity];
p4 =
  Plot[
   Evaluate[
    Table[sola2[[1, 1, 2]] /.
      {y0 -> 0, v0 -> -i, a0 -> i},
     {i, -5, 5}]], {t, 0, 2},
   PlotStyle -> grays,
   DisplayFunction -> Identity];
```

```
Show[GraphicsArray[{{p1,
```

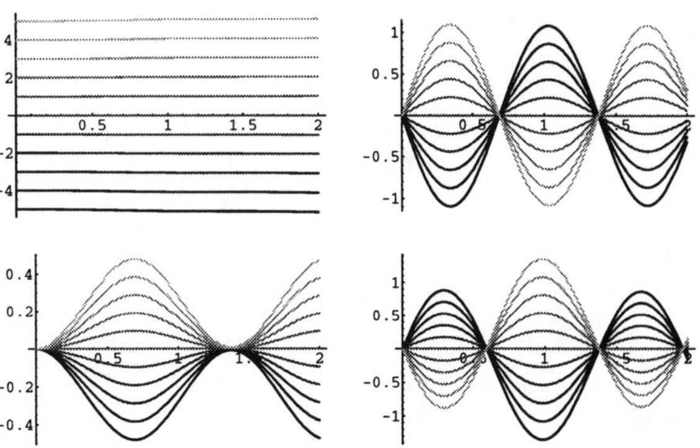

(c)

```
sola1 =
 DSolve[
  {8.9 y^(4) [t] - 2.5 y''[t] + 32.07 y'[t] +
     0.773 y[t] == 0}, y[t], t]
```

$$\{\{y[t] \to e^{-1.58665\,t}\,C[1] + e^{-0.0240585\,t}\,C[2] + e^{(0.805354-1.27543\,i)\,t}\,C[3] + e^{(0.805354+1.27543\,i)\,t}\,C[4]\}\}$$

```
sola2 =
 DSolve[
  {8.9 y^(4) [t] - 2.5 y''[t] + 32.07 y'[t] +
     0.773 y[t] == 0, y[0] == y0,
    y'[0] == v0, y''[0] == a0, y^(3) [0] == b0},
   y[t], t]
```

$$\begin{aligned}\{\{y[t] \to\ & (-0.0946981 + 0.00823387\,i)\\ & e^{(0.805354-1.27543\,i)\,t}\\ & ((0.805354 - 1.27543\,i)\,a0 +\\ & \quad b0 - (1.25902 + 2.05434\,i)\,v0 -\\ & \quad (0.0307423 + 0.0486861\,i)\,y0) -\\ & (0.0946981 + 0.00823387\,i)\\ & e^{(0.805354+1.27543\,i)\,t}\\ & ((0.805354 + 1.27543\,i)\,a0 +\\ & \quad b0 - (1.25902 - 2.05434\,i)\,v0 -\\ & \quad (0.0307423 - 0.0486861\,i)\,y0) -\\ & (0.0870888 - 3.26212\times10^{-20}\,i)\\ & e^{-1.58665\,t}\,(-1.58665\,a0 + b0 +\\ & \quad (2.23656 + 1.53123\times10^{-17}\,i)\,v0 +\\ & \quad (0.0547405 - 2.15354\times10^{-18}\,i)\,y0) +\\ & (0.276485 - 5.55548\times10^{-18}\,i)\\ & e^{-0.0240585\,t}\,(-0.0240585\,a0 + b0 -\\ & \quad (0.28032 - 1.53123\times10^{-17}\,i)\,v0 +\\ & \quad (3.61011 + 8.18856\times10^{-17}\,i)\,y0)\}\}\end{aligned}$$

```
grays = Table[GrayLevel[i],
   {i, 0, 0.7, 0.7 / 10}];
Show[GraphicsArray[{{p1,
```

```
p1 =
  Plot[
   Evaluate[
    Table[sola2[[1, 1, 2]] /.
       {y0 -> i, v0 -> 0, a0 -> 0, b0 -> 0},
      {i, -5, 5}]], {t, 0, 8},
   PlotStyle -> grays,
   DisplayFunction -> Identity];
p2 =
  Plot[
   Evaluate[
    Table[sola2[[1, 1, 2]] /.
       {y0 -> 0, v0 -> i, a0 -> 0, b0 -> 0},
      {i, -5, 5}]], {t, 0, 4},
   PlotStyle -> grays,
   DisplayFunction -> Identity];
p3 =
  Plot[
   Evaluate[
    Table[sola2[[1, 1, 2]] /.
       {y0 -> 0, v0 -> 0, a0 -> i, b0 -> 0},
      {i, -5, 5}]], {t, 0, 4},
   PlotStyle -> grays,
   DisplayFunction -> Identity];
p4 =
  Plot[
   Evaluate[
    Table[sola2[[1, 1, 2]] /.
       {y0 -> 0, v0 -> 0, a0 -> 0, b0 -> i},
      {i, -5, 5}]], {t, 0, 4},
   PlotStyle -> grays,
   DisplayFunction -> Identity];
```

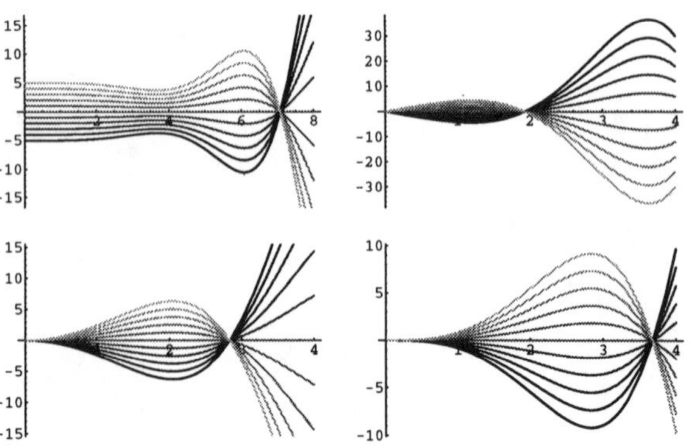

45. (a)

```
Solve[r^3 + 3 r^2 + 2 r + 6 == 0]
```

$\{\{r \to -3\}, \{r \to -i\sqrt{2}\}, \{r \to i\sqrt{2}\}\}$

```
y[t_] = c1 Exp[-3] + c2 Cos[Sqrt[2] t] +
   c3 Sin[Sqrt[2] t]
```

$\frac{c1}{e^3} + c2\,\mathrm{Cos}[\sqrt{2}\ t] + c3\,\mathrm{Sin}[\sqrt{2}\ t]$

```
cvals =
  Solve[{y[0] == 0, y'[0] == 1,
    y''[0] == -1}]
```

$\{\{c1 \to -\frac{e^3}{2},\ c3 \to \frac{1}{\sqrt{2}},\ c2 \to \frac{1}{2}\}\}$

```
y[t] /. cvals
```

$\{-\frac{1}{2} + \frac{1}{2}\mathrm{Cos}[\sqrt{2}\ t] + \frac{\mathrm{Sin}[\sqrt{2}\ t]}{\sqrt{2}}\}$

```
Plot[y[t] /. cvals, {t, 0, 16}]
```

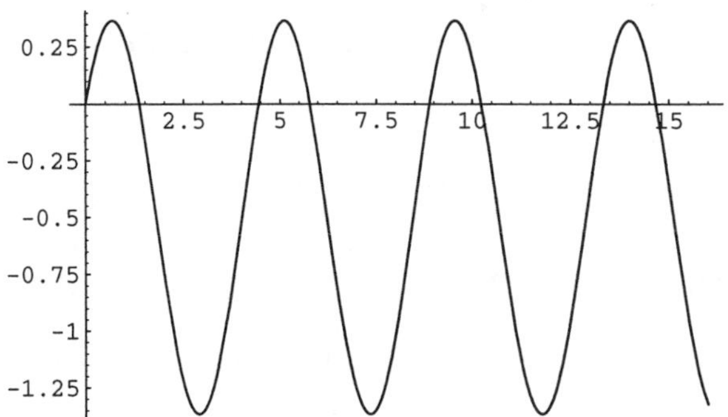

(b)

```
Solve[r^4 - 8 r^3 + 30 r^2 - 56 r + 49 == 0]
```

$\{\{r \to 2 - i\sqrt{3}\}, \{r \to 2 - i\sqrt{3}\}, \{r \to 2 + i\sqrt{3}\}, \{r \to 2 + i\sqrt{3}\}\}$

```
y[t_] = Exp[2 t] (c1 + c2 t) Cos[Sqrt[3] t] +
   (c3 + c4 t) Sin[Sqrt[3] t]
```

$e^{2t}(c1 + c2\,t)\,\mathrm{Cos}[\sqrt{3}\ t] + (c3 + c4\,t)\,\mathrm{Sin}[\sqrt{3}\ t]$

```
vals =
  Solve[{y[0] == 1, y'[0] == 2, y''[0] == ·
    y'''[0] == -1}]
```

$\{\{c4 \to -\frac{4}{\sqrt{3}},\ c3 \to -\frac{\sqrt{3}}{2},\ c2 \to \frac{3}{2},\ c1 \to 1\}\}$

```
y[t] /. vals
```

$\{e^{2t}\left(1 + \frac{3t}{2}\right)\mathrm{Cos}[\sqrt{3}\ t] + \left(-\frac{\sqrt{3}}{2} - \frac{4t}{\sqrt{3}}\right)\mathrm{Sin}[\sqrt{3}\ t]\}$

```
Plot[y[t] /. vals, {t, 0, 3},
 PlotRange -> {-10, 10}]
```

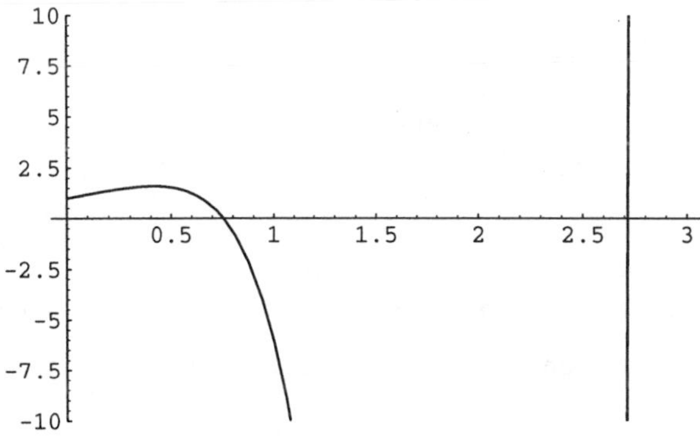

(c)

```
Solve[0.31 r^3 + 11.2 r^2 - 9.8 r + 5.3 == 0]

{{r → -36.996}, {r → 0.433493 - 0.523649 i},
 {r → 0.433493 + 0.523649 i}}

y[t_] = c1 Exp[-36.996 t] +
   Exp[0.433493 t]
    (c2 Cos[.523649 t] + c3 Sin[.523649 t]

c1 e^(-36.996 t) + e^(0.433493 t)
  (c2 Cos[0.523649 t] + c3 Sin[0.523649 t

cvals =
 Solve[{y[0] == -1, y'[0] == -1,
   y''[0] == 0}]

{{c1 → 0.000288931,
  c2 → -1.00029, c3 → -1.06119}}

y[t] /. cvals

{0.000288931 e^(-36.996 t) +
  e^(0.433493 t) (-1.00029 Cos[0.523649 t
    1.06119 Sin[0.523649 t])}

Plot[y[t] /. cvals, {t, 0, 12}]
```

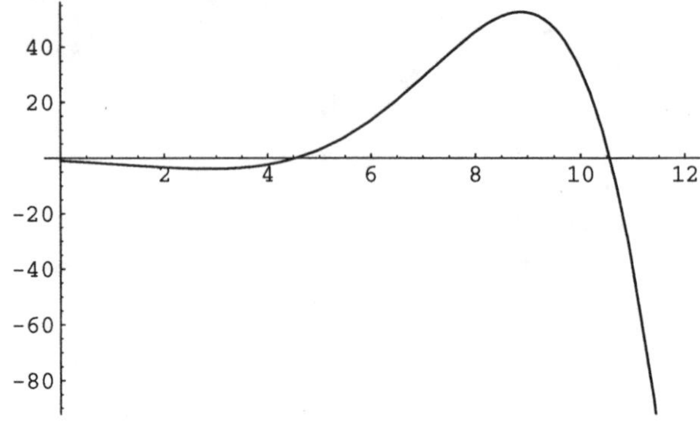

*46.

```
Clear[k, y]
gensol = DSolve[y[x] -
     k^4 D[y[x], {x, 4}] == 0, y[x], x]

{{y[x] → e^(-x/k) C[1] +
     e^(-i x/k) C[2] + e^(i x/k) C[3] + e^(x/k) C[4]}}

s2 = ExpToTrig[gensol[[1, 1, 2]]];
s3 = Collect[s2, {Cos[x/k], Sin[x/k]}]

(C[2] + C[3]) Cos[x/k] + C[1] Cosh[x/k] +
 C[4] Cosh[x/k] + (-i C[2] + i C[3]) Sin[x/k] -
 C[1] Sinh[x/k] + C[4] Sinh[x/k]

cvals = Solve[{C[2] + C[3] == c2,
     -I C[2] + I C[3] == c3},
    {C[2], C[3]}];
s3 /. cvals // Expand

{c2 Cos[x/k] + C[1] Cosh[x/k] +
  C[4] Cosh[x/k] + c3 Sin[x/k] -
```

Typo on page 168. Immediately after $y = Ae^{-x/k} + Be^{x/k} + C\sin\frac{x}{k} + D\cos\frac{x}{k}$ should be the next problem: Find conditions on a and b, if possible, so that the solution to the initial-value problem

$$\begin{cases} y'' + 4y' + 3y = 0 \\ y(0) = a,\ y'(0) = b \end{cases}$$

has (a) neither local maxima nor local minima; (b) exactly one local maximum; and (c) exactly one local minimum on the interval $[1, \infty)$.

```
Clear[x,y,a,b]
partsol=DSolve[{y''[x]+4y'[x]+3y[x]==0,
    y[0]==a,y'[0]==b},y[x],x]
dsol=D[partsol[[1,1,2]],x]//Simplify
Solve[dsol==0,x]
```

```
x:='x':y:='y':a:='a':b:='b':
partsol:=dsolve({diff(y(x),x$2)+
   4*diff(y(x),x)+3*y(x)=0,
   y(0)=a,D(y)(0)=b},y(x));
assign(partsol):
dsol:=diff(y(x),x);
solve(dsol=0,x);
```

The local extrema occur at the values of x where the derivative of the solution is zero:

$$3a+3b-3ae^{2x}-be^{2x}=0$$

$$-(3a+b)e^{2x}=-(3a+3b)$$

$$e^{2x}=\frac{3a+3b}{3a+b}$$

$$x=\ln\sqrt{\frac{3a+3b}{3a+b}}.$$

Because the domain of the natural logarithm function is $(0,\infty)$, there will be no local extrema if

$$\frac{3a+3b}{3a+b}\le 0 \Rightarrow a\le -b \text{ and } a>-\tfrac{1}{3}b \text{ OR } a\ge -b \text{ and } a<-\tfrac{1}{3}b.$$

*47. We find a general solution of the equation by solving the characteristic equation. The system of equations $\begin{cases} y(0)=0 \\ y'(0)=y_0 \\ y(4)=0 \end{cases}$ has a unique solution.

```
Clear[x,y,m]
Solve[4.02063m^3-0.224975m^2+4.486m-
    2.48493==0]
y[x_]=c1*Exp[0.4712630669x]+
    Exp[-0.2076539528x](c2*Cos[1.1262088x]+
        c3*Sin[1.126208088x]);
eqs={y[0]==0,y'[0]==y0,y[4]==0}
cvals=Solve[eqs,{c1,c2,c3}]
toplot=Table[y[x] /. cvals[[1]],{y0,-3,3}]
grays=Table[GrayLevel[i],{i,0,.7,.7/6}];
Plot[Evaluate[toplot],{x,-1,5},
    PlotStyle->grays,
    PlotRange->{-3,3},AspectRatio->1]
```

```
solve(4.02063*m^3-0.224975*m^2+
   4.486*m-2.48493=0);
y:=x->c1*exp(0.4712630669*x)+
   exp(-0.2076539528*x)*(
      c2*cos(1.1262088*x)+
      c3*sin(1.126208088*x));
eqs:={y(0)=0,D(y)(0)=y0,y(4)=0};
cvals:=solve(eqs,{c1,c2,c3});
toplot:={seq(subs(cvals,y(x)),
   y0=-3..3)};
plot(toplot,x=-1..5);
```

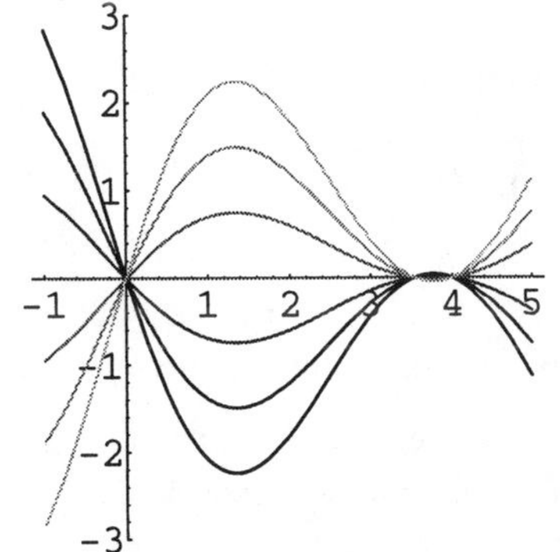

EXERCISES 4.5

5. f involves both exponential and polynomial terms. Associated with the term e^{-t} we have $S_1 = \{e^{-t}\}$ while associated with the t^4 and t terms we have $S_2 = \{t^4, t^3, t^2, t, 1\}$.

11. f involves both exponential and polynomial terms. Associated with the term $e^{-t}\cos t$ we have $S_1 = \{e^{-t}\cos t, e^{-t}\sin t\}$ while associated with the 1 term $S_2 = \{1\}$.

15. The characteristic equation for the corresponding homogeneous equation is

$$r^2 - 4r + 3 = (r-3)(r-1) = 0 \Rightarrow r = 1, 3$$

so $y_h = c_1 e^t + c_2 e^{3t}$. $S = \{e^{3t}\}$ and because e^{3t} is a solution to the corresponding homogeneous equation we multiply S by t: $tS = \{te^{3t}\}$. Because te^{3t} is not a solution to the corresponding homogeneous equation we assume that $y_p = Ate^{3t}$.

19. The characteristic equation for the corresponding homogeneous equation is

$$r^2 + 1 = 0 \Rightarrow r_{1,2} = \pm i$$

so $y_h = c_1 \cos t + c_2 \sin t$. $S = \{\cos t, \sin t\}$ and no element of S is a solution of the corresponding homogeneous equation so we assume that

$$y_p = A\cos t + B\sin t.$$

23. The characteristic equation of the corresponding homogeneous equation is $r^2 = 0 \Rightarrow r_{1,2} = 0$ so $y_h = c_1 + c_2 t$. $S = \{t^4, t^3, t^2, t, 1\}$. Elements of S are solutions to the corresponding homogeneous equation so we multiply by t: $tS = \{t^5, t^4, t^3, t^2, t\}$. Elements of tS are solutions to the corresponding homogeneous equation so we multiply tS by t: $t^2 S = \{t^6, t^5, t^4, t^3, t^2\}$. No element of $t^2 S$ is a solution to the corresponding homogeneous equation so we assume that

$$y_p = At^6 + Bt^5 + Ct^4 + Dt^3 + Ct^2.$$

27. The characteristic equation for the corresponding homogeneous equation is

$$r^2 - 2r - 8 = (r-4)(r+2) = 0 \Rightarrow r = 4, -2$$

so $y_h = c_1 e^{4t} + c_2 e^{-2t}$. The associated set of functions is $S = \{t, 1\}$ and because none of these are solutions to the corresponding homogeneous equations we assume $y_p = At + B \Rightarrow y_p' = A, y_p'' = 0$. Substitution into the nonhomogeneous equation yields

$$-2A - 8(At + B) = 32t$$
$$-2A - 8B - 8At = 32t$$

and equating coefficients gives us

$$\begin{cases} -2A - 8B = 0 \\ -8A = 32 \end{cases} \Rightarrow A = -4,\ B = 1$$

so $y_p = -4t + 1$ and

$$y = y_h + y_p = c_1 e^{4t} + c_2 e^{-2t} - 4t + 1.$$

31. The characteristic equation of the corresponding homogeneous equation is

$$8r^2 + 6r + 1 = (4r+1)(2r+1) = 0 \Rightarrow r = -1/4, -1/2$$

so $y_h = c_1 e^{-t/4} + c_2 e^{-t/2}$. The associated set of functions is $S = \{t^2, t, 1\}$ and because none of these are solutions to the corresponding homogeneous equation we assume that

$$y_p = At^2 + Bt + C \Rightarrow y_p' = 2At + B, y_p'' = 2A.$$

Substitution into the nonhomogeneous equation gives us

$$8 \cdot 2A + 6(2At + B) + At^2 + Bt + C = 5t^2$$
$$16A + 6B + C + (12A + B)t + At^2 = 5t^2$$

and equating coefficients results in

$$\begin{cases} 16A + 6B + C = 0 \\ 12A + B = 0 \\ A = 5 \end{cases} \Rightarrow A = 5,\ B = -60,\ C = 280$$

so $y_p = 5t^2 - 60t + 280$ and

$$y = c_1 e^{-t/4} + c_2 e^{-t/2} + 5t^2 - 60t + 280.$$

35. The characteristic equation for the corresponding homogeneous equation is $r^2 - 9 = (r+3)(r-3) = 0 \Rightarrow r = -3, 3$ so $y_h = c_1 e^{3t} + c_2 e^{-3t}$. The associated set of functions is $S = \{t\cos 3t, t\sin 3t, \cos 3t, \sin 3t\}$ and because none of these are solutions to the corresponding homogeneous equation we assume that

$$y_p = At\cos 3t + Bt\sin 3t + C\cos 3t + D\sin 3t \Rightarrow y_p' = (A + 3D + 3Bt)\cos 3t + (B - 3C - 3At)\sin 3t$$

and $y_p'' = (6B - 9C - 9At)\cos 3t + (-6A - 9D - 9Bt)\sin 3t$. Substituting y_p into the nonhomogeneous equation yields

$$(6B - 9C - 9At)\cos 3t + (-6A - 9D - 9Bt)\sin 3t - 9(At\cos 3t + Bt\sin 3t + C\cos 3t + D\sin 3t) = 54t\sin 3t$$
$$(6B - 18C - 18At)\cos 3t + (-6A - 18D - 18Bt)\sin 3t = 54t\sin 3t.$$

Equating coefficients gives us

$$\begin{cases} 6B-18C=0 \\ -18A=0 \\ -6A-18D=0 \\ -18B=54 \end{cases} \Rightarrow A=0,\ B=-3,\ C=-1,\ D=0$$

so $y_p = -3t\sin 3t - \cos 3t$ and

$$y = y_h + y_p = c_1e^{3t} + c_2e^{-3t} - \cos 3t - 3t\sin 3t.$$

39. A general solution of the corresponding homogeneous equation $y'' - 4y' - 5y = 0$ with charactertistic equation $r^2 - 4r - 5 = (r+1)(r-5) = 0$ is $y_h = c_1e^{-t} + c_2e^{5t}$. In this case, $S = \{t^2e^{5t}, te^{5t}, e^{5t}\}$. Because e^{5t} is a solution of the corresponding homogeneous equation, we multiply each member of S by t and assume that there is a particular solution of the nonhomogeneous equation of the form $y_p = A\,te^{5t} + B\,t^2e^{5t} + C\,t^3e^{5t}$. Substituting y_p into the nonhomogeneous equation yields

$$(6A+2B)e^{5t} + (12B+6C)te^{5t} + 18Ct^2e^{5t} = -648t^2e^{5t}.$$

Equating coefficients results in the system of equations

$$\begin{cases} 6A+2B=0 \\ 12B+6C=0 \\ 18C=-648 \end{cases},$$

which has solution $A=-6$, $B=18$, and $C=-36$. Thus, a general solution of the nonhomogeneous equation is

$$y = c_1e^{-t} + c_2e^{5t} - 36t^3e^{5t} + 18t^2e^{5t} - 6te^{5t}.$$

43. The characteristic equation for the corresponding homogeneous equation is

$$r^3 + 6r^2 - 14r - 104 = (r-4)\left(r^2 + 10r + 26\right) = 0 \Rightarrow r = 4$$

or

$$r = \frac{-10 \pm \sqrt{10^2 - 4\cdot 1\cdot 26}}{2} = -5 \pm i$$

so a general solution of the corresponding homogeneous equation is $y = c_1e^{4t} + c_2e^{-5t}\cos t + c_3e^{-5t}\sin t$. The associated set of function is $S = \{e^t\}$ and because e^t is not a solution of the corresponding homogeneous equation we assume that $y_p = Ae^t \Rightarrow y_p' = Ae^t, y_p'' = Ae^t, y_p''' = Ae^t$. Substituting y_p into the nonhomogeneous equation yields $-111Ae^t = -111e^t \Rightarrow A = 1$ so $y_p = e^t$ and

$$y = y_h + y_p = c_1e^{4t} + c_2e^{-5t}\cos t + c_3e^{-5t}\sin t + e^t.$$

49. The characteristic equation of the corresponding homogeneous equation is

$$r^2 + 2r - 3 = (r+3)(r-1) = 0 \Rightarrow r = -3, 1$$

so $y_h = c_1e^{-3t} + c_2e^t$. The associated set of functions is $S = \{1\}$ and 1 is not a solution tothe corresponding homogeneous equaiton so we assume that $y_p = A$. Substituting y_p into the nonhomogeneous equation yields $-3A = -2 \Rightarrow A = 2/3 \Rightarrow y_p = 2/3$ so $y = c_1e^{-3t} + c_2e^t + 2/3 \Rightarrow y' = -3c_1e^{-3t} + c_2e^t$. Application of the initial conditions gives us $\{c_1 + c_2 + 2/3 = 2/3, -3c_1 + c_2 = 8\} \Rightarrow c_1 = -2, c_2 = 2$ so $y = -2e^{-3t} + 2e^t + 2/3$.

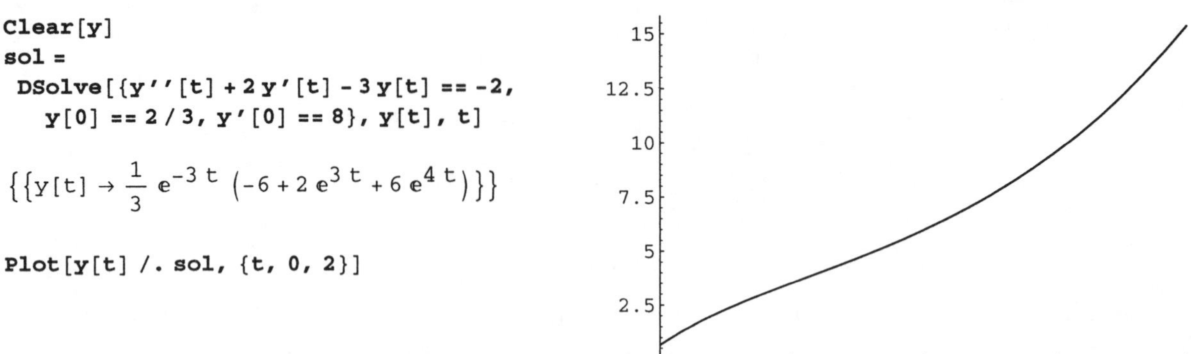

53. The characteristic equation for the corresponding homogeneous equation is

$$r^2+6r+25=0 \Rightarrow r=\frac{-6\pm\sqrt{6^2-4\cdot1\cdot25}}{2}=-3\pm4i$$

so $y_h=e^{-3t}\left(c_1\cos4t+c_2\sin4t\right)$. The associated set of functions is $S=\{1\}$ and because 1 is not a solution to the corresponding homogeneous equation we assume that $y_p=A$. Substituting y_p into the nonhomogeneous equation results in $25A=-1\Rightarrow A=-1/25\Rightarrow y_p=-1/25$ and $y=e^{-3t}\left(c_1\cos4t+c_2\sin4t\right)-1/25\Rightarrow$

$$y'=e^{-3t}\left[\left(-3c_1+4c_2\right)\cos4t+\left(-4c_1-3c_2\right)\sin4t\right].$$

Application of the initial conditions results in $\left\{c_1-1/25=-1/25,-3c_1+4c_2=7\right\}\Rightarrow c_1=0, c_2=7/4$ so

$$y=\frac{7}{4}e^{-3t}\sin4t-\frac{1}{25}.$$

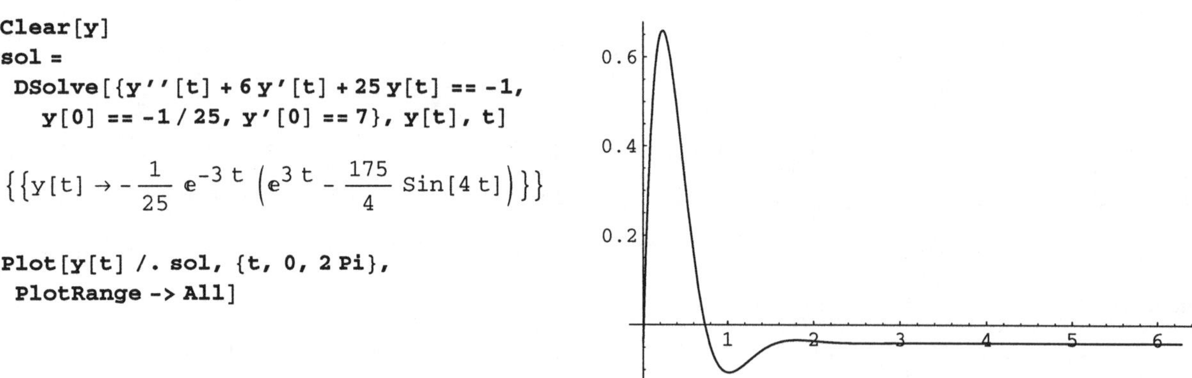

57. The corresponding homogeneous equation has characteristic equation

$$r^3+5r^2=r^2(r+5)=0\Rightarrow r_{1,2}=0,\ r_3=-5$$

so $y_h=c_1+c_2t+c_3e^{-5t}$. The associated set of functions is $S=\{t,1\}\Rightarrow t^2S=\left\{t^3,t^2\right\}$ so we search for a particular solution of the form $y_p=At^3+Bt^2\Rightarrow y_p'=3At^2+2Bt,\ y_p''=6At+2B,\ y_p'''=6A$. Substitution of y_p into the nonhomogeneous equation results in

$$\begin{array}{c}6A+5(2B+6At)=125t\\6A+10B+30At=125t\end{array}\Rightarrow\begin{cases}6A+10B=0\\30A=125\end{cases}\Rightarrow A=\frac{25}{6},\ B=-\frac{5}{2}$$

so $y_p=\frac{25}{6}t^3-\frac{5}{2}t^2$ and $y=y_h+y_p=c_1+c_2t+c_3e^{-5t}+\frac{25}{6}t^3-\frac{5}{2}t^2$. Application of the initial conditions yields

$$\begin{cases} c_1 + c_3 = 0 \\ c_2 - 5c_3 = 0 \Rightarrow c_1 = -\frac{1}{5}, c_2 = 1, c_3 = \frac{1}{5} \\ 25c_3 - 5 = 0 \end{cases}$$

so

$$y = \frac{25}{6}t^3 - \frac{5}{2}t^2 + t + \frac{1}{5}e^{-5t} - \frac{1}{5}.$$

```
Clear[y]
sol =
 DSolve[{y'''[t] + 5 y''[t] == 125 t,
   y[0] == 0, y'[0] == 0, y''[0] == 0},
  y[t], t]
```

$\{\{y[t] \to -\frac{1}{5} + \frac{e^{-5t}}{5} + t - \frac{5t^2}{2} + \frac{25t^3}{6}\}\}$

```
Plot[y[t] /. sol, {t, -1, 1}]
```

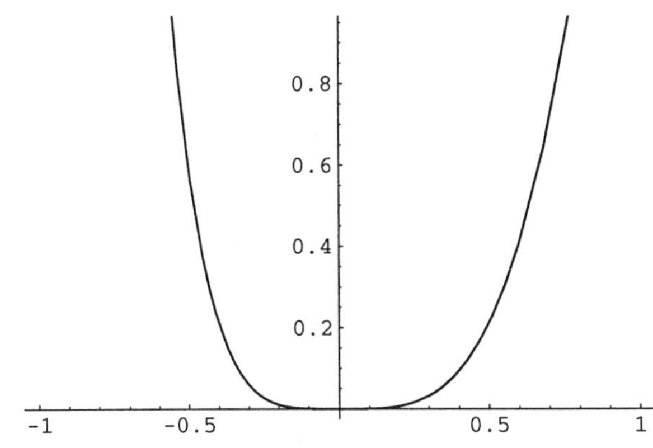

69. (a)

```
rvals = Solve[2 r^3 - 6 r^2 + 18 r - 56 == 0]
```

{{r → 3.05455}, {r → -0.0272772 + 3.02752 i}
{r → -0.0272772 - 3.02752 i}}

```
yh[t_] = c1 Exp[3.05455 t] +
   Exp[-0.027272 t]
    (c2 Cos[3.02752 t] + c3 Sin[3.02752 t]);
yp[t_] = a Exp[-2 t] + b Exp[3 t]
```

$a e^{-2t} + b e^{3t}$

```
Collect[2 yp'''[t] - 6 yp''[t] + 18 yp'[t] -
  56 yp[t], {a, b}]
```

$-132 a e^{-2t} - 2 b e^{3t}$

```
vals = Solve[{-132 a == 1, -2 b == 4}]
```

$\{\{a \to -\frac{1}{132}, b \to -2\}\}$

```
yp[t_] = yp[t] /. vals[[1]];
y[t_] = yh[t] + yp[t]
```

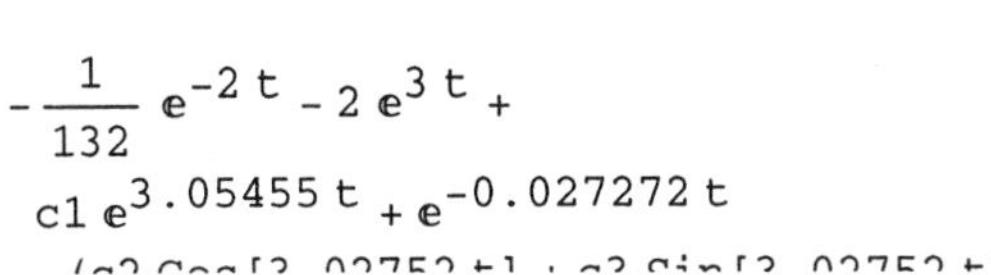

$-\frac{1}{132}e^{-2t} - 2e^{3t} +$
$c1\, e^{3.05455 t} + e^{-0.027272 t}$

```
cvals =
 Solve[{y[0] == 1, y'[0] == 0, y''[0] == -1}]
```

{{c1 → 2.40716,

```
Plot[y[t] /. cvals, {t, -2, 2}]
```

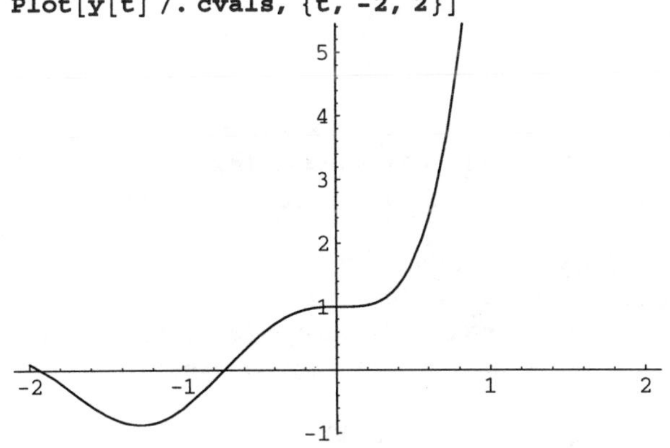

(b)

```
Solve[r^4 - 1/4 r^3 - 21 r^2 - 50/4 r + 5 ==
    0] // N // Chop
```

{{r → -4.07594}, {r → -0.903589},
 {r → 0.273954}, {r → 4.95557}}

```
yh[t_] = c1 Exp[-4.07594 t] +
   c2 Exp[-0.903589 t] + c3 Exp[-.273954 t] +
   c4 Exp[4.95557 t];
yp[t_] = a t + b;
vals = SolveAlways[
  yp''''[t] - 1/4 yp'''[t] - 21 yp''[t] -
    50/4 yp'[t] + 5 yp[t] == t, t]
```

$\{\{a \to \frac{1}{5},\ b \to \frac{1}{2}\}\}$

```
yp[t_] = yp[t] /. vals[[1]]
```

$\frac{1}{2} + \frac{t}{5}$

```
y[t_] = yh[t] + yp[t]
```

$\frac{1}{2} + c1\, e^{-4.07594\, t} + c2\, e^{-0.903589\, t}$
$c3\, e^{-0.273954\, t} + c4\, e^{4.95557\, t} + \frac{t}{5}$

```
cvals =
 Solve[{y[0] == 1, y'[0] == 0, y''[0] == 0,
   y'''[0] == 0}]
```

{{c1 → -0.00462875, c2 → 0.11289,
  c3 → 0.393564, c4 → -0.00182468}}

```
Plot[y[t] /. cvals, {t, -2, 2}]
```


70.

```
Clear[x, y, a, b, c]
sol = DSolve[{y'''[x] + y''[x] + 4 y'[x] +
       4 y[x] == Exp[-x] Cos[2 x],
    y[0] == a, y'[0] == b, y''[0] == c},
   y[x], x] // Simplify
convals = Solve[{-10 + 17 a - 17 c == 0,
    5 + 68 a + 85 b + 17 c == 0}, {a, b}]
toplot = sol[[1, 1, 2]] /. convals[[1]] /.
  c -> 0
Plot[toplot, {x, 0, Pi}]
convals = Solve[{-10 + 17 a - 17 c == 85,
    5 + 68 a + 85 b + 17 c == 0}, {a, b}]
toplot = sol[[1, 1, 2]] /. convals[[1]] /.
  c -> 0
Plot[toplot, {x, 0, 4 Pi}]
convals = Solve[{-10 + 17 a - 17 c == 0,
    5 + 68 a + 85 b + 17 c == 170}, {a, b}]
toplot = sol[[1, 1, 2]] /. convals[[1]] /.
  c -> 0
Plot[toplot, {x, 0, 4 Pi}]
```

```
sol:=dsolve({diff(y(x),x$3)+
      diff(y(x),x$2)+
      4*diff(y(x),x)+4*y(x)=
      exp(-x)*cos(2*x),
   y(0)=a,D(y)(0)=b,(D@@2)(y)(0)=c},
   y(x));
convals:=solve({-2/17+1/5*a-1/5*c=0,
   1/34+1/2*b+1/10*c+2/5*a=0},{a,b});
s1:=subs(convals,rhs(sol));
toplot:=subs(c=0,s1);
plot(toplot,x=0..Pi);
convals:=solve({-2/17+1/5*a-1/5*c=1,
   1/34+1/2*b+1/10*c+2/5*a=0},{a,b});
s1:=subs(convals,rhs(sol));
toplot:=subs(c=0,s1);
plot(toplot,x=0..3*Pi);
convals:=solve({-2/17+1/5*a-1/5*c=0,
   1/34+1/2*b+1/10*c+2/5*a=1},{a,b});
s1:=subs(convals,rhs(sol));
toplot:=subs(c=0,s1);
plot(toplot,x=0..3*Pi);
```

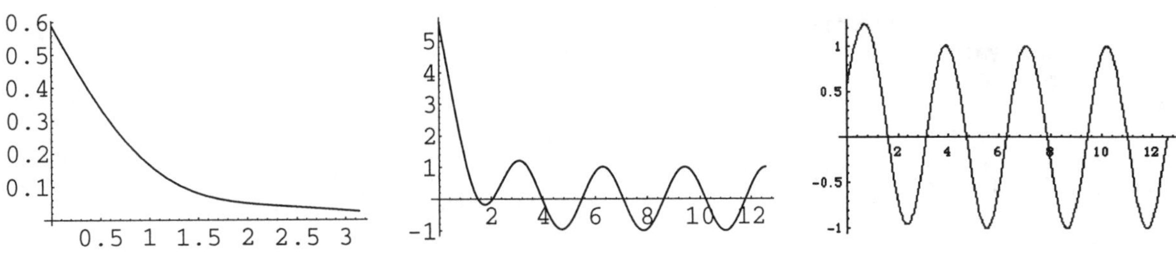

71. (a) A general solution of the corresponding homogeneous equation $y^{(4)} + a_3 y''' + a_2 y'' + a_1 y' + a_0 y = 0$ must be $y(t) = c_1 e^{-2t} + c_2 e^{-t} + c_3 e^t + c_4 e^{2t}$. To have general solution $y(t) = c_1 e^{-2t} + c_2 e^{-t} + c_3 e^t + c_4 e^{2t}$, we must have characteristic equation $(r+2)(r+1)(r-1)(r-2) = 0$ while the characteristic equation of $y^{(4)} + a_3 y''' + a_2 y'' + a_1 y' + a_0 y = 0$ is $r^4 + a_3 r^3 + a_2 r^2 + a_1 r + a_0 = 0$. Hence, we must have

$$(r+2)(r+1)(r-1)(r-2) = r^4 + a_3 r^3 + a_2 r^2 + a_1 r + a_0,$$

and equating coefficients of like terms results in $a_0 = 4$, $a_1 = 0$, $a_2 = -5$, and $a_3 = 0$.

```
chareqn=Expand[(m+2)(m+1)(m-1)(m-2)]
avals=SolveAlways[chareqn==
    m^4+a3 m^3+a2 m^2+a1 m+ a0,m]
```

```
chareqn:=expand((m+2)*(m+1)*(m-1)*
    (m-2));
match(chareqn=
    m^4+a3*m^3+a2*m^2+a1*m+a0,m,
    'avals');
avals;
```

For these values, we see that $y(t) = c_1 e^{-2t} + c_2 e^{-t} + c_3 e^t + c_4 e^{2t} + \frac{1}{10}\cos t$ is a general solution of the nonhomogeneous equation $y^{(4)} + a_3 y''' + a_2 y'' + a_1 y' + a_0 y = \cos t$.

```
Clear[y]
y[t_]=c1 Exp[-2t]+c2 Exp[-t]+c3 Exp[t]+
    c4 Exp[2t]+1/10 Cos[t];
step1=D[y[t],{t,4}]+a3 y'''[t]+a2 y''[t]+
    a1 y'[t]+a0 y[t]-Cos[t] /.
    avals[[1]] //Expand
```

```
y:=t->c1*exp(-2*t)+c2*exp(-t)+
    c3*exp(t)+c4*exp(2*t)+
    1/10*cos(t);
subs(avals,diff(y(t),t$4)+
    a3*diff(y(t),t)+
    a2*diff(y(t),t$2)+
    a1*diff(y(t),t)+a0*y(t)-cos(t));
```

For (b)-(d), we proceed in the same manner as in (a). For (b) and (c) the answer is "yes"; for (d), the answer is "no."

```
chareqn=Expand[(m+1)(m-1)(m^2+1)]
avals=SolveAlways[chareqn==
    m^4+a3 m^3+a2 m^2+a1 m+ a0,m]
Clear[y]
y[t_]=c1 Exp[-t]+c2 Exp[t]+c3 Cos[t]+
    c4 Sin[t]-3/8Cos[t]-1/4 t Sin[t];
step1=D[y[t],{t,4}]+a3 y'''[t]+a2 y''[t]+
    a1 y'[t]+a0 y[t]-Cos[t] /.
    avals[[1]] //Expand
chareqn=Expand[(m^2+1)^2]
avals=SolveAlways[chareqn==
    m^4+a3 m^3+a2 m^2+a1 m+ a0,m]
Clear[y]
y[t_]=c1 Cos[t]+c2 Sin[t]+c3 t Cos[t]+
    c4 t Sin[t]+3/8Cos[t]+
    3/8 Sin[t]-1/8 t^2 Cos[t]
step1=D[y[t],{t,4}]+a3 y'''[t]+a2 y''[t]+
    a1 y'[t]+a0 y[t]-Cos[t] /.
    avals[[1]] //Expand
Clear[y]
y[t_]=c1 Cos[t]+c2 Sin[t]+c3 t Cos[t]+
    c4 t Sin[t]+t^3 Sin[t]
step1=D[y[t],{t,4}]+a3 y'''[t]+a2 y''[t]+
    a1 y'[t]+a0 y[t]-Cos[t] /.
    avals[[1]] //Expand
```

```
chareqn:=expand((m+1)*(m-1)*(m^2+1));
match(chareqn=
    m^4+a3*m^3+a2*m^2+a1*m+a0,m,
    'avals');
avals;
y:=t->c1*exp(-t)+c2*exp(t)+
    c3*cos(t)+c4*sin(t)-3/8*cos(t)-
    1/4*t*sin(t);
subs(avals,diff(y(t),t$4)+
    a3*diff(y(t),t)+
    a2*diff(y(t),t$2)+
    a1*diff(y(t),t)+a0*y(t)-cos(t));
chareqn:=expand((m^2+1)^2);
match(chareqn=
    m^4+a3*m^3+a2*m^2+a1*m+a0,m,
    'avals');
avals;
y:=t->c1*cos(t)+c2*sin(t)+
    c3*t*cos(t)+c4*t*sin(t)+
    3/8*cos(t)+3/8*sin(t)-
    1/8*t^2*cos(t);
subs(avals,diff(y(t),t$4)+
    a3*diff(y(t),t)+
    a2*diff(y(t),t$2)+
    a1*diff(y(t),t)+a0*y(t)-cos(t));
y:=t->c1*cos(t)+c2*sin(t)+
    c3*t*cos(t)+c4*t*sin(t)+
    t^3*sin(t);
subs(avals,diff(y(t),t$4)+
    a3*diff(y(t),t)+
    a2*diff(y(t),t$2)+
    a1*diff(y(t),t)+a0*y(t)-cos(t));
```

72. For (i), we assume that the solution to the equation is periodic; we choose $y(x)=a\sin x$, which satisfies the initial conditions, and then calculate $f(x)=y''+y'-2y$. We then see that for this $f(x)$, $y(x)=a\sin x$ is the solution tto the initial-value problem. For (ii) and (iii), we proceed in the same way.

```
Clear[x,y,f,t]
sol=DSolve[{y''[x]+y'[x]-2y[x]==f[x],
   y[0]==0,y'[0]==a},y[x],x]
Clear[x,y,f,t]
sol=DSolve[{y''[x]+y'[x]-2y[x]==Sin[x],
   y[0]==0,y'[0]==0},y[x],x]
Clear[y]
y[x_]=a Sin[x]
y[0]
y'[0]
f[x_]=y''[x]+y'[x]-2 y[x]
Clear[y]
sol=DSolve[{y''[x]+y'[x]-2y[x]==f[x],
   y[0]==0,y'[0]==a},y[x],x]
Clear[y]
y[x_]=a (Exp[-x]-Exp[-2x])
y[0]
y'[0]
f[x_]=y''[x]+y'[x]-2 y[x]//Simplify
Clear[y]
sol=DSolve[{y''[x]+y'[x]-2y[x]==f[x],
   y[0]==0,y'[0]==a},y[x],x]
Clear[y]
y[x_]=a (Exp[x]-1)
y'[0]
y[0]
f[x_]=y''[x]+y'[x]-2 y[x]//Simplify
Clear[y]
sol=DSolve[{y''[x]+y'[x]-2y[x]==f[x],
   y[0]==0,y'[0]==a},y[x],x]
```

```
x:='x':y:='y':f:='f':t:='t':
sol:=dsolve({diff(y(x),x$2)+
   diff(y(x),x)-2*y(x)=f(x),
   y(0)=0,D(y)(0)=a},y(x));
y:=x->a*sin(x):
f:=simplify(diff(y(x),x$2)+
   diff(y(x),x)-2*y(x));
x:='x':y:='y':t:='t':
dsolve({diff(y(x),x$2)+
   diff(y(x),x)-2*y(x)=f,
   y(0)=0,D(y)(0)=a},y(x));
y:=x->a*(exp(-x)-exp(-2*x)):
f:=simplify(diff(y(x),x$2)+
   diff(y(x),x)-2*y(x));
x:='x':y:='y':t:='t':
dsolve({diff(y(x),x$2)+
   diff(y(x),x)-2*y(x)=f,
   y(0)=0,D(y)(0)=a},y(x));
y:=x->a*(exp(x)-1):
f:=simplify(diff(y(x),x$2)+
   diff(y(x),x)-2*y(x));
x:='x':y:='y':t:='t':
dsolve({diff(y(x),x$2)+
   diff(y(x),x)-2*y(x)=f,
   y(0)=0,D(y)(0)=a},y(x));
```

73. How is the solution for c = 3 different from the others? (*Hint:* A finite sum of periodic functions is periodic.)

```
cvals = Table[c, {c, 0, 6, 1/2}];
Clear[x, y, t, c, sol]
sol[c_] := {c, DSolve[{x''[t] +
     9 x[t] == Sin[c t],
     x[0] == 0, x'[0] == 0}, x[t], t] //
    Simplify}
sols = Map[sol, cvals];
sols // TableForm


toshow =
  Table[Plot[sols[[i, 2, 1, 1, 2]],
    {t, 0, 8 Pi}, DisplayFunction ->
     Identity], {i, 2, 13}];


Show[GraphicsArray[Partiti
```

```
for c from 0 to 6 by 1/2 do
   dsolve({diff(x(t),t$2)+9*x(t)=
         sin(c*t),
      x(0)=0,D(x)(0)=0},x(t))
   od;
```

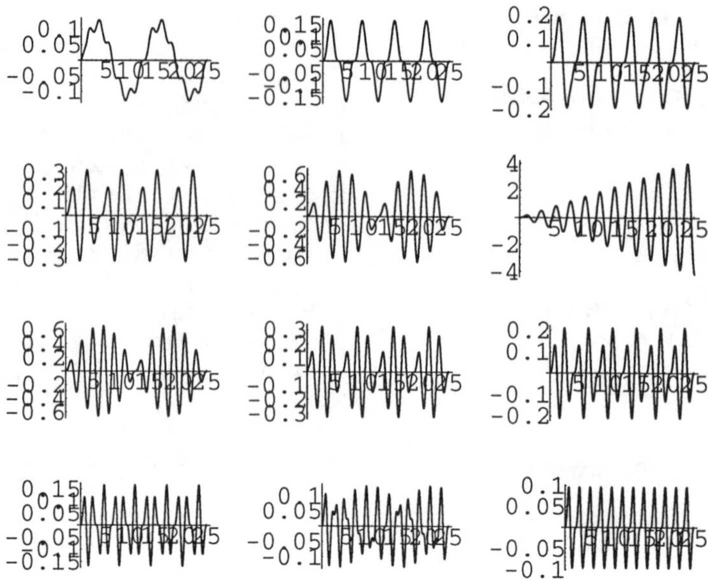

74.

```
sol =
 DSolve[
   {4 y''[t] + 4 y'[t] + 37 y[t] == Cos[3 t],
    y[0] == y[Pi]}, y[t], t] // Simplify
```

$$\left\{\left\{y[t] \to \frac{1}{145}\left(\left(1 - \frac{2\,e^{-t/2}}{1 + e^{-\pi/2}}\right) \text{Cos}[3\,t] + \left(12 - 145\,e^{-t/2}\,C[1]\right) \text{Sin}[3\,t]\right)\right\}\right\}$$

```
toplot = Table[sol[[1, 1, 2]] /. C[1] -> i,
   {i, -5, 5}];
grays = Table[GrayLevel[i],
   {i, 0, 0.7, .7 / 10}];
```

```
Plot[Evaluate[toplot], {t, 0, 2 Pi},
 PlotStyle -> grays]
```

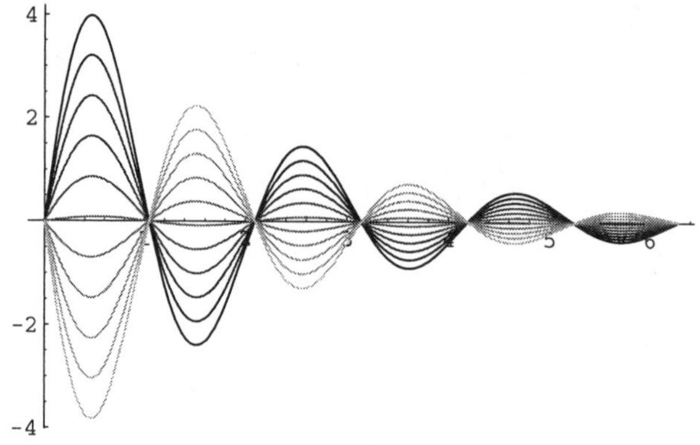

(b) Choosing $y(0) = \frac{1}{145} - \frac{2e^{\pi/2}}{145\left(1 + e^{\pi/2}\right)}$ and $y(\pi) = -\frac{1}{145} + \frac{2}{145\left(1 + e^{\pi/2}\right)}$ leads to a unique solution.

```
solb =
 DSolve[
   {4 y''[t] + 4 y'[t] + 37 y[t] == Cos[3
   y[t], t] // Simplify
```

$$\left\{\left\{y[t] \to \left(\frac{1}{145} + e^{-t/2}\, C[2]\right) \mathrm{Cos}[3\,t] + \frac{1}{145}\left(12 - 145\, e^{-t/2}\, C[1]\right) \mathrm{Sin}[3\,t]\right\}\right\}$$

```
solb[[1, 1, 2]] /. t -> 0
solb[[1, 1, 2]] /. t -> Pi
```

$$\frac{1}{145} + C[2]$$

$$-\frac{1}{145} - e^{-\pi/2}\, C[2]$$

```
cvals =
```

$$\mathrm{Solve}\left[\frac{1}{145} + C[2] == -\frac{1}{145} - e^{-\pi/2}\, C[2]\right]$$

$$\left\{\left\{C[2] \to -\frac{2\, e^{\pi/2}}{145\,(1 + e^{\pi/2})}\right\}\right\}$$

```
solb[[1, 1, 2]] /. {t -> 0, cvals[[1, 1]]}
```

$$\frac{1}{145} - \frac{2\, e^{\pi/2}}{145\,(1 + e^{\pi/2})}$$

```
solb[[1, 1, 2]] /. {t -> Pi, cvals[[1, 1]]}
```

$$-\frac{1}{145} + \frac{2}{145\,(1 + e^{\pi/2})}$$

EXERCISES 4.6

3. A general solution of the corresponding homogeneous equation is

$$y_h = c_1 e^{-2t} + c_2 t\, e^{-2t}.$$

Then,

$$W\left(\left\{e^{-2t}, t\, e^{-2t}\right\}\right) = \begin{vmatrix} e^{-2t} & t\, e^{-2t} \\ -2e^{-2t} & e^{-2t} - 2te^{-2x} \end{vmatrix}$$

$$= e^{-4t}$$

so

$$u_1(t) = -\int t\, dt = -\tfrac{1}{2}t^2 + C_1$$

and

$$u_2(t) = \int dt = t + C_2.$$

A particular solution to the nonhomogeneous equation is then $y_p(t) = \frac{1}{2}t^2 e^{-2t}$ and a general solution is then $y = c_1 e^{-2t} + c_2 te^{-2t} + \frac{1}{2}t^2 e^{-2t}$.

7. A general solution of the corresponding homogeneous equation is $y_h = c_1 \cos 4t + c_2 \sin 4t$ and $W(\{\cos 4t, \sin 4t\}) = \begin{vmatrix} \cos 4t & \sin 4t \\ -4\sin 4t & 4\cos 4t \end{vmatrix} = 4$. Then,

$$u_1(t) = -\tfrac{1}{4}\int dt = -\tfrac{1}{4}t + C_1 \qquad \text{and} \qquad u_2(t) = \tfrac{1}{4}\int \cot 4t\, dt = \tfrac{1}{16}\ln|\sin 4t| + C_2$$

so a particular solution to the nonhomogeneous equation is $y_p = -\frac{1}{4}t\cos 4t + \frac{1}{16}\sin 4t \ln|\sin 4t|$ and a general solution is $y = c_1 \cos 4t + c_2 \sin 4t - \frac{1}{4}t\cos 4t + \frac{1}{16}\sin 4t \ln|\sin 4t|$.

11. A general solution of the corresponding homogeneous equation is $y_h = e^t\left(c_1 \cos 5t + c_2 \sin 5t\right)$ and

$$W\left(\left\{e^t \cos 5t, e^t \sin 5t\right\}\right) = \begin{vmatrix} e^t \cos 5t & e^t \sin 5t \\ e^t(\cos 5t - 5\sin 5t) & e^t(5\cos 5t + \sin 5t) \end{vmatrix} = 5e^{2t}.$$

Then,

$$u_1(t) = -\tfrac{1}{5}\int (\csc 5t + \sec 5t)\sin 5t\, dx = -\tfrac{1}{5}t + \tfrac{1}{25}\ln|\cos 5t| + C_1$$

and

$$u_2(t) = \tfrac{1}{5}\int (\csc 5t + \sec 5t)\cos 5t\, dt = \tfrac{1}{5}t + \tfrac{1}{25}\ln|\sin 5t| + C_2$$

so a particular solution to the nonhomogeneous equation is

$$y_p = \tfrac{1}{25}e^t\left(-5t\cos 5t + \cos 5t \ln|\cos 5t| + 5t\sin 5t + \sin 5t \ln|\sin 5t|\right)$$

and a general solution is

$$y = e^t\left(c_1 \cos 5t + c_2 \sin 5t\right) + \tfrac{1}{25}e^t\left(-5t\cos 5t + \cos 5t \ln|\cos 5t| + 5t\sin 5t + \sin 5t \ln|\sin 5t|\right).$$

15. A general solution of the corresponding homogeneous equation is $y_h = e^{6t}\left(c_1 \cos t + c_2 \sin t\right)$ and

$$W\left(\left\{e^{6t} \cos t, e^{6t} \sin t\right\}\right) = \begin{vmatrix} e^{6t} \cos t & e^{6t} \sin t \\ e^{6t}(6\cos t - \sin t) & e^{6t}(\cos t + 6\sin t) \end{vmatrix} = e^{12t}.$$

Then, $u_1(t) = -\int \tan t\, dt = \ln|\cos t| + C_1$ and $u_2(t) = \int dt = t + C_2$ so a particular solution to the nonhomogeneous equation is $y_p = e^{6t}\left(\cos t \ln|\cos t| + t\sin t\right)$ and a general solution is

$$y = e^{6t}\left(c_1 \cos t + c_2 \sin t\right) + e^{6t}\left(\cos t \ln|\cos t| + t\sin t\right).$$

19. A general solution of the corresponding homogeneous equation $y'' - y = 0$ with characteristic equation $r^2 - 1 = (r+1)(r-1) = 0$ is

$$y_h = c_1 e^{-t} + c_2 e^t$$

so a fundamental set of solutions is given by $S = \left\{e^{-t}, e^t\right\}$. Then,

$$W(s) = \begin{vmatrix} e^{-t} & e^t \\ -e^{-t} & e^t \end{vmatrix} = 2.$$

Using the variation of parameters formula, we obtain

$$u_1(t) = -\int e^t \sinh t\,dt = \frac{1}{2}t - \frac{1}{4}e^{2t}$$

and

$$u_2(t) = \int e^{-t} \sinh t\,dt = \frac{1}{2}t + \frac{1}{4}e^{-2t}$$

so a particular solution to the nonhomogeneous equation is given by

$$\begin{aligned} y_p &= u_1 y_1 + u_2 y_2 \\ &= \left(\frac{1}{4} + \frac{1}{2}t\right)e^{-t} + \left(\frac{1}{2}t - \frac{1}{4}\right)e^t \end{aligned}$$

and a general solution is given by

$$y = c_1 e^{-t} + c_2 e^t + \left(\frac{1}{4} + \frac{1}{2}t\right)e^{-t} + \left(\frac{1}{2}t - \frac{1}{4}\right)e^t$$

or

$$y = c_1 e^{-t} + c_2 e^t + \frac{1}{2}te^{-t} + \frac{1}{2}te^t.$$

23. A general solution of the corresponding homogeneous equation is $y_h = c_1 e^{-3t} + c_2 te^{-3t}$ and

$$W\left(\left\{e^{-3t}, te^{-3t}\right\}\right) = \begin{vmatrix} e^{-3t} & te^{-3t} \\ -3e^{-3t} & e^{2t}(1-3t) \end{vmatrix} = e^{-6t}.$$

Then, $u_1(t) = -\int dt = -t + C_1$ and $u_2(t) = \int \frac{1}{t}\,dt = \ln|t| + C_2$ so a particular solution to the nonhomogeneous equation is $y_p = t\,e^{-3t}(\ln|t| - 1)$ or $y_p = te^{-3t}\ln|t|$ because $-te^{-3t}$ is a solution of the corresponding homogeneous equation. Therefore, a general solution is $y = c_1 e^{-3t} + c_2 te^{-3t} + te^{-3t}\ln|t|$.

27. A general solution of the corresponding homogeneous equation is $y_h = c_1 e^t + c_2 te^t$ and

$$W\left(\left\{e^t, te^t\right\}\right) = \begin{vmatrix} e^t & te^t \\ e^t & e^t(t+1) \end{vmatrix} = e^{2t}.$$

Then,

$$u_1(t) = -\int t\sqrt{1-t^2}\,dt = \frac{1}{3}\left(1-t^2\right)^{3/2} + C_1$$

and

$$u_2(t) = \int \sqrt{1-t^2}\,dt \underbrace{=}_{\substack{\text{trigonometric substitution with} \\ t=\sin\theta \Rightarrow dt=\cos\theta\,d\theta}} \int \cos^2\theta\,d\theta$$

$$= \frac{1}{4}(2\theta + \sin 2\theta) + C_2 = \frac{1}{2}t\sqrt{1-t^2} + \frac{1}{2}\sin^{-1}t + C_2$$

so a particular solution to the nonhomogeneous equation is given by

$$y_p = \frac{1}{6}e^t\left[\left(2+t^2\right)\sqrt{1-t^2} + 3t\sin^{-1}t\right]$$

and a general solution is

$$y = c_1 e^t + c_2 te^t + \sqrt{1-t^2}\left(\frac{1}{3}e^t + \frac{1}{6}t^2 e^t\right) + \frac{1}{2}te^t\sin^{-1}t.$$

31. A general solution of the corresponding homogeneous equation is $y_h = c_1\cos\frac{t}{2} + c_2\sin\frac{t}{2}$. Then,

$$W\left(\left\{\cos\frac{t}{2}, \sin\frac{t}{2}\right\}\right) = \begin{vmatrix} \cos\frac{t}{2} & \sin\frac{t}{2} \\ -\frac{1}{2}\sin\frac{t}{2} & \frac{1}{2}\cos\frac{t}{2} \end{vmatrix} = \frac{1}{2}$$

so

$$u_1(t) = -2\int\left(\csc\frac{t}{2} + \sec\frac{t}{2}\right)\sin\frac{t}{2}\,dt = -2t + 4\ln\left|\cos\frac{t}{2}\right| + C_1$$

and

$$u_2(t)=2\int\left(\csc\tfrac{t}{2}+\sec\tfrac{t}{2}\right)\cos\tfrac{t}{2}\,dt=2t+4\ln\left|\sin\tfrac{t}{2}\right|+C_2.$$

Thus, a particular solution of the nonhomogeneous equation is

$$y_p(t)=\left(-2t+4\ln\left|\cos\tfrac{t}{2}\right|\right)\cos\tfrac{t}{2}+\left(2t+4\ln\left|\sin\tfrac{t}{2}\right|\right)\sin\tfrac{t}{2}$$

and a general solution is

$$y(t)=c_1\cos\tfrac{t}{2}+c_2\sin\tfrac{t}{2}+\left(-2t+4\ln\left|\cos\tfrac{t}{2}\right|\right)\cos\tfrac{t}{2}+\left(2t+4\ln\left|\sin\tfrac{t}{2}\right|\right)\sin\tfrac{t}{2}.$$

35. A general solution of the corresponding homogeneous equation $y'''-3y''+3y'-y=0$ is $y_h=c_1e^t+c_2te^t+c_3t^2e^t$ and

$$W\left(\left\{e^t,te^t,t^2e^t\right\}\right)=\begin{vmatrix}e^t & te^t & t^2e^t\\ e^t & e^t(t+1) & te^t(t+2)\\ e^t & e^t(t+2) & e^t\left(t^2+4t+2\right)\end{vmatrix}=2e^{3t}.$$

Then, $u_1(t)=\frac{1}{2}\int t\,dt=\frac{1}{4}t^2+C_1$, $u_2(t)=-\int dt=t+C_2$, and $u_3(t)=\frac{1}{2}\int\frac{1}{t}\,dt=\frac{1}{2}\ln|t|+C_3$ so a particular solution to the nonhomogeneous equation is $y_p=\frac{1}{4}t^2e^t(2\ln|t|-3)$ and a general solution is

$$y=c_1e^t+c_2te^t+c_3t^2e^t-\tfrac{3}{4}t^2e^t+\tfrac{1}{2}t^2e^t\ln|t|.$$

41. A general solution of the corresponding homogeneous equation is $y_h=c_1e^{-3t}+c_2e^{-2t}$ and $W\left(\left\{e^{-3t},e^{-2t}\right\}\right)=e^{-5t}$. Then,

$$u_1(t)=\int e^{3t}(3+\sin 4t)\,dt=\tfrac{1}{25}e^{3t}(25-4\cos 4t+3\sin 4t)+C_1$$

and

$$u_2(t)=-\int e^{2t}(3+\sin 4t)dt=\tfrac{1}{10}e^{2t}(-15+2\cos 4t-\sin 4t)+C_2$$

so a particular solution to the nonhomogeneous equation is $y_p=\frac{1}{50}(-25+2\cos 4t+\sin 4t)$ and a general solution is $y=c_1e^{-3t}+c_2e^{-2t}+\frac{1}{50}(-25+2\cos 4t+\sin 4t)$. Application of the initial conditions yields

$$\begin{cases}y(0)=-\frac{23}{50}\\ y'(0)=\frac{27}{25}\end{cases}\Rightarrow\begin{cases}-\frac{23}{50}+c_1+c_2=-\frac{23}{50}\\ \frac{2}{25}-3c_1-2c_2=\frac{27}{25}\end{cases}\Rightarrow(c_1,c_2)=(-1,1)$$

so the solution to the initial-value problem is $y=-\frac{1}{2}+\frac{1}{25}\cos 4t+\frac{1}{50}\sin 4t+e^{-2t}-e^{-3t}$.

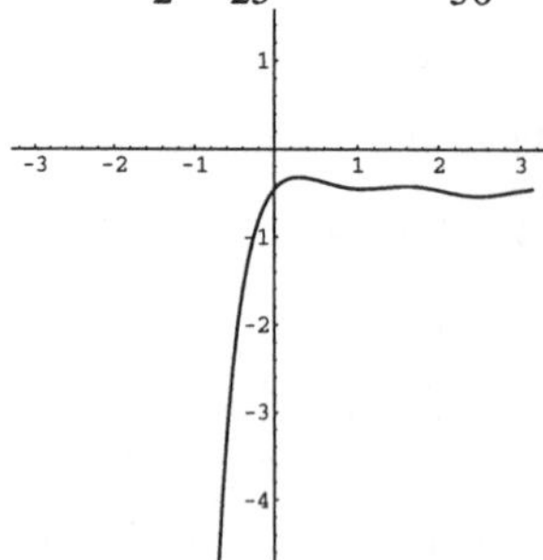

45. A general solution of the corresponding homogeneous equation is $y_h=c_1+c_2t+c_3e^t$ and $W\left(\left\{1,t,e^t\right\}\right)=e^t$. Then, $u_1(t)=3\int(t-1)t^2dt=\frac{3}{4}t^4-t^3+C_1$, $u_2(t)=-\int 3t^2dt=-t^3+C_2$, and $u_3(t)=3\int t^2e^{-t}dt=\left(-3t^2-6t-6\right)e^{-t}+C_3$ so a particular solution to the nonhomogeneous equation is $y_p=-\frac{1}{4}t^4-t^3-3t^2-6t-6$ and a general solution is $y=c_1+c_2t+c_3e^t-\frac{1}{4}t^4-t^3-3t^2-6t-6$. (Note that

$-6t-6$ can be omitted from the particular solution because $-6t-6$ is a solution of the corresponding homogeneous equation.) Application of the initial conditions yields

$$\begin{cases} -6+c_1+c_3=0 \\ -6+c_2+c_3=0 \Rightarrow (c_1,c_2,c_3)=(0,0,6) \\ -6+c_3=0 \end{cases}$$

so the solution to the initial-value problem is $y=-\frac{1}{4}t^4-t^3-3t^2-6t-6+6e^t$.

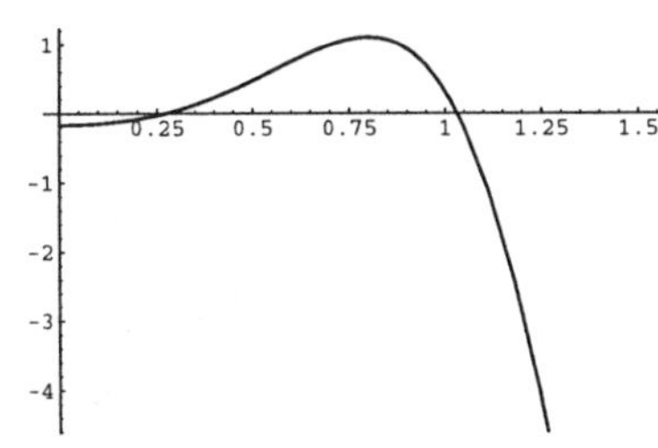

Graph for Exercise 44

Graph for Exercise 45

53. $r^2-k^2=(r+k)(r-k)=0 \Rightarrow r=\pm k$ so $y_h=c_1e^{-kt}+c_2e^{kt}$. Then, $W\left(\left\{e^{-kt},e^{kt}\right\}\right)=\begin{vmatrix} e^{-kt} & e^{kt} \\ -ke^{-kt} & ke^{kt} \end{vmatrix}=2k$

and $y=\int_0^t \frac{e^{kt}e^{-kx}-e^{-kt}e^{kx}}{2k} f(x)\,dx=\frac{1}{2k}\int_0^t \left(e^{kt-kx}-e^{-kt+kx}\right)f(x)dx$.

56. Before applying the variation of parameters formula, first rewrite the equation: $y''+\frac{1}{t}y'+\frac{4}{t^2}y=\underbrace{\frac{1}{t}}_{f(t)}$. Then,

$$W(\{\sin(2\ln t),\cos(2\ln t)\})=\begin{vmatrix} \sin(2\ln t) & \cos(2\ln t) \\ 2t^{-1}\cos(2\ln t) & -2t^{-1}\sin(2\ln t) \end{vmatrix}=-2t^{-1} \text{ so}$$

$$u_1(t)=\frac{1}{2}\int \cos(2\ln t)dt \underbrace{=}_{t=e^x \Rightarrow dt=e^x dx} \frac{1}{2}\int e^x\cos(2x)dx \underbrace{=}_{\text{integration by parts}} \frac{1}{10}e^x(\cos 2x+2\sin 2x)+C_1$$

$$=\frac{1}{10}t(\cos(2\ln t)+2\sin(2\ln t))+C_1$$

and

$$u_2(t)=-\frac{1}{2}\int \sin(2\ln t)\,dt=\frac{1}{10}t(2\cos(2\ln t)-\sin(2\ln t))+C_2.$$

Then, a particular solution to the nonhomogeneous equation is $y_p=\frac{1}{5}t$ and a general solution is $y=c_1\sin(2\ln t)+c_2\cos(2\ln t)+\frac{1}{5}t$.

58. The following code uses random number commands so your results will almost certainly look somewhat different from those shown here.

```
sol=DSolve[{4y''[x]+4y'[x]+
        y[x]==Exp[-x/2],
    y[0]==a,y'[0]==b},y[x],x]//Simplify
dsol=D[sol[[1,1,2]],x]//Simplify
extrema=Solve[dsol==0,x]//Simplify
Solve[1-2a+a^2+4a b+4 b^2==0,b]//Simplify
extrema[[1,1,2]]
extrema[[2,1,2]]
Solve[{extrema[[1,1,2]]==1,
    extrema[[2,1,2]]==3},{a,b}]
solc=DSolve[{4y''[x]+4y'[x]+
        y[x]==Exp[-x/2],
    y[0]==3/8,y'[0]==-3/16},
    y[x],x]//Simplify
Plot[y[x] /. solc,{x,0,5}]
Solve[{2(1-a-2b)==3,
    1-2a+a^2+4a b+4 b^2==0},
    {a,b}]//Simplify
sold=DSolve[{4y''[x]+4y'[x]+
        y[x]==Exp[-x/2],
    y[0]==5/8,y'[0]==-9/16},
    y[x],x]//Simplify
Solve[D[sold[[1,1,2]],x]==0,x]
Plot[y[x] /. sold,{x,0,5},PlotRange->All]
Solve[{extrema[[1,1,2]]==x0,
    extrema[[2,1,2]]==x1},{a,b}]
sole=DSolve[{4y''[x]+4y'[x]+y[x]==
    Exp[-x/2],
    y[0]==(8-2x0-2x1+x0 x1)/8,
    y'[0]==-x0 x1/16},y[x],x]//Simplify
{x0,x1}=Table[Random[],{2}]
sole
Plot[y[x] /. sole,{x,0,2}]
```

```
x:='x':y:='y':
sol:=dsolve({4*diff(y(x),x$2)+
   4*diff(y(x),x)+y(x)=exp(-x/2),
   y(0)=a,D(y)(0)=b},y(x));
dsol:=collect(diff(rhs(sol),x),
   exp(-x/2));
ext:=solve(dsol=0,x);
ext[1];
ext[2];
solve({ext[1]=1,ext[2]=3},{a,b});
solc:=dsolve({4*diff(y(x),x$2)+
   4*diff(y(x),x)+y(x)=exp(-x/2),
   y(0)=3/8,D(y)(0)=-3/16},y(x));
plot(rhs(solc),x=0..5);
solve({2*(1-a-2*b)=3,
   1-2*a+a^2+4*a*b+4*b^2=0},{a,b});
sold:=dsolve({4*diff(y(x),x$2)+
   4*diff(y(x),x)+y(x)=exp(-x/2),
   y(0)=5/8,D(y)(0)=-9/16},y(x));
plot(rhs(sold),x=0..5);
solve({ext[1]=x0,ext[2]=x1},{a,b});
sole:=dsolve({4*diff(y(x),x$2)+
   4*diff(y(x),x)+y(x)=exp(-x/2),
   y(0)=(8-2*x0-2*x1+x0*x1)/8,
   D(y)(0)=-x0*x1/16},y(x));
rand();
x0:=0.5:x1:=0.6:
plot(rhs(sole),x=0..5);
```

59.

```
Clear[x,y,u]
y[x_]=Exp[u[x]]
y''[x]
y'[x]
step1=Expand[Exp[-2x](y[x] y''[x]-
    y'[x]^2)-2x (1+x)y[x]^2]
step2=Factor[step1]
step2[[2]]
usol = DSolve[2 e^(2 x) x + 2 e^(2 x) x^2 - u''[x] ==
    u[x], x]

Clear[y]
y[x_]=Exp[usol[[1,1,2]]]
toplot=Table[y[x] /. {C[1]->i,C[2]->j},
    {i,-1,1},{j,-1,1}]//Flatten
grays=Table[GrayLevel[i],{i,0,.7,.7/8}];
Plot[Evaluate[toplot],{x,-1,1},
    PlotStyle->grays]
```

```
x:='x':y:='y':u:='u':
y:=x->exp(u(x));
diff(y(x),x$2);
diff(y(x),x);
step1:=expand(exp(-2*x)*
    (y(x)*diff(y(x),x$2)-
    diff(y(x),x)^2)-2*x*(1+x)*y(x)^2);
step2:=factor(step1);
usol:=dsolve(2*x*exp(2*x)+
    2*x^2*exp(2*x)-diff(u(x),x$2)=0,
    u(x));
y:='y':
y:=exp(rhs(usol));
toplot:={seq(seq(subs({_C1=i,_C2=j},
    y),i=-1..1),j=-1..1)};
plot(toplot,x=-1..1);
```

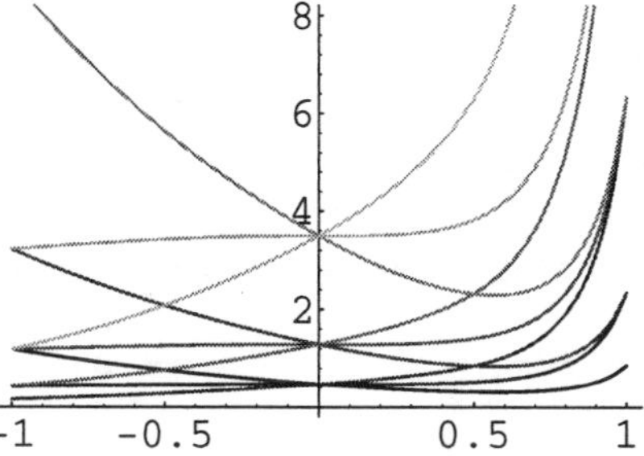

60.

```
Clear[x,y,f]
sol=DSolve[{y''[x]+4y[x]==f[x],y[0]==0,
    y'[0]==2},y[x],x]//Simplify
Clear[y]
y[x_]=Sin[2x]+1/2 Sin[2x]*
        Integrate[f[t]Cos[2t],{t,0,x}]-
    1/2Cos[2x]*
        Integrate[f[t]Sin[2t],{t,0,x}]
step1=y''[x]+4y[x]
step2=Simplify[%]
y[x] /. f[t]->c
Simplify[%]
Clear[x,y]
notperiodic=DSolve[{y''[x]+4y[x]==Cos[2x],
    y[0]==0,y'[0]==2},y[x],x]//Simplify
```

```
sol:=dsolve({diff(y(x),x$2)+4*y(x)=
    f(x),y(0)=0,D(y)(0)=2},y(x));
y:='y':
y:=x->sin(2*x)+1/2*sin(2*x)*
        Int(f(t)*cos(2*t),t=0..x)-
    1/2*cos(2*x)*
        Int(f(t)*sin(2*t),t=0..x);
step1:=diff(y(x),x$2)+4*y(x);
simplify(step1);
step2:=subs(f(t)=c,y(x));
with(student);
step3:=value(step2);
simplify(step3);
y:='y':
notperiodic:=dsolve({diff(y(x),x$2)+
    4*y(x)=cos(2*x),
    y(0)=0,D(y)(0)=2},y(x));
expand(notperiodic,trig);
simplify(",trig);
```

61. (a) We must first find a second linearly independent solution to the corresponding homogeneous equation

$$t^2y'' - 4ty' + \left(t^2+6\right)y = 0$$

$$y'' - \frac{4}{t}y' + \frac{t^2+6}{t^2}y = 0.$$

Using reduction of order we find a second linearly independent solution of the form $y = v(t)t^2 \cos t$ where

$$v(t) = \int \frac{e^{-\int -4/t\,dt}}{\left(t^2 \cos t\right)^2} dt = \int \frac{t^4}{t^4 \cos^2 t} dt = \int \sec^2 t \, dt = \tan t$$

and thus a second linearly independent solution is $y = \tan t \cdot t^2 \cos t = t^2 \sin t$.

We now use variation of parameters to find a particular solution of the nonhomogeneous equation First,

$$W(S) = \begin{vmatrix} t^2 \cos t & t^2 \sin t \\ 2t\cos t - t^2 \sin t & 2t \sin t + t^2 \cos t \end{vmatrix} = 2t^3 \cos t \sin t + t^4 \cos^2 t - 2t^3 \cos t \sin t + t^4 \sin^2 t = t^4.$$

Then,

$$u_1 = -\int \frac{y_2 f(t)}{W(S)} dt = -\int \frac{t^2 \sin t \cdot \left(t + \frac{2}{t}\right)}{t^4} dt = -\int \left(\frac{1}{t} + \frac{2}{t^3}\right) \sin t \, dt = -\int \frac{\sin t}{t} dt - 2\int \frac{\sin t}{t^3} dt.$$

To evaluate $-2\int \frac{\sin t}{t^3} dt$ we use integration by parts twice. First we let $u = \sin t \Rightarrow du = \cos t \, dt$ and $dv = 1/t^3 \, dt \Rightarrow v = -1/\left(2t^2\right)$:

$$-2\int \frac{\sin t}{t^3} dt = -2\left(-\frac{1}{2t^2} \sin t + \frac{1}{2}\int \frac{\cos t}{t^2} dt\right)$$

and then let $u = \cos t \Rightarrow du = -\sin t \, dt$ and $dv = 1/t^2 \, dt \Rightarrow v = -1/t$:

$$-2\int \frac{\sin t}{t^3} dt = -2\left(-\frac{1}{2t^2} \sin t + \frac{1}{2}\int \frac{\cos t}{t^2} dt\right) = -2\left(-\frac{1}{2t^2} \sin t + \frac{1}{2}\left(-\frac{1}{t} \cos t - \int \frac{\sin t}{t} dt\right)\right)$$

$$= \frac{1}{t^2} \sin t + \frac{1}{t} \cos t + \int \frac{\sin t}{t} dt.$$

Thus,

$$u_1 = -\int \frac{y_2 f(t)}{W(S)} dt = -\int \frac{\sin t}{t} dt - 2\int \frac{\sin t}{t^3} dt$$

$$= -\int \frac{\sin t}{t} dt + \frac{1}{t^2} \sin t + \frac{1}{t} \cos t + \int \frac{\sin t}{t} dt = \frac{1}{t^2} \sin t + \frac{1}{t} \cos t.$$

In the same way, u_2 is found to be

$$u_2 = \int \frac{y_1 f(t)}{W(S)} dt = \int \frac{t^2 \cos t \cdot \left(t + \frac{2}{t}\right)}{t^4} dt = -\frac{\cos t}{t^2} + \frac{\sin t}{t}.$$

Then a particular solution to the nonhomogeneous equation is

$$y_p = u_1 y_1 + u_2 y_2 = \left(\frac{1}{t^2} \sin t + \frac{1}{t} \cos t\right) t^2 \cos t + \left(-\frac{\cos t}{t^2} + \frac{\sin t}{t}\right) t^2 \sin t$$

$$= \cos t \sin t + t \cos^2 t - \cos t \sin t + t \sin^2 t = t$$

and a general solution to the nonhomogeneous equation is

$$y = c_1 t^2 \cos t + c_2 t^2 \sin t + t.$$

(b) Note that $y' = 1 + 2c_1 t \cos t + c_2 t^2 \cos t + 2c_2 t \sin t - c_1 t^2 \sin t$ and $y(0) = 0$ and $y(0) = 1$ for all values of c_1 and c_2.

(c) No.

EXERCISES 4.7

3. Method 1. If $y = x^r$,

$$\begin{aligned} 2r(r-1)x^r - 8rx^r + 8x^r &= 0 \\ \left(2r^2 - 10r + 8\right)x^r &= 0 \\ (r-1)(r-4) &= 0 \end{aligned}$$

so $r = 1$ or $r = 4$, and a general solution is given by $y = c_1 x + c_2 x^4$.

Method 2. Let $x = e^t$. Then, $\dfrac{dy}{dx} = \dfrac{1}{x}\dfrac{dy}{dt}$ and $\dfrac{d^2y}{dx^2} = \dfrac{1}{x^2}\left(\dfrac{d^2y}{dt^2} - \dfrac{dy}{dt}\right)$. Substituting into the equation gives us

$$\begin{aligned} 2x^2 \cdot \frac{1}{x^2}\left(\frac{d^2y}{dt^2} - \frac{dy}{dt}\right) - 8x \cdot \frac{1}{x}\frac{dy}{dt} + 8y &= 0 \\ 2\left(\frac{d^2y}{dt^2} - 5\frac{dy}{dt} + 4y\right) &= 0. \end{aligned}$$

The corresponding characteristic equation is $r^2 - 5r + 4 = (r-4)(r-1) = 0 \Rightarrow r = 1, 4$ so $y = c_1 e^t + c_2 e^{4t}$. Using $t = \ln x$ to return to the original variable gives us

$$y = c_1 e^{\ln x} + c_2 e^{4\ln x} = c_1 x + c_2 x^4.$$

9. If $y = x^r$,

$$\begin{aligned} r(r-1)x^r + 2rx^r - 6x^r &= 0 \\ \left(r^2 + r - 6\right)x^r &= 0 \\ (r+3)(r-2)x^r &= 0 \end{aligned}$$

so $r = -3$ or $r = 2$, and a general solution is $y = c_1 x^2 + c_2 x^{-3}$.

15. Method 1. If $y = x^r$,

$$\begin{aligned} r(r-1)(r-2)x^r + 22r(r-1)x^r + 124rx^r + 140x^r &= 0 \\ \left(r^3 + 19r^2 + 104r + 140\right)x^r &= 0 \\ (r+10)(r+7)(r+2) &= 0 \end{aligned}$$

$r = -10$, $r = -7$, and $r = -2$ so a general solution is $y = c_1 x^{-10} + c_2 x^{-7} + c_3 x^{-2}$.

Method 2. Let $x = e^t$. Then, $\dfrac{dy}{dx} = \dfrac{1}{x}\dfrac{dy}{dt}$, $\dfrac{d^2y}{dx^2} = \dfrac{1}{x^2}\left(\dfrac{d^2y}{dt^2} - \dfrac{dy}{dt}\right)$, and $\dfrac{d^3y}{dx^3} = \dfrac{1}{x^3}\left(\dfrac{d^3y}{dt^3} - 3\dfrac{d^2y}{dt^2} + 2\dfrac{dy}{dt}\right)$.
Substituting into the equation gives us

$$\begin{aligned} x^3 \cdot \frac{1}{x^3}\left(\frac{d^3y}{dt^3} - 3\frac{d^2y}{dt^2} + 2\frac{dy}{dt}\right) + 22x^2 \cdot \frac{1}{x^2}\left(\frac{d^2y}{dt^2} - \frac{dy}{dt}\right) + 124x \cdot \frac{1}{x}\frac{dy}{dt} + 140y &= 0 \\ \frac{d^3y}{dt^3} + 19\frac{d^2y}{dt^2} + 104\frac{dy}{dt} + 140y &= 0. \end{aligned}$$

The characteristic equation is $r^3 + 19r^2 + 104r + 140 = (r+2)(r+7)(r+10) = 0 \Rightarrow \quad r = -2, -7, -10$ so $y = c_1 e^{-2t} + c_2 e^{-7t} + c_3 e^{-10t}$. Returning to the original variable with $t = \ln x$ gives us

$$y = c_1 e^{-2\ln x} + c_2 e^{-7\ln x} + c_3 e^{-10\ln x} = c_1 x^{-2} + c_2 x^{-7} + c_3 x^{-10}.$$

19. $r(r-1)(r-2)x^r + 2rx^r - 2x^r = \left(r^3 - 3r^2 + 4r - 2\right)x^r = (r-1)\left(r^2 - 2r + 2\right)x^r = 0 \Rightarrow$
$r = 1$ or $r = 1 \pm i$ so $y = c_1 x + c_2 x \sin(\ln x) + c_3 x \cos(\ln x)$.

23. Method 1. The corresponding homogeneous equation is $x^2y'' + xy' + y = 0$ and $y = x^r \Rightarrow (r^2+1)x^r = 0$ so $r = \pm i$. Then, a general solution of the corresponding homogeneous equation is $y_h = c_1\cos(\ln x) + c_2\sin(\ln x)$ and $W(\{\cos(\ln x), \sin(\ln x)\}) = \begin{vmatrix} \cos(\ln x) & \sin(\ln x) \\ -x^{-1}\sin(\ln x) & x^{-1}\cos(\ln x) \end{vmatrix} = x^{-1}$. To apply the variation of parameters formula, we first rewrite the equation as $y'' + \frac{1}{x}y' + \frac{1}{x^2}y = \underbrace{x^{-4}}_{f(x)}$. Then,

$$u_1(x) = -\int x^{-3}\sin(\ln x)\,dx \underset{\substack{x=e^t \Rightarrow dx=e^t dt \\ \text{and} \\ x^{-3}=e^{-3t}}}{=} -\int e^{-2t}\sin t\,dt \underset{\substack{\text{integration by} \\ \text{parts twice}}}{=} e^{-2t}\left(\tfrac{1}{5}\cos t + \tfrac{2}{5}\sin t\right) + C_1$$

$$= x^{-2}\left(\tfrac{1}{5}\cos(\ln x) + \tfrac{2}{5}\sin(\ln x)\right) + C_1$$

and

$$u_2(x) = \int x^{-3}\cos(\ln x)dx \underset{\substack{x=e^t \Rightarrow dx=e^t dt \\ \text{and} \\ x^{-3}=e^{-3t}}}{=} \int e^{-2t}\cos t\,dt \underset{\substack{\text{integration by} \\ \text{parts twice}}}{=} e^{-2t}\left(-\tfrac{2}{5}\cos t + \tfrac{1}{5}\sin t\right) + C_2$$

$$= x^{-2}\left(-\tfrac{2}{5}\cos(\ln x) + \tfrac{1}{5}\sin(\ln x)\right) + C_1$$

so a particular solution of the nonhomogeneous equation is $y_p = \frac{1}{5}x^{-2}$ and a general solution is $y = \frac{1}{5x^2}\left(1 - 5c_1x^2\sin(\ln x) + 5c_2x^2\cos(\ln x)\right)$.

Method 2. Let $x = e^t$. Then, $\frac{dy}{dx} = \frac{1}{x}\frac{dy}{dt}$ and $\frac{d^2y}{dx^2} = \frac{1}{x^2}\left(\frac{d^2y}{dt^2} - \frac{dy}{dt}\right)$. Substituting into the equation gives us

$$x^2 \cdot \frac{1}{x^2}\left(\frac{d^2y}{dt^2} - \frac{dy}{dt}\right) + x \cdot \frac{1}{x}\frac{dy}{dt} + y = \left(e^t\right)^{-2}$$

$$\frac{d^2y}{dt^2} + y = e^{-2t}.$$

The corresponding homogeneous equation $y'' + y = 0$ has general solution $y_h = c_1\cos t + c_2\sin t$ so we search for a particular solution to the nonhomogeneous equation of the form $y_p = Ae^{-2t}$. Substituting y_p into the nonhomogeneous equation yields $5Ae^{-2t} = e^{-2t} \Rightarrow A = 1/5$ so $y_p = \frac{1}{5}e^{-2t}$ and $y = c_1\cos t + c_2\sin t + \frac{1}{5}e^{-2t}$. Returning the original variable using $t = \ln x$ gives us

$$y = c_1\cos(\ln x) + c_2\sin(\ln x) + \tfrac{1}{5}e^{-2\ln x} = c_1\cos(\ln x) + c_2\sin(\ln x) + \tfrac{1}{5}x^{-2}.$$

27. Method 1. The corresponding homgeneous equation is $x^2y'' + xy' + 4y = 0$ and $y = x^r \Rightarrow (r^2+4)x^r = 0 \Rightarrow r = \pm 2i$ so a general solution of the corresponding homogeneous equation is $y_h = c_1\cos(2\ln x) + c_2\sin(2\ln x)$ and $W(\{\cos(2\ln x), \sin(2\ln x)\}) = \begin{vmatrix} \cos(2\ln x) & \sin(2\ln x) \\ -2x^{-1}\sin(2\ln x) & 2x^{-1}\cos(2\ln x) \end{vmatrix} = 2x^{-1}$. To apply the variation of parameters formula, we first rewrite the equation as $y'' + \frac{1}{x}y' + \frac{4}{x^2} = \underbrace{8x^{-2}}_{f(x)}$. Then,

$$u_1(x) = -4\int x^{-1}\sin(2\ln x)\,dx \underset{\substack{x=e^t \Rightarrow dx=e^t dt \\ \text{and} \\ x^{-1}=e^{-t}}}{=} 2\cos(2\ln x) + C_1$$

and $u_2(x)=4\int x^{-1}\cos(2\ln x)x\,dx=2\sin(2\ln x)+C_2$ so a particular solution of the nonhomogeneous equation is $y_p=2$ and a general solution is $y=2+c_1\sin(2\ln x)+c_2\cos(2\ln x)$.

Method 2. Let $x=e^t$. Then, $\dfrac{dy}{dx}=\dfrac{1}{x}\dfrac{dy}{dt}$ and $\dfrac{d^2y}{dx^2}=\dfrac{1}{x^2}\left(\dfrac{d^2y}{dt^2}-\dfrac{dy}{dt}\right)$. Substituting into the equation gives us

$$x^2\cdot\frac{1}{x^2}\left(\frac{d^2y}{dt^2}-\frac{dy}{dt}\right)+x\cdot\frac{1}{x}\frac{dy}{dt}+4y=8$$

$$\frac{d^2y}{dt^2}+4y=8.$$

The corresponding homogeneous equation has general solution $y_h=c_1\cos 2t+c_2\sin 2t$ so we search for a particular solution to the nonhomogeneous equation of the form $y_p=A$. Substituting into the nonhomogeneous equation gives us $4A=8\Rightarrow A=2\Rightarrow y_p=2$ and $y=c_1\cos 2t+c_2\sin 2t+2$. Returning to the original variable with $t=\ln x$ gives us $y=c_1\cos(2\ln x)+c_2\sin(2\ln x)+2$.

29. Method 1. The corresponding homogeneous equation is $x^3y'''+3x^2y''-11xy'+16y=0$ and $y=x^r\Rightarrow (r+4)(r-2)^2x^r=0$ so a general solution of the corresponding homogeneous equation is $y_h=c_1x^{-4}+c_2x^2+c_3x^2\ln x$ and

$$W\left(\left\{x^{-4},x^2,x^2\ln x\right\}\right)=\begin{vmatrix} x^{-4} & x^2 & x^2\ln x\\ -4x^{-5} & 2x & x(1+2\ln x)\\ 20x^{-6} & 2 & 3+2\ln x\end{vmatrix}=36x^{-3}.$$

To apply the variation of parameters formula, we first rewrite the equation as $y'''+\frac{3}{x}y''-\frac{11}{x^2}y'+\frac{16}{x^3}y=x^{-6}$.

Then, $u_1(x)=\frac{1}{36}\int dx=\frac{1}{36}x+C_1$,

$$u_2(x)=-\tfrac{1}{46}\int x^{-6}(1+6\ln x)dx=\tfrac{1}{900}x^{-5}(11+30\ln x)+C_2$$

(to evaluate $\int x^{-6}\ln x\,dx$, use integration by parts with $u=\ln x\Rightarrow du=\frac{1}{x}dx$ and $dv=x^{-6}dx\Rightarrow v=-\frac{1}{5}x^{-5}$:

$$\int x^{-6}\ln x\,dx=-\tfrac{1}{5}x^{-5}\ln x+\tfrac{1}{5}\int x^{-6}dx=-\tfrac{1}{5}x^{-5}\ln x-\tfrac{1}{25}x^{-5}+C_4),$$

and $u_3(x)=\frac{1}{6}\int x^{-6}dx=-\frac{1}{30}x^{-5}+C_3$ so a particular solution of the nonhomogeneous equation is $y_p=\frac{1}{25}x^{-3}$ and a general solution is $y=\dfrac{1}{25x^4}\left(25c_1+25c_2x^6+25c_3x^6\ln x+x\right)$.

Method 2. Let $x=e^t$. Then, $\dfrac{dy}{dx}=\dfrac{1}{x}\dfrac{dy}{dt}$, $\dfrac{d^2y}{dx^2}=\dfrac{1}{x^2}\left(\dfrac{d^2y}{dt^2}-\dfrac{dy}{dt}\right)$, and $\dfrac{d^3y}{dx^3}=\dfrac{1}{x^3}\left(\dfrac{d^3y}{dt^3}-3\dfrac{d^2y}{dt^2}+2\dfrac{dy}{dt}\right)$.

Substituting into the equation gives us

$$x^3\cdot\frac{1}{x^3}\left(\frac{d^3y}{dt^3}-3\frac{d^2y}{dt^2}+2\frac{dy}{dt}\right)+3x^2\cdot\frac{1}{x^2}\left(\frac{d^2y}{dt^2}-\frac{dy}{dt}\right)-11x\cdot\frac{1}{x}\frac{dy}{dt}+16y=\left(e^t\right)^{-3}$$

$$\frac{d^3y}{dt^3}-12\frac{dy}{dt}+16y=e^{-3t}.$$

The corresponding homogeneous equation has characteristic equation

$$r^3-12r+16=(r-2)^2(r+4)=0\Rightarrow r_{1,2}=2,\ r_3=-4$$

so $y_h=c_1e^{2t}+c_2te^{2t}+c_3e^{-4t}$. We search for a particular solution to the nonhomogeneous equation of the form $y_p=Ae^{-3t}$. Substituting y_p into the nonhomogenous equation gives us $25Ae^{-3t}=e^{-3t}\Rightarrow A=1/25$ so

$y_p = \frac{1}{25}e^{-3t}$ and $y = c_1e^{2t} + c_2te^{2t} + c_3e^{-4t} + \frac{1}{25}e^{-3t}$. Returning to the original variable with $t = \ln x$ gives us $y = c_1e^{2\ln x} + c_2 \ln x\, e^{2\ln x} + c_3e^{-4\ln x} + \frac{1}{25}e^{-3\ln x} = c_1x^2 + c_2x^2 \ln x + c_3x^{-4} + \frac{1}{25}x^{-3}$.

33. $y = x^r \Rightarrow x^2y'' + xy' + 4y = x^2 \cdot r(r-1)x^{r-2} + x \cdot rx^{r-1} + 4x^r = \left(r^2 + 4\right)x^r = 0$ so $r = \pm 2i$ and a general solution of the equation is $y = c_1 \cos(2\ln x) + c_2 \sin(2\ln x)$ and application of the initial conditions yields

$$\begin{cases} c_1 = 1 \\ 2c_2 = 0 \end{cases} \Rightarrow (c_1, c_2) = (1, 0)$$

so $y = \cos(2\ln x)$.

```
sol =
 DSolve[{x^2 y''[x] + x y'[x] + 4 y[x] =
   y[1] == 1, y'[1] == 0}, y[x], x]

{{y[x] → Cos[2 Log[x]]}}

Plot[y[x] /. sol, {x, 0.01, 16}]
```

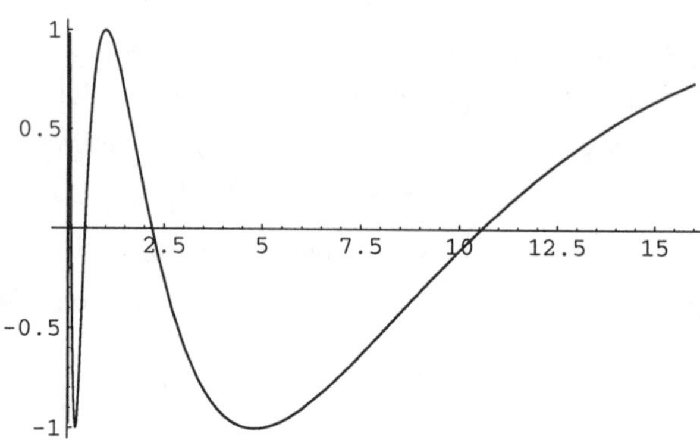

37. $y = x^r \Rightarrow$

$$x^3y''' - 2x^2y'' + 5xy' - 5y = x^3r(r-1)(r-2)x^{r-3} - 2x^2r(r-1)x^{r-2} + 5xrx^{r-1} - 5x^r = \left(r^3 - 5r^2 + 9r - 5\right)x^r$$

$$= (r-1)\left(r^2 - 4r + 5\right)x^r = 0 \Rightarrow r = 1, 2 \pm i$$

so general solution of the equation is

$$y = c_1x + x^2\left(c_2 \cos(\ln x) + c_3 \sin(\ln x)\right)$$

and application of the initial conditions yields

$$\begin{cases} c_1 + c_2 = 5 \\ c_1 + 2c_2 + c_3 = -1 \\ c_2 + 3c_3 = 0 \end{cases}$$

so $c_1 = \frac{3}{2}$, $c_2 = -\frac{3}{2}$, and $c_3 = \frac{1}{2}$, and $y = \frac{3}{2}x - \frac{3}{2}x^2 \cos(\ln x) + \frac{1}{2}x^2 \sin(\ln x)$.

```
sol =
 DSolve[
  {x^3 y'''[x] - 2 x^2 y''[x] + 5 x y'[x] -
     5 y[x] == 0, y[1] == 0, y'[1] == -1,
   y''[1] == 0}, y[x], x]

{{y[x] → 3 x/2 -
     3/2 x^2 Cos[Log[x]] + 1/2 x^2 Sin[Log[x]]}}

Plot[y[x] /. sol, {x, 0, 5}]
```

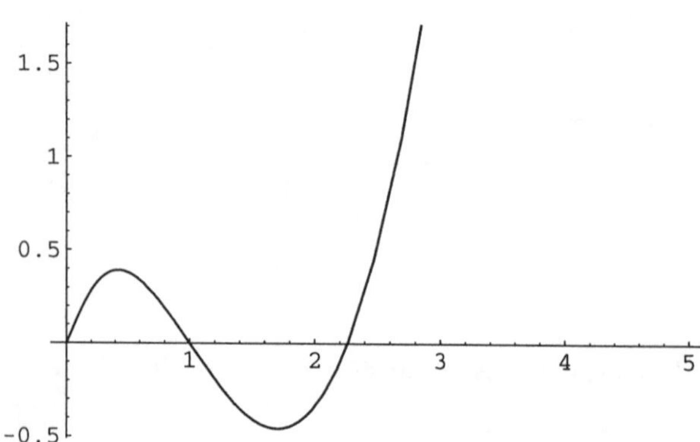

41. The corresponding homogeneous equation is $4x^2y'' + y = 0$ and $y = x^r \Rightarrow$

$$4x^2y''+y=4x^2r(r-1)x^{r-2}+x^r=(2r-1)^2x^r=0\Rightarrow r_{1,2}=1/2$$

so a general solution of the corresponding homogeneous equation is $y_h=c_1\sqrt{x}+c_2\sqrt{x}\ln x$. A fundamental set of solutions is $S=\left\{\sqrt{x},\sqrt{x}\ln x\right\}$ and $W(S)=\begin{vmatrix}\sqrt{x} & \sqrt{x}\ln x\\ \frac{1}{2\sqrt{x}} & \frac{2+\ln x}{2\sqrt{x}}\end{vmatrix}=1$. In standard form, the nonhomogeneous equation is $y''+\frac{1}{4x^2}y=\frac{x^3}{4x^2}=\frac{1}{4}x$. Using the method of variation of parameters, a particular solution is given by $y_p=u_1\sqrt{x}+u_2\sqrt{x}\ln x$ where

$$u_1=-\int\frac{\sqrt{x}\ln x\cdot\frac{1}{4}x}{1}dx=\frac{1}{50}x^{5/2}(2-5\ln x)$$

and

$$u_2=\int\frac{\sqrt{x}\cdot\frac{1}{4}x}{1}dx=\frac{1}{10}x^{5/2}.$$

Thus,

$$y_p=\frac{1}{50}x^{5/2}(2-5\ln x)\sqrt{x}+\frac{1}{10}x^{5/2}\sqrt{x}\ln x=\frac{1}{25}x^3$$

and a general solution of the equation is

$$y=c_1\sqrt{x}+c_2\sqrt{x}\ln x+\tfrac{1}{25}x^3$$

Application of the initial conditions yields

$$\begin{cases}\frac{1}{25}+c_1=1\\ \frac{3}{25}+\frac{1}{2}c_1+c_2=-1\end{cases}\Rightarrow(c_1,c_2)=\left(\tfrac{24}{25},-\tfrac{8}{5}\right)$$

so $y=\frac{1}{25}x^3+\frac{24}{25}\sqrt{x}-\frac{8}{5}\sqrt{x}\ln x$.

```
sol =
  DSolve[{4 x^2 y''[x] + y[x] == x^3,
    y[1] == 1, y'[1] == -1}, y[x], x]

{{y[x] → 1/25 (24 √x + x^3 - 40 √x Log[x])}}

Plot[y[x] /. sol, {x, 0, 5}]
```

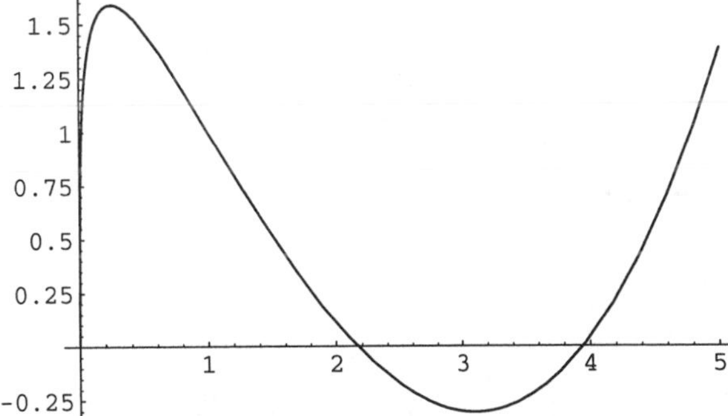

47. The equation should be $x^2y''+Axy'+By=0$, not $y''+Ay'+By=0$. (a) $y=c_1x\cos(\ln x)+c_2x\sin(\ln x)$; (b) $y=c_1x^{-1}+c_2x^{-2}$; (c) $y=c_1\cos(\ln x)+c_2\sin(\ln x)$

```
Map[
 DSolve[
   x^2 y''[x] + #[[1]] x y'[x] + #[[2]] y[x] ==
    0, y[x], x] &, {{-1, 2}, {4, 2}, {1, 1}}]
```

```
{{{y[x] →
    x C[2] Cos[Log[x]] - x C[1] Sin[Log[x]]}},
 {{y[x] → C[1]/x + C[2]/x^2}}, {{y[x] →
    C[2] Cos[Log[x]] - C[1] Sin[Log[x]]}}}
```

57. If $x=-e^t$, substituting into the original equation yields

$$x^2 \cdot \frac{1}{x^2}\left(\frac{d^2y}{dt^2}-\frac{dy}{dt}\right)+x\cdot\frac{1}{x}\frac{dy}{dt}+4y=0$$

$$\frac{d^2y}{dt^2}+4y=0$$

which has general solution $y(t)=c_1\cos 2t+c_2\sin 2t$. Because $x=-e^t$, $t=\ln(-x)$ so

$$y(x)=c_1\cos(2\ln(-x))+c_2\sin(2\ln(-x)).$$

Application of the initial conditions yields $\begin{cases} c_1=0 \\ -2c_2=2 \end{cases} \Rightarrow (c_1,c_2)=(0,-1)$ so $y=-\sin(2\ln(-x))$.

```
sol =
 DSolve[{x^2 y''[x] + x y'[x] + 4 y[x] == 0,
    y[-1] == 0, y'[-1] == 2}, y[x], x] //
  Simplify
```

```
{{y[x] → -Cosh[2 π] Sin[2 Log[x]] +
    i Cos[2 Log[x]] Sinh[2 π]}}
```

```
Plot[y[x] /. sol, {x, -4, 0}]
```

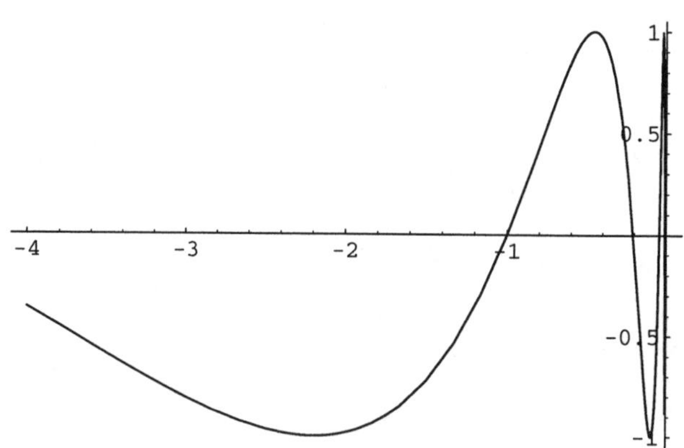

62. We can solve all five equations at the same time.

```
eqs={x^3 y'''[x]+16x^2 y''[x]+
      79x y'[x]+125y[x]==0,
   x^4 y''''[x]+5x^3y'''[x]-
      12x^2y''[x]-
      12x y'[x]+48y[x]==0,
   x^4 y''''[x]+14x^3y'''[x]+
      55x^2y''[x]+
      65x y'[x]+15y[x]==0,
   x^4 y''''[x]+8x^3y'''[x]+
      27x^2y''[x]+
      35x y'[x]+45y[x]==0,
   x^4 y''''[x]+10x^3y'''[x]+
      27x^2y''[x]+
      21x y'[x]+4y[x]==0}
Map[DSolve[#,y[x],x]&,eqs]
```

```
eqs:=[x^3*diff(y(x),x$3)+
       16*x^2*diff(y(x),x$2)+
       79*x*diff(y(x),x)+125*y(x)=0,
    x^4*diff(y(x),x$4)+
       5*x^3*diff(y(x),x$3)-
       12*x^2*diff(y(x),x$2)-
       12*x*diff(y(x),x)+48*y(x)=0,
    x^4*diff(y(x),x$4)+
       14*x^3*diff(y(x),x$3)+
       55*x^2*diff(y(x),x$2)+
       65*x*diff(y(x),x)+15*y(x)=0,
    x^4*diff(y(x),x$4)+
       8*x^3*diff(y(x),x$3)+
       27*x^2*diff(y(x),x$2)+
       35*x*diff(y(x),x)+45*y(x)=0,
    x^4*diff(y(x),x$4)+
       10*x^3*diff(y(x),x$3)+
       27*x^2*diff(y(x),x$2)+
       21*x*diff(y(x),x)+4*y(x)=0];
map(dsolve,eqs,y(x));
```

63. (a) After solving the equation, we graph the solution on the interval [0.2,1.8]. From the graph, we see that the solution function has four local minima and maxima on this interval.

```
partsol=DSolve[{x^3 y'''[x]+
    9x^2 y''[x]+44x y'[x]+58y[x]==0,
    y[1]==2,y'[1]==10,
    y''[1]==-2},y[x],x]
sol=partsol[[1,1,2]]
Plot[sol,{x,0.2,1.8}]
```

```
partsol:=dsolve({x^3*diff(y(x),x$3)+
   9*x^2*diff(y(x),x$2)+
   44*x*diff(y(x),x)+58*y(x)=0,
   y(1)=2,D(y)(1)=10,
   (D@@2)(y)(1)=-2},y(x));
assign(partsol):
plot(y(x),x=0.2..1.8);
```

The x-coordinates of these four points correspond to the zeros of the derivative of the solution which we approximate using numerical methods.

```
dsol=D[sol,x]
Plot[dsol,{x,0.2,1.8}]
xcoords=Map[FindRoot[dsol==0,{x,#}]&,
    {0.25,0.38,0.87,1.3}]
```

```
plot(diff(y(x),x),x=0.2..1.8);
c1:=fsolve(diff(y(x),x)=0,
   x=0.2..0.3);
c2:=fsolve(diff(y(x),x)=0,
   x=0.35..0.45);
c3:=fsolve(diff(y(x),x)=0,
   x=0.55..1.0);
c4:=fsolve(diff(y(x),x)=0,
   x=1.0..1.5);
```

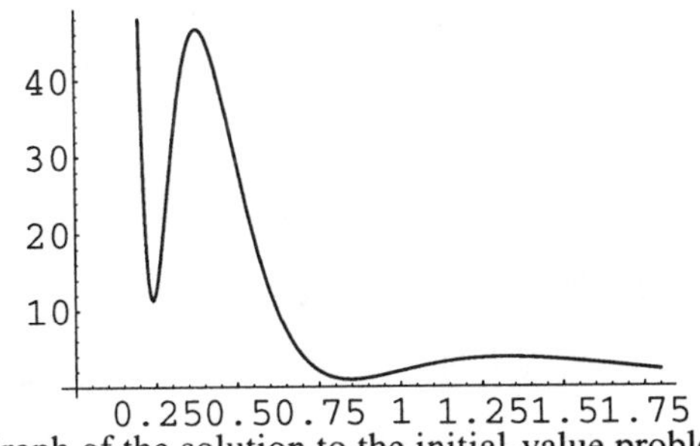

Graph of the solution to the initial-value problem

Graph of the derivative of the solution to the initial-value problem

We then use these values to approximate the local minima and maxima.

```
{x,sol} /. xcoords
```

```
evalf(subs(x=c1,y(x)));
evalf(subs(x=c2,y(x)));
evalf(subs(x=c3,y(x)));
evalf(subs(x=c4,y(x)));
```

64. First, we solve the equation to see that the solution to the initial-value problem is

$$y=\frac{3}{5}(a-2b)x^{-1/3}+\frac{1}{5}(2a+6b)x^{1/2}.$$

```
Clear[x,y,a,b]
partsol=DSolve[{6x^2 y''[x]+5x y'[x]
    -y[x]==0,y[1]==a,y'[1]==b},y[x],x]
```

```
x:='x':y:='y':a:='a':b:='b':
partsol:=dsolve({
    6*x^2*diff(y(x),x$2)+
       5*x*diff(y(x),x)-y(x)=0,
    y(1)=a,D(y)(1)=b},y(x));
```

(b) We see that $\lim_{x\to 0^+} y(x)=0$ if

$$a-2b=0$$
$$a=2b.$$

```
toplot=Table[partsol[[1,1,2]] /.
    a->2b,{b,-5,5}];
grays=Table[GrayLevel[i],
    {i,0,.6,.6/10}];
Plot[Evaluate[toplot],{x,0,10},
    PlotStyle->grays]
```

```
assign(partsol):
y1:=subs(a=2*b,y(x)):
toplot1:={seq(y1,b=-5..5)};
plot(toplot1,x=0..10);
```

(c) We see that $\lim_{x\to\infty} y(x)=0$ if

$$2a+6b=0$$
$$a=-3b.$$

```
toplot=Table[partsol[[1,1,2]] /.
    a->-3b,{b,-5,5}];
grays=Table[GrayLevel[i],
    {i,0,.6,.6/10}];
Plot[Evaluate[toplot],{x,0,40},
    PlotRange->{-20,20},
    AspectRatio->1,
    PlotStyle->grays]
```

```
b:='b':a:='a':
y2:=subs(a=-3*b,y(x));
toplot2:={seq(y2,b=-5..5)};
plot(toplot2,x=0..40,-20..20);
```

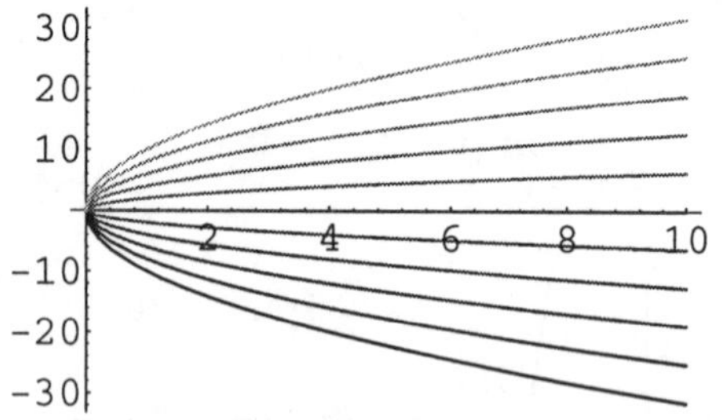

Graph for 3 (b)

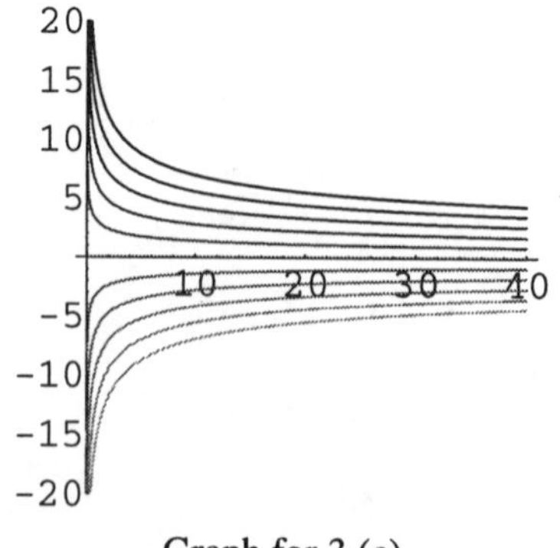

Graph for 3 (c)

(d) No, because the only solution to the system of equations $\begin{cases} a-2b=0 \\ 2a+6b=0 \end{cases}$ is $(a,b)=(0,0)$.

EXERCISES 4.8

5. First, we find the Maclaurin series for e^{-x}: $e^x = \sum_{n=0}^{\infty} \frac{1}{n!} x^n \Rightarrow e^{-x} = \sum_{n=0}^{\infty} \frac{(-1)^n}{n!} x^n$. Then, $y = \sum_{n=0}^{\infty} a_n x^n \Rightarrow$

$$y'' - y' - 2y = e^{-x}$$

$$\sum_{n=0}^{\infty} (n+1)(n+2)a_{n+2}x^n - \sum_{n=0}^{\infty} (n+1)a_{n+1}x^n - 2\sum_{n=0}^{\infty} a_n x^n = \sum_{n=0}^{\infty} \frac{(-1)^n}{n!} x^n$$

$$\sum_{n=0}^{\infty} [(n+1)(n+2)a_{n+2} - (n+1)a_{n+1} - 2a_n]x^n = \sum_{n=0}^{\infty} \frac{(-1)^n}{n!} x^n$$

so $(n+1)(n+2)a_{n+2} - (n+1)a_{n+1} - 2a_n = \frac{(-1)^n}{n!}$ and solving for a_{n+2} yields

$$a_{n+2} = \frac{2a_n + (n+1)a_{n+1}}{(n+2)(n+1)} - \frac{(-1)^n}{n!} \Rightarrow a_n = \frac{2a_{n-2} + (n-1)a_{n-1}}{n(n-1)} - \frac{(-1)^n}{(n-2)!} \textit{ for } n \ge 2.$$

Thus, $y = a_0 + a_1 x + \left(a_0 + \frac{1}{2}a_1 + \frac{1}{2}\right)x^2 + \left(\frac{1}{3}a_0 + \frac{1}{2}a_1\right)x^3 + \left(\frac{1}{8} + \frac{1}{4}a_0 + \frac{5}{24}a_1\right)x^4 + \cdots$.

16. (Maple V only) $y = \sum_{n=0}^{\infty} a_n x^n \Rightarrow y' = \sum_{n=0}^{\infty} (n+1)a_{n+1}x^n$ and $y'' = \sum_{n=0}^{\infty} (n+1)(n+2)a_{n+2}x^n$. Then,

```
with(powseries);
powcreate(y(n)=a(n));
tpsform(y,x);
yp:=powdiff(y);
tpsform(yp,x);
ypp:=powdiff(yp);
ser1:=evalpow(yp^2);
tpsform(ser1,x);
ser2:=evalpow(1/3*yp^2);
tpsform(ser2,x);
ser3:=evalpow((1/3*yp^2-1)*yp);
tpsform(ser3,x);
```

$$(y')^2 = a_1^2 + 4a_1a_2x + \left(6a_1a_3 + 4a_2^2\right)x^2 + (8a_1a_4 + 12a_2a_3)x^3 + \cdots,$$

$$\tfrac{1}{3}(y')^2 - 1 = \tfrac{1}{3}a_1^2 - 1 + \tfrac{4}{3}a_1a_2x + \left(2a_1a_3 + \tfrac{4}{3}a_2^2\right)x^2 + \left(\tfrac{8}{3}a_1a_4 + 4a_2a_3\right)x^3 + \cdots, \textit{ and}$$

$$\left(\tfrac{1}{3}(y')^2 - 1\right)y' = \tfrac{1}{3}a_1\left(a_1^2 - 3\right) + \left(\tfrac{4}{3}a_1^2 + \tfrac{2}{3}a_2\left(a_1^2 - 3\right)\right)x +$$
$$\left(\tfrac{1}{3}a_1\left(6a_1a_3 + 4a_2^2\right) + \tfrac{8}{3}a_2^2a_1 + a_3\left(a_1^2 - 3\right)\right)x^2 + \cdots.$$

Thus,

```
ser4:=evalpow(ypp+(1/3*yp^2-1)*yp+y);
tpsform(ser4,x);
```

$$y'' + \left(\tfrac{1}{3}(y')^2 - 1\right)y' + y = 2a_2 + \tfrac{1}{3}a_1\left(a_1^2 - 3\right) + a_0 + \left(6a_3 + \tfrac{4}{3}a_1^2a_2 + \tfrac{2}{3}a_2\left(a_1^2 - 3\right) + a_1\right)x +$$
$$\left(12a_4 + \tfrac{1}{3}a_1\left(6a_1a_3 + 4a_2^2\right) + \tfrac{8}{3}a_2^2a_1 + a_3\left(a_1^2 - 3\right) + a_2\right)x^2 + \cdots = 0$$

The initial conditions imply that $a_0 = 1$ and $a_1 = 0$. Equating coefficients of like terms results in $y = 1 - \frac{1}{2}x^2 - \frac{1}{6}x^3 + \frac{1}{40}x^5 + \cdots$.

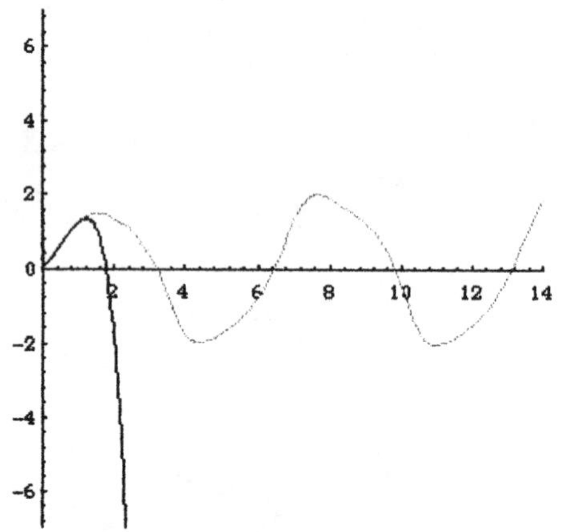

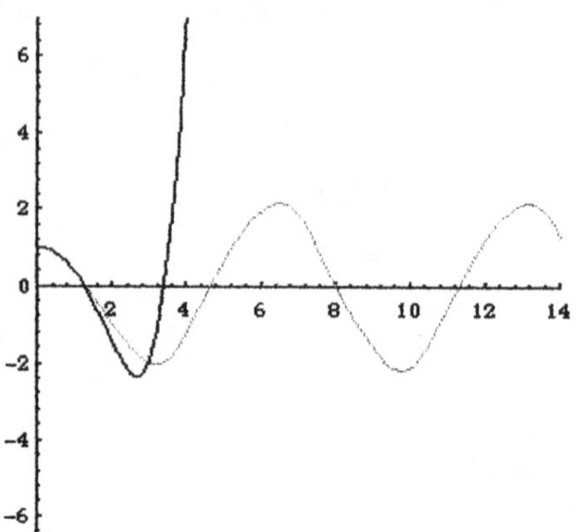

Exercise 15: The graph of a numerical solution (in gray) is shown together with the approximate solution $y \approx x + \frac{1}{2}x^2 - \frac{1}{8}x^4 - \frac{1}{8}x^5$.

Exercise 16: The graph of a numerical solution (in gray) is shown together with the approximate solution $y \approx 1 - \frac{1}{2}x^2 - \frac{1}{6}x^3 + \frac{1}{40}x^5$.

17. $x(x+1)y'' - \frac{1}{x^2}y' + 5y = 0 \Rightarrow y'' - \frac{1}{x^3(x+1)}y' - \frac{5}{x(x+1)}y = 0$ so $x = 0$ is irregular and $x = -1$ is regular.

19. We rewrite the equation in the form $y'' - \frac{5}{2x}y' - \frac{3}{2x}y = 0$ to see that the indicial equation is

$$r(r-1) - \frac{5}{2}r = \frac{1}{2}r(2r-7) = 0 \Rightarrow r = 0 \text{ or } r = \frac{7}{2}.$$

Now, $y = \sum_{n=0}^{\infty} a_n x^n \Rightarrow y' = \sum_{n=1}^{\infty} n a_n x^{n-1}$ and $y'' = \sum_{n=2}^{\infty} n(n-1)a_n x^{n-2}$ so

$$2xy'' - 5y' - 3y = 0$$

$$2x\sum_{n=2}^{\infty} n(n-1)a_n x^{n-2} - 5\sum_{n=1}^{\infty} n a_n x^{n-1} - 3\sum_{n=0}^{\infty} a_n x^n = 0$$

$$2\sum_{n=2}^{\infty} n(n-1)a_n x^{n-1} - 5\sum_{n=1}^{\infty} n a_n x^{n-1} - 3\sum_{n=0}^{\infty} a_n x^n = 0$$

$$-3a_0 - 5a_1 + 2\sum_{n=2}^{\infty} n(n-1)a_n x^{n-1} - 5\sum_{n=2}^{\infty} n a_n x^{n-1} - 3\sum_{n=1}^{\infty} a_n x^n = 0$$

$$-3a_0 - 5a_1 + 2\sum_{n=2}^{\infty} n(n+1)a_{n+1} x^n - 5\sum_{n=1}^{\infty} (n+1)a_{n+1} x^n - 3\sum_{n=1}^{\infty} a_n x^n = 0$$

$$-3a_0 - 5a_1 + \sum_{n=1}^{\infty}\left[\left(2n^2 - 3n - 5\right)a_{n+1} - 3a_n\right]x^n = 0.$$

Thus, $-3a_0 - 5a_1 = 0 \Rightarrow a_1 = -\frac{3}{5}a_0$ and

$$\left(2n^2 - 3n - 5\right)a_{n+1} - 3a_n = 0 \Rightarrow a_{n+1} = \frac{3a_n}{2n^2 - 3n - 5} \Rightarrow a_n = \frac{3a_{n-1}}{2n^2 - 7n} \text{ for } n \ge 2.$$

Choosing $a_0 = 1$ yields

n	0	1	2	3	4	5
a_n	1	-3/5	3/10	-3/10	-9/40	-9/200

so one solution is $y=1-\frac{3}{5}x+\frac{3}{10}x^2-\frac{3}{10}x^3+\cdots$. A second linearly independent solution has the form

$$y=\sum_{n=0}^{\infty}a_nx^{n+7/2}\Rightarrow y'=\sum_{n=0}^{\infty}\left(n+\tfrac{7}{2}\right)a_nx^{n+5/2} \text{ and } y''=\sum_{n=0}^{\infty}\left(n+\tfrac{7}{2}\right)\left(n+\tfrac{5}{2}\right)a_nx^{n+3/2}:$$

$$2\sum_{n=0}^{\infty}\left(n+\tfrac{7}{2}\right)\left(n+\tfrac{5}{2}\right)a_nx^{n+5/2}-5\sum_{n=0}^{\infty}\left(n+\tfrac{7}{2}\right)a_nx^{n+5/2}-3\sum_{n=0}^{\infty}a_nx^{n+7/2}=0$$

$$2\sum_{n=1}^{\infty}\left(n+\tfrac{7}{2}\right)\left(n+\tfrac{5}{2}\right)a_nx^{n+5/2}-5\sum_{n=1}^{\infty}\left(n+\tfrac{7}{2}\right)a_nx^{n+5/2}-3\sum_{n=0}^{\infty}a_nx^{n+7/2}=0$$

$$2\sum_{n=0}^{\infty}\left(n+\tfrac{7}{2}\right)\left(n+\tfrac{9}{2}\right)a_{n+1}x^{n+7/2}-5\sum_{n=0}^{\infty}\left(n+\tfrac{9}{2}\right)a_{n+1}x^{n+7/2}-3\sum_{n=0}^{\infty}a_nx^{n+7/2}=0$$

$$\sum_{n=0}^{\infty}\left[\left(2n^2+11n+9\right)a_{n+1}-3a_n\right]x^{n+7/2}=0.$$

Then, $\left(2n^2+11n+9\right)a_{n+1}-3a_n=0\Rightarrow a_{n+1}=\dfrac{3a_n}{2n^2+11n+9}\Rightarrow a_n=\dfrac{3a_{n-1}}{7n^2+2n}$ and choosing $a_0=1$ yields

n	0	1	2	3	4	5
a_n	1	1/3	1/22	1/286	1/5720	3/486200

so a second linearly independent solution is $y=x^{7/2}\left(1+\frac{1}{3}x+\frac{1}{22}x^2+\frac{1}{286}x^3+\cdots\right)$ and a general solution is $y=a_0x^{7/2}\left(1+\frac{1}{3}x+\frac{1}{22}x^2+\frac{1}{286}x^3+\cdots\right)+a_1\left(1-\frac{3}{5}x+\frac{3}{10}x^2-\frac{3}{10}x^3+\cdots\right)$.

23. The indicial equation is $r(r-1)+\frac{1}{2}r-\frac{35}{16}=\frac{1}{16}(4r+5)(4r-7)=0\Rightarrow r=-\frac{5}{4}$ or $r=\frac{7}{4}$. We now find a solution of the form

$$y_1(x)=\sum_{n=0}^{\infty}a_nx^{n+7/4}\Rightarrow y_1'(x)=\sum_{n=0}^{\infty}\left(n+\tfrac{7}{4}\right)a_nx^{n+3/4} \quad \text{and} \quad y_1''(x)=\sum_{n=0}^{\infty}\left(n+\tfrac{7}{4}\right)\left(n+\tfrac{3}{4}\right)a_nx^{n-1/4}:$$

$$\sum_{n=0}^{\infty}\left(n+\tfrac{7}{4}\right)\left(n+\tfrac{3}{4}\right)a_n x^{n-1/4}+\left(\tfrac{1}{2x}-2\right)\sum_{n=0}^{\infty}\left(n+\tfrac{7}{4}\right)a_n x^{n+3/4}-\frac{35}{16x^2}\sum_{n=0}^{\infty}a_n x^{n+7/4}=0$$

$$\sum_{n=0}^{\infty}\left(n+\tfrac{7}{4}\right)\left(n+\tfrac{3}{4}\right)a_n x^{n-1/4}+\tfrac{1}{2}\sum_{n=0}^{\infty}\left(n+\tfrac{7}{4}\right)a_n x^{n-1/4}-2\sum_{n=0}^{\infty}\left(n+\tfrac{7}{4}\right)a_n x^{n+3/4}-\tfrac{35}{16}\sum_{n=0}^{\infty}a_n x^{n-1/4}=0$$

$$\sum_{n=1}^{\infty}\left(n+\tfrac{7}{4}\right)\left(n+\tfrac{3}{4}\right)a_n x^{n-1/4}+\tfrac{1}{2}\sum_{n=1}^{\infty}\left(n+\tfrac{7}{4}\right)a_n x^{n-1/4}-2\sum_{n=0}^{\infty}\left(n+\tfrac{7}{4}\right)a_n x^{n+3/4}-\tfrac{35}{16}\sum_{n=1}^{\infty}a_n x^{n-1/4}=0$$

$$\sum_{n=0}^{\infty}\left(n+\tfrac{7}{4}\right)\left(n+\tfrac{11}{4}\right)a_{n+1} x^{n+3/4}+\tfrac{1}{2}\sum_{n=0}^{\infty}\left(n+\tfrac{11}{4}\right)a_{n+1} x^{n+3/4}-2\sum_{n=0}^{\infty}\left(n+\tfrac{7}{4}\right)a_n x^{n+3/4}-\tfrac{35}{16}\sum_{n=0}^{\infty}a_{n+1} x^{n+3/4}=0$$

$$\sum_{n=0}^{\infty}\left[\left(n^2+5n+4\right)a_{n+1}-\left(2n+\tfrac{7}{2}\right)a_n\right]x^{n+3/4}=0.$$

Then, $\left(n^2+5n+4\right)a_{n+1}-\left(2n+\tfrac{7}{2}\right)a_n=0\Rightarrow a_{n+1}=\dfrac{(4n+7)a_n}{2\left(n^2+5n+4\right)}\Rightarrow a_n=\dfrac{(4n+3)a_{n-1}}{2\left(n^2+3n\right)}$. Choosing $a_0=1$ yields

n	0	1	2	3	4	5
a_n	1	7/8	77/160	77/384	209/3072	4807/245760

so one solution is $y_1(x)=x^{7/4}\left(1+\frac{7}{8}x+\frac{77}{160}x^2+\frac{77}{384}x^3+\cdots\right)$. We now find a second linearly independent solution of the form $y_2(x)=cy_1(x)\ln x+\sum_{n=0}^{\infty}b_n x^{n-5/4}$:

$$-\tfrac{1}{2}cx^{-2}y_1(x)-2cx^{-1}y_1(x)+2cx^{-1}y_1'(x)+\sum_{n=0}^{\infty}\left(n-\tfrac{5}{4}\right)\left(n-\tfrac{9}{4}\right)b_n x^{n-13/4}+$$

$$\tfrac{1}{2}\sum_{n=0}^{\infty}\left(n-\tfrac{5}{4}\right)b_n x^{n-13/4}-2\sum_{n=0}^{\infty}\left(n-\tfrac{5}{4}\right)b_n x^{n-9/4}-\tfrac{35}{16}\sum_{n=0}^{\infty}b_n x^{n-13/4}=0$$

$$-\tfrac{1}{2}cx^{-2}y_1(x)-2cx^{-1}y_1(x)+2cx^{-1}y_1'(x)+\sum_{n=0}^{\infty}\left(n-\tfrac{5}{4}\right)\left(n-\tfrac{1}{4}\right)b_{n+1} x^{n-9/4}+$$

$$\tfrac{1}{2}\sum_{n=0}^{\infty}\left(n-\tfrac{1}{4}\right)b_{n+1} x^{n-9/4}-2\sum_{n=0}^{\infty}\left(n-\tfrac{5}{4}\right)b_n x^{n-9/4}-\tfrac{35}{16}\sum_{n=0}^{\infty}b_{n+1} x^{n-9/4}=0$$

$$-\tfrac{1}{2}cx^{-2}y_1(x)-2cx^{-1}y_1(x)+2cx^{-1}y_1'(x)+\sum_{n=0}^{\infty}\left[\left(n^2-n-2\right)b_{n+1}-\left(2n-\tfrac{5}{2}\right)b_n\right]x^{n-9/4}=0$$

and choosing $c=\frac{15}{8}$ yields (Remember that if you choose a different value of c, your second linearly independent solution will look different from that obtained here--but be correct, as long as your arithmetic is correct. Indeed, it would be a good exercise to show that the general solution you obtain is the same as that obtained here.)

$$\tfrac{45}{8}x^{-1/4}+\tfrac{285}{64}x^{3/4}+\tfrac{777}{256}x^{7/4}+\tfrac{1617}{1024}x^{11/4}+\tfrac{5335}{8192}x^{15/4}+\cdots+\sum_{n=0}^{\infty}\left[\left(n^2-n-2\right)b_{n+1}-\left(2n-\tfrac{5}{2}\right)b_n\right]x^{n-9/4}=0.$$

Equating coefficients of like terms then yields

$$\tfrac{5}{2}b_0 - 2b_1 = 0 \Rightarrow b_0 = \tfrac{4}{5}b_1$$
$$\tfrac{1}{2}b_1 - 2b_2 = 0 \Rightarrow b_1 = 4b_2$$
$$\tfrac{45}{8} - \tfrac{3}{2}b_2 = 0 \Rightarrow b_2 = \tfrac{15}{4} \Rightarrow b_1 = 15 \Rightarrow b_0 = 12$$
$$\tfrac{285}{64} - \tfrac{7}{2}b_3 + 4b_4 = 0 \Rightarrow b_4 = \tfrac{1}{256}(-285 + 224b_3) \underset{\text{choose } b_3=0}{=} -\tfrac{285}{256}$$
$$\tfrac{777}{256} - \tfrac{11}{2}b_4 + 10b_5 = 0 \Rightarrow b_5 = \tfrac{1}{2560}(-777 + 1408b_4) = -\tfrac{4689}{5120}$$
$$\vdots$$

Thus, a second linearly independent solution is

$$y = x^{-5/4}\ln x\left(\frac{15}{8}x^3 + \frac{105}{64}x^4 + \cdots\right) + x^{-5/4}\left(12 + 15x + \frac{15}{4}x^2 - \frac{13}{2}x^3 + \cdots\right).$$

and a general solution is

$$y = a_0x^{7/4}\left(1 + \frac{7}{8}x + \frac{77}{160}x^2 + \frac{77}{384}x^3 + \cdots\right) + a_1x^{-5/4}\ln x\left(\frac{15}{8}x^3 + \frac{105}{64}x^4 + \cdots\right)$$
$$+a_1x^{-5/4}\left(12 + 15x + \frac{15}{4}x^2 - \frac{13}{2}x^3 + \cdots\right).$$

33. The equations are equivalent because $\frac{d}{dx}\left[\left(1-x^2\right)y'\right] + k(k+1)y = \left(1-x^2\right)y'' - 2xy' + k(k+1)y$.

Multiplying $\frac{d}{dx}\left[\left(1-x^2\right)P_n'(x)\right] + n(n+1)P_n(x) = 0$ by $P_m(x)$ yields

$$\frac{d}{dx}\left[\left(1-x^2\right)P_n'(x)\right]P_m(x) + n(n+1)P_n(x)P_m(x) = 0 \text{ (*)};$$

multiplying $\frac{d}{dx}\left[\left(1-x^2\right)P_m'(x)\right] + m(m+1)P_m(x) = 0$ by $P_n(x)$ yields

$$\frac{d}{dx}\left[\left(1-x^2\right)P_m'(x)\right]P_n(x) + m(m+1)P_n(x)P_m(x) = 0 \text{ (**)}.$$

Subtracting Equation (**) from Equation (*) gives us

$$\frac{d}{dx}\left[\left(1-x^2\right)P_n'(x)\right]P_m(x) - \frac{d}{dx}\left[\left(1-x^2\right)P_m'(x)\right]P_n(x) + \left[n(n+1) - m(m+1)\right]P_n(x)P_m(x) = 0. \text{ (***)}$$

Integrating each side of this equation with respect to x from $x = -1$ to $x = 1$, we have

$$\int_{-1}^{1}\frac{d}{dx}\left[\left(1-x^2\right)P_n'(x)\right]P_m(x)dx - \int_{-}^{1}\frac{d}{dx}\left[\left(1-x^2\right)P_m'(x)\right]P_n(x)dx$$
$$+\int_{-1}^{1}\left[n(n+1) - m(m+1)\right]P_n(x)P_m(x)dx = 0$$

Through integration by parts, we find that

$$\int_{-1}^{1}P_n(x)\frac{d}{dx}\left[\left(1-x^2\right)P_m'(x)\right]dx = \left[P_n(x)\left(1-x^2\right)P_m'(x)\right]_{-1}^{1} - \int_{-1}^{1}\left(1-x^2\right)P_m'(x)P_n'(x)dx$$
$$= 0 - \int_{-1}^{1}\left(1-x^2\right)P_m'(x)P_n'(x)dx$$

Similarly, $\int_{-1}^{1}P_m(x)\frac{d}{dx}\left[\left(1-x^2\right)P_n'(x)\right]dx = 0 - \int_{-1}^{1}\left(1-x^2\right)P_m'(x)P_n'(x)dx$. Therefore,

$$\int_{-1}^{1}\left[n(n+1) - m(m+1)\right]P_n(x)P_m(x)dx = \left[n(n+1) - m(m+1)\right]\int_{-1}^{1}P_n(x)P_m(x)dx = 0.$$

Because $m \neq n$, $\int_{-1}^{1}P_m(x)P_n(x)dx = 0$.

39. $y_1(x) = F(1,0,1;x) = 1$; Second lin. indep. soln.:

$$y_2(x) = y_1(x)\int \frac{e^{\int dx/(x(1-x))}}{[y_1(x)]^2} = \int \frac{x/(1-x)}{(1)^2}dx = x + \ln|x-1|$$

43.

$$\Gamma(p+1) = \int_0^{\infty} e^{-u}u^p du = \lim_{M\to\infty}\int_0^M e^{-u}u^p du = \lim_{M\to\infty}\left\{\left[-u^p e^{-u}\right]_{u=0}^M + \int_0^M e^{-u}u^{p-1}du\right\}$$

$$= 0 + \int_0^{\infty} e^{-u}u^{p-1}du = \Gamma(p)$$

45. $x = 0$ is an ordinary point because

$$f(x) = \begin{cases} \frac{\sin x}{x}, & x \neq 0 \\ 1, & x = 0 \end{cases} = 1 - \frac{1}{6}x^2 + \frac{1}{120}x^4 - \frac{1}{5040}x^6 + \frac{1}{362880}x^8 - \frac{1}{39916800}x^{10} + \cdots$$

```
Series[Sin[x]/x,{x,0,10}]
```

```
series(sin(x)/x,x=0,11);
```

To avoid error messages that result when dividing by zero, first multiply through by x: $y'' + \frac{\sin x}{x}y' + y = 0 \Rightarrow xy'' + y'\sin x + x\,y = 0$. Note, however, that Mathematica is unable to generate a numerical solution because of division by zero, so we slightly "move" the initial conditions to approximate the solution.

```
Clear[x,y,f,eq]
lhseq=x y''[x]+Sin[x] y'[x]+x y[x]
step1=Series[lhseq,{x,0,7}]
step2=step1 /. {y[0]->1,y'[0]->-1}
eqs=LogicalExpand[step2==0]
vals=Solve[eqs]
step3=Series[y[x],{x,0,5}]
step4=step3 /. {y[0]->1,y'[0]->-1} /.
    vals[[1]]
step5=Normal[step4]
Plot[step5,{x,0,2}]
numsol=NDSolve[{x y''[x]+
    Sin[x] y'[x]+x y[x]==0,
    y[0.001]==1,y'[0.001]==-1},y[x],
    {x,0.001,2}]
Plot[y[x] /. numsol,{x,0.001,2}]
```

```
x:='x':y:='y':
step1:=dsolve({x*diff(y(x),x$2)+
    sin(x)*diff(y(x),x)+x*y(x)=0,
    y(0)=1,D(y)(0)=-1},
    y(x),series);
p5:=convert(rhs(step1),polynom);
plot(p5,x=0..2);
step2:=dsolve({x*diff(y(x),x$2)+
    sin(x)*diff(y(x),x)+x*y(x)=0,
    y(0)=1,D(y)(0)=-1},
    y(x),numeric);
with(plots):
odeplot(step2,[x,y(x)],0..2);
```

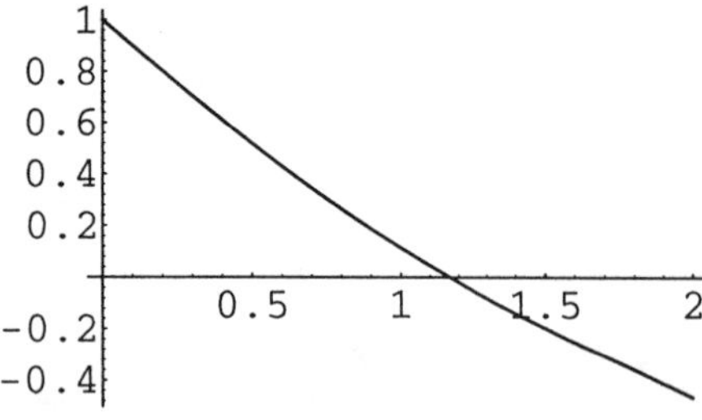

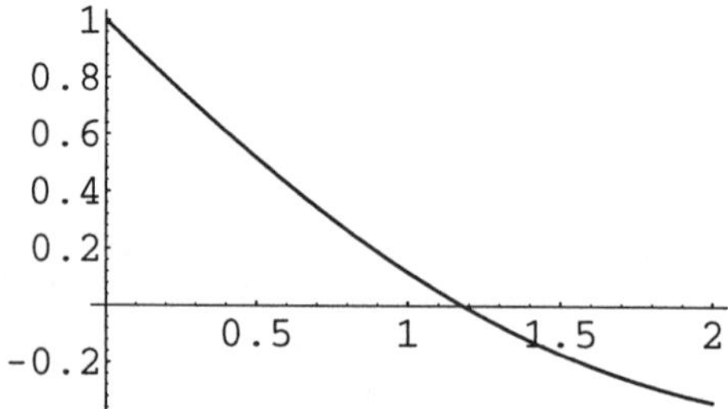

46.

```
Clear[x,y,y,eq]
lhseq=y''[x]-Cos[x] y[x]
rhseq=Sin[x]
step1=Series[lhseq,{x,0,15}]
step2=step1 /. {y[0]->1,y'[0]->0}
step3=Series[rhseq,{x,0,15}]
eqs=LogicalExpand[step2==step3]
vals=Solve[eqs]
step4=Series[y[x],{x,0,15}]
step5=step4 /. {y[0]->1,y'[0]->0} /.
    vals[[1]]
step6=Normal[step5]
p4=Take[step6,4]
p7=Take[step6,7]
p10=Take[step6,10]
p13=Take[step6,13]
Plot[{p4,p7,p10,p13},{x,0,4},
    PlotRange->{{0,8},{-1,7}},
    AspectRatio->1,
    PlotStyle->{GrayLevel[0],
        GrayLevel[.2],GrayLevel[.4],
            GrayLevel[.6]}]
numsol=NDSolve[{y''[x]-
    Cos[x] y[x]==Sin[x],
    y[0]==1,y'[0]==0},y[x],{x,0,8}]
Plot[y[x] /. numsol,{x,0,8},
    PlotRange->{-2.5,5.5},
    AspectRatio->1]
```

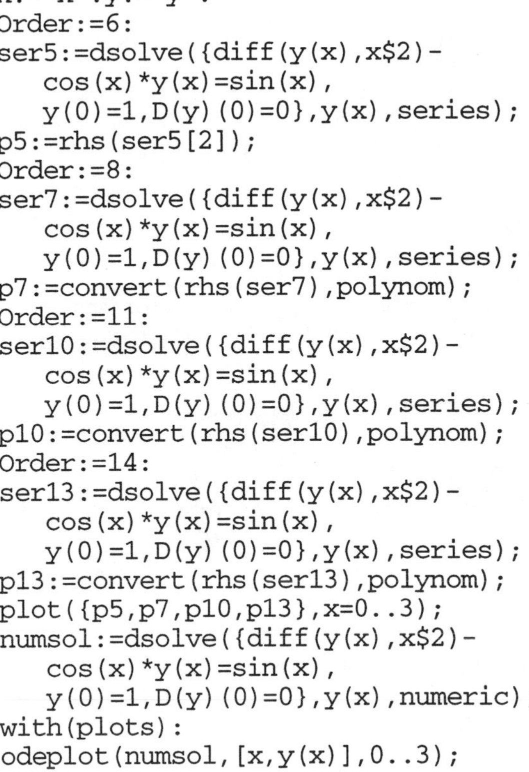

```
x:='x':y:='y':
Order:=6:
ser5:=dsolve({diff(y(x),x$2)-
    cos(x)*y(x)=sin(x),
    y(0)=1,D(y)(0)=0},y(x),series);
p5:=rhs(ser5[2]);
Order:=8:
ser7:=dsolve({diff(y(x),x$2)-
    cos(x)*y(x)=sin(x),
    y(0)=1,D(y)(0)=0},y(x),series);
p7:=convert(rhs(ser7),polynom);
Order:=11:
ser10:=dsolve({diff(y(x),x$2)-
    cos(x)*y(x)=sin(x),
    y(0)=1,D(y)(0)=0},y(x),series);
p10:=convert(rhs(ser10),polynom);
Order:=14:
ser13:=dsolve({diff(y(x),x$2)-
    cos(x)*y(x)=sin(x),
    y(0)=1,D(y)(0)=0},y(x),series);
p13:=convert(rhs(ser13),polynom);
plot({p5,p7,p10,p13},x=0..3);
numsol:=dsolve({diff(y(x),x$2)-
    cos(x)*y(x)=sin(x),
    y(0)=1,D(y)(0)=0},y(x),numeric);
with(plots):
odeplot(numsol,[x,y(x)],0..3);
```

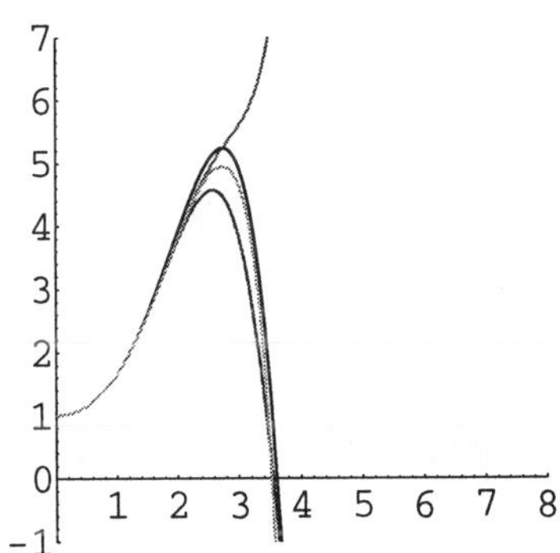

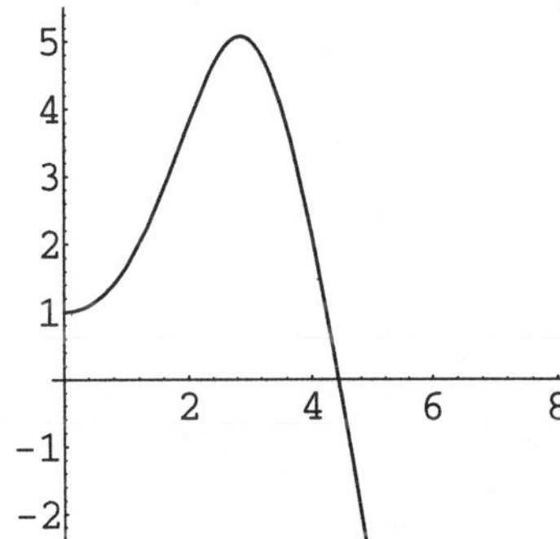

47. (Mathematica only for (c))

```
Plot[{AiryAi[x],AiryBi[x]},{x,-15,5},
    PlotStyle->{GrayLevel[0],
    GrayLevel[.5]}]
```

```
plot({Ai(x),Bi(x)},x=-15..5,-1..1);
```

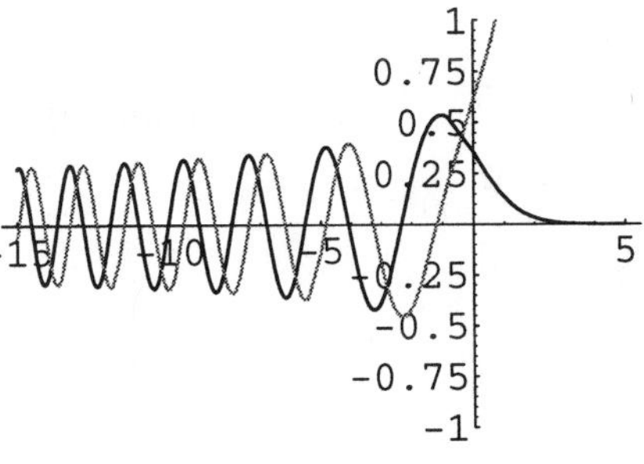

```
seriesai=Series[AiryAi[x],{x,0,45}];
Short[seriesai]
p45=Normal[seriesai];
p6=Take[p45,2]
p15=Take[p45,5]
p30=Take[p45,10]
Plot[{p6,p15,p30,p45},{x,-15,5},PlotRange->{-1,1},
    PlotStyle->{GrayLevel[0],GrayLevel[.2],GrayLevel[.4],GrayLevel[.6]}]
seriesbi=Series[AiryBi[x],{x,0,45}];
Short[seriesai]
p45=Normal[seriesbi];
p6=Take[p45,2]
p15=Take[p45,5]
p30=Take[p45,10]
Plot[{p6,p15,p30,p45},{x,-15,5},PlotRange->{-1,1},
    PlotStyle->{GrayLevel[0],GrayLevel[.2],GrayLevel[.4],GrayLevel[.6]}]
```

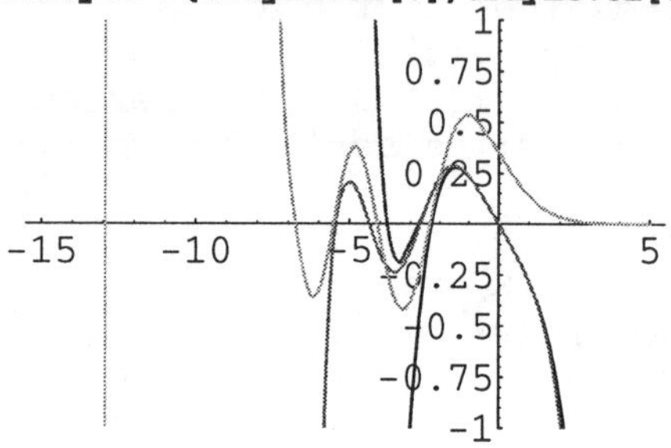

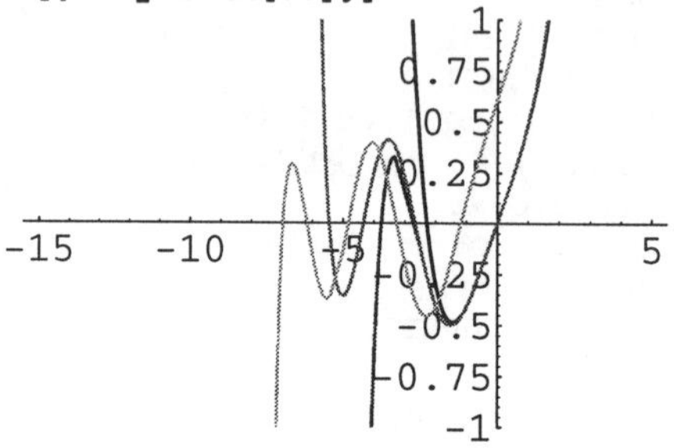

48. Here, we use an approximation.

```
y1=x^(-1/2)-1/2x^(1/2)-
    1/16x^(3/2)-1/96x^(5/2)-
    5/3072x^(7/2)-7/30720x^(9/2)
y2=y1 Log[x]+2x^(1/2)+
    3/16x^(3/2)+1/32x^(5/2)+
    31/6144x^(7/2)+3/4096x^(9/2)
approxsol=c1 y1+c2 y2;
s1=approxsol /. x->1;
eq1=s1==1
s2=D[approxsol,x] /. x->1
eq2=s2==-1
cvals=Solve[{eq1,eq2}]
sersol=approxsol /. cvals[[1]]
numsol=NDSolve[{x y''[x]+
    (2-x)y'[x]+1/(4x)y[x]==0,
    y[1]==1,y'[1]==-1},y[x],{x,0.1,2}]
```

```
y1:=x^(-1/2)-1/2*x^(1/2)-
    1/16*x^(3/2)-1/96*x^(5/2)-
    5/3072*x^(7/2)-7/30720*x^(9/2);
y2:=y1*ln(x)+2*x^(1/2)+3/16*x^(3/2)+
    1/32*x^(5/2)+
    31/6144*x^(7/2)+3/4096*x^(9/2);
approxsol:=c1*y1+c2*y2;
eq1:=eval(subs(x=1,approxsol))=1;
eq2:=eval(subs(x=1,
    diff(approxsol,x)))=-1;
cvals:=solve({eq1,eq2});
sersol:=subs(cvals,approxsol);
```

Note that the series approximation agrees quite well with the numerical approximation.

```
p1=Plot[sersol,{x,0,2},
    PlotStyle->GrayLevel[.4],
    DisplayFunction->Identity]
p2=Plot[y[x] /. numsol,{x,0.1,2},
    DisplayFunction->Identity];
Show[p1,p2,
    DisplayFunction->$DisplayFunction]
```

```
plot(sersol,x=0..2);
numsol:=dsolve({x*diff(y(x),x$2)+
    (2-x)*diff(y(x),x)+1/(4*x)*y(x)=0,
    y(1)=1,D(y)(1)=-1},y(x),numeric);
odeplot(numsol,[x,y(x)],x=0.1..2);
```

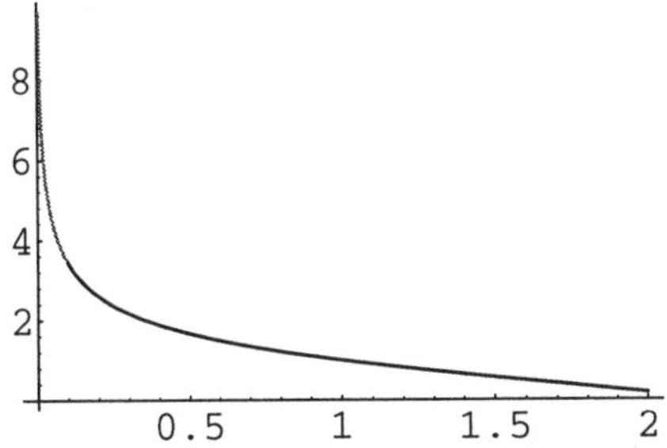

49.

```
Clear[x,y,Y,eq]
lhseq=x y''[x]+(1-x)y'[x]+n y[x]
step1=Series[lhseq,{x,0,15}]
step2=step1 /. y[0]->1
eqs=LogicalExpand[step2==0]
vars=Table[D[y[x],{x,n}] /. x->0,
    {n,1,16}]
vals=Solve[eqs,vars]
step4=Series[y[x],{x,0,15}];
step5=step4 /. y[0]->1 /. vals[[1]]
```

```
Order:=15:
dsolve({x*diff(y(x),x$2)+
    (1-x)*diff(y(x),x)+n*y(x)=0,
    y(0)=1},y(x),series);
```

```
Table[{n,LaguerreL[n,x]},
    {n,1,8}]//TableForm
```

```
with(orthopoly);
seq(L(n,x),n=1..8);
```

```
Table[{n,Simplify[LaguerreL[n,x]-
        Exp[x]/n!*
            D[x^n Exp[-x],{x,n}]]},
    {n,1,8}]//TableForm
```

```
array([seq([n,
    simplify(L(n,x)-exp(x)/n!*
        diff(x^n*exp(-x),x$n))],
    n=1..8)]);
```

```
Table[Integrate[Exp[-x]LaguerreL[n,x]
    LaguerreL[m,x],{x,0,Infinity}],
    {n,1,8},{m,1,8}]//TableForm
```

```
array([seq([seq(
    int(exp(-x)*L(n,x)*L(m,x),
        x=0..infinity),
    n=1..8)],m=1..8)]);
```

CHAPTER 4 REVIEW EXERCISES

3.
$$W(S)=\begin{vmatrix} t & t\ln t \\ 1 & 1+\ln t \end{vmatrix}$$
$= t \neq 0$; Linearly independent

13. The characteristic equation is
$$6r^2+5r-4=(3r+4)(2r-1)=0$$
so a general solution is
$$y=c_1e^{-4t/3}+c_2e^{t/2}.$$

15. The characteristic equation is
$$r^2+3r+2=(r+2)(r+1)=0$$
so a general solution is $y=c_1e^{-t}+c_2e^{-2t}$.

17. The characteristic equation is
$$2r^2-5r+2=(2r-1)(r-2)=0$$
so a general solution is $y=c_1e^{t/2}+c_2e^{2t}$.

19. The characteristic equation is
$$20r^2+r-1=(5r-1)(4r+1)=0$$
so a general solution is $y=c_1e^{-t/4}+c_2e^{t/5}$.

21. The characteristic equation is
$$2r^3+3r^2+r=r(r+1)(2r+1)=0$$
so a general solution is $y=c_1+c_2e^{-t}+c_3e^{-t/2}$.

23. The characteristic equation is
$$9r^2+12r+13=0 \Rightarrow r=-\tfrac{2}{3}\pm i$$
so a general solution is
$$y=e^{-2t/3}\left(c_1\sin t+c_2\cos t\right).$$

25. Note that any of the methods (undetermined coefficients, annihilators, and variation of paramters) discussed can be used to solve this equation. We use the method of undetermined coefficients. A general solution of the

corresponding homogeneous equation, with characteristic equation $r^2+5r=r(r+5)=0$, is $y_h=c_1+c_2e^{-5r}$. Then, there is a particular solution to the nonhomogeneous equation of the form $y_p=At^3+Bt^2+Ct$. Substituting into the nonhomogeneous equation yields

$$15At^2+(6A+10B)t+5C+2B=5t^2 \Rightarrow \begin{cases} 15A=5 \\ 6A+10B=0 \\ 5C+2B=0 \end{cases} \Rightarrow (A,B,C)=\left(\tfrac{1}{3},-\tfrac{1}{5},\tfrac{2}{25}\right)$$

so a general solution of the nonhomogeneous equations is $y=c_1+c_2e^{-5t}+\frac{2}{25}t-\frac{1}{5}t^2+\frac{1}{3}t^3$.

27. A general solution of the corresponding homogeneous equation is $y=e^{-t}(c_1\cos 2t+c_2\sin 2t)$ so, using the method of undetermined coefficients, we assume that there is a particular solution to the nonhomogeneous equation of the form $y_p=A\cos 2t+B\sin 2t$. Substituting into the nonhomogeneous equation yields

$$(A+4B)\cos 2t+(-4A+B)\sin 2t=3\sin 2t \Rightarrow \begin{cases} A+4B=0 \\ -4A+B=3 \end{cases} \Rightarrow (A,B)=\left(-\tfrac{12}{17},\tfrac{3}{17}\right)$$

so $y=c_1e^{-t}\cos 2t+c_2e^{-t}\sin 2t-\frac{12}{17}\cos 2t+\frac{3}{17}\sin 2t$.

31. A general solution of the corresponding homogeneous equation is $y_h=e^{3t}(c_1\cos 2t+c_2\sin 2t)$ so, using the method of undetermined coefficients, we assume that there is a particular solution of the nonhomogeneous equation of the form $y_p=Ae^{-2t}$. Substitution into the nonhomogeneous equation yields $29Ae^{-2t}=3e^{-2t}\Rightarrow A=\frac{3}{29}$ so $y=c_1e^{3t}\cos 2t+c_2e^{3t}\sin 2t+\frac{3}{29}e^{-2t}$.

35. The characteristic equation of the corresponding homogeneous equation is $r^3-12r-16=(r+2)^2(r-4)=0$ so a general solution of the corresponding homogeneous equation is $y_h=c_1e^{-2t}+c_2te^{-2t}+c_3e^{4t}$. Using the method of undetermined coefficients, we assume that there is a particular solution of the nonhomogeneous equation of the form $y_p=At^2e^{-2t}+Bte^{4t}$. Substitution into the nonhomogeneous equation yields $-12Ae^{-2t}+36Be^{4t}=e^{4t}-e^{-2t}\Rightarrow(A,B)=\left(\frac{1}{12},\frac{1}{36}\right)$ so a general solution of the nonhomogeneous equation is $y=c_1e^{-2t}+c_2te^{-2t}+c_3e^{4t}+\frac{1}{36}te^{4t}+\frac{1}{12}t^2e^{-2t}$.

41. The characteristic equation is $r^2+25=0\Rightarrow r=\pm 5i$ so a general solution of the equation is $y=c_1\cos 5t+c_2\sin 5t$. Application of the initial conditions yields $\begin{cases} c_1=1 \\ 5c_2=0 \end{cases}\Rightarrow(c_1,c_2)=(1,0)$ so the solution to the initial-value problem is $y=\cos 5t$.

```
sol =
  DSolve[{y''[t] + 25 y[t] == 0, y[0] == 1,
    y'[0] == 0}, y[t], t]

{{y[t] → Cos[5 t]}}

Plot[y[t] /. sol, {t, 0, 2 Pi}]
```

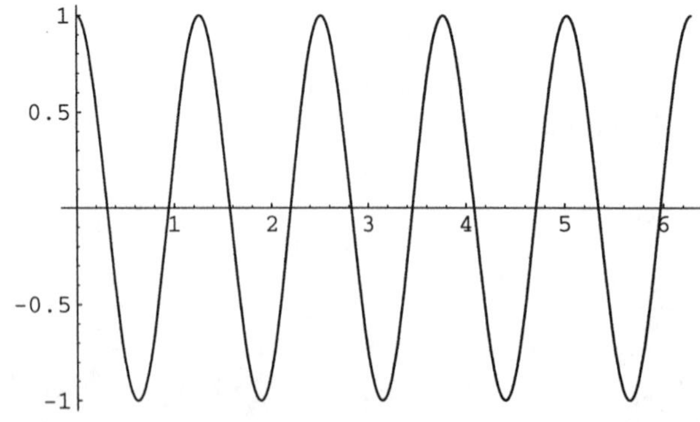

49. A general solution of the corresponding homogeneous equation is $y_h=c_1e^{4t}+c_2te^{4t}$ and $W\left(\left\{e^{4t},te^{4t}\right\}\right)=\begin{vmatrix} e^{4t} & te^{4t} \\ 4e^{4t} & e^{4t}(4t+1) \end{vmatrix}=e^{8t}$. Then,

$$u_1(t) = -\int t^{-2}dt = t^{-1} + C_1 \qquad \text{and} \qquad u_2(t) = \int t^{-3}dt = -\tfrac{1}{2}t^{-2} + C_2$$

so a particular solution of the nonhomogeneous equation is $y_p = \frac{1}{2}t^{-1}e^{4t}$ and a general solution is $y = c_1e^{4t} + c_2te^{4t} + \dfrac{1}{2t}e^{4t}$. Application of the initial conditions yields

$$\begin{cases} \frac{1}{2}e^4 + c_1e^4 + c_2e^4 = 0 \\ \frac{3}{2}e^4 + 4c_1e^4 + 5c_2e^4 = 0 \end{cases} \Rightarrow (c_1, c_2) = \left(-1, \tfrac{1}{2}\right)$$

so the solution to the initial-value problem is $y = \dfrac{1}{2t}e^{4t} - e^{4t} + \dfrac{t}{2}e^{4t}$.

```
sol =
 DSolve[
  {y''[t] - 8 y'[t] + 16 y[t] == t^-3 Exp[4 t]
   y[1] == 0, y'[1] == 0}, y[t], t]
```

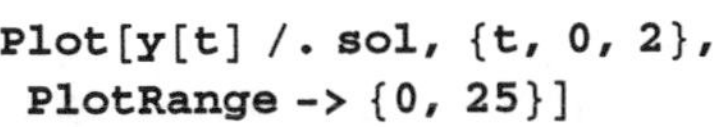

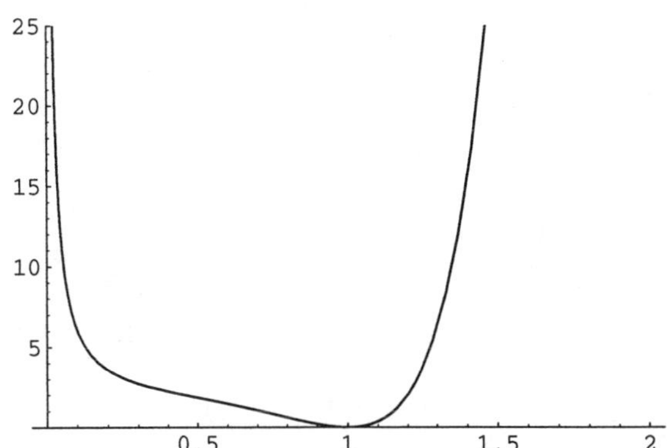


```
Plot[y[t] /. sol, {t, 0, 2},
 PlotRange -> {0, 25}]
```

54. First, we calculate

$$\begin{aligned} \frac{d}{dt}W(y_1(t), y_2(t)) &= \frac{d}{dt}\left[y_1(t)y_2'(t) - y_1'(t)y_2(t)\right] \\ &= y_1(t)y_2''(t) + y_1'(t)y_2'(t) - y_1'(t)y_2'(t) - y_1''(t)y_2(t) \\ &= y_1(t)y_2''(t) - y_1''(t)y_2(t). \end{aligned}$$

Then, $y_1''(t) = -p(t)y_1'(t) - q(t)y_1(t)$ and $y_2''(t) = -p(t)y_2'(t) - q(t)y_2(t)$, so

$$\begin{aligned} \frac{d}{dt}W(y_1(t), y_2(t)) &= y_1(t)y_2''(t) - y_1''(t)y_2(t) \\ &= y_1(t)\left[-p(t)y_2'(t) - q(t)y_2(t)\right] - y_2(t)\left[-p(t)y_1'(t) - q(t)y_1(t)\right] \\ &= -p(t)\left[y_1(t)y_2'(t) - y_1'(t)y_2(t)\right] = -p(t)W(y_1(t), y_2(t)). \end{aligned}$$

Therefore, $W(y_1(t), y_2(t))$ satisfies the first-order linear separable equation $\dfrac{dW}{dt} = -p(t)W$ or $\dfrac{dW}{W} = -p(t)dt$ with solution $\ln|W| = -\int p(t)dt$ or $W(y_1(t), y_2(t)) = Ce^{-\int p(t)dt}$.

56. (a) $W(y_1, y_2) = Ce^{-\int 3dt} = Ce^{-3t}$; $W\left(e^{-4t}, e^t\right) = \begin{vmatrix} e^{-4t} & e^t \\ -4e^{-4t} & e^t \end{vmatrix} = e^{-3t} + 4e^{-3t} = 5e^{-3t}$

(b) $W(y_1, y_2) = Ce^{-\int 4dt} = Ce^{-4t}$; $W\left(e^{-2t}\cos 3t, e^{-2t}\sin 3t\right) = 3e^{-4t}$

(c) $W(y_1, y_2) = Ce^{-\int 4dt} = Ce^{-4t}$; $W\left(e^{-2t}, te^{-2t}\right) = e^{-4t}$

(d) $W(y_1, y_2) = Ce^{-\int 0dt} = C$; $W(\cos 3t, \sin 3t) = 3$

58. (a) $y = \dfrac{1}{2}e^{t-1} - e^{-1-t}$

(b) $\lambda = 0 \Rightarrow y = 0$; $\lambda < 0 \Rightarrow y = 0$; $\lambda > 0 \Rightarrow \lambda_n = (n\pi/p)^2 \Rightarrow y_n = \sin(n\pi t/p), n = 1, 2, \ldots$

(c) $\lambda = 0 \Rightarrow y = 1$; $\lambda < 0 \Rightarrow y = 0$; $\lambda > 0 \Rightarrow \lambda_n = (n\pi/p)^2 \Rightarrow y_n = \cos(n\pi t/p), n = 1, 2, \ldots$

(d) $\lambda = 0 \Rightarrow y = 0$; $\lambda > 0 \Rightarrow y = 0$; $\lambda < 0 \Rightarrow \lambda_n = -(n\pi/2)^2 \Rightarrow y_n = e^{-t}\sin(n\pi t/2), n = 1, 2, \ldots$

60. $y = x^r \Rightarrow r(r-1)x^r + 7rx^r + 8x^r = 0$ so $(r+4)(r+2) = 0 \Rightarrow r = -4$ or $r = -2$; $y = c_1x^{-4} + c_2x^{-2}$

62. $y = x^r \Rightarrow r(r-1)x^r + rx^r + x^r = 0$ and simplifying yields $r^2 + 1 = 0 \Rightarrow r = \pm i$ so

$$y = c_1\sin(\ln x) + c_2\cos(\ln x).$$

64. $y = x^r \Rightarrow$

$$5r(r-1)x^r - rx^r + 2x^r = 0$$

and simplifying yields

$$5r^2 - 6r + 2 = 0 \Rightarrow r = \tfrac{3}{5} \pm \tfrac{1}{5}i$$

so $y = x^{3/5}\left[c_1\sin\left(\frac{1}{5}\ln x\right) + c_2\cos\left(\frac{1}{5}\ln x\right)\right]$.

65. A general solution of the corresponding homogeneous equation is $y_h = c_1x^3 + c_2x^5$ and $W\left(\{x^3, x^5\}\right) = \begin{vmatrix} x^3 & x^5 \\ 3x^2 & 5x^4 \end{vmatrix} = 2x^7$. To apply the variation of parameters formula, we first rewrite the equation as $y'' - \frac{7}{x}y' + \frac{15}{x^2}y = \frac{8}{x}$ to see that $f(x) = \frac{8}{x}$. Then,

$$u_1(x) = -4\int x^{-3}dx = 2x^{-2} + C_1 \quad \text{and} \quad u_2(x) = 4\int x^{-5}dx = -x^{-4} + C_2$$

so a particular solution of the nonhomogeneous equation is $y_p(x) = x$ and a general solution is $y = c_1x^3 + c_2x^5 + x$.

67. Because $e^x = \sum_{n=0}^{\infty} \frac{1}{n!}x^n$, $xe^x = \sum_{n=0}^{\infty} \frac{1}{n!}x^{n+1}$. Then, $y = \sum_{n=0}^{\infty} a_nx^n \Rightarrow$

$$y'' + 2y' - 3y = xe^x$$

$$\sum_{n=2}^{\infty} n(n-1)a_nx^{n-2} + 2\sum_{n=1}^{\infty} na_nx^{n-1} - 3\sum_{n=0}^{\infty} a_nx^n = \sum_{n=0}^{\infty} \frac{1}{n!}x^{n+1}$$

$$-3a_0 + 2a_1 + 2a_2 + \sum_{n=3}^{\infty} n(n-1)a_nx^{n-2} + 2\sum_{n=2}^{\infty} na_nx^{n-1} - 3\sum_{n=1}^{\infty} a_nx^n = \sum_{n=0}^{\infty} \frac{1}{n!}x^{n+1}$$

$$-3a_0 + 2a_1 + 2a_2 + \sum_{n=0}^{\infty} (n+3)(n+2)a_{n+3}x^{n+1} + 2\sum_{n=0}^{\infty} (n+2)a_{n+2}x^{n+1} - 3\sum_{n=0}^{\infty} a_{n+1}x^{n+1} = \sum_{n=0}^{\infty} \frac{1}{n!}x^{n+1}$$

$$-3a_0 + 2a_1 + 2a_2 + \sum_{n=0}^{\infty} \left[(n+3)(n+2)a_{n+3} + 2(n+2)a_{n+2} - 3a_{n+1}\right]x^{n+1} = \sum_{n=0}^{\infty} \frac{1}{n!}x^{n+1}.$$

Equating coefficients of like terms results in $-3a_0 + 2a_1 + 2a_2 = 0 \Rightarrow a_2 = -\frac{1}{2}(3a_0 - 2a_1)$ and

$$(n+3)(n+2)a_{n+3}+2(n+2)a_{n+2}-3a_{n+1}=\frac{1}{n!}\Rightarrow a_{n+3}=\frac{1}{n!}+\frac{-2(n+2)a_{n+2}+3a_{n+1}}{(n+3)(n+2)}$$

$$\Rightarrow a_n=\frac{1}{(n-3)!}+\frac{-2(n-1)a_{n-1}+3a_{n-2}}{n(n-1)}\ \textit{for}\ n\geq 3.$$

Therefore,

$$y=a_0+a_1x+\left(\frac{3}{2}a_0-a_1\right)x^2+\left(\frac{7}{6}a_1-a_0+\frac{1}{6}\right)x^3+\left(\frac{7}{8}a_0-\frac{5}{6}a_1\right)x^4+\left(\frac{1}{20}-\frac{1}{2}a_0+\frac{61}{120}a_1\right)x^5+\cdots.$$

70. (Note that this is a Cauchy-Euler equation so can also be solved by the methods used in Section 4.7.) We write the equation in the form $y'' + \frac{5}{x}y' - \frac{1}{x^2}y = 0$ to see that the indicial equation is

$$r(r-1) + \tfrac{5}{2}r - 1 = \tfrac{1}{2}(r+2)(2r-1) = 0 \Rightarrow r = -2 \text{ or } r = \tfrac{1}{2}.$$

We now find a solution of the form $y = \sum_{n=0}^{\infty} a_n x^{n-2}$:

$$2\sum_{n=0}^{\infty}(n-2)(n-3)a_n x^{n-2} + 5\sum_{n=0}^{\infty}(n-2)a_n x^{n-2} - 2\sum_{n=0}^{\infty}a_n x^{n-2} = 0$$

$$\sum_{n=0}^{\infty} n(2n-5)a_n x^{n-2} = 0$$

and equating coefficients of like terms indicates that $n(2n-5)a_n = 0 \Rightarrow a_n = 0$ for $n \ge 1$. Choosing $a_0 = 1$ results in $y = x^{-2}$. Because the indicial roots do not differ by an integer, we can also find a solution of the form $y = \sum_{n=0}^{\infty} a_n x^{n+1/2}$:

$$2\sum_{n=0}^{\infty}\left(n+\tfrac{1}{2}\right)\left(n-\tfrac{1}{2}\right)a_n x^{n+1/2} + 5\sum_{n=0}^{\infty}\left(n+\tfrac{1}{2}\right)a_n x^{n+1/2} - 2\sum_{n=0}^{\infty}a_n x^{n+1/2} = 0$$

$$\sum_{n=0}^{\infty} n(2n+5)a_n x^{n+1/2} = 0$$

and equating coefficients of like terms results in $n(2n+5)a_n = 0 \Rightarrow a_n = 0$ for $n \ge 1$. Choosing $a_0 = 1$ yields $y = x^{1/2}$ so a general solution is $y = a_0 x^{-2} + a_1 x^{1/2}$.

Differential Equations at Work: A. Testing for Diabetes

1.

```
eq1=g'[t]==-a g[t]-b h[t]+j[t]
eq2=h'[t]==-c h[t]+d g[t]
step1=Solve[eq1,h[t]]
step2=D[step1[[1,1,2]],t]
```

```
EQ1:=diff(g(t),t)=
   -a*g(t)-b*h(t)+J(t);
EQ2:=diff(h(t),t)=-c*h(t)+d*g(t);
step_1:=solve(EQ1,h(t));
step_2:=diff(step_1,t);
```

2.

```
step3=eq2 /. {step1[[1,1]],h'[t]->step2}
step4=step3 /. {j'[t]->0,j[t]->0}
```

```
step_3:=subs({h(t)=step_1,
   diff(h(t),t)=step_2},EQ2);
step_4:=expand(b*step_3);
step_5:=subs({diff(J(t),t)=0,J(t)=0},
   step_4);
```

5.

```
sol=DSolve[step4,g[t],t]
step5=sol[[1,1,2]] //.
    {(a+c)^2-4(a c+b d)->-4 omega^2,
        -a-c->-2 alpha}
step6=PowerExpand[step5]
<<Algebra`Trigonometry`
step7=ComplexToTrig[step6]//Simplify
step8=Collect[step7,{Cos[omega t],
    Sin[omega t]}]
toapply=Solve[{C[1]+C[2]==c1,
    -I C[1]+I C[2]==c2},{C[1],C[2]}]
model=step8 /. toapply[[1]]//Simplify
```

```
assume((a-c)^2-4*b*d<0):
Sol:=dsolve(step_5,g(t));
```

6. (Mathematica only) In each case, we must find α, ω, c_1, and c_2 so that $G(t) = G_0 + e^{-\alpha t}(c_1 \cos\omega t + c_2 \sin\omega t)$ agrees with the data as closely as possible.
To accomplish this, we take advantage of the `NonlinearFit` command that is contained in the **NonlinearFit** package which is located in the **Statistics** folder (or directory).
First, we load the **NonlinearFit** package.

```
<<Statistics`NonlinearFit`
```

For the first patient, we use `NonlinearFit` to find values of α, ω, c_1, and c_2

```
p1=NonlinearFit[{{1,85.32},{2,82.54},{3,78.25},{4,76.61}},
    model+80,t,{c1,c2,omega,alpha}]
```

and then evaluate $2\pi/\omega$ for the value of ω obtained.

```
2Pi/omega /. p1 // N
```

Similarly, we use `NonlinearFit` to determine if the other patients 3 have diabetes.

```
p2=NonlinearFit[{{1,91.77},{2,85.69},{3,92.39},{4,91.13}},
    model+90,t,{c1,c2,omega,alpha}]
2Pi/omega /. p2 // N

p3=NonlinearFit[{{1,103.35},{2,98.26},{3,96.59},{4,99.47}},
    model+100,t,{c1,c2,omega,alpha}]
2Pi/omega /. p3 // N

p4=NonlinearFit[{{1,114.64},{2,105.89},{3,108.14},{4,113.76}},
    model+110,t,{c1,c2,omega,alpha}]
2Pi/omega /. p4 // N
```

Differential Equations at Work: B. Modeling the Motion of a Skier

1.

```
Clear[v]
step1=DSolve[v'[t]==k^2-h^2 v[t]^2,v[t],t]
```

```
k:='k':h:='h':v:='v':s:='s':
step1:=dsolve(diff(v(t),t)=
    k^2-h^2*v^2,v(t));
```

```
Clear[h,k,v,t]
Solve[(h v[t]+k)/(h v[t]-k)==
    Exp[2 h k c+2 h k t],v[t]]
```

```
solve((h*v(t)+k)/(h*v(t)-k)=
    exp(2*h*k*c+2*h*k*t),v(t));
```

```
Clear[v,k,h];
v[t_]=k/h Tanh[h k (t+c)]
v'[t]-k^2+h^2v[t]^2//Simplify
```

```
v:='v':k:='k':h:='h':
v:=t->k/h*tanh(h*k*(t+c));
diff(v(t),t)-k^2+h^2*v(t)^2;
simplify(");
```

```
cval=Solve[v[0]==v0,c]
v[t_,h_,k_,v0_]=v[t] /. cval[[1]] //
    Simplify
```

```
cval:=solve(v(0)=v0,c);
v:=subs(c=cval,v(t));
```

2.

```
step2=Integrate[v[t,h,k,v0],t]+
    anotherconstant
step3=step2 /. t->0
cvaltwo=Solve[step3==0,anotherconstant]
s[t_,h_,k_,v0_]=step2 /.
    cvaltwo[[1]] //Simplify
```

```
step2:=int(subs(c=cval,v),t)+
   anotherconstant;
step3:=eval(subs(t=0,step2));
cvaltwo:=solve(step3=0,
   anotherconstant);
s:='s':
s:=subs(anotherconstant=cvaltwo,
   step2);
```

3.

```
m=75;
rho=1.29;
mu=0.06;
cda=0.16;
constants=Table[{alpha,
    9.8 (Sin[alpha Pi/180]-
        mu Cos[alpha Pi/180]),
        cda rho/(2 m)}//N,{alpha,30,50,5}];
TableForm[constants]
```

```
m:=75:
rho:=1.29:
mu:=0.06:
cda:=0.16:
contants:=evalf(array([seq(
   [alpha,9.8*(sin(alpha*Pi/180)-
         mu*cos(alpha*Pi/180)),
      cda*rho/(2*m)],
   alpha={30,40,45,50})]));
```

```
graphs[v0_,alpha_]:=Module[{},
    k=Sqrt[9.8 (Sin[alpha Pi/180]-
        mu Cos[alpha Pi/180])]//N;
    h=Sqrt[cda rho/(2 m)]//N;
    p1=Plot[v[t,h,k,v0],{t,0,40},
        DisplayFunction->Identity];
    p2=Plot[s[t,h,k,v0],{t,0,40},
        DisplayFunction->Identity];
    Show[GraphicsArray[{p1,p2}]];
    ParametricPlot[{s[t,h,k,v0],
        v[t,h,k,v0]},{t,0,40},
        Compiled->False];
                                    ]
```

```
graph:='graph':
graph:=proc(initvel,alpha)
   local k0, h0,vel,dis;
   k0:=evalf(sqrt(9.8*
      (sin(alpha*Pi/180)-
         mu*cos(alpha*Pi/180))));
   h0:=evalf(sqrt(cda*rho/(2*m)));
   vel:=subs({v0=initvel,k=k0,h=h0},
      v);
   dis:=subs({v0=initvel,k=k0,h=h0},
      s);
   plot({vel,dis},t=0..40)
   end:
```

```
Map[graphs[#,30]&,{0,5,10}];
```

```
graph(0,30);
graph(5,30);
graph(10,30);
```

5.

```
Clear[h,k,w,s]
step1=DSolve[w'[s]+2h^2 w[s]==2k^2,w[s],s]
```

```
w:='w':k:='k':h:='h':s:='s':v:='v':
step1:=dsolve(diff(w(s),s)+2*h^2*w=
   2*k^2,w(s));
```

6.

```
step2=step1[[1,1,2]] /. s->0
cval=Solve[step2==v0^2,C[1]]
Clear[v]
v[s_,h_,k_,v0_]=Sqrt[step1[[1,1,2]] /.
    cval[[1]]] // Simplify
```

```
step2:=eval(subs(s=0,rhs(step1)));
cval:=solve(step2=v0^2,_C1);
v:=subs(_C1=cval,sqrt(rhs(step1)));
```

7.

```
m=75;
rho=1.29;
mu=0.06;
cda=0.16;
graphs[v0_,alpha_]:=Module[{},
    k=Sqrt[9.8 (Sin[alpha Pi/180]-
        mu Cos[alpha Pi/180])]//N;
    h=Sqrt[cda rho/(2 m)]//N;
    Plot[v[s,h,k,v0],{s,0,1500}];
                                  ]

Map[graphs[#,30]&,{0,5,10}];
```

```
graph:='graph':
graph:=proc(initvel,alpha)
   local k0, h0,vel;
   k0:=evalf(sqrt(9.8*
      (sin(alpha*Pi/180)-
         mu*cos(alpha*Pi/180))));
   h0:=evalf(sqrt(cda*rho/(2*m)));
   vel:=subs({v0=initvel,k=k0,h=h0},
      v);
   plot(vel,s=0..1500)
   end:

for i in {0,5,10} do graph(i,30) od;
```

8. (Mathematica only)

```
Clear[mu,cda]
m=75;
rho=1.29;
graphs[v0_,alpha_,mu_,cda_,opts___]:=Module[{},
        k=Sqrt[9.8 (Sin[alpha Pi/180]-mu Cos[alpha Pi/180])]//N;
        h=Sqrt[cda rho/(2 m)]//N;
        Plot[v[s,h,k,v0],{s,0,1500},opts]
                                  ]
```

Differential Equations at Work: C. The Schrödinger Equation

1.

```
Clear[a0]
capr[r_]:=a Exp[-r/a0]
sub=capr''[r]+2/r capr'[r]+
2 mu/h^2 (capE+smalle^2/(4 Pi eps0 r)) capr[r]
```

A series solutions can be computed directly with Maple V.

```
eq:=diff(u(r),r$2)+2*(1/r-p)*diff(u(r),r)+(2*(z-p/r)-L*(L+1)/r^2)*u(r)=0:
sol:=dsolve(eq,u(r),series);
```

6.

```
serone=Normal[Series[Exp[-p r],{r,0,3}]]
prod=Expand[uapprox serone]
Collect[prod,r]
```

Applications of Higher-Order Equations

5

EXERCISES 5.1

3. $m = 1/4$ slugs, $k = 16$ lb/ft; Released 0.75 ft (8 inches) below eq. with an upward initial vel. of 2 ft/s.

```
sol =
  DSolve[{1 / 4 x''[t] + 16 x[t] == 0,
    x[0] == 3 / 4, x'[0] == -2}, x[t], t]
```

$\{\{x[t] \to \frac{3}{4} \text{Cos}[8\,t] - \frac{1}{4} \text{Sin}[8\,t]\}\}$

```
Plot[x[t] /. sol, {t, 0, 2 Pi}]
```

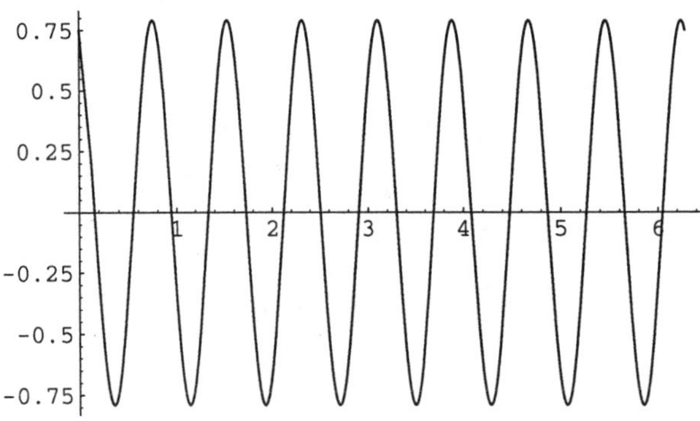

7. A general solution of $x'' + 16x = 0$ is $x = c_1 \cos 4t + c_2 \sin 4t$. Then, $x'(t) = -4c_1 \sin 4t + 4c_2 \cos 4t$; $x(0) = c_1 = -2$, $x'(0) = 4c_2 = 1 \Rightarrow c_2 = 1/4$ so $x(t) = -2\cos 4t + \frac{1}{4}\sin 4t$ which we can rewrite as

$$x(t) = \sqrt{(-2)^2 + \left(\frac{1}{4}\right)^2}\cos(4t - \phi) = \sqrt{\frac{65}{16}}\cos(4t - \phi) = \frac{\sqrt{65}}{4}\cos(4t - \phi),$$

where $\phi = \cos^{-1}\left(-\frac{8}{\sqrt{65}}\right) \approx 3.02$ rads; per. $= \frac{\pi}{2}$; amp $= \frac{\sqrt{65}}{4}$.

13. We first determine the spring constant k and the mass m:

$$F = k\,s$$
$$6 = \underbrace{\frac{1}{2}}_{6\text{ in}=\frac{1}{2}\text{ ft}} \cdot k$$
$$k = 12$$

and

$$F = m\,g$$
$$6 = m \cdot 32$$
$$m = 3/16.$$

Thus, we solve

$$\frac{3}{16}x'' + 12x = 0 \Leftrightarrow x'' + 64x = 0.$$

A general solution of this homogeneous equation is $x = c_1 \cos 8t + c_2 \sin 8t$. If the object is lifted 3 in. $= \frac{1}{4}$ ft. above the equilibrium position and released, we find the solution that satisfies the initial conditions $x(0) = -1/4$ and $x'(0) = 0$:

$$\left.\begin{array}{r} x(0) = c_1 = -\frac{1}{4} \\ x'(0) = 8c_2 = 0 \end{array}\right\} \Rightarrow x(t) = -\frac{1}{4}\cos 8t.$$

The time required for the mass to its equilibrium position is found by finding the first positive value of t for which $x(t)=0$:

$$x(t)=-\tfrac{1}{4}\cos 8t=0$$
$$8t=(2n+1)\pi/2,\ n \text{ any integer}$$
$$t=(2n+1)\pi/16,\ n \text{ any integer; } t=\pi/16 \text{ is the first positive solution.}$$

After five seconds the mass is $x(5)\approx 0.167\ ft$ below the equilibrium position. If the object is released from its equilibrium position with a downward velocity of 1 ft/s, we find that the solution that satisfies the initial conditions $x(0)=0$ and $x'(0)=1$ is $x(t)=\frac{1}{8}\sin 8t$; the time required for the object to return to its equilibrium position is $t=\pi/8$ s.

```
sola =
  DSolve[{3 / 16 x''[t] + 12 x[t] == 0,
    x[0] == -1 / 4, x'[0] == 0}, x[t], t]
```

$\{\{x[t]\to -\frac{1}{4}\,\mathrm{Cos}[8\,t]\}\}$

```
solb =
  DSolve[{3 / 16 x''[t] + 12 x[t] == 0,
    x[0] == 0, x'[0] == 1}, x[t], t]
```

$\{\{x[t]\to \frac{1}{8}\,\mathrm{Sin}[8\,t]\}\}$

```
Plot[Evaluate[x[t] /. {sola, solb}],
  {t, 0, Pi},
  PlotStyle -> {GrayLevel[0], GrayLevel[
```

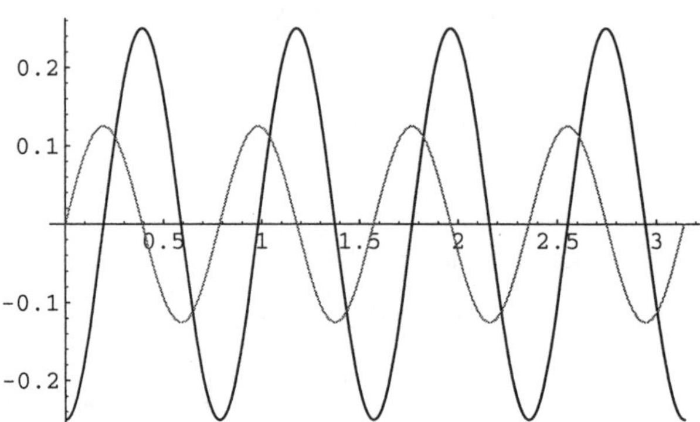

17. Here, $1=k\cdot 1/8\Rightarrow k=8$ and $1=m\cdot 32\Rightarrow m=1/32$ so we solve $\left\{\frac{1}{32}x''+8x=0, x(0)=b, x'(0)=-1\right\}$. A general solution of $x''+256x=0$ is $x=c_1\cos 16t+c_2\sin 16t$ and application of the initial conditions yields $x(t)=b\cos 16t-\frac{1}{16}\sin 16t$. The amplitude of the solution is $A=\sqrt{b^2+\frac{1}{k/m}}=\sqrt{b^2+\frac{1}{256}}$ and $\sqrt{b^2+\frac{1}{256}}=2\Rightarrow b=\frac{\sqrt{1023}}{16}$.

```
sol =
  DSolve[{1 / 32 x''[t] + 8 x[t] == 0, x[
    x'[0] == -1}, x[t], t]
```

$\{\{x[t]\to b\,\mathrm{Cos}[16\,t]-\frac{1}{16}\,\mathrm{Sin}[16\,t]\}\}$

```
bval = Solve[Sqrt[b^2 + 1 / 256] == 2, b]
```

$\{\{b\to -\frac{\sqrt{1023}}{16}\},\ \{b\to \frac{\sqrt{1023}}{16}\}\}$

```
Plot[Evaluate[x[t] /. sol /. bval[[2]]],
  {t, 0, Pi / 2}]
```

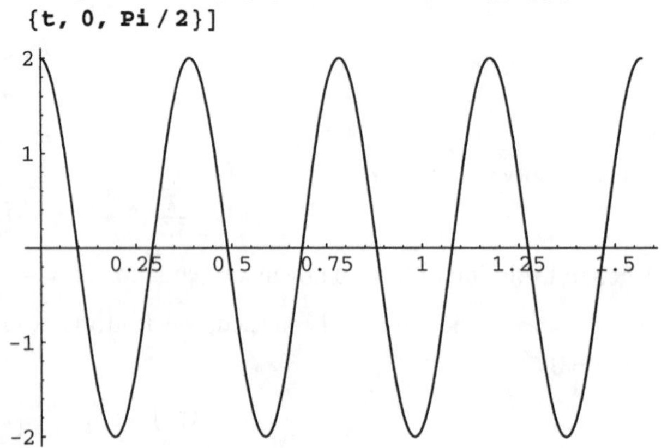

25. First, we find the mass of the cylinder: $512 = m \cdot 32 \Rightarrow m = 16$ slugs. Also, $h = \dfrac{mg}{\pi R^2 \rho} = \dfrac{512}{62.5\pi} \approx 2.61$ ft $\Rightarrow$ Eq. Position: $4 - 2.61 \approx 1.39$ ft and $y(0) = 3 - 1.39 = 1.61$ ft so we solve the initial value problem $\left\{ \dfrac{d^2y}{dt^2} + \dfrac{\pi \cdot 62.5}{16} y = 0, y(0) = 1.61, y'(0) = -3 \right\}$. A general solution of $\dfrac{d^2y}{dt^2} + \dfrac{\pi \cdot 62.5}{16} y = 0$ is $y(t) = c_1 \cos 3.5t + c_2 \sin 3.5t$ and applying the initial conditions results in $y(t) = 1.61 \cos 3.5t - 0.856379 \sin 3.5t$; max. displacement= $\sqrt{(1.61)^2 + (0.856444)^2} \approx 1.8236$ ft.

27. For (a) we solve $\begin{cases} \dfrac{d^2x}{dt^2} + 4x = 0 \\ x(0) = \alpha, x'(0) = 0 \end{cases}$, $\alpha = 1, 4, -2$.

```
sol=DSolve[{x''[t]+4x[t]==0,x[0]==alpha,
    x'[0]==0},x[t],t]
toplot=Table[sol[[1,1,2]] /.
    alpha->{1,4,-2}];
grays=Table[GrayLevel[i],{i,0,.5,.5/2}];
Plot[Evaluate[toplot],{t,0,3Pi},
    PlotRange->{-3Pi/2,3Pi/2},
    AspectRatio->1,PlotStyle->grays]
```

```
x:='x':alpha:='alpha':
sol:=dsolve({diff(x(t),t$2)+4*x(t)=0,
   x(0)=alpha,D(x)(0)=0},x(t));
toplot:={seq(rhs(sol),
   alpha={1,4,-2})};
plot(toplot,t=0..3*Pi);
```

From the graph, we see that varying α affects the amplitude of the displacement but not the times at which the mass passes through the equilibrium position.

For (b) we solve $\begin{cases} \dfrac{d^2x}{dt^2} + 4x = 0 \\ x(0) = 0, x'(0) = \beta \end{cases}$, $\beta = 1, 4, -2$.

```
sol=DSolve[{x''[t]+4x[t]==0,x[0]==0,
    x'[0]==beta},x[t],t];
toplot=Table[sol[[1,1,2]] /.
    beta->{1,4,-2}];
grays=Table[GrayLevel[i],{i,0,.5,.5/2}];
Plot[Evaluate[toplot],{t,0,3Pi},
    PlotRange->{-3Pi/2,3Pi/2},
    AspectRatio->1,PlotStyle->grays]
```

```
sol:=dsolve({diff(x(t),t$2)+4*x(t)=0,
   x(0)=0,D(x)(0)=beta},x(t));
toplot:={seq(rhs(sol),
   beta={1,4,-2})};
plot(toplot,t=0..3*Pi);
```

Graph for (a)

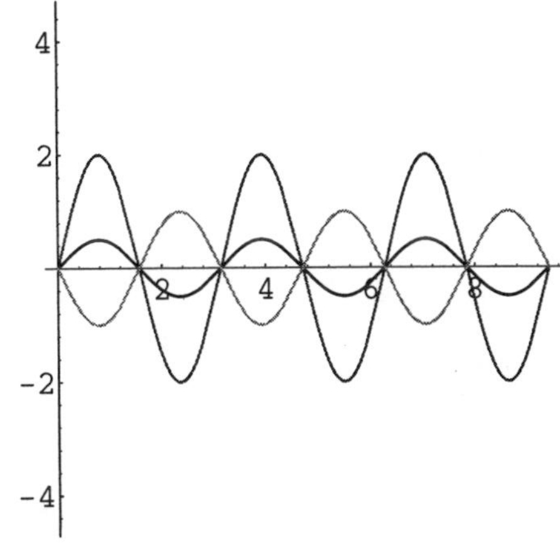

Graph for (b)

From the graph we see that varying β not only affects the initial velocity but also the amplitude of the displacement. The time at which the object passes through the equilibrium position is not affected.

28. In this case we solve $\begin{cases} 4\dfrac{d^2x}{dt^2} + 20x = 0 \\ x(0) = -7/12,\, x'(0) = 5/2 \end{cases}$

```
sol = DSolve[{4 x''[t] + 20 x[t] == 0,
    x[0] == -7 / 12, x'[0] == 5 / 2}, x[t], t]
```

$\{\{x[t] \to -\frac{7}{12} \text{Cos}[\sqrt{5}\, t] + \frac{1}{2} \sqrt{5}\, \text{Sin}[\sqrt{5}\, t]\}\}$

```
Plot[x[t] /. sol, {t, 0, 3 Pi}]
```

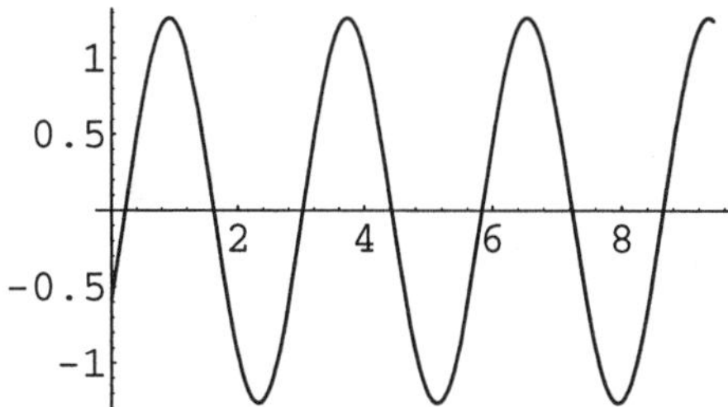


```
x:='x':
sol:=dsolve({4*diff(x(t),t$2)+
   20*x(t)=0,x(0)=-7/12,D(x)(0)=5/2},
   x(t));
plot(rhs(sol),t=0..3*Pi);
```

We compute the period and amplitude quickly.

```
alpha=-7/12;
beta=5/2;
omega=Sqrt[20/4]
capt=2Pi/omega
amplitude=Sqrt[alpha^2+beta^2/omega^2]
N[amplitude]
```

```
alpha:=-7/12:
beta:=5/2:
omega:=sqrt(20/4);
capt:=2*Pi/omega;
amplitude:=sqrt(
   alpha^2+beta^2/omega^2);
evalf(amplitude);
```

Notice that we can approximate the amplitude in part (a) by observing the graph of each solution and determining the maximum displacement from $x = 0$. Similarly, we can approximate the time at which the mass first passes through its equilibrium position in part (b) by finding the first value of t for which $x(t) = 0$. In part (c), the period can be approximated by measuring the distance between successive "peaks" or successive "valleys". `FindRoot` may be helpful in these calculations in *Mathematica*; `fsolve` may be useful with *Maple*.

30. Here we solve $\begin{cases} 3\dfrac{d^2x}{dt^2}+15x=0 \\ x(0)=3/4,\ x'(0)=1 \end{cases}$

```
sol=DSolve[{3 x''[t]+15x[t]==0,
    x[0]==3/4,x'[0]==1},x[t],t]
Plot[x[t] /. sol,{t,0,3Pi}]
alpha=3/4;
beta=1;
omega=Sqrt[15/3]
capt=2Pi/omega
amplitude=Sqrt[alpha^2+beta^2/omega^2]
N[amplitude]
```

```
x:='x':
sol:=dsolve({3*diff(x(t),t$2)+
    15*x(t)=0,x(0)=3/4,D(x)(0)=1},
    x(t));
plot(rhs(sol),t=0..3*Pi);
alpha:=3/4:
beta:=1:
omega:=sqrt(15/3);
capt:=2*Pi/omega;
amplitude:=sqrt(
    alpha^2+beta^2/omega^2);
evalf(amplitude);
```

EXERCISES 5.2

3. $m=1/4, c=2, k=1$; Released 6 inches above equil. with a downward init. vel. of 1 ft/s.

```
sol =
  DSolve[{1 / 4 x''[t] + 2 x'[t] + x[t] == 0,
    x[0] == -1 / 2, x'[0] == 1}, x[t], t]
```

$$\left\{\left\{x[t] \to \frac{1}{12}\left(-3+\sqrt{3}\right) e^{2\left(-2-\sqrt{3}\right)t} + \frac{1}{12}\left(-3-\sqrt{3}\right) e^{2\left(-2+\sqrt{3}\right)t}\right\}\right\}$$

```
Plot[x[t] /. sol, {t, 0, 10}]
```

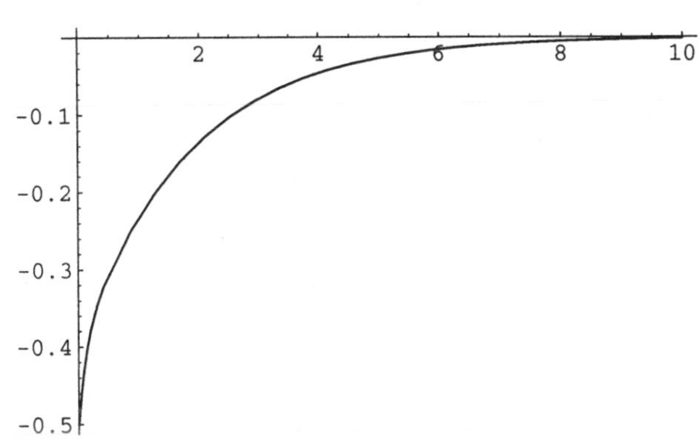

7. The characteristic equation of $x''+2x'+26x=0$ is $r^2+2r+26=0 \Rightarrow$

$$r=\frac{-2\pm\sqrt{2^2-4\cdot 26}}{2}=\frac{-2\pm\sqrt{-100}}{2}=\frac{-2\pm 10i}{2}=-1\pm 5i$$

so a general solution of $x''+2x'+26x=0$ is $x=e^{-t}\left(c_1\cos 5t+c_2\sin 5t\right)$ and application of the initial conditions yields

$$\left\{c_1=1,-c_1+5c_2=1\right\}\Rightarrow\left(c_1,c_2\right)=(1,2/5)\Rightarrow x(t)=e^{-t}\left(\cos 5t+\tfrac{2}{5}\sin 5t\right).$$

Then, $x(t)=\frac{\sqrt{29}}{5}e^{-t}\cos(5t-\phi)$, $\phi=\cos^{-1}\left(5/\sqrt{29}\right)\approx 0.38$ rads., the quasiperiod is $2\pi/5$, and the mass first passes through its equilibrium position when $t=\frac{1}{5}\left(\pi/2+\cos^{-1}\left(5/\sqrt{29}\right)\right)\approx 0.390261$.

```
sol =
 DSolve[{x''[t] + 2 x'[t] + 26 x[t] == 0,
   x[0] == 1, x'[0] == 1}, x[t], t]
```

$\{\{x[t] \to e^{-t} (\text{Cos}[5\,t] + \frac{2}{5}\,\text{Sin}[5\,t])\}\}$

```
Plot[x[t] /. sol, {t, 0, Pi}]
```

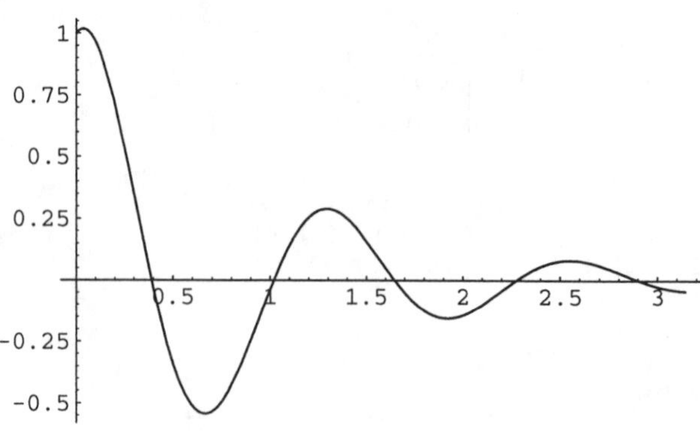

11. The characteristic equation is $m^2 + \frac{3}{2}m + \frac{1}{2} = \left(m + \frac{1}{2}\right)(m+1) = 0$ so a general solution is given by $x = c_1e^{-t} + c_2e^{-t/2}$. Application of the initial conditions yields the system of equations

$$\begin{cases} c_1 + c_2 = -1 \\ -c_1 - \frac{1}{2}c_2 = 2 \end{cases},$$

which has solution $c_1 = -3$ and $c_2 = 2$. Thus, $x(t) = -3e^{-t} + 2e^{-t/2}$; the system is overdamped. $x(t) = 0 \Rightarrow t = 2\ln\frac{3}{2} \approx 0.811$. The maximum displacement is 1 at $t = 0$.

```
sol =
 DSolve[{x''[t] + 3 / 2 x'[t] + 1 / 2 x[t] == 0,
   x[0] == -1, x'[0] == 2}, x[t], t]
```

$\{\{x[t] \to e^{-t} (-3 + 2\,e^{t/2})\}\}$

```
Plot[x[t] /. sol, {t, 0, 10}, Pl
```

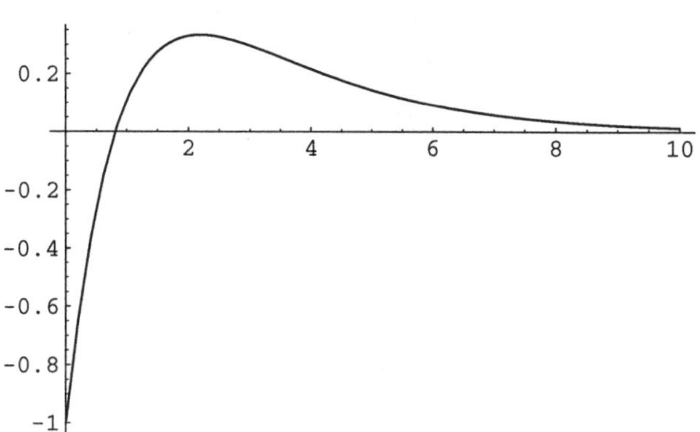

15. The characteristic equation of $x'' + 10x' + 25x = 0$ is $r^2 + 10r + 25 = 0 \Rightarrow (r+5)^2 = 0 \Rightarrow r_{1,2} = -5$ so a general solution of $x'' + 10x' + 25x = 0$ is $x = c_1e^{-5t} + c_2te^{-5t}$ and application of the initial conditions yields $\{c_1 = -5, -5c_1 + c_2 = 1\} \Rightarrow (c_1, c_2) = (-5, -24) \Rightarrow x(t) = -5e^{-5t} - 24te^{-5t}$; critically damped; does not pass through equilibrium; Max. Dis. = 5 at $t = 0$.

```
sol =
 DSolve[{x''[t] + 10 x'[t] + 25 x[t] == 0,
   x[0] == -5, x'[0] == 1}, x[t], t]
```

$\{\{x[t] \to e^{-5\,t}\,(-5 - 24\,t)\}\}$

```
Plot[x[t] /. sol, {t, 0, 3 / 2}, Pl
```

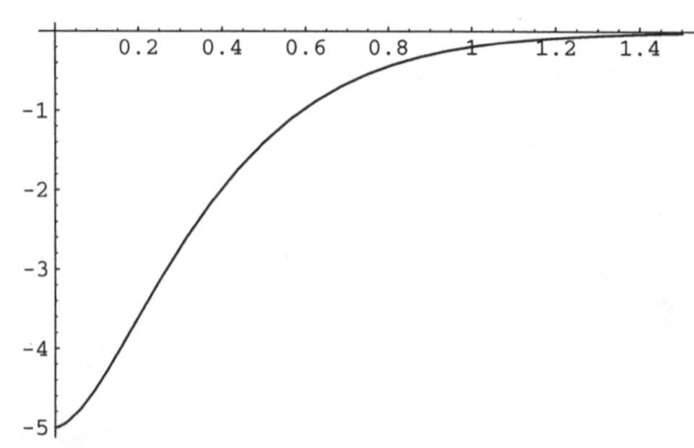

19. The object's mass is given by

$$F = m\,g$$
$$32 = m \cdot 32$$
$$m = 1.$$

We solve the initial-value problem

$$\begin{cases} x'' + 10x' + 24x = 0 \\ x(0) = -1/2,\ x'(0) = 0 \end{cases}.$$

The characteristic equation is $r^2 + 10r + 24 = 0 \Rightarrow (r+6)(r+4) = 0 \Rightarrow r = -4, -6$ so general solution of the differential equation is $x = c_1 e^{-4t} + c_2 e^{-6t}$. Application of the initial conditions yields the system of equations

$$\begin{cases} c_1 + c_2 = -1/2 \\ -4c_1 - 6c_2 = 0 \end{cases},$$

which has solution $c_1 = -3/2$ and $c_2 = 1$ so $x(t) = -\frac{3}{2}e^{-4t} + e^{-6t}$. The object passes through its equilibrium position when

$$x(t) = -\frac{3}{2}e^{-4t} + e^{-6t} = 0$$
$$\frac{3}{2}e^{-4t} = e^{-6t}$$
$$-4t + \ln\frac{3}{2} = -6t$$
$$2t = -\ln\frac{3}{2}$$
$$t = -\frac{1}{2}\ln\frac{3}{2},$$

which is a negative number; thus the object does not pass through the equilibrium position. The maximum displacement is 1/2 at $t = 0$.

```
sol =
  DSolve[{x''[t] + 10 x'[t] + 24 x[t] == 0,
    x[0] == -1 / 2, x'[0] == 0}, x[t], t]
```

$$\left\{\left\{x[t] \to e^{-6t}\left(1 - \frac{3 e^{2t}}{2}\right)\right\}\right\}$$

```
Plot[x[t] /. sol, {t, 0, 1}]
```

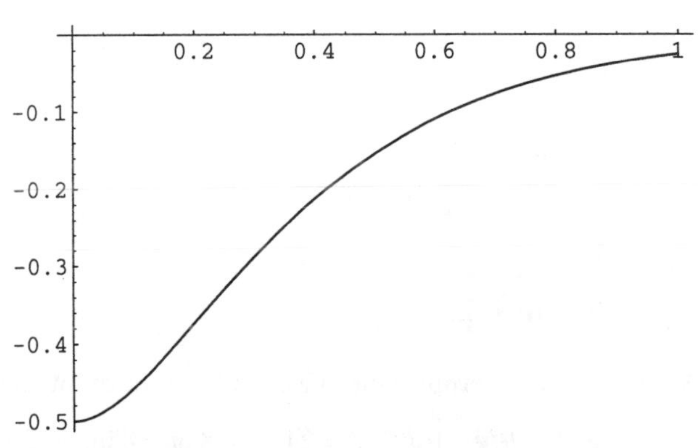

23. The motion is critically damped if

$$c^2 - 4mk = 0$$
$$c^2 - 4 \cdot 4 \cdot 64 = 0$$
$$c = 32.$$

The motion is underdamped if

$$c^2 - 4mk < 0$$
$$c^2 - 4 \cdot 4 \cdot 64 < 0$$
$$0 < c < 32.$$

29. $x(t)=c_1e^{-\rho t}+c_2te^{-\rho t}, \rho=\dfrac{c}{2m}$; $x(0)=\alpha, x'(0)=0 \Rightarrow x(t)=\alpha e^{-\rho t}(1+\rho t)$; $x(t)=0 \Rightarrow t=-\frac{1}{\rho}<0$

35. Note that $c=2\sqrt{6}$ results in critical damping, $c=4\sqrt{6}$ in overdamping, and $c=-\sqrt{6}$ in underdamping.

```
sols =
 Map[
  DSolve[{x''[t] + # x'[t] + 6 x[t] == 0,
     x[0] == 0, x'[0] == 1}, x[t], t] &,
  {2 Sqrt[6], 4 Sqrt[6], Sqrt[6]}]
```

$\{\{\{x[t] \to e^{-\sqrt{6}\,t}\,t\}\}, \{\{x[t] \to -\dfrac{e^{(-3\sqrt{2}-2\sqrt{6})t}}{6\sqrt{2}} + \dfrac{e^{(3\sqrt{2}-2\sqrt{6})t}}{6\sqrt{2}}\}\},$
$\{\{x[t] \to \dfrac{1}{3}\sqrt{2}\,e^{-\sqrt{\frac{3}{2}}\,t}\,\mathrm{Sin}[\dfrac{3t}{\sqrt{2}}]\}\}\}$

```
sol1:=dsolve({diff(x(t),t$2)+
    2*sqrt(6)*diff(x(t),t)+
    6*x(t)=0,x(0)=0,D(x)(0)=1},x(t));
sol2:=dsolve({diff(x(t),t$2)+
    4*sqrt(6)*diff(x(t),t)+
    6*x(t)=0,x(0)=0,D(x)(0)=1},x(t));
sol3:=dsolve({diff(x(t),t$2)+
    sqrt(6)*diff(x(t),t)+
    6*x(t)=0,x(0)=0,D(x)(0)=1},x(t));
plot({rhs(sol1),rhs(sol2),
    rhs(sol3)},t=0..2*Pi);
```

```
Plot[Evaluate[x[t] /. sols], {t, 0, 4},
  PlotStyle -> {GrayLevel[0], GrayLevel[.5]
   Dashing[{0.01}]}]
```

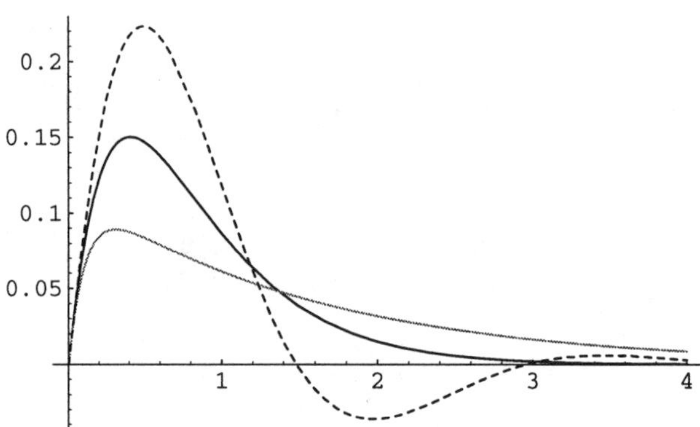

EXERCISES 5.3

3. First, we determine the mass and the value of the spring constant: $F=16 \Rightarrow m=F/g=16/32=1/2$ and $F=16 \Rightarrow k=F/s=16/(2/3)=24$ ($s=8\text{ in.}=8\text{ in.}\cdot\frac{1\text{ ft.}}{12\text{ in.}}=\frac{2}{3}\text{ ft.}$). Then, we solve the initial-value problem

$$\begin{cases}\frac{1}{2}x''+24x=2\cos t\\ x(0)=\frac{1}{3},\ x'(0)=0\end{cases} \Leftrightarrow \begin{cases}x''+48x=4\cos t\\ x(0)=\frac{1}{3},\ x'(0)=0\end{cases}.$$

The characteristic equation of the corresponding homogeneous equation $x''+48x=0$ is

$$m^2+48=0 \Rightarrow m=\pm\sqrt{48}i=\pm4\sqrt{3}i$$

so a general solution of the corresponding homogeneous equation is $x_h=c_1\cos(4\sqrt{3}t)+c_2\sin(4\sqrt{3}t)$. Using the method of undetermined coefficients, we assume that there is a particular solution to the nonhomogeneous equation of the form $x_p=A\cos t+B\sin t$. Substitution of this function and its derivatives into the nonhomogeneous equation yields $47A\cos t+47B\sin t=4\cos t$ so $A=4/47$ and $B=0$. Thus, a general solution of the nonhomogeneous equation is $x(t)=c_1\cos\left(4\sqrt{3}t\right)+c_2\sin\left(4\sqrt{3}t\right)+\frac{4}{47}\cos t$. Applying the initial conditions results in the system of equations

$$\begin{cases} c_1+\frac{4}{47}=\frac{1}{3} \\ 4\sqrt{3}c_2=0 \end{cases},$$

which has solution $c_1=35/141$ and $c_2=0$. Thus, $x(t)=\frac{35}{141}\cos\left(4\sqrt{3}t\right)+\frac{4}{47}\cos t$.

```
sol =
  DSolve[{1 / 2 x''[t] + 24 x[t] == 2 Cos[t],
     x[0] == 1 / 3, x'[0] == 0}, x[t], t] //
   Simplify

{{x[t] →
   1
  ─── (12 Cos[t] + 35 Cos[4 √3 t])}}
  141

Plot[x[t] /. sol, {t, 0, 4 Pi}]
```

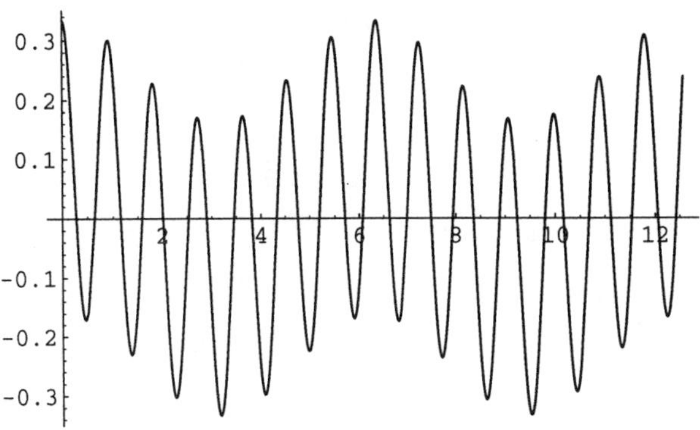

7. We solve $\begin{cases} x''+8x'+25x=\cos t-\sin t \\ x(0)=0,\ x'(0)=0 \end{cases}$ A general solution of the corresponding homgoeneous equation, $x''+8x'+25x=0$, which has characteristic equation

$$m^2+8m+25=0 \Rightarrow m=\frac{-8\pm\sqrt{8^2-4\cdot1\cdot25}}{2\cdot1}=-4\pm3i,$$

is $x_h=e^{-4t}\left(c_1\cos 3t+c_2\sin 3t\right)$. Using the method of undetermined coefficients, we assume that there is a particular solution to the nonhomogeneous equation of the form $x_p=A\cos t+B\sin t$. Substitution of this function into the nonhomogeneous equation yields

$$(24A+8B)\cos t+(-8A+24B)\sin t=\cos t-\sin t \Rightarrow \begin{cases} 24A+8B=1 \\ -8A+24B=-1 \end{cases} \Rightarrow (A,B)=\left(\tfrac{1}{20},-\tfrac{1}{40}\right)$$

so a general solution of the nonhomogeneous equation is $x=e^{-4t}\left(c_1\cos 3t+c_2\sin 3t\right)+\frac{1}{20}\cos t-\frac{1}{40}\sin t$. Application of the initial conditions yields

$$\begin{cases} \frac{1}{20}+c_1=0 \\ -\frac{1}{40}-4c_1+3c_2=0 \end{cases} \Rightarrow (c_1,c_2)=\left(-\tfrac{1}{20},-\tfrac{7}{120}\right)$$

so the solution to the initial-value problem is $x(t)=-e^{-4t}\left(\dfrac{1}{20}\cos 3t+\dfrac{7}{120}\sin 3t\right)+\dfrac{1}{40}(2\cos t-\sin t)$.

```
sol =
  DSolve[
    {x''[t] + 8 x'[t] + 25 x[t] ==
      Cos[t] - Sin[t], x[0] == 0,
     x'[0] == 0}, x[t], t] // Simplify
```

$$\left\{\left\{x[t] \to \frac{1}{120} e^{-4t}\left(6 e^{4t}\,\mathrm{Cos}[t] - 6\,\mathrm{Cos}[3t] - 3 e^{4t}\,\mathrm{Sin}[t] - 7\,\mathrm{Sin}[3t]\right)\right\}\right\}$$

```
Plot[x[t] /. sol, {t, 0, 4 Pi}]
```

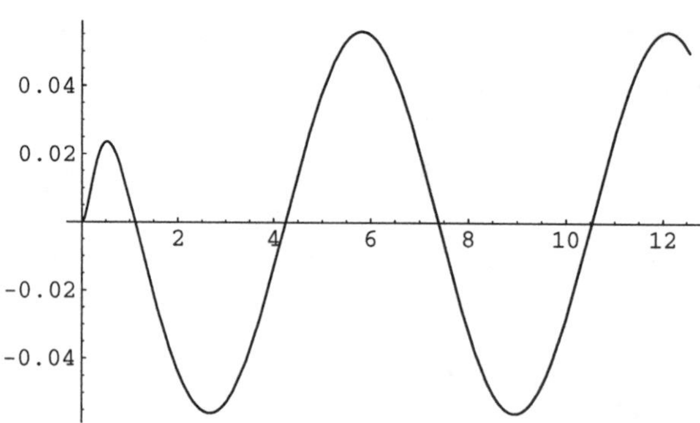

11. We solve $\{4x'' + 4x' + 26x = 250\sin t, x(0) = 0, x'(0) = 0\}$. The characteristic equation of the corresponding homogeneous equation is $4m^2 + 4m + 26 = 0 \Rightarrow m = -\frac{1}{2} \pm \frac{5}{2}i$ so a general solution of the corresponding homogeneous equation is $x_h = e^{-t/2}\left(c_1 \cos\frac{5t}{2} + c_2 \sin\frac{5t}{2}\right)$. Using the method of undetermined coefficients, we assume that there is a particular solution to the nonhomogeneous equation of the form $x_p = A\cos t + B\sin t$. Substituting this function into the nonhomogeneous equation yields

$$(22A + 4B)\cos t + (-4A + 22B)\sin t = 250\sin t \Rightarrow \begin{cases} 22A + 4B = 0 \\ -4A + 22B = 250 \end{cases} \Rightarrow (A, B) = (-2, 11)$$

so a general solution of the nonhomogeneous equation is $x = e^{-t/2}\left(c_1 \cos\frac{5t}{2} + c_2 \sin\frac{5t}{2}\right) - 2\cos t + 11\sin t$. Application of the initial conditions yields

$$\begin{cases} -2 + c_1 = 0 \\ 11 - \frac{1}{2}c_1 + \frac{5}{2}c_2 = 0 \end{cases} \Rightarrow (c_1, c_2) = (2, -4)$$

so the solution to the initial-value problem is $x(t) = e^{-t/2}\left(2\cos\frac{5t}{2} - 4\sin\frac{5t}{2}\right) - 2\cos t + 11\sin t$. The transient solution is $e^{-t/2}\left(2\cos\frac{5t}{2} - 4\sin\frac{5t}{2}\right)$ and the steady-state solution is $-2\cos t + 11\sin t$.

```
sol =
  DSolve[
    {4 x''[t] + 4 x'[t] + 26 x[t] ==
      250 Sin[t], x[0] == 0, x'[0] == 0},
    x[t], t] // Simplify
```

$$\left\{\left\{x[t] \to -2\,\mathrm{Cos}[t] + 2 e^{-t/2}\,\mathrm{Cos}\left[\frac{5t}{2}\right] + 11\,\mathrm{Sin}[t] - 4 e^{-t/2}\,\mathrm{Sin}\left[\frac{5t}{2}\right]\right\}\right\}$$

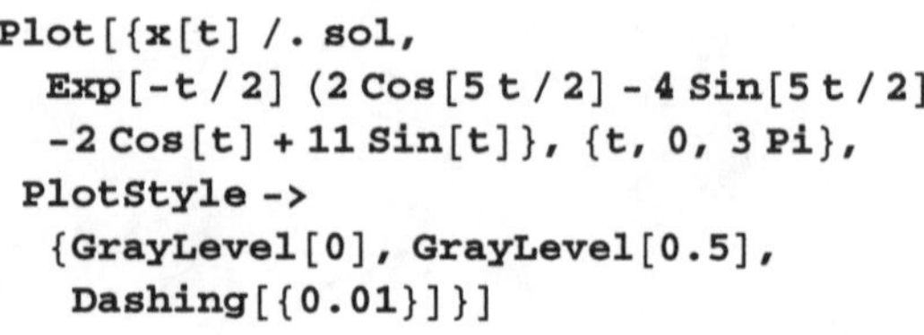

```
Plot[{x[t] /. sol,
  Exp[-t / 2] (2 Cos[5 t / 2] - 4 Sin[5 t / 2]
  -2 Cos[t] + 11 Sin[t]}, {t, 0, 3 Pi},
 PlotStyle ->
  {GrayLevel[0], GrayLevel[0.5],
   Dashing[{0.01}]}]
```

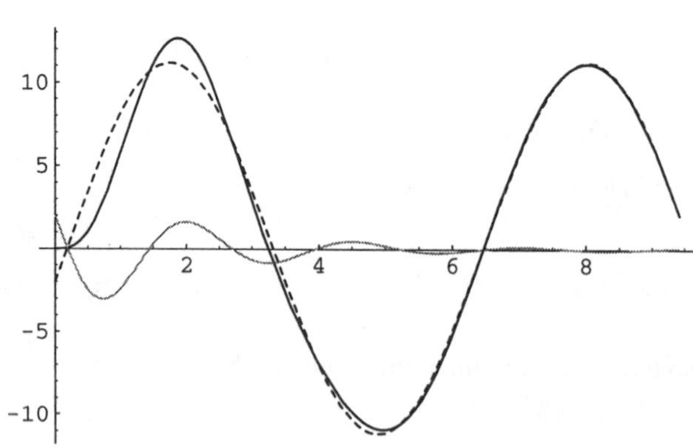

15. First, we solve $\{x''+x=1, x(0)=0, x'(0)=0\}$. A general solution of the corresponding homogeneous equation is $x_h=c_1\cos t+c_2\sin t$ and using the method of undetermined coefficients we find that a particular solution to the nonhomogeneous equation is $x_p=1$ so a general solution of the nonhomogeneous equation is $x=c_1\cos t+c_2\sin t+1$. Application of the initial conditions yields $\begin{cases}1+c_1=0\\ c_2=0\end{cases}\Rightarrow(c_1,c_2)=(-1,0)$ so the solution to the initial-value problem is $x(t)=1-\cos t$. Because $x(\pi)=2$ and $x'(\pi)=0$, we now solve $\{x''+x=0, x(\pi)=2, x'(\pi)=0\}$. A general solution of this homogeneous equation is $x_h=c_1\cos t+c_2\sin t$ and application of the initial conditions yields $\begin{cases}-c_1=2\\ -c_2=0\end{cases}\Rightarrow(c_1,c_2)=(-2,0)$ so the solution to this initial-value problem is $x(t)=-2\cos t$. Thus, the solution to the original problem is $x(t)=\begin{cases}1-\cos t, & 0\le t\le\pi\\ -2\cos t, & t>\pi\end{cases}$

```
sol =
  DSolve[{x''[t] + x[t] == UnitStep[Pi -
    x[0] == 0, x'[0] == 0}, x[t], t]

{{x[t] → 1 - Cos[t] - UnitStep[-π + t] -
    Cos[t] UnitStep[-π + t]}}

Plot[x[t] /. sol, {t, 0, 3 Pi}]
```

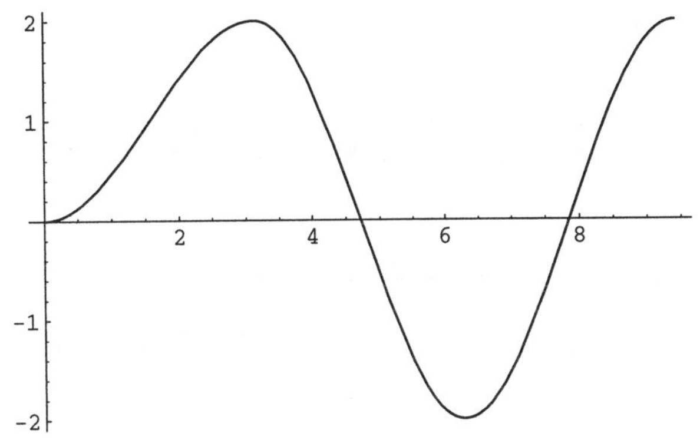

21.

```
sols =
 Map[
   DSolve[{x''[t] + x[t] == Cos[t],
      x[0] == 0, x'[0] == #}, x[t], t] &,
   {-2, -1, 0, 1, 2}] // Simplify
```

$\{\{\{x[t]\to\frac{1}{2}(-4+t)\,\mathrm{Sin}[t]\}\},$
$\{\{x[t]\to\frac{1}{2}(-2+t)\,\mathrm{Sin}[t]\}\},$
$\{\{x[t]\to\frac{1}{2}t\,\mathrm{Sin}[t]\}\},$
$\{\{x[t]\to\frac{1}{2}(2+t)\,\mathrm{Sin}[t]\}\},$
$\{\{x[t]\to\frac{1}{2}(4+t)\,\mathrm{Sin}[t]\}\}\}$

```
sol1:=dsolve({diff(x(t),t$2)+
   x(t)=cos(t),x(0)=0,D(x)(0)=0},
   x(t))
sol2:=dsolve({diff(x(t),t$2)+
   x(t)=cos(t),x(0)=0,D(x)(0)=1},
   x(t))
plot({rhs(sol1),rhs(sol2)},
   t=0..4*Pi);
```

```
grays = Table[GrayLevel[i],
    {i, 0, 0.7, .7 / 4}];
Plot[Evaluate[x[t] /. sols], {t, 0, 4
 PlotStyle -> grays]
```

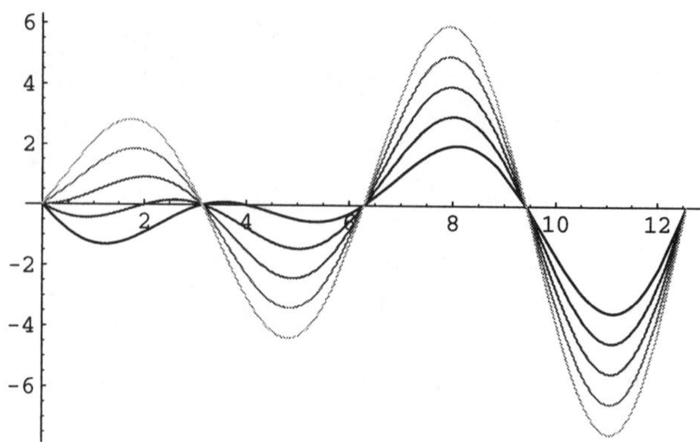

23.

```
sols =
 Map[
   DSolve[{x''[t] + 4 x[t] == #,
      x[0] == 0, x'[0] == 0}, x[t], t] &,
   {Cos[19 / 10 t], Cos[21 / 10 t]}] //
  Simplify
```

$$\left\{\left\{\left\{x[t] \to \frac{100}{39}\left(\mathrm{Cos}\left[\frac{19\,t}{10}\right] - \mathrm{Cos}[2\,t]\right)\right\}\right\}, \left\{\left\{x[t] \to \frac{100}{41}\left(\mathrm{Cos}[2\,t] - \mathrm{Cos}\left[\frac{21\,t}{10}\right]\right)\right\}\right\}\right\}$$

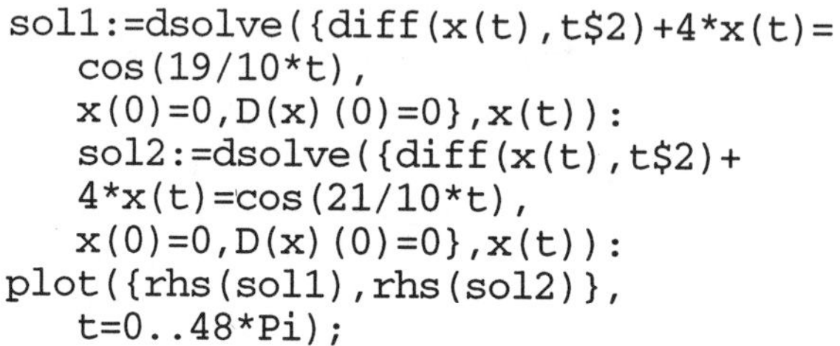

```
sol1:=dsolve({diff(x(t),t$2)+4*x(t)=
    cos(19/10*t),
    x(0)=0,D(x)(0)=0},x(t)):
    sol2:=dsolve({diff(x(t),t$2)+
    4*x(t)=cos(21/10*t),
    x(0)=0,D(x)(0)=0},x(t)):
plot({rhs(sol1),rhs(sol2)},
    t=0..48*Pi);
```

```
Plot[Evaluate[x[t] /. sols], {t, 0, 32
 PlotStyle ->
  {GrayLevel[0], GrayLevel[0.5]},
 PlotPoints -> 1000]
```

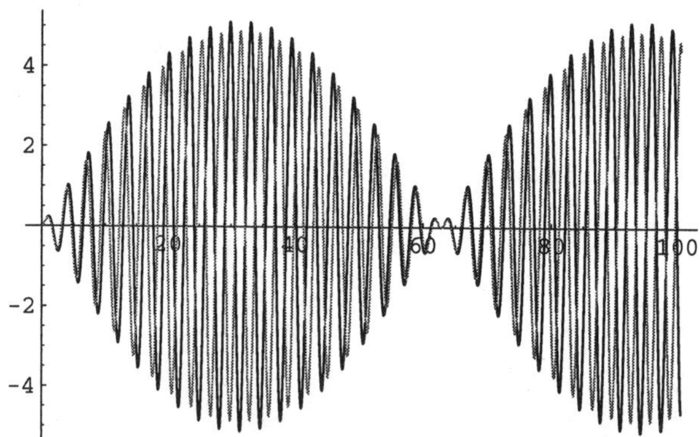

26.

```
g[gamma_]=capf/Sqrt[(omega^2-gamma^2)^2+
    4 lambda^2 gamma^2]
omega=Sqrt[k/m];
k=4;
m=1;
capf=2;
toplot1=g[gamma] /.
    lambda->{1,.5,.375,.25,.125}
grays=Table[GrayLevel[i],{i,0,.5,.5/4}]
Plot[Evaluate[toplot1],{gamma,-5,5},
    PlotRange->All,PlotStyle->grays]
```

```
g:=gamma->capf/sqrt(
    (omega^2-gamma^2)^2+
       4*lambda^2*gamma^2);
omega:=sqrt(k/m);
k:=4;
m:=1;
capf:=2;
lvals:={1,.5,.375,.25,.125}:
toplot1:={seq(g(gamma),
    lambda=lvals)};
plot(toplot1,gamma=-5..5);
```

```
omega=Sqrt[k/m];
k=49;
m=10;
capf=20;
toplot2=g[gamma] /.
    lambda->{1,.5,.375,.25,.125};
Plot[Evaluate[toplot2],{gamma,-5,5},
    PlotRange->All,PlotStyle->grays]
```

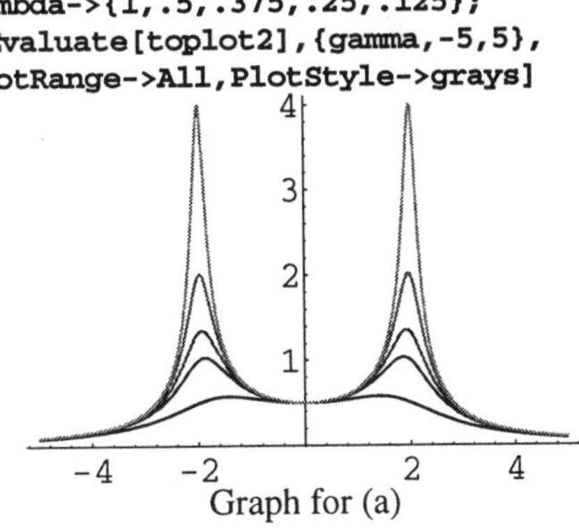

Graph for (a)

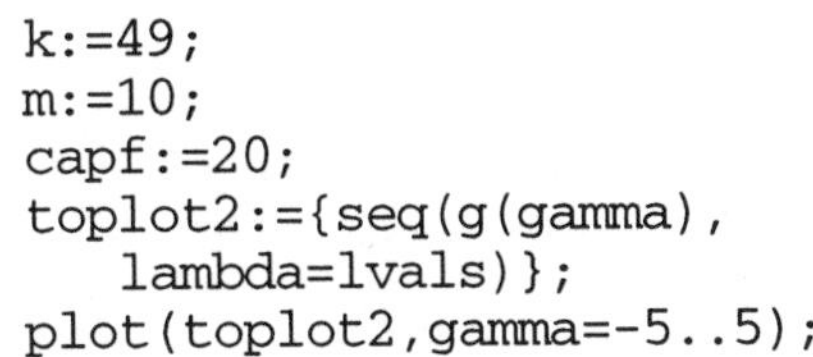

```
k:=49;
m:=10;
capf:=20;
toplot2:={seq(g(gamma),
   lambda=lvals)};
plot(toplot2,gamma=-5..5);
```

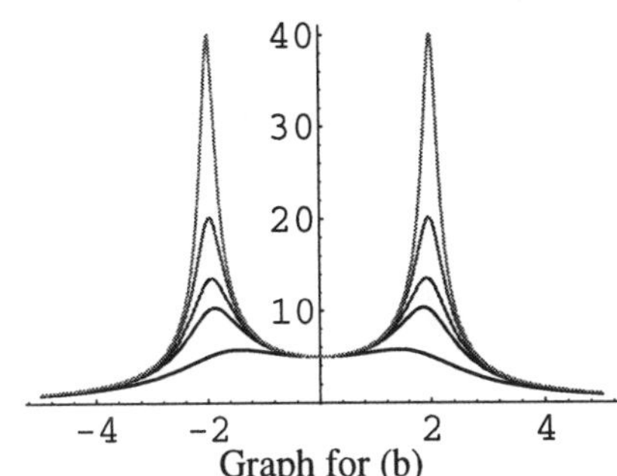

Graph for (b)

27 and 28.

```
Clear[x,sol,omegas,f,beta,sol]
sol[omegas_,beta_,f_]:=DSolve[{x''[t]+omegas x[t]==f Cos[beta t],
    x[0]==0,x'[0]==0},x[t],t][[1,1,2]]//Expand
a=sol[6000^2,5991.62,2]
Play[a,{t,0,6}]
b=sol[6000^2,6000,2]
Play[b,{t,0,6}]
```

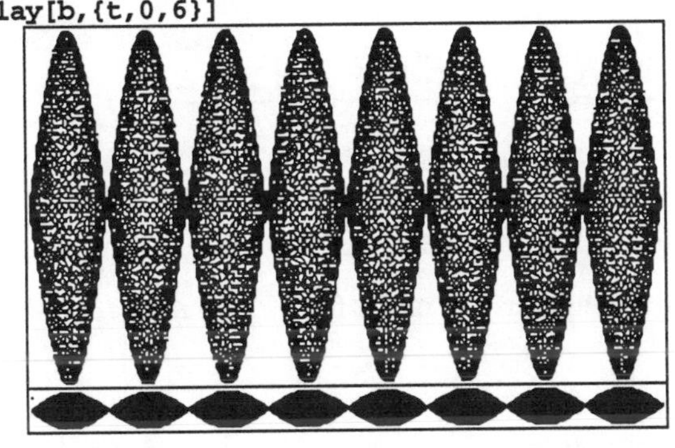

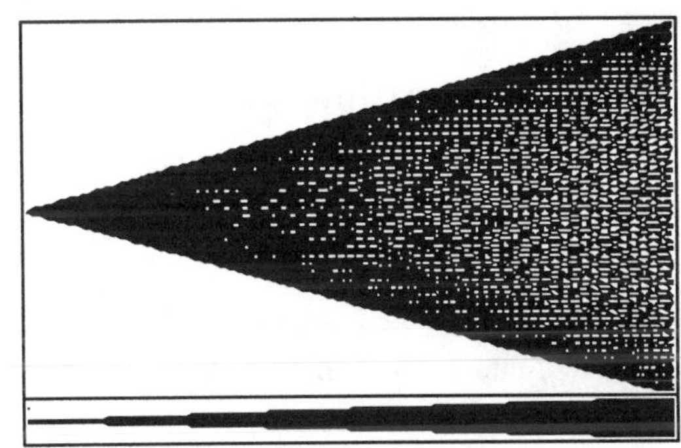

EXERCISES 5.4

3. We solve the initial-value problem $\begin{cases}\frac{1}{4}Q''+64Q=16t\\ Q(0)=0,\ Q'(0)=0\end{cases} \Leftrightarrow \begin{cases}Q''+256Q=64t\\ Q(0)=0,\ Q'(0)=0\end{cases}$. The corresponding homogeneous equation $Q''+256Q=0$ has characteristic equation $r^2+256=0 \Rightarrow r=\pm\sqrt{-256}=\pm 16i$ so a general solution of the corresponding homogeneous equation is $Q_h=c_1\cos 16t+c_2\sin 16t$. Using the method of undertermined coefficients, we assume that a particular solution has the form $Q_p=At+B \Rightarrow Q_p'=A, Q_p''=0$. Substituting into the nonhomogeneous equation yields

$$0+256(At+B)=64t \Rightarrow B=0, A=64/256=1/4 \Rightarrow Q_p=t/4$$

Thus, a general solution of the equation is $Q=c_1\cos 16t+c_2\sin 16t+t/4$ and application of the initial conditions yields $\{c_1=0, 1/4+16c_2=0\} \Rightarrow (c_1,c_2)=(0,-1/64)$ so $Q(t)=\frac{1}{4}t-\frac{1}{64}\sin 16t$ and $I(t)=Q'(t)=\frac{1}{4}-\frac{1}{4}\cos 16t$.

```
sol =
 DSolve[{1 / 4 q''[t] + 64 q[t] == 64 t,
    q[0] == 0, q'[0] == 0}, q[t], t] //
  Simplify
```

$\{\{q[t] \to t - \frac{1}{16}\,\text{Sin}[16\,t]\}\}$

```
TrigReduce[sol[[1, 1, 2]]]
```

$\frac{1}{16}\,(16\,t - \text{Sin}[16\,t])$

```
Plot[
 Evaluate[
  {q[t] /. sol, D[q[t] /. sol, t]}],
 {t, 0, 2 Pi},
 PlotStyle ->
  {GrayLevel[0], GrayLevel[0.5]}]
```

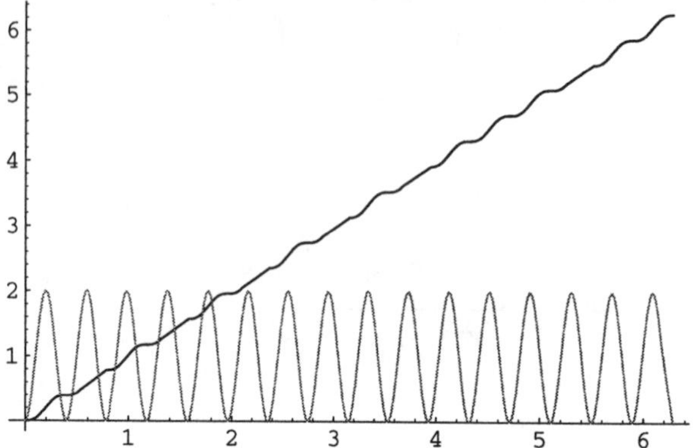

7. We solve the initial-value problem

$$\begin{cases} 0.2\dfrac{d^2Q}{dt^2} + 25\dfrac{dQ}{dt} + \dfrac{1}{0.001}Q = 126\cos t + 5000\sin t \\ Q(0) = 0,\ Q'(0) = 4 \end{cases} \Leftrightarrow \begin{cases} \dfrac{d^2Q}{dt^2} + 125\dfrac{dQ}{dt} + 5000Q = 630\cos t + 25000\sin t \\ Q(0) = 0,\ Q'(0) = 4 \end{cases}$$

The characteristic equation for the corresponding homogeneous equation is

$$r^2 + 125r + 5000 = 0 \Rightarrow r_{1,2} = \frac{-125 \pm \sqrt{125^2 - 4\cdot 1\cdot 5000}}{2} = \frac{-125 \pm \sqrt{-4375}}{2} = -\frac{125}{2} + \frac{25\sqrt{7}}{2}i$$

so a general solution of the corresponding homogeneous equation $Q'' + 125\,Q' + 5000Q = 0$ is

$$Q_h(t) = e^{-125t/2}\left(c_1 \cos\left(\frac{25\sqrt{7}}{2}t\right) + c_2 \sin\left(\frac{25\sqrt{7}}{2}t\right)\right)$$

so we assume that there is a particular solution to the nonhomogeneous equation of the form $Q_p = A\cos t + B\sin t$. Substituting Q_p into the nonhomogeneous equation results in

$$(4999A + 125B)\cos t + (-125A + 4999B)\sin t = 630\cos t + 25000\sin t$$

and equating coefficients results in the system of equations

$$\begin{cases} 4999A + 125B = 630 \\ -125A + 4999B = 5000 \end{cases},$$

which has solution $A = \frac{12185}{12502813}$ and $B = \frac{62526875}{12502813}$ so

$$Q(t) = e^{-125t/2}\left(c_1 \cos\left(\frac{25\sqrt{7}}{2}t\right) + c_2 \sin\left(\frac{25\sqrt{7}}{2}t\right)\right) + \frac{12185}{12502813}\cos t + \frac{62526875}{12502813}\sin t.$$

Application of the initial conditions results in the system of equations

$$\begin{cases} \dfrac{12185}{12502813} + c_1 = 0 \\ \dfrac{62526875}{12502813} - \dfrac{125}{2}c_1 + \dfrac{25\sqrt{7}}{2}c_2 = 4 \end{cases},$$

which has solution $c_1 = -\frac{12185}{12502813}$ and $c_2 = -\frac{26554371}{312570325\sqrt{7}}$ so

$$Q(t) = e^{-125t/2}\left(-\frac{12185}{12502813}\cos\left(\frac{25\sqrt{7}}{2}t\right) - \frac{26554371}{312570325\sqrt{7}}\sin\left(\frac{25\sqrt{7}}{2}t\right)\right) + \frac{12185}{12502813}\cos t + \frac{62526875}{12502813}\sin t.$$

```
sol =
 DSolve[
   {q''[t] + 125 q'[t] + 5000 q[t] ==
     630 Cos[t] + 25000 Sin[t],
    q[0] == 0, q'[0] == 4}, q[t], t] //
  Simplify
```

$$\left\{\left\{q[t] \to \left(e^{-125\,t/2}\left(2132375\, e^{125\,t/2}\,\text{Cos}[t] - 2132375\,\text{Cos}\left[\frac{25\sqrt{7}\,t}{2}\right] + 10942203125\, e^{125\,t/2}\,\text{Sin}[t] - 26554371\sqrt{7}\,\text{Sin}\left[\frac{25\sqrt{7}\,t}{2}\right]\right)\right)\Big/ 2187992275\right\}\right\}$$

```
Plot[
  Evaluate[
   {q[t] /. sol, D[q[t] /. sol, t]}],
  {t, 0, 4Pi},
  PlotStyle ->
   {GrayLevel[0], GrayLevel[0.5]}]
```

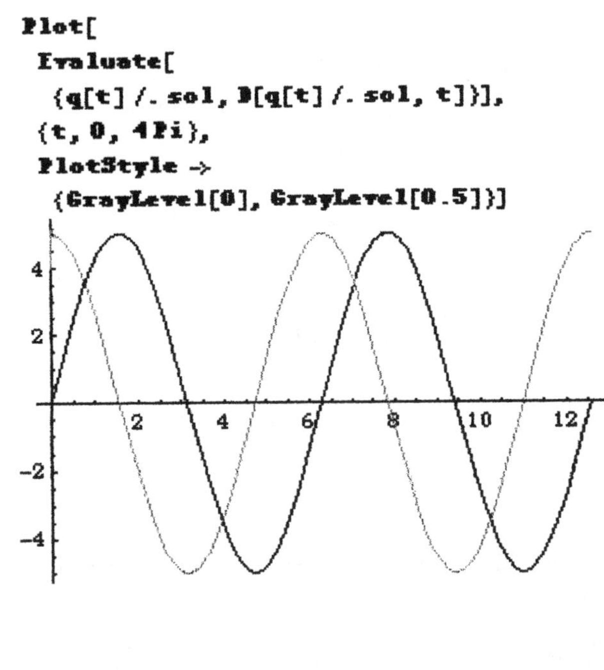

The charge is first equal to zero when $t \approx 3.1414$. The steady-state charge is $\lim_{t\to\infty} Q(t) = \frac{12185}{12502813}\cos t + \frac{62526875}{12502813}\sin t$ and the steady-state current is the derivative of the steady-state charge:

$$\frac{d}{dt}\left(\frac{12185}{12502813}\cos t + \frac{62526875}{12502813}\sin t\right) = -\frac{12185}{12502813}\sin t + \frac{62526875}{12502813}\cos t.$$

11. In each case, we consider the boundary value problem $EI\frac{d^4s}{dx^4} = 8$, $s(0) = 0, s''(0) = 0$, $s(10) = 0$, $s''(10) = 0$.

(a) If $EI = 100$, we consider the differential equation $\frac{d^4s}{dx^4} = \frac{8}{100} = \frac{2}{25}$. Integration gives us $\frac{d^3s}{dx^3} = \frac{2}{25}x + c_1$, $\frac{d^2s}{dx^2} = \frac{1}{25}x^2 + c_1x + c_2$, $\frac{ds}{dx} = \frac{1}{75}x^3 + \frac{1}{2}c_1x^2 + c_2x + c_3$, and $s = \frac{1}{300}x^4 + \frac{1}{6}c_1x^3 + \frac{1}{2}c_2x^2 + c_3x + c_4$. Application of the boundary conditions yields the system of equations $s(0) = c_4 = 0$, $s''(0) = c_2 = 0$, $s(10) = \frac{1}{300}(10)^4 + \frac{1}{6}c_1(10)^3 + c_3(10) = 0$, $s''(10) = \frac{1}{25}(10)^2 + c_1(10) = 0$ which has solution $\{c_1 = -2/5, c_2 = 0, c_3 = 10/3, c_4 = 0\}$. Therefore, $s(x) = \frac{1}{300}x^4 - \frac{1}{15}x^3 + \frac{10}{3}x$. In a similar manner, we find (b) $s(x) = \frac{1}{30}x^4 + \frac{100}{3}x - \frac{2}{3}x^3$ and (c) $s(x) = \frac{1}{3}x^4 + \frac{1000}{3}x - \frac{20}{3}x^3$; Simple support leads to larger max. displacement.

```
sols =
 Map[
  DSolve[{# s(4) [x] == 8, s[0] == 0,
     s''[0] == 0, s[10] == 0, s''[10]
    s[x], x] &, {100, 10, 1}]
```

$$\left\{\left\{\left\{s[x] \to \frac{1}{300}\left(1000\,x - 20\,x^3 + x^4\right)\right\}\right\},\right.$$
$$\left\{\left\{s[x] \to \frac{1}{30}\left(1000\,x - 20\,x^3 + x^4\right)\right\}\right\},$$
$$\left.\left\{\left\{s[x] \to \frac{1000\,x}{3} - \frac{20\,x^3}{3} + \frac{x^4}{3}\right\}\right\}\right\}$$

```
Plot[Evaluate[s[x] /. sols], {x, 0, 10},
 PlotStyle -> {GrayLevel[0], GrayLevel[0.5]
   Dashing[{0.01}]}]
```

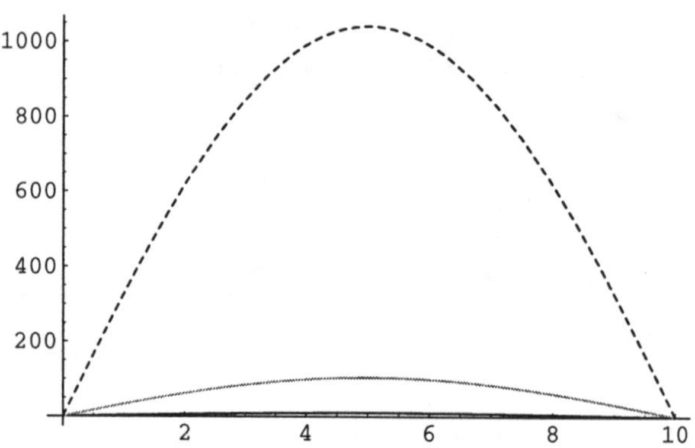

13 and 22. (a) $s(x) = \frac{1}{360}x^6 + \frac{10000}{9}x - \frac{125}{9}x^3$; (b) $\frac{d^4s}{dx^4} = x^2$, $s(0) = 0$, $s''(0) = 0$, $s''(10) = 0$, $s'''(10) = 0$. Integration yields $\frac{d^3s}{dx^3} = \frac{1}{3}x^3 + c_1$, $\frac{d^2s}{dx^2} = \frac{1}{12}x^4 + c_1x + c_2$, $\frac{ds}{dx} = \frac{1}{60}x^5 + \frac{1}{2}c_1x^2 + c_2x + c_3$, and $s = \frac{1}{360}x^6 + \frac{1}{6}c_1x^3 + \frac{1}{2}c_2x^2 + c_3x + c_4$. Applying the boundary conditions, we obtain $s(0) = c_4 = 0$, $s''(0) = c_2 = 0$, $\frac{d^2s}{dx^2}(10) = \frac{1}{12}(10)^4 + c_1(10) = 0$, and $\frac{d^3s}{dx^3}(10) = \frac{1}{3}(10)^3 + c_1 = 0$. The third equation indicates that $c_1 = -\frac{10^3}{12}$ while the fourth implies that $c_1 = -\frac{10^3}{3}$. This contradiction means that there is no solution. (c) $s(x) = \frac{1}{360}x^6 - \frac{500}{9}x^3 + 15000x$ (d) $s(x) = \frac{1}{360}x^6 + \frac{1250}{3}x - \frac{125}{18}x^3$

```
sols =
 Map[
  DSolve[Evaluate[
     Join[{s⁽⁴⁾[x] == x^2}, #]], s[x], x]
   {{s[0] == 0, s''[0] == 0, s[10] == 0,
     s''[10] == 0},
    {s[0] == 0, s''[0] == 0, s''[10] == 0,
     s'''[10] == 0},
    {s[0] == 0, s''[0] == 0, s'[10] == 0,
     s'''[10] == 0},
    {s[0] == 0, s''[0] == 0, s[10] == 0,
     s'[10] == 0}}]
```

$$\left\{\left\{\left\{s[x]\to\frac{10000\,x}{9}-\frac{125\,x^3}{9}+\frac{x^6}{360}\right\}\right\},\ \{\},\ \left\{\left\{s[x]\to 15000\,x-\frac{500\,x^3}{9}+\frac{x^6}{360}\right\}\right\},\ \left\{\left\{s[x]\to\frac{1250\,x}{3}-\frac{125\,x^3}{18}+\frac{x^6}{360}\right\}\right\}\right\}$$

```
grays = Table[GrayLevel[i],
    {i, 0, 0.7, 0.7/3}];
Plot[Evaluate[s[x] /. sols], {x, 0, 10},
  PlotStyle -> grays]
```

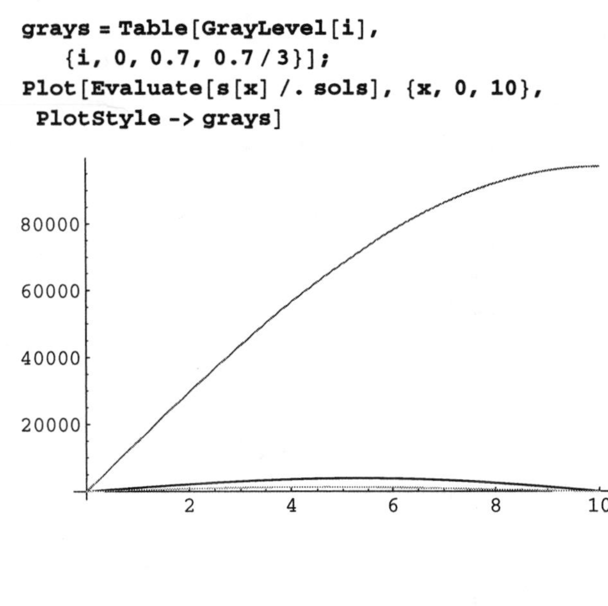

14 and 23. (a) $s(x)=\frac{480000}{\pi^4}\sin\frac{\pi x}{10}$; (b) No Solution; (c) $s(x)=\frac{480000}{\pi^4}\sin\frac{\pi x}{10}+\frac{24000}{\pi^3}\left(2+\pi^2\right)x-\frac{80}{\pi}x^3$; (d) $s(x)=\frac{480000}{\pi^4}\sin\frac{\pi x}{10}-\frac{24000}{\pi^3}\left(2+\pi^2\right)x+\frac{240}{\pi}x^3$; Max. displacement occurs in (c). The smallest max. displacement occurs in (d).

```
sols =
 Map[
  DSolve[Evaluate[
     Join[{s^(4)[x] == 48 Sin[Pi x / 10]},
      #]], s[x], x] &,
   {{s[0] == 0, s''[0] == 0, s[10] == 0,
     s''[10] == 0},
    {s[0] == 0, s''[0] == 0, s''[10] == 0,
     s'''[10] == 0},
    {s[0] == 0, s''[0] == 0, s'[10] == 0,
     s'''[10] == 0},
    {s[0] == 0, s''[0] == 0, s[10] == 0,
     s'[10] == 0}}]
```

$$\left\{\left\{\left\{s[x] \to \frac{480000\,\mathrm{Sin}\left[\frac{\pi x}{10}\right]}{\pi^4}\right\}\right\}, \{\},\right.$$

$$\left\{\left\{s[x] \to \frac{24000\,(2+\pi^2)\,x}{\pi^3} - \frac{80\,x^3}{\pi} + \frac{480000\,\mathrm{Sin}\left[\frac{\pi x}{10}\right]}{\pi^4}\right\}\right\}, \left\{\left\{s[x] \to\right.\right.$$

$$\left. -\frac{24000\,x}{\pi^3} + \frac{240\,x^3}{\pi^3} + \frac{480000\,\mathrm{Sin}\left[\frac{\pi x}{10}\right]}{\pi^4} \right\}$$

```
grays = Table[GrayLevel[i],
    {i, 0, 0.7, 0.7 / 3}];
Plot[Evaluate[s[x] /. sols], {x, 0, 10},
  PlotStyle -> grays]
```

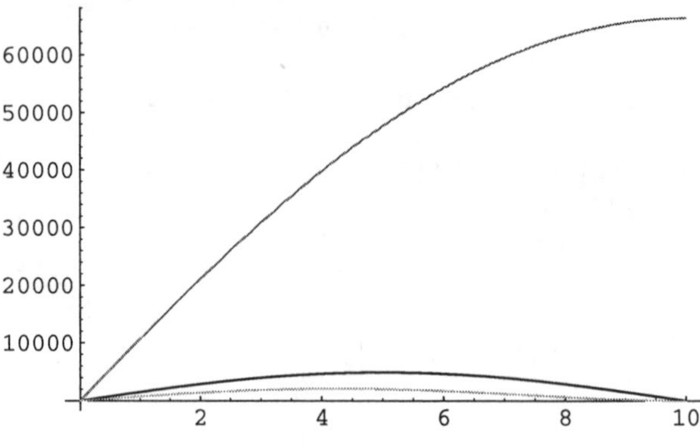

17. The characteristic equation of the corresponding homogeneous equation $L\frac{d^2Q}{dt^2} + R\frac{dQ}{dt} + \frac{1}{C}Q = 0$ is

$$Lr^2 + Rr + 1/C = 0 \Rightarrow r_{1,2} = \frac{-R\sqrt{C} \pm \sqrt{CR^2 - 4L}}{2L\sqrt{C}} = -\frac{R}{2L} \pm \frac{\sqrt{R^2C - 4L}}{2L\sqrt{C}}$$

so a general solution of the corresponding homogeneous equation is given by

$$Q_h = c_1 \exp\left(\frac{-R\sqrt{C} + \sqrt{CR^2 - 4L}}{2L\sqrt{C}}t\right) + c_2 \exp\left(\frac{-R\sqrt{C} - \sqrt{CR^2 - 4L}}{2L\sqrt{C}}t\right) \left(CR^2 - 4L > 0\right),$$

$$Q_h = c_1 \exp\left(-\frac{R}{2L}t\right) + c_2 t \exp\left(-\frac{R}{2L}t\right) \left(CR^2 - 4L = 0\right), \text{ or}$$

$$Q_h = \exp\left(-\frac{R}{2L}t\right)\left(c_1 \cos\left(\frac{\sqrt{4L - R^2C}}{2L\sqrt{C}}t\right) + c_2 \sin\left(\frac{\sqrt{4L - R^2C}}{2L\sqrt{C}}t\right)\right) \left(CR^2 - 4L < 0\right)$$

Using the method of undetermined coefficients, we assume that there is a particular solution of the nonhomogeneous equation of the form $Q_p(t) = A\cos\omega t + B\sin\omega t$. Substituting Q_p into the nonhomogeneous equation yields

$$\left(A/C - AL\omega^2 + BR\omega\right)\cos\omega t + \left(B/C - BL\omega^2 - A\omega R\right)\sin\omega t = E_0 \sin\omega t$$

and equating coefficients results in the system of equations

$$\begin{cases} A/C - AL\omega^2 + BR\omega = 0 \\ B/C - BL\omega^2 - A\omega R = E_0 \end{cases}$$

which has solution $A=-\dfrac{C^2E_0\omega R}{1-2CL\omega^2+C^2\omega^2\left(L^2\omega^2+R^2\right)}$ and $B=-\dfrac{CE_0\left(CL\omega^2-1\right)}{1-2CL\omega^2+C^2\omega^2\left(L^2\omega^2+R^2\right)}$. Thus,

$$Q_p(t)=-\frac{C^2E_0\omega R}{1-2CL\omega^2+C^2\omega^2\left(L^2\omega^2+R^2\right)}\cos\omega t-\frac{CE_0\left(CL\omega^2-1\right)}{1-2CL\omega^2+C^2\omega^2\left(L^2\omega^2+R^2\right)}\sin\omega t$$

and

$$I_p(t)=\frac{E_0}{\sqrt{\left(L\omega-\frac{1}{C\omega}\right)^2+R^2}}\left[\frac{R\sin\omega t}{\sqrt{\left(L\omega-\frac{1}{C\omega}\right)^2+R^2}}-\frac{\left(L\omega-\frac{1}{C\omega}\right)\cos\omega t}{\sqrt{\left(L\omega-\frac{1}{C\omega}\right)^2+R^2}}\right].$$

20. (a) We solve $\{\theta''+4\theta'+13\theta=0,\theta(0)=\theta_0,\theta'(0)=0\}$. The characteristic equation is $r^2+4r+13=0\Rightarrow r_{1,2}=-2\pm3i$ so a general solution is $\theta(t)=e^{-2t}(c_1\cos3t+c_2\sin3t)$. Applying the initial conditions yields $\{\theta(0)=c_1=\theta_0,\theta'(0)=-2c_1+3c_2=0\}\Rightarrow c_1=\theta_0, c_2=2\theta_0/3$ so the solution to the initial-value problem is

$$\theta(t)=\theta_0e^{-2t}\left(\frac{2}{3}\sin3t+\cos3t\right).$$

```
sols =
  Map[
   DSolve[{θ''[t] + 4 θ'[t] + 13 θ[t] == 0,
      θ[0] == #, θ'[0] == 0}, θ[t], t] &,
   Table[i, {i, -4, 4}]];

grays = Table[GrayLevel[i],
   {i, 0, 0.7, 0.7/8}];

Plot[Evaluate[θ[t] /. sols], {t, 0, π/2},
  PlotRange -> All, PlotStyle -> grays]
```

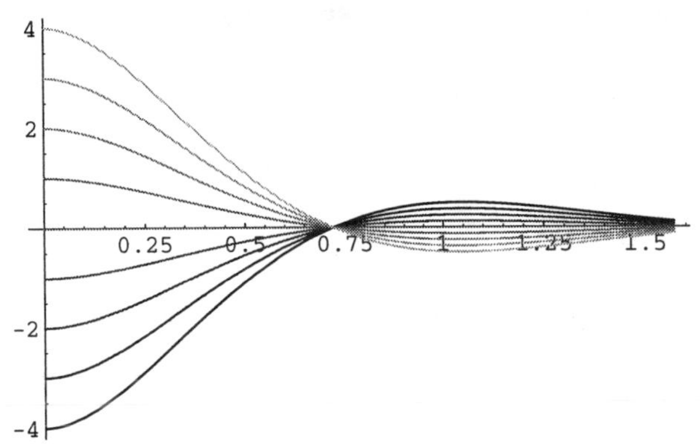

(b) We use the results of (a) and see that a general solution of the corresponding homogeneous equation is $\theta_h(t)=e^{-2t}(c_1\cos3t+c_2\sin3t)$ so we search for a particular solution of the nonhomogeneous equation of the form

$$\theta_p(t)=A\cos\pi t+B\sin\pi t\Rightarrow\theta_p'(t)=-A\pi\sin\pi t+B\pi\cos\pi t, \theta_p''(t)=-A\pi^2\cos\pi t-B\pi^2\sin\pi t.$$

Substituting into the nonhomogeneous equation and simplifying results in

$$\left(13A+4B\pi-A\pi^2\right)\cos\pi t+\left(13B-4A\pi-b\pi^2\right)\sin\pi t=\sin\pi t$$

so

$$\begin{cases}13A+4B\pi-A\pi^2=0\\ 13B-4A\pi-b\pi^2=1\end{cases}\Rightarrow A=-\frac{4\pi}{\pi^4-10\pi^2+169}, B=-\frac{\pi^2+13}{\pi^4-10\pi^2+169}$$

and thus

$$\begin{aligned}\theta(t)&=\theta_h(t)+\theta_p(t)\\&=e^{-2t}(c_1\cos3t+c_2\sin3t)-\frac{4\pi}{\pi^4-10\pi^2+169}\cos\pi t-\frac{\pi^2+13}{\pi^4-10\pi^2+169}\sin\pi t\end{aligned}$$

Application of the initial conditions results in

$$\begin{cases} c_1 - \dfrac{4\pi}{\pi^4 - 10\pi^2 + 169} = \theta_0 \\ -2c_1 + 3c_2 - \dfrac{\pi\left(\pi^2 - 13\right)}{\pi^4 - 10\pi^2 + 169} = 0 \end{cases} \Rightarrow c_1 = \theta_0 + \frac{4\pi}{\pi^4 - 10\pi^2 + 169},\ c_2 = \frac{5\pi - \pi^3 - 338\theta_0 + 20\pi^2\theta_0 - 2\pi^4\theta_0}{\pi^4 - 10\pi^2 + 169}$$

and thus

$$\theta(t) = \frac{1}{3\left(169 - 10\pi^2 + \pi^4\right)}\left[e^{-2t}\left(\pi^3 - 5\pi\right)\sin 3t + 12\pi e^{-2t}\cos 3t + (39 - 3\pi^2)\sin \pi t - 12\pi \cos \pi t\right].$$

```
sols =
  (DSolve[{θ''[t] + 4 θ'[t] + 13 θ[t] == Sin[π
      θ[0] == #1, θ'[0] == 0}, θ[t], t] &) /@
   Table[i, {i, -4, 4}];

grays = Table[GrayLevel[i],
   {i, 0, 0.7, 0.7/8}];

Plot[Evaluate[θ[t] /. sols], {t, 0, π/2},
 PlotRange → All, PlotStyle → grays]
```

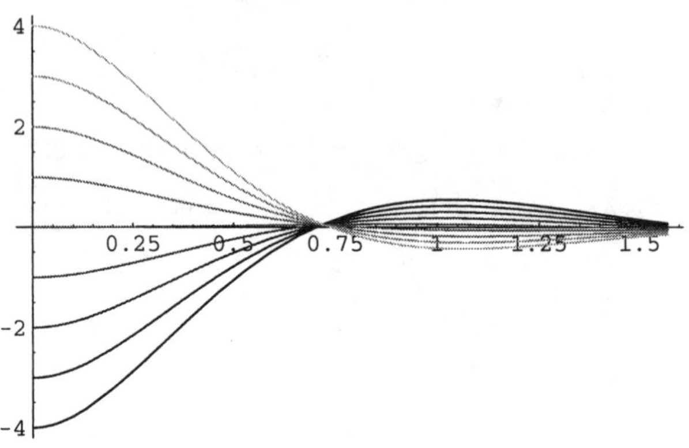

The motion stops in (a) because $\lim_{t\to\infty} \theta(t) = 0$; it eventually oscillates in (b).

EXERCISES 5.5

3. We consider the equation $\theta'' + 2\sqrt{7}\theta' + 16\theta = 0$ with characteristic roots $r_{1,2} = \frac{1}{2}\left(-2\sqrt{7} \pm 6i\right) = -\sqrt{7} \pm 3i$. Therefore, a general solution is $\theta(t) = e^{-t\sqrt{7}}\left(c_1 \cos 3t + c_2 \sin 3t\right)$. Applying the initial conditions where $\theta'(t) = -\sqrt{7}e^{-t\sqrt{7}}\left(c_1 \cos 3t + c_2 \sin 3t\right) + 3e^{-t\sqrt{7}}\left(-c_1 \sin 3t + c_2 \cos 3t\right)$, we find:

(a) $\theta(t) = \frac{\sqrt{7}}{60}e^{-t\sqrt{7}}\sin 3t + \frac{1}{20}e^{-t\sqrt{7}}\cos 3t$; (b) $\theta(t) = \left(\frac{\sqrt{7}}{60} + \frac{1}{3}\right)e^{-t\sqrt{7}}\sin 3t + \frac{1}{20}e^{-t\sqrt{7}}\cos 3t$;

(c) $\theta(t) = \left(\frac{\sqrt{7}}{60} - \frac{1}{3}\right)e^{-t\sqrt{7}}\sin 3t + \frac{1}{20}e^{-t\sqrt{7}}\cos 3t$

```
sols =
 Map[
   DSolve[
    {θ''[t] + 2 Sqrt[7] θ'[t] + 16 θ[t] == 0,
     θ[0] == #[[1]],
     θ'[0] == #[[2]]}, θ[t], t] &,
   {{1/20, 0}, {1/20, 1},
    {1/20, -1}}] // Simplify
```

$\{\{\{\theta[t] \to \frac{1}{60} e^{-\sqrt{7}\,t} (3\,\mathrm{Cos}[3\,t] + \sqrt{7}\,\mathrm{Sin}[3\,t])\}\},$

$\{\{\theta[t] \to \frac{1}{60} e^{-\sqrt{7}\,t} (3\,\mathrm{Cos}[3\,t] + (20 + \sqrt{7})\,\mathrm{Sin}[3\,t])\}\},$

$\{\{\theta[t] \to \frac{1}{60} e^{-\sqrt{7}\,t} (3\,\mathrm{Cos}[3\,t] + (-20 + \sqrt{7})\,\mathrm{Sin}[3\,t])\}\}\}$

```
Plot[Evaluate[θ[t] /. sols], {t, 0, Pi},
 PlotStyle -> {GrayLevel[0], GrayLevel[0.5],
   Dashing[{0.01}]}, PlotRange -> All]
```

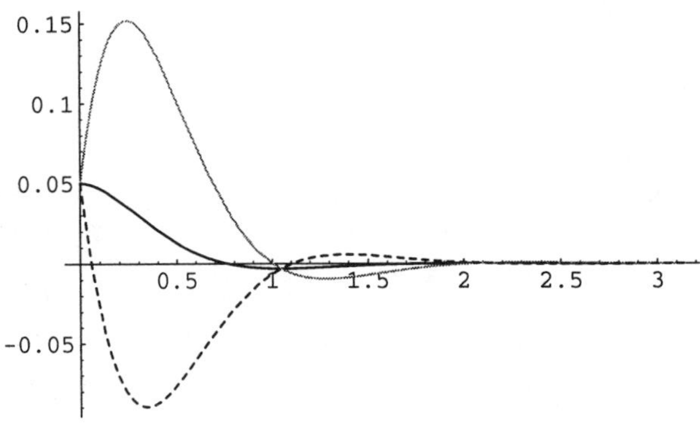

7. Because $\theta(t) = \theta_0 \cos\omega t + \frac{v_0}{\omega}\sin\omega t$ and $\theta(t) = A\cos(\omega t - \phi) = A\cos\omega t\cos\phi + A\sin\omega t\sin\phi$, we have $A\cos\phi = \theta_0$ and $A\sin\phi = \frac{v_0}{\omega}$. Therefore, $\cos\phi = \frac{\theta_0}{A}$ and $\sin\phi = \frac{v_0}{\omega A}$, so $\left(\frac{\theta_0}{A}\right)^2 + \left(\frac{v_0}{\omega A}\right)^2 = 1$. Solving for A, we find that $A = \sqrt{\theta_0^2 + \frac{v_0^2}{\omega^2}}$. The phase angle is determined by finding the angle ϕ that satisfies $\cos\phi = \frac{\theta_0}{A}$ and $\sin\phi = \frac{v_0}{\omega A}$.

11. $T = 2\pi\sqrt{8/32} \approx 3.14$ s

15. Max. Dis.: $\sqrt{\theta_0^2 + v_0^2/\omega^2}$ when $\cos(\omega t - \phi) = \pm 1 \Rightarrow \omega t - \phi = n\pi\ (n = 0, 1, 2, \ldots) \Rightarrow t = \frac{1}{\omega}(n\pi + \phi)$; $\omega = \sqrt{g/L}, \phi = \cos^{-1}\left(\theta_0 / \sqrt{\theta_0^2 + v_0^2/\omega^2}\right)$

19. (a) $x(t) = A\cos\omega t \Rightarrow \frac{dx}{dt} = -A\omega\sin\omega t$; Substitution into the nonlinear term yields

$$\varepsilon\left(x^2 - 1\right)\frac{dx}{dt} = \varepsilon\left(A^2\cos^2\omega t - 1\right)\left(-A\omega\sin\omega t\right) = -\varepsilon A^3\omega\cos^2\omega t\sin\omega t + \varepsilon A\omega\sin\omega t.$$

However, $\cos^2\omega t = 1 - \sin^2\omega t$, so we have

$$\varepsilon\left(x^2 - 1\right)\frac{dx}{dt} = -\varepsilon A^3\omega\left(1 - \sin^2\omega t\right)\sin\omega t + \varepsilon A\omega\sin\omega t$$

$$= -\varepsilon A^3\omega\sin\omega t + \varepsilon A^3\omega\sin^3\omega t + \varepsilon A\omega\sin\omega t$$

With the identity $\sin^3\omega t = -\frac{1}{4}\sin 3\omega t + \frac{3}{4}\sin\omega t$, $\varepsilon\left(x^2 - 1\right)\frac{dx}{dt} = \left(-\frac{1}{4}\varepsilon A^3\omega + \varepsilon A\omega\right)\sin\omega t - \frac{1}{4}\varepsilon A^3\omega\sin 3\omega t$.

(b) If $A=2$, the differential equation becomes $\frac{d^2x}{dt^2}+x=0$ which has a periodic general solution $x(t)=c_1\cos t+c_2\sin t$. (c) If $A\neq 2$, the differential equation involves a damping term, so solutions are not periodic.

21.

```
sols =
 Map[
   DSolve[{θ''[t] + θ[t] == 0, θ[0] == #[[1]],
      θ'[0] == #[[2]]}, θ[t], t] &,
   {{0, 2}, {2, 0}, {-2, 0}, {0, -1},
    {0, -2}, {1, -1}, {-1, 1}}] //
  Simplify

{{{θ[t] → 2 Sin[t]}},
 {{θ[t] → 2 Cos[t]}}, {{θ[t] → -2 Cos[t]}},
 {{θ[t] → -Sin[t]}}, {{θ[t] → -2 Sin[t]}},
 {{θ[t] → Cos[t] - Sin[t]}},
 {{θ[t] → -Cos[t] + Sin[t]}}}

grays = Table[GrayLevel[i],
   {i, 0, 0.7, .7 / 6}];
Plot[Evaluate[θ[t] /. sols], {t, 0, 3 Pi},
 PlotStyle -> grays]
```

```
sol:=proc(ics)
   local theta;
   rhs(dsolve({diff(theta(t),t$2)+
      theta(t)=0,
      theta(0)=ics[1],
      D(theta)(0)=ics[2]},theta(t)))
   end:
toplot:=map(sol,{[0,2],[2,0],
   [-2,0],[0,-1],[0,-2],
   [1,-1],[-1,1]});

plot(toplot,t=0..2*Pi);
```

22 and 23.

```
sols =
  Map[
    DSolve[{θ''[t] + 1 / 2 θ'[t] + θ[t] == 0,
        θ[0] == #[[1]],
        θ'[0] == #[[2]]}, θ[t], t] &,
     {{1, 0}, {-1, 0}, {0, 1}, {0, -1},
      {1, 1}, {1, -1}, {-1, 1}, {-1, -1},
      {1, 2}, {1, 3}, {-1, 2}, {-1, 3}}] //
   Simplify;
```

```
sol:=proc(ics)
   local theta;
   rhs(dsolve({diff(theta(t),t$2)+
      1/2*diff(theta(t),t)+
      theta(t)=0,theta(0)=ics[1],
      D(theta)(0)=ics[2]},
      theta(t)))
   end:
initconds:={[1,0],[-1,0],[0,1],
   [0,-1],[1,1],[1,-1],[-1,1],
   [-1,-1],[1,2],[1,3],[-1,2],
   [-1,3]};
toplot:=map(sol,initconds);
```

```
toshow =
  Table[Plot[Evaluate[θ[t] /. sols[[i]]],
    {t, 0, 3 Pi}, DisplayFunction -> Identity]
   {i, 1, 12}];
```

```
Show[GraphicsArray[
  Partition[toshow, 3]]]
```

```
plot(toplot,t=0..3*Pi);
```

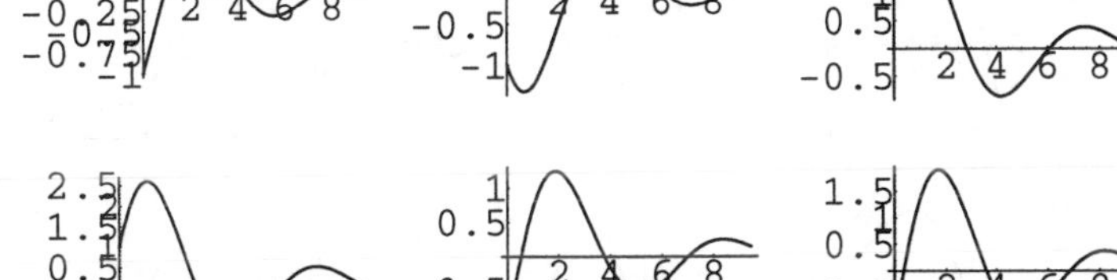

```
grays = Table[GrayLevel[i],
    {i, 0, 0.7, .7 / 11}];
Plot[Evaluate[θ[t] /. sols], {t, 0, 3 Pi},
  PlotStyle -> grays]
```

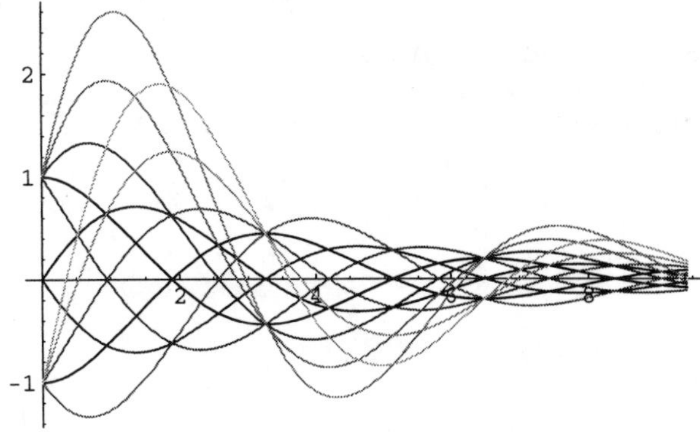

24.

```
sols =
  Map[
   NDSolve[{θ''[t] + 1 / 2 θ'[t] + Sin[θ[t]] =
      θ[0] == #[[1]],
      θ'[0] == #[[2]]}, θ[t], {t, 0, 3 Pi}] &
  {{1, 0}, {-1, 0}, {0, 1}, {0, -1}, {1, 1}
   {1, -1}, {-1, 1}, {-1, -1}, {1, 2},
   {1, 3}, {-1, 2}, {-1, 3}}];
```

```
toshow =
  Table[Plot[Evaluate[θ[t] /. sols[[i]]],
    {t, 0, 3 Pi}, DisplayFunction -> Identity]
   {i, 1, 12}];
```

```
Show[GraphicsArray[
  Partition[toshow, 3]]]
```

```
with(plots):
sol:=proc(ics)
   local theta,numsol;
   numsol:=dsolve({diff(theta(t),t$2)
      +1/2*diff(theta(t),t)+
      sin(theta(t))=0,
      theta(0)=ics[1],
      D(theta)(0)=ics[2]},
      theta(t),numeric);
   odeplot(numsol,[t,theta(t)],
      0..10);
   end:
for cond in initconds do
   sol(cond)
   od;
```

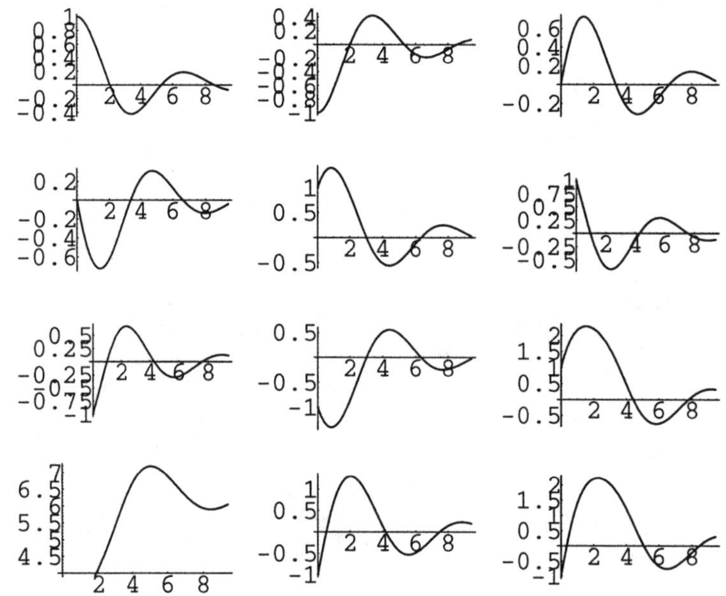

```
grays = Table[GrayLevel[i],
   {i, 0, 0.7, .7 / 11}];
Plot[Evaluate[θ[t] /. sols], {t, 0, 3 Pi},
 PlotStyle -> grays]
```

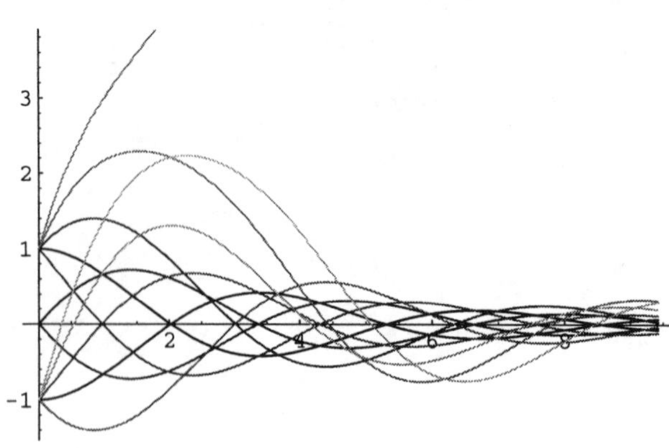

25. (Note that each solution is graphed individually with Maple V.)

```
sols =
  Map[
   NDSolve[
     {θ''[t] + 1 / 2 (θ[t] ^ 2 - 1) θ'[t] +
        θ[t] == 0, θ[0] == #[[1]],
       θ'[0] == #[[2]]}, θ[t],
     {t, 0, 3 Pi}] &,
   {{1, 0}, {-1, 0}, {0, 1}, {0, -1},
    {1, 1}, {1, -1}, {-1, 1}, {-1, -1},
    {1, 2}, {1, 3}, {-1, 2}, {-1, 3}}];

toshow =
  Table[Plot[Evaluate[θ[t] /. sols[[i]]],
    {t, 0, 3 Pi}, DisplayFunction -> Identity]
   {i, 1, 12}];

Show[GraphicsArray[
  Partition[toshow, 3]]]
```

```
sol:=proc(ics)
   local theta,numsol;
   numsol:=dsolve({diff(theta(t),t$2)
      +1/2*(theta(t)^2-1)*
         diff(theta(t),t)+
      theta(t)=0,
      theta(0)=ics[1],
      D(theta)(0)=ics[2]},
      theta(t),numeric);
   odeplot(numsol,[t,theta(t)],
      0..10);
   end:
for cond in initconds do
   sol(cond)
   od;
```

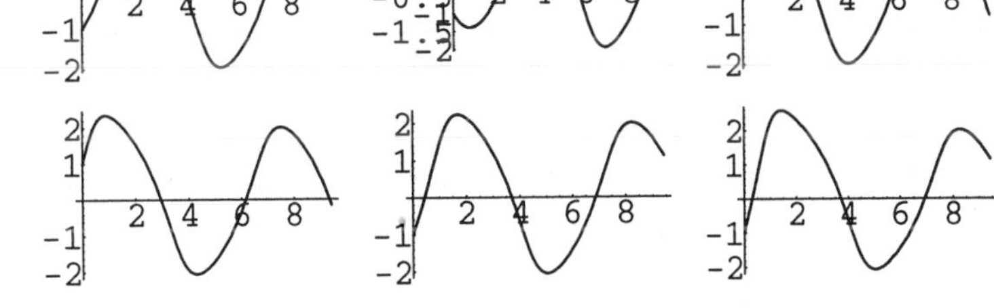

```
grays = Table[GrayLevel[i],
   {i, 0, 0.7, .7 / 11}];
Plot[Evaluate[θ[t] /. sols], {t, 0, 3 Pi},
  PlotStyle -> grays]
```

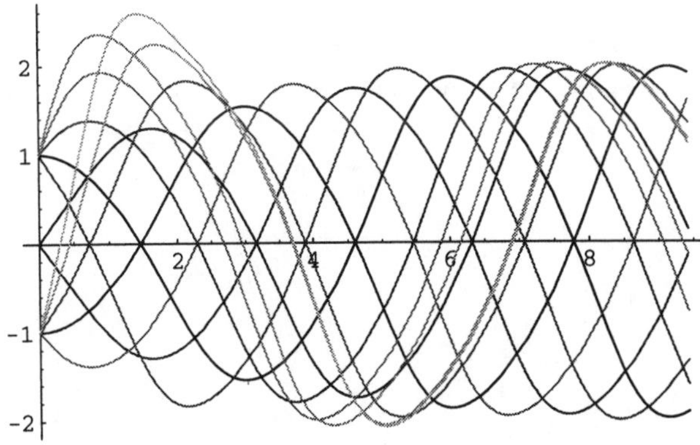

CHAPTER 5 REVIEW EXERCISES

3. We solve $\{5x''+20x'+65x=0,\ x(0)=0, x'(0)=-1\}$. The characteristic equation is

$$5r^2+20r+65=0 \Rightarrow r_{1,2}=\frac{-20\pm\sqrt{20^2-4\cdot5\cdot65}}{2\cdot5}=\frac{-20\pm\sqrt{-900}}{2\cdot5}=-2\pm3i$$

so a general solution of the equation is $x=e^{-2t}(c_1\cos3t+c_2\sin3t)$. Application of the initial conditions yields

$$\begin{cases} x(0)=c_1=0 \\ x'(0)=-2c_1+3c_2=-1 \end{cases} \Rightarrow c_1=0, c_2=-1/3$$

so

$$x(t)=-\frac{1}{3}e^{-2t}\sin3t,\ \lim_{t\to\infty}x(t)=0;\ \text{quasiper.}=\frac{2\pi}{3};\ \text{max. dis.}=\left|x\left(\frac{1}{3}\tan^{-1}\frac{3}{2}\right)\right|\approx0.144;\ t=\frac{\pi}{3}.$$

```
Clear[x]
sol =
  DSolve[{5 x''[t] + 20 x'[t] + 65 x[t] == 0,
    x[0] == 0, x'[0] == -1}, x[t], t]

{{x[t] → -1/3 e^(-2 t) Sin[3 t]}}

Plot[x[t] /. sol, {t, 0, Pi}]
```

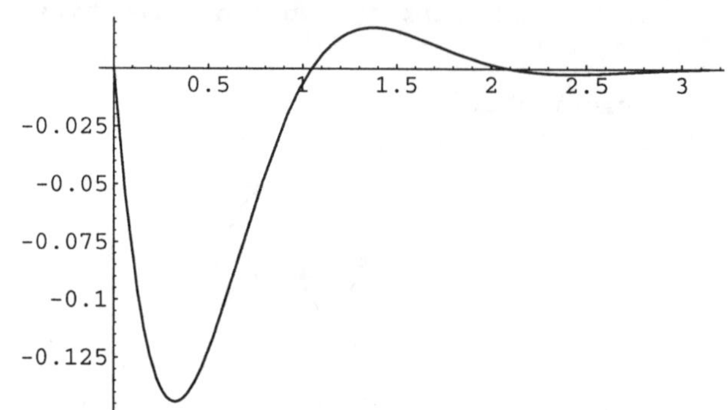

7. We solve the initial-value problem

$$\begin{cases} 4x''+16x=4\cos t \\ x(0)=x'(0)=0 \end{cases} \Leftrightarrow \begin{cases} x''+4x=\cos t \\ x(0)=x'(0)=0 \end{cases}.$$

A general solution of the corresponding homogeneous equation $x''+4x=0$ is $x_h(t)=c_1\cos2t+c_2\sin2t$. Using the method of undetermined coefficients, we assume that there is a particular solution to the nonhomogeneous equation of the form $x_p(t)=A\cos t+B\sin t$. Substituting x_p into the nonhomogeneous equation yields

$$3A\cos t+3B\sin t=\cos t$$

so $A=1/3$ and $B=0$. Thus, $x(t)=c_1\cos2t+c_2\sin2t+\frac{1}{3}\cos t$. Application ofthe initial conditions results in the system of equations

$$\begin{cases} 1/3+c_1=0 \\ 2c_2=0 \end{cases} \Rightarrow c_1=-1/3, c_2=0,$$

so $x(t)=\frac{1}{3}\cos t-\frac{1}{3}\cos2t$. The envelope functions are $\pm\frac{2}{3}\sin\frac{t}{2}$, which is confirmed by the graphs of $x(t)$ together and the envelope functions (in gray).

```
sol =
 DSolve[{x''[t] + 4 x[t] == Cos[t], x
   x'[0] == 0}, x[t], t]
```

$\{\{x[t] \to \frac{1}{3}(\text{Cos}[t] - \text{Cos}[2\,t])\}\}$

```
Plot[{x[t] /. sol, 2 / 3 Sin[t / 2],
  -2 / 3 Sin[t / 2]}, {t, 0, 8 Pi},
  PlotStyle -> {GrayLevel[0], GrayLev
   GrayLevel[0.5]}]
```

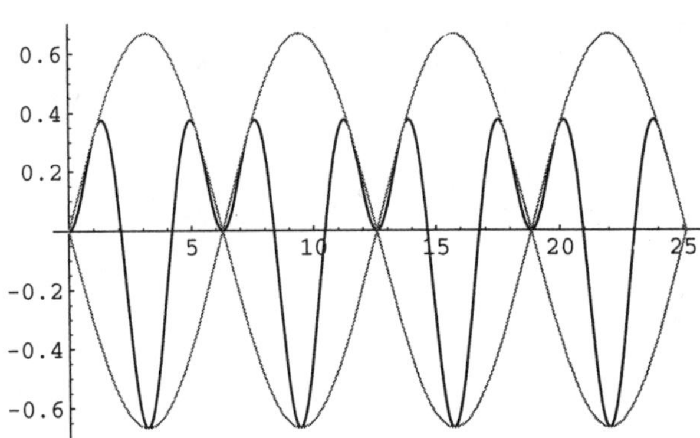

11. We solve $\{Q'' + 10^4 Q = 220, Q(0) = 0, Q'(0) = 0\}$. The characteristic equation for the corresponding homogeneneous equation is $r^2 + 10^4 = 0 \Rightarrow r_{1,2} = \pm 100i$ so a general solution of the corresponding homogeneous equation is $Q_h = c_1 \cos 100t + c_2 \sin 100t$. Using the method of undetermined coefficients, we find a particular solution to the nonhomogeneous equation of the form $Q_p = A$. Substitution into the nonhomogeneous equation gives us

$$0 + 10^4 A = 220 \Rightarrow A = 220/10^4 = 11/500$$

so $Q_p = 11/500$ and $Q = c_1 \cos 100t + c_2 \sin 100t + 11/500$. Application of the initial conditions results in

$$\begin{cases} Q(0) = 11/500 + c_1 = 0 \\ Q'(0) = 100c_2 = 0 \end{cases} \Rightarrow c_1 = -11/500, c_2 = 0$$

so $Q(t) = \frac{11}{500} - \frac{11}{500}\cos 100t$; $I(t) = \frac{11}{5}\sin 100t$; $\lim_{t\to\infty} Q(t)$ and $\lim_{t\to\infty} I(t)$ *DNE*.

```
Clear[q]
sol =
 DSolve[{q''[t] + 10 ^ 4 q[t] == 220,
   q[0] == 0, q'[0] == 0}, q[t], t]
```

$\{\{q[t] \to \frac{1}{500}(11 - 11\,\text{Cos}[100\,t])\}\}$

```
Plot[q[t] /. sol, {t, 0, 1}]
```

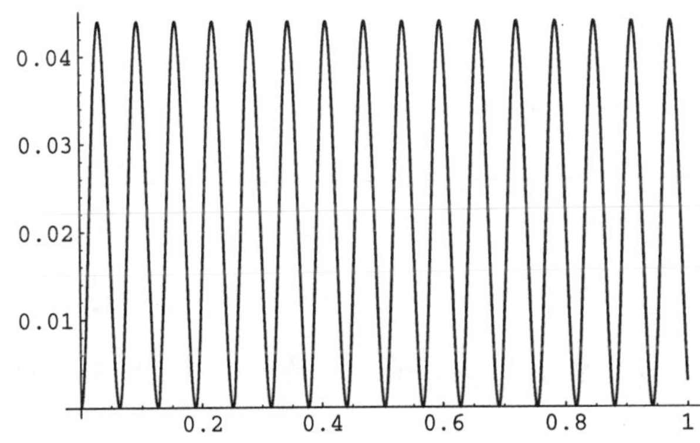

15. We solve $\{s^{(4)} = x(10 - x), s(0) = 0, s'(0) = 0, s''(10) = 0, s'''(10) = 0\}$ with solution

$$s(x) = \frac{1250}{3}x^2 - \frac{250}{9}x^3 + \frac{1}{12}x^5 - \frac{1}{360}x^6$$

```
sol =
  DSolve[{s''''[x] == x (10 - x), s[0] == 0,
    s'[0] == 0, s''[10] == 0, s'''[10] == 0},
   s[x], x]
```

$$\left\{\left\{s[x] \to \frac{1250\,x^2}{3} - \frac{250\,x^3}{9} + \frac{x^5}{12} - \frac{x^6}{360}\right\}\right\}$$

```
Plot[s[x] /. sol, {x, 0, 10}]
```

19. We solve $\left\{\frac{1}{2}\theta'' + 8\theta' + 32\theta = 0, \theta(0) = 1, \theta'(0) = 0\right\}$. The characteristic equation is

$$\frac{1}{2}r^2 + 8r + 32 = 0 \Leftrightarrow r^2 + 16r + 64 = 0 \Rightarrow (r+8)^2 = 0 \Rightarrow r_{1,2} = -8$$

so a general solution is $\theta(t) = c_1 e^{-8t} + c_2 t e^{-8t}$. Application of the initial conditions gives us

$$\begin{cases} \theta(0) = c_1 = 1 \\ \theta'(0) = -8c_1 + c_2 = 0 \end{cases} \Rightarrow c_1 = 1, c_2 = 8$$

so the solution to the initial-value problem is $\theta(t) = e^{-8t} + 8te^{-8t}$; The motion is not periodic.

```
sol =
  DSolve[{θ''[t] + 16 θ'[t] + 64 θ[t] == 0,
    θ[0] == 1, θ'[0] == 0}, θ[t], t]
```

$$\{\{\theta[t] \to e^{-8\,t}\,(1 + 8\,t)\}\}$$

```
Plot[θ[t] /. sol, {t, 0, 1}]
```

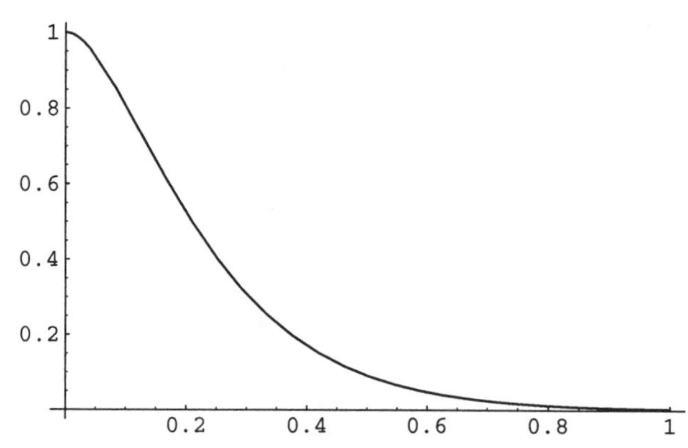

Differential Equations at Work: B. Soft Springs

```
numsolsa1 =
  Map[
   NDSolve[
     {x''[t] + 0.2 x'[t] + 10 x[t] - 0.2 x[t
       -9.8, x[0] == #, x'[0] == 0}, x[t],
     {t, 0, 10}] &, Table[i, {i, -3, 3}]];
```

```
grays = Table[GrayLevel[i],
   {i, 0, 0.7, 0.7 / 6}];
```

```
Plot[Evaluate[x[t] /. numsolsa1], {t,
 PlotStyle -> grays]
```

Alternatively, we can rewrite this second-order equation as a system.

```
<< Graphics`PlotField`
pvfa = PlotVectorField[{y,
    -0.2 y - 10 x + 0.2 x^3 - 9.8},
    {x, -20, 20}, {y, -40, 40},
    Axes -> Automatic,
    AxesOrigin -> {0, 0},
    ScaleFunction -> (0.5 &),
    PlotPoints -> 20];
```

```
Clear[x, y, t, solgraph]
solgraph[xy_, opts___] :=
 Module[{numsol},
  numsol =
   NDSolve[{x'[t] == y[t],
     y'[t] == -0.2 y[t] - 10 x[t] +
       0.2 x[t]^3 - 9.8, x[0] == xy[[1]],
     y[0] == xy[[2]]}, {x[t], y[t]},
    {t, 0, 10}];
  ParametricPlot[
   Evaluate[{x[t], y[t]} /. numsol],
   {t, 0, 5}, opts,
   PlotStyle ->
    {{GrayLevel[.5],
      Thickness[0.0075]}},
   PlotRange -> {{-20, 20}, {-40, 40}},
   AspectRatio -> 1,
   DisplayFunction -> Identity]]
```

```
l1 = Table[{i, 40}, {i, -20, 20, 2}];
l2 = Table[{i, -40}, {i, -20, 20, 2}];
l3 = Table[{i, 0}, {i, -20, 20, 2}];
l4 = Table[{0, i}, {i, -40, 40, 4}];
ll = Union[l1, l2, l3, l4];
toshowa = Map[solgraph, ll];
```

```
Show[{toshowa, pvfa},
 DisplayFunction -> $DisplayFunction,
 AspectRatio -> 2]
```

Differential Equations at Work: C. Soft Springs

```
numsolsa1 =
  Map[
   NDSolve[
     {x''[t] + 0.3 x[t] + 0.04 x[t]^3 == 0,
      x[0] == #, x'[0] == 0}, x[t],
     {t, 0, 12}] &, Table[i, {i, -5, 5}]];
```

```
grays = Table[GrayLevel[i],
    {i, 0, 0.7, 0.7 / 10}];
Plot[Evaluate[x[t] /. numsolsa1],
 {t, 0, 10}, PlotStyle -> grays]
```

```
numsolsb1 =
  Map[
   NDSolve[
     {x''[t] + 0.3 x[t] + 0.04 x[t] ^3 == 0,
      x[0] == 0, x'[0] == #}, x[t],
     {t, 0, 10}] &, Table[i, {i, -5, 5}]];
```

```
Plot[Evaluate[x[t] /. numsolsb1],
 {t, 0, 10}, PlotStyle -> grays]
```

Differential Equations at Work: E. Bode Plots

2.

```
Clear[x]
sol =
 DSolve[
   {x''[t] + 2 x'[t] + 4 x[t] == Sin[2 t],
    x[0] == 1 / 2, x'[0] == 1}, x[t], t] //
  Simplify
```

$$\left\{\left\{x[t] \to \frac{1}{12} e^{-t} \left(-3 e^{t} \text{Cos}[2 t] + 9 \text{Cos}[\sqrt{3} t] + 7 \sqrt{3} \text{Sin}[\sqrt{3} t]\right)\right\}\right\}$$

```
Plot[{x[t] /. sol, Sin[2 t]}, {t, 0, 3]
 PlotStyle ->
  {GrayLevel[0], GrayLevel[0.5]}]
```

3

```
k = 2; c = 1;
```

$$\text{m[w\_]} = \frac{1}{\sqrt{(k^2 - w^2) + 4 c^2 w^2}};$$

```
<< Graphics`Graphics`
LogLogPlot[m[w] ^20, {w, 0.1, 10}]
```

```
N[m[2]]
```

0.25

The branch of $\cot^{-1} x$ used by Mathematics is not continuous at $x = 0$ as seen the following graph.

```
Plot[ArcCot[x], {x, -10, 10}]
```

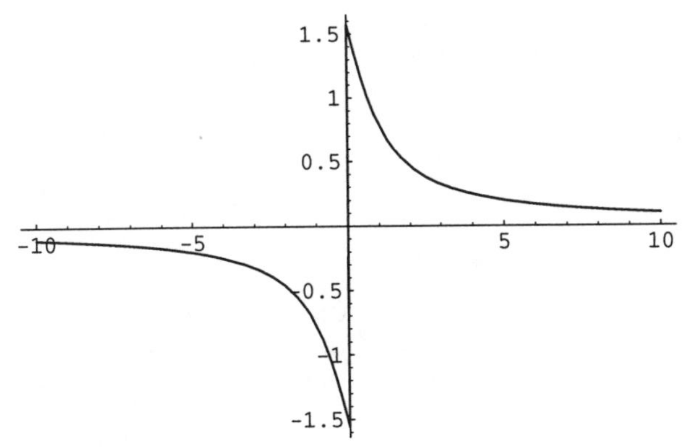

However, we can construct a function continuous at $x = 0$ as we do in `newarccot`.

```
Clear[newarccot]
newarccot[x_] := ArcCot[x] /; x >= 0;
newarccot[x_] := ArcCot[x] + Pi /; x < 0;
Plot[newarccot[x], {x, -10, 10}]
```

6

Systems of Differential Equations

EXERCISES 6.1

3. By integrating, $x(t)=C_1$. Solving the characteristic equation, $r+2=0$, so $r=-2$ and $y(t)=C_2e^{-2t}$; $\left\{x(t)=C_1, y(t)=C_2e^{-2t}\right\}$

9. $x''=-x'-2y'=-x'-2(x+y)=-x'-2x-2[-1/2(x'+x)]=-x$, $x''+x=0$, Solving $r^2+1=0$, we have $x(t)=C_1\cos t+C_2\sin t$; Then, $y=-1/2(x'+x)$, so $y(t)=-\frac{1}{2}(C_1+C_2)\cos t+\frac{1}{2}(C_1-C_2)\sin t$; $\left\{x(t)=C_1\cos t+C_2\sin t, y(t)=-\frac{1}{2}(C_1+C_2)\cos t+\frac{1}{2}(C_1-C_2)\sin t\right\}$

15. $x'=y$, $x''=y'=t\sin t-16x$, $\begin{cases} x'=y \\ y'=-16x+t\sin t \end{cases}$

19. (a) Let $x'=y$ so that $x''=y'=-\mu\left[\frac{1}{3}(x')^2-1\right]x'-x$. The system is $\begin{cases} x'=y \\ y'=-\mu\left[\frac{1}{3}y^2-1\right]y-x \end{cases}$.

(b) $x'''+\mu\left[\frac{2}{3}x'\cdot x''\right]x'+\mu\left[\frac{1}{3}(x')^2-1\right]x''+x'=x'''+\mu(x')^2x''-\mu x''+x'=z''+\mu\left(z^2-1\right)z'+z=0$.

25. In operator notation the system is

$$\begin{cases} \left(D^2+1\right)x+\left(D^2+1\right)y=0 \\ 2(D-1)x+\left(D^2-1\right)y=0 \end{cases}.$$

We eliminate y by first applying D^2-1 to the first equation and $-\left(D^2+1\right)$ to the second

$$\begin{cases} \left(D^2-1\right)\left(D^2+1\right)x+\left(D^2-1\right)\left(D^2+1\right)y=0 \\ -2\left(D^2+1\right)(D-1)x-\left(D^2+1\right)\left(D^2-1\right)y=0 \end{cases}$$

and then adding to obtain

$$\left[\left(D^2-1\right)\left(D^2+1\right)-2\left(D^2+1\right)(D-1)\right]x=0$$

$$\left(D^2+1\right)\left[\left(D^2-1\right)-2(D-1)\right]x=0$$

$$\left(D^2+1\right)(D-1)^2x=0,$$

which has general solution $x=c_1\cos t+c_2\sin t+c_3e^t+c_4te^t$. Substituting x into the second equation (we use the second equation to find y because it is simpler than the first) gives us the nonhomogeneous equation

$$y''+2x'-2x-y=0$$

$$y''-y=2x-2x'$$

$$y''-y=-2c_4e^t+2(c_1-c_2)\cos t+2(c_1+c_2)\sin t.$$

A general solution of the homogeneous equation $y''-y=0$ is $y_h(t)=c_5+c_6e^t$ and a particular solution of the nonhomogeneous equation $y''-y=-2c_4e^t+2(c_1-c_2)\cos t+2(c_1+c_2)\sin t$ has the form $y_p(t)=A\,t\,e^t+B\cos t+C\sin t$. Substituting $y_p(t)$ into the nonhomogeneous equation leads to

$$2Ae^t-2B\cos t-2C\sin t=-2c_4e^t+2(c_1-c_2)\cos t+2(c_1+c_2)\sin t$$

and equating coefficients results in $A=-c_4$, $B=c_2-c_1$, and $C=-(c_1+c_2)$ so

$$y=c_5+c_6e^t-c_4t\,e^t+(c_2-c_1)\cos t-(c_1+c_2)\sin t.$$

Because y must also satisfy the first equation, we have

$$x''+y''+x+y=0$$

$$c_5+2(c_3+c_6)e^t=0$$

so $c_5=0$ and $c_6=-c_3$. Thus,

$$y=-c_3e^t-c_4t\,e^t+(c_2-c_1)\cos t-(c_1+c_2)\sin t.$$

Another form of the solution is

$$\begin{cases} x(t)=(c_2+c_1)\sin t+c_1\cos t-c_3e^t-c_4te^t \\ y(t)=c_2\cos t+(-c_2-2c_1)\sin t+c_3e^t+c_4te^t. \end{cases}$$

29. In operator notation, the system is

$$\begin{cases} (D-3)x+2y=0 \\ -2x+(D+1)y=10 \end{cases} \Leftrightarrow \begin{cases} (D-3)(D+1)x+2(D+1)y=0 \\ 4x-2(D+1)y=-20 \end{cases}$$

Adding, we have $\left(D^2-2D+1\right)x=-20$ with general solution $x(t)=c_1e^t+c_2te^t-20$. Solving the first differential equation for y, we have $y(t)=-\frac{1}{2}(x'(t)-3x(t))=c_1e^t-\frac{1}{2}c_2e^t+c_2te^t-30$. Application of the initial conditions yields $\left\{x(0)=c_1-20=0, y(0)=c_1-\frac{1}{2}c_2-30=0\right\}$ with solution $c_1=20$ and $c_2=-20$. Therefore, $\begin{cases} x(t)=20e^t-20te^t-20 \\ y(t)=30e^t-20te^t-30 \end{cases}$.

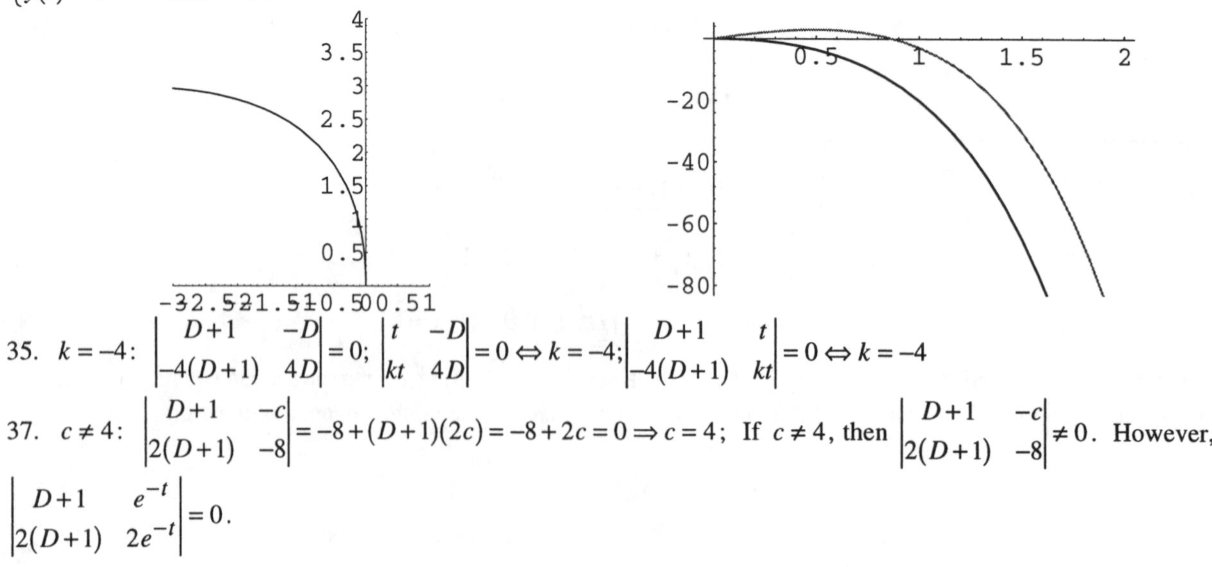

35. $k=-4$: $\begin{vmatrix} D+1 & -D \\ -4(D+1) & 4D \end{vmatrix}=0$; $\begin{vmatrix} t & -D \\ kt & 4D \end{vmatrix}=0\Leftrightarrow k=-4$; $\begin{vmatrix} D+1 & t \\ -4(D+1) & kt \end{vmatrix}=0\Leftrightarrow k=-4$

37. $c\neq 4$: $\begin{vmatrix} D+1 & -c \\ 2(D+1) & -8 \end{vmatrix}=-8+(D+1)(2c)=-8+2c=0\Rightarrow c=4$; If $c\neq 4$, then $\begin{vmatrix} D+1 & -c \\ 2(D+1) & -8 \end{vmatrix}\neq 0$. However, $\begin{vmatrix} D+1 & e^{-t} \\ 2(D+1) & 2e^{-t} \end{vmatrix}=0$.

41.

We use the command `PlotVectorField` *to graph the direction field associated with the system*

$$\begin{cases} \frac{dx}{dt} = f(x,y) \\ \frac{dy}{dt} = g(x,y) \end{cases}$$

```
<<Graphics`PlotField`
pvf1=PlotVectorField[
    {2x-x y,-3y+x y},{x,0,6},{y,0,6},
    ScaleFunction->(1&),
    PlotPoints->30,
    DefaultColor->GrayLevel[.4]]
num1=NDSolve[{x'[t]==2x[t]-x[t]y[t],
    y'[t]==-3y[t]+x[t]y[t],x[0]==2,
    y[0]==3},{x[t],y[t]},{t,0,6}]
pp1=ParametricPlot[{x[t],y[t]} /.
    num1,{t,0,6},
    PlotStyle->GrayLevel[0],
    Compiled->False]
num2=NDSolve[{x'[t]==2x[t]-x[t]y[t],
    y'[t]==-3y[t]+x[t]y[t],x[0]==3,
    y[0]==2},{x[t],y[t]},{t,0,6}]
pp2=ParametricPlot[{x[t],y[t]} /.
    num2,{t,0,6},
    PlotStyle->GrayLevel[0],
    Compiled->False]
Show[pvf1,pp1,pp2,Axes->Automatic,
    AxesOrigin->{0,0},
    PlotRange->{{0,6},{0,6}},
    AspectRatio->1]
```

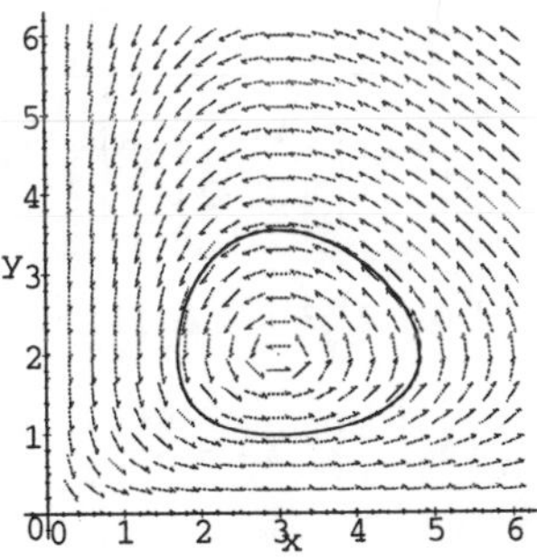

The solution that satisfies (b) is a constant solution; the solution that satisfies (a) is periodic.

EXERCISES 6.2

11.

$$\mathbf{AB} = \begin{pmatrix} 1 & 2 \\ 3 & 5 \end{pmatrix}\begin{pmatrix} 1 & -1 \\ -1 & -5 \end{pmatrix} = \begin{pmatrix} (1)(1)+(2)(-1) & (1)(-1)+(2)(-5) \\ (3)(1)+(5)(-1) & (3)(-1)+(5)(-5) \end{pmatrix} = \begin{pmatrix} -1 & -11 \\ -2 & -28 \end{pmatrix};$$

$$\mathbf{BA} = \begin{pmatrix} 1 & -1 \\ -1 & -5 \end{pmatrix}\begin{pmatrix} 1 & 2 \\ 3 & 5 \end{pmatrix} = \begin{pmatrix} (1)(1)+(-1)(3) & (1)(2)+(-1)(5) \\ (-1)(1)+(-5)(3) & (-1)(2)+(-5)(5) \end{pmatrix} = \begin{pmatrix} -2 & -3 \\ -16 & -27 \end{pmatrix}.$$

and

17. $$|\mathbf{A}| = \begin{vmatrix} 3 & -2 & 0 \\ -1 & 3 & 0 \\ 3 & -2 & 0 \end{vmatrix} = 3\begin{vmatrix} 3 & 0 \\ -2 & 0 \end{vmatrix} - (-2)\begin{vmatrix} -1 & 0 \\ 3 & 0 \end{vmatrix} + 0\begin{vmatrix} -1 & 3 \\ 3 & -2 \end{vmatrix} = 0$$

19. Compute the determinant by expanding along the fourth row (it contains the most zeros):

$$|\mathbf{A}| = \begin{vmatrix} 2 & 0 & -2 & -1 \\ -3 & -2 & 0 & 1 \\ -3 & -1 & -3 & -3 \\ 1 & 0 & 0 & 1 \end{vmatrix} = 1\begin{vmatrix} 0 & -2 & -1 \\ -2 & 0 & 1 \\ -1 & -3 & -3 \end{vmatrix} + 1\begin{vmatrix} 2 & 0 & -2 \\ -3 & -2 & 0 \\ -3 & -1 & -3 \end{vmatrix}$$

$$= \left(0\begin{vmatrix} 0 & 1 \\ -3 & -3 \end{vmatrix} - (-2)\begin{vmatrix} 2 & 1 \\ -1 & -3 \end{vmatrix} - 1\begin{vmatrix} -2 & 0 \\ -1 & -3 \end{vmatrix}\right) + \left(2\begin{vmatrix} -2 & 0 \\ -1 & -3 \end{vmatrix} - 0\begin{vmatrix} -3 & 0 \\ -3 & -3 \end{vmatrix} - 2\begin{vmatrix} -3 & -2 \\ -3 & -1 \end{vmatrix}\right)$$

$$= 10.$$

21. $$\mathbf{A} = \begin{pmatrix} 0 & -1 \\ 2 & 2 \end{pmatrix} \Rightarrow \mathbf{A}^c = \begin{pmatrix} 2 & -2 \\ 1 & 0 \end{pmatrix} \Rightarrow \mathbf{A}^a = \left(\mathbf{A}^c\right)^T = \begin{pmatrix} 2 & 1 \\ -2 & 0 \end{pmatrix} \Rightarrow \mathbf{A}^{-1} = \frac{1}{|\mathbf{A}|}\mathbf{A}^a = \frac{1}{2}\begin{pmatrix} 2 & 1 \\ -2 & 0 \end{pmatrix} = \begin{pmatrix} 1 & 1/2 \\ -1 & 0 \end{pmatrix}$$

23.

$$\mathbf{A}^{-1} = \frac{1}{|\mathbf{A}|}\mathbf{A}^a = \frac{1}{-2}\begin{pmatrix} \begin{vmatrix} 2 & 2 \\ 0 & -2 \end{vmatrix} & -\begin{vmatrix} 3 & 2 \\ 0 & -2 \end{vmatrix} & \begin{vmatrix} 3 & 2 \\ 0 & 0 \end{vmatrix} \\ -\begin{vmatrix} -1 & -2 \\ 0 & -2 \end{vmatrix} & \begin{vmatrix} -1 & -2 \\ 0 & -2 \end{vmatrix} & -\begin{vmatrix} -1 & -1 \\ 0 & 0 \end{vmatrix} \\ \begin{vmatrix} -1 & -2 \\ 2 & 2 \end{vmatrix} & -\begin{vmatrix} -1 & -2 \\ 3 & 2 \end{vmatrix} & \begin{vmatrix} -1 & -1 \\ 3 & 2 \end{vmatrix} \end{pmatrix}^T = \begin{pmatrix} 2 & 1 & -1 \\ -3 & -1 & 2 \\ 0 & 0 & -\frac{1}{2} \end{pmatrix}$$

25. $$\begin{vmatrix} -6-\lambda & 1 \\ -2 & -3-\lambda \end{vmatrix} = \lambda^2 + 9\lambda + 20 = (\lambda+4)(\lambda+5) = 0 \Rightarrow \lambda_1 = -4, \lambda_2 = -5;$$

$$\lambda_1 = -4: \begin{pmatrix} -2 & 1 \\ -2 & 1 \end{pmatrix}\begin{pmatrix} x_1 \\ y_1 \end{pmatrix} = \begin{pmatrix} 0 \\ 0 \end{pmatrix} \Rightarrow y_1 = 2x_1; \mathbf{v}_1 = x_1\begin{pmatrix} 1 \\ 2 \end{pmatrix}; \quad \lambda_2 = -5: \begin{pmatrix} -1 & 1 \\ -2 & 2 \end{pmatrix}\begin{pmatrix} x_2 \\ y_2 \end{pmatrix} = \begin{pmatrix} 0 \\ 0 \end{pmatrix} \Rightarrow y_2 = x_2; \mathbf{v}_2 = x_2\begin{pmatrix} 1 \\ 1 \end{pmatrix}$$

29. $$\begin{vmatrix} -7-\lambda & 4 \\ -1 & -3-\lambda \end{vmatrix} = \lambda^2 + 10\lambda + 25 = (\lambda+5)^2 = 0 \Rightarrow \lambda_1 = \lambda_2 = -5;$$

$$\begin{pmatrix} -2 & 4 \\ -1 & 2 \end{pmatrix}\begin{pmatrix} x_1 \\ y_1 \end{pmatrix} = \begin{pmatrix} 0 \\ 0 \end{pmatrix} \Rightarrow x_1 = 2y_1; \mathbf{v} = x_1\begin{pmatrix} 2 \\ 1 \end{pmatrix}$$

33. $$\begin{vmatrix} -41-\lambda & -38 & 18 \\ 48 & 44-\lambda & -21 \\ 8 & 6 & -3-\lambda \end{vmatrix} = -\lambda^3 + 7\lambda + 6 = (3-\lambda)(\lambda+1)(\lambda+2) = 0; \ \lambda_1 = 3, \lambda_2 = -1, \lambda_3 = -2; \ \lambda_1 = 3:$$

$$\begin{pmatrix}-44 & -38 & 18\\ 48 & 41 & -21\\ 8 & 6 & -6\end{pmatrix}\begin{pmatrix}x_1\\ y_1\\ z_1\end{pmatrix}=\begin{pmatrix}0\\0\\0\end{pmatrix}\Rightarrow\begin{pmatrix}1 & 0 & -3\\ 0 & 1 & 3\\ 0 & 0 & 0\end{pmatrix}\begin{pmatrix}x_1\\ y_1\\ z_1\end{pmatrix}=\begin{pmatrix}0\\0\\0\end{pmatrix},\quad x_1=3z_1, y_1=-3z_1,\quad \mathbf{v}_1=z_1\begin{pmatrix}3\\-3\\1\end{pmatrix};\quad \lambda_2=-1:$$

$$\begin{pmatrix}-40 & -38 & 18\\ 48 & 45 & -21\\ 8 & 6 & -2\end{pmatrix}\begin{pmatrix}x_2\\ y_2\\ z_2\end{pmatrix}=\begin{pmatrix}0\\0\\0\end{pmatrix}\Rightarrow\begin{pmatrix}1 & 0 & 1/2\\ 0 & 1 & -1\\ 0 & 0 & 0\end{pmatrix}\begin{pmatrix}x_2\\ y_2\\ z_2\end{pmatrix}=\begin{pmatrix}0\\0\\0\end{pmatrix},\quad x_2=-\tfrac{1}{2}z_2, y_2=z_2,\quad \mathbf{v}_2=z_2\begin{pmatrix}-1/2\\1\\1\end{pmatrix};\quad \lambda_3=-2:$$

$$\begin{pmatrix}-39 & -38 & 18\\ 48 & 46 & -21\\ 8 & 6 & -1\end{pmatrix}\begin{pmatrix}x_3\\ y_3\\ z_3\end{pmatrix}=\begin{pmatrix}0\\0\\0\end{pmatrix}\Rightarrow\begin{pmatrix}1 & 0 & 1\\ 0 & 1 & 3\\ 0 & 0 & 0\end{pmatrix}\begin{pmatrix}x_3\\ y_3\\ z_3\end{pmatrix}=\begin{pmatrix}0\\0\\0\end{pmatrix},\quad x_3=-z_3, y_3=\tfrac{3}{2}z_3,\quad \mathbf{v}_3=z_3\begin{pmatrix}-1\\3/2\\1\end{pmatrix}$$

37. $u=t, dv=\cos t dt \Rightarrow du=dt, v=\sin t;\quad \int t\cos t dt = t\sin t-\int \sin t dt = t\sin t+\cos t;$

$u=t, dv=\sin t dt \Rightarrow du=dt, v=-\cos t,\quad \int t\sin t dt=-t\cos t+\int \cos t dt=\sin t-t\cos t$

39. $\begin{pmatrix}4e^{4t}\\ -3\sin 3t\\ 3\cos 3t\end{pmatrix};\begin{pmatrix}\frac{1}{4}e^{4t}+c_1\\ \frac{1}{3}\sin 3t+c_2\\ -\frac{1}{3}\cos 3t+c_3\end{pmatrix}$

41. $\mathbf{A}^c=\begin{pmatrix}d & -c\\ -b & a\end{pmatrix}; \mathbf{A}^{-1}=\frac{1}{|\mathbf{A}|}\mathbf{A}^c=\frac{1}{ad-bc}\begin{pmatrix}d & -c\\ -b & a\end{pmatrix}$

43.

```
capa={{-4,4,-4},{2,3,-4},{5,0,-1}};
eigs=Eigensystem[capa]
(capa-eigs[[1,1]]IdentityMatrix[3]).
   eigs[[2,1]]
(capa-eigs[[1,3]]IdentityMatrix[3]).
   eigs[[2,3]]
capb={{1,-3,5},{5,5,0},{-5,-2,3}};
eigs=Eigensystem[capb]
(capb-eigs[[1,1]]IdentityMatrix[3]).
   eigs[[2,1]]
(capb-eigs[[1,2]]IdentityMatrix[3]).
   eigs[[2,2]]
```

```
capa:=array([[-4,4,-4],
   [2,3,-4],[5,0,-1]]):
eigs:=eigenvects(capa);
capb:=array([[1,-3,5],[5,5,0],
   [-5,-2,3]]):
eigs:=eigenvects(capb);
```

45. For the first matrix, we see that the eigenvalues are complex unless $k = 0$. For the second matrix, the eigenvalues are always complex and purely imaginary if $k = 0$.

```
m1={{0,-1-k^2},{1,2}};
Eigenvalues[m1]
m2={{0,-1-k^2},{1,2k}};
Eigenvalues[m2]
```

```
m1:=array([[0,-1-k^2],[1,2]]);
eigenvals(m1);
m2:=array([[0,-1-k^2],[1,2*k]]);
eigenvals(m2);
```

46. The exact eigenvalues and eigenvectors are rather complicated. The approximations make more sense to most people.

```
capa={{3,1,1,1},{-2,-3,2,-3},
   {-3,3,-2,1},{0,1,2,0}};
eigs=Eigensystem[capa]
N[eigs]
```

```
capa:=array([[3,1,1,1],[-2,-3,2,-3],
   [-3,3,-2,1],[0,1,2,0]]);
eigsa:=eigenvects(capa);
evalf(eigsa);
```

EXERCISES 6.3

11. $\begin{vmatrix} \cos 2t & \sin 2t \\ -2\sin 2t & 2\cos 2t \end{vmatrix} = 2\cos^2 2t + 2\sin^2 2t = 2 \neq 0$; Lin. Indep.

13. $\begin{vmatrix} e^{2t} & e^t & e^{-t} \\ e^{2t} & -e^t & e^{-t} \\ 2e^{2t} & 3e^t & e^{-t} \end{vmatrix} = e^{2t}\begin{vmatrix} -e^t & e^{-t} \\ 3e^t & e^{-t} \end{vmatrix} - e^t\begin{vmatrix} e^{2t} & e^{-t} \\ 2e^{2t} & e^{-t} \end{vmatrix} + e^{-t}\begin{vmatrix} e^{2t} & -e^t \\ 2e^{2t} & 3e^t \end{vmatrix} = 2e^{2t} \neq 0$; Lin. Indep.

17. $\begin{pmatrix} \cos t + 2\sin t \\ \sin t \end{pmatrix}' = \begin{pmatrix} -\sin t + 2\cos t \\ \cos t \end{pmatrix} = \begin{pmatrix} 2(\cos t + 2\sin t) - 5\sin t \\ (\cos t + 2\sin t) - 2\sin t \end{pmatrix}$;

$\begin{pmatrix} 2\cos t - \sin t \\ \cos t \end{pmatrix}' = \begin{pmatrix} -2\sin t - \cos t \\ -\sin t \end{pmatrix} = \begin{pmatrix} 2(2\cos t - \sin t) - 5\cos t \\ (2\cos t - \sin t) - 2\cos t \end{pmatrix}$; $W = \cos^2 t + \sin^2 t = 1 \neq 0$; Yes.

19. $\begin{pmatrix} 27e^{-6t} \\ -29e^{-6t} \\ 36e^{-6t} \end{pmatrix}' = \begin{pmatrix} -162e^{-6t} \\ 174e^{-6t} \\ -216e^{-6t} \end{pmatrix} = \begin{pmatrix} -2\left(27e^{-6t}\right) - 3\left(36e^{-6t}\right) \\ 3\left(27e^{-6t}\right) + 3\left(-29e^{-6t}\right) + 5\left(36e^{-6t}\right) \\ -4\left(27e^{-6t}\right) - 3\left(36e^{-6t}\right) \end{pmatrix}$

$\begin{pmatrix} -e^{-t} \\ -e^{-t} \\ e^{-t} \end{pmatrix}' = \begin{pmatrix} e^{-t} \\ e^{-t} \\ -e^{-t} \end{pmatrix} \neq \begin{pmatrix} 2e^{-t} - 3e^{-t} \\ -3e^{-t} - 3e^{-t} + 5e^{-t} \\ 4e^{-t} - 3e^{-t} \end{pmatrix}$; No.

26. (a) - (c)

```
fm={{Cos[3t]-Sin[3t],
    -Sin[3t]-Cos[3t]},
    {Sin[3t],Cos[3t]}};
Simplify[Det[fm]]
capxa=fm.{c1,c2}
capa={{-3,-6},{3,3}}
D[capxa,t]-capa.capxa//Simplify

fm={{Exp[t](Sin[3t]-3Cos[3t]),
        5Exp[t]Cos[3t]},
    {2Exp[t]Sin[3t],
    Exp[t](Cos[3t]-3Sin[3t])}};
Simplify[Det[fm]]
capxb=fm.{c1,c2}
capa={{0,5},{-2,2}}
D[capxb,t]-capa.capxb//Simplify

fm={{Exp[-t](Sin[3t]+
    3Cos[3t]),5Exp[-t]Cos[3t]},
    {2Exp[-t]Sin[3t],
    Exp[-t](Cos[3t]+3Sin[3t])}};
Simplify[Det[fm]]
capxc=fm.{c1,c2}
capa={{0,-5},{2,-2}}
D[capxc,t]-capa.capxc//Simplify
```

```
fm:=array([[cos(3*t)-sin(3*t),
   -sin(3*t)-cos(3*t)],
   [sin(3*t),cos(3*t)]]);
with(linalg):
simplify(det(fm));
capxa:=evalm(fm &* [[c1],[c2]]);
capa:=array([[-3,-6],[3,3]]);
evalm(map(diff,capxa,t)-
   capa &* capxa);

fm:=array([
   [exp(t)*(sin(3*t)-3*cos(3*t)),
      5*exp(t)*cos(3*t)],
   [2*exp(t)*sin(3*t),
   exp(t)*(cos(3*t)-3*sin(3*t))]]);
simplify(det(fm));
capxb:=evalm(fm &* [[c1],[c2]]);
capa:=array([[0,5],[-2,2]]);
evalm(map(diff,capxb,t)-
   capa &* capxb);

fm:=array([[exp(-t)*(sin(3*t)+
   3*cos(3*t)),5*exp(-t)*cos(3*t)],
   [2*exp(-t)*sin(3*t),
   exp(-t)*(cos(3*t)+3*sin(3*t))]]);
simplify(det(fm));
capxc:=evalm(fm &* [[c1],[c2]]);
capa:=array([[0,-5],[2,-2]]);
evalm(map(diff,capxc,t)-
   capa &* capxx);
```

(d)

```
Clear[x,y,t]
partsol=DSolve[{x'[t]==-3x[t]-6y[t],
    y'[t]==3x[t]+3y[t],x[0]==0,
    y[0]==1},{x[t],y[t]},t]


<<Algebra`Trigonometry`
partsol=ComplexToTrig[partsol]//
    Simplify
Plot[Evaluate[{x[t],y[t]} /.
    partsol],{t,0,2Pi},
    PlotRange->{-Pi,Pi},
    AspectRatio->1,
    PlotStyle->{GrayLevel[0],
        GrayLevel[.4]}]
ParametricPlot[{x[t],y[t]} /.
    partsol,{t,0,2Pi},
    Compiled->False,
    PlotRange->{{-2,2},{-2,2}},
    AspectRatio->1]
```

```
x:=capxa[1,1];
y:=capxa[2,1];
eqs:=eval(subs(t=0,{x=0,y=1}));
vals:=solve(eqs);
toplot:=subs(vals,[x,y,t=0..2*Pi]);
plot(toplot);
```

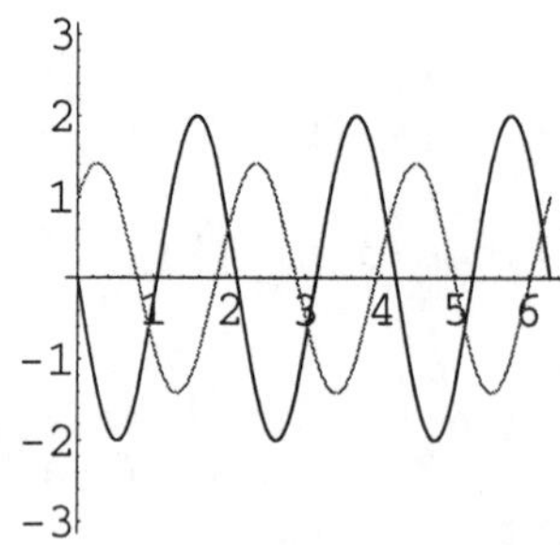

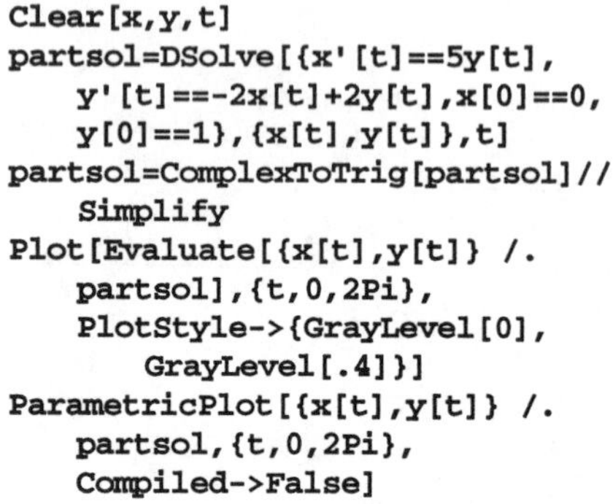

```
Clear[x,y,t]
partsol=DSolve[{x'[t]==5y[t],
    y'[t]==-2x[t]+2y[t],x[0]==0,
    y[0]==1},{x[t],y[t]},t]
partsol=ComplexToTrig[partsol]//
    Simplify
Plot[Evaluate[{x[t],y[t]} /.
    partsol],{t,0,2Pi},
    PlotStyle->{GrayLevel[0],
       GrayLevel[.4]}]
ParametricPlot[{x[t],y[t]} /.
    partsol,{t,0,2Pi},
    Compiled->False]
```

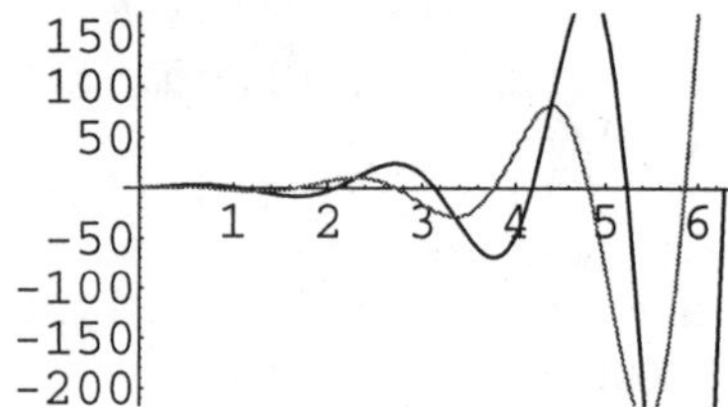


```
Clear[x,y,t]
partsol=DSolve[{x'[t]==-5y[t],
    y'[t]==2x[t]-2y[t],x[0]==0,
    y[0]==1},{x[t],y[t]},t]
partsol=ComplexToTrig[partsol]//
    Simplify
Plot[Evaluate[{x[t],y[t]} /.
    partsol],{t,0,2Pi},
    PlotRange->{-Pi,Pi},
    AspectRatio->1,
    PlotStyle->{GrayLevel[0],
       GrayLevel[.4]}]
ParametricPlot[{x[t],y[t]} /.
    partsol,{t,0,2Pi},
    Compiled->False,
    PlotRange->{{-2,2},{-2,2}},
    AspectRatio->1]
```

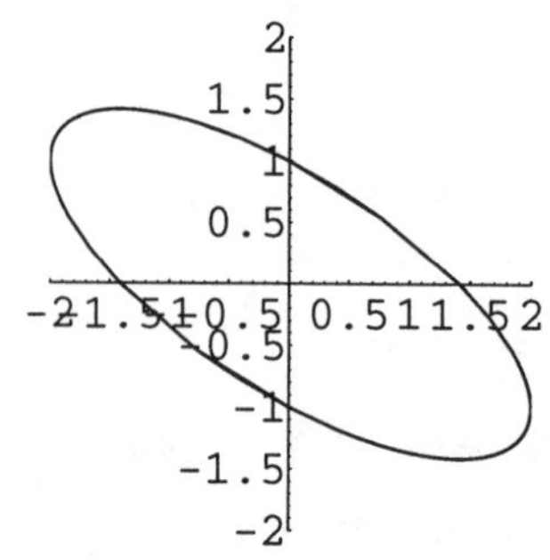


```
x:=capxb[1,1];
y:=capxb[2,1];
eqs:=eval(subs(t=0,{x=0,y=1}));
vals:=solve(eqs);
toplot:=subs(vals,[x,y,t=0..2*Pi]);
plot(toplot);
```

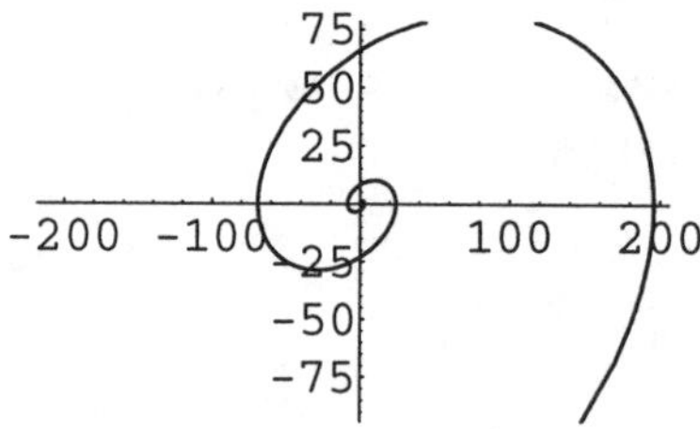


```
x:=capxc[1,1];
y:=capxc[2,1];
eqs:=eval(subs(t=0,{x=0,y=1}));
vals:=solve(eqs);
toplot:=subs(vals,[x,y,t=0..2*Pi]);
plot(toplot);
```

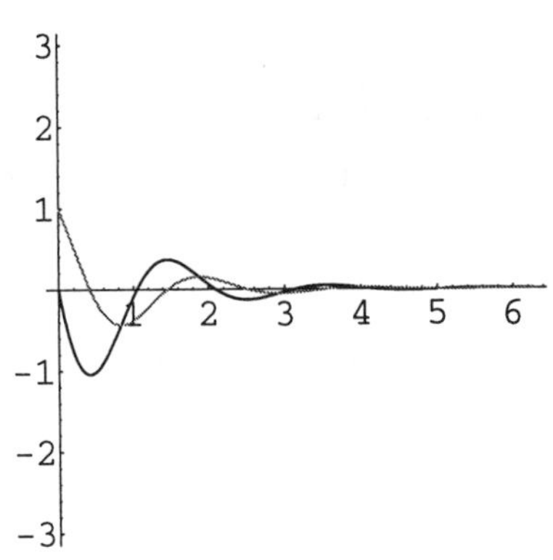

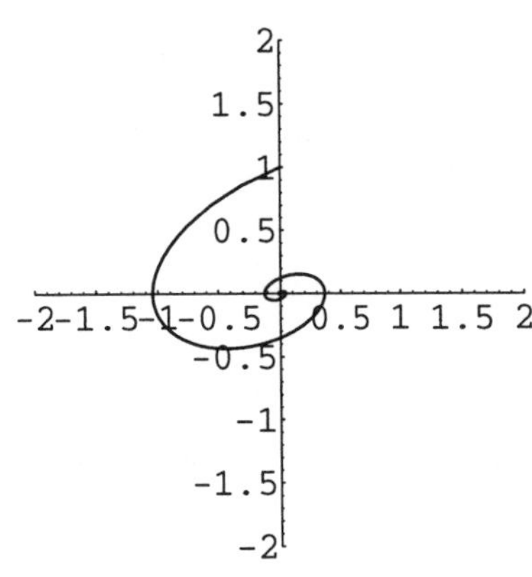

EXERCISES 6.4

5. Repeated; Solve $(\mathbf{A}-2\mathbf{I})\mathbf{w}=\begin{pmatrix}1\\1\end{pmatrix}$ for $\mathbf{w}$ to obtain $\mathbf{w}=\begin{pmatrix}0\\1\end{pmatrix}$; $\mathbf{X}(t)=c_1\begin{pmatrix}1\\1\end{pmatrix}e^{2t}+c_2\left[\begin{pmatrix}1\\1\end{pmatrix}t+\begin{pmatrix}0\\1\end{pmatrix}\right]e^{2t}$

9. Repeated; Solve $(\mathbf{A}-2\mathbf{I})\mathbf{w}=\begin{pmatrix}0\\1\\0\end{pmatrix}$ for $\mathbf{w}$ to obtain $\mathbf{w}=\begin{pmatrix}0\\1\\0\end{pmatrix}$; $\mathbf{X}(t)=c_1\begin{pmatrix}0\\1\\0\end{pmatrix}e^{2t}+c_2\left[\begin{pmatrix}0\\1\\0\end{pmatrix}t+\begin{pmatrix}0\\1\\0\end{pmatrix}\right]e^{2t}+c_3\begin{pmatrix}1\\0\\0\end{pmatrix}e^{3t}$

15. The eigenvalues and corresponding eigenvectors of $\begin{pmatrix}6&-1\\5&0\end{pmatrix}$ are $\lambda_1=1, \mathbf{v}_1=\begin{pmatrix}1\\5\end{pmatrix}$ and $\lambda_2=5, \mathbf{v}_2=\begin{pmatrix}1\\1\end{pmatrix}$. A general solution is $\begin{pmatrix}x(t)\\y(t)\end{pmatrix}=c_1\begin{pmatrix}1\\5\end{pmatrix}e^{t}+c_2\begin{pmatrix}1\\1\end{pmatrix}e^{5t}=\begin{pmatrix}c_1e^{t}+c_2e^{5t}\\5c_1e^{t}+c_2e^{5t}\end{pmatrix}$.

19. The eigenvalues and corresponding eigenvectors of $\begin{pmatrix}-5&3\\2&-10\end{pmatrix}$ are $\lambda_1=-11, \mathbf{v}_1=\begin{pmatrix}1\\-2\end{pmatrix}$ and $\lambda_2=-4, \mathbf{v}_2=\begin{pmatrix}3\\1\end{pmatrix}$. A general solution is $\begin{pmatrix}x(t)\\y(t)\end{pmatrix}=c_1\begin{pmatrix}1\\-2\end{pmatrix}e^{-11t}+c_2\begin{pmatrix}3\\1\end{pmatrix}e^{-4t}=\begin{pmatrix}c_1e^{-11t}+3c_2e^{-4t}\\-2c_1e^{-11t}+c_2e^{-4t}\end{pmatrix}$.

23. $\begin{pmatrix}-6&2\\-2&-10\end{pmatrix}$ has the repeated eigenvalue $\lambda_1=\lambda_2=-8$ with only one linearly independent eigenvector $\mathbf{v}_1=\begin{pmatrix}-1\\1\end{pmatrix}$. One solution is $\begin{pmatrix}-1\\1\end{pmatrix}e^{-8t}$. A second linearly independent solution has the form $\left[\begin{pmatrix}-1\\1\end{pmatrix}t+\mathbf{w}_2\right]e^{-8t}$ where $(A-(-8)I)\mathbf{w}_2=\begin{pmatrix}2&2\\-2&-2\end{pmatrix}\begin{pmatrix}x_2\\y_2\end{pmatrix}=\begin{pmatrix}-1\\1\end{pmatrix}$. A solution to $2x_2+2y_2=-1$ is $\begin{pmatrix}-1/2\\0\end{pmatrix}$, so a second solution is $\left[\begin{pmatrix}-1\\1\end{pmatrix}t+\begin{pmatrix}-1/2\\0\end{pmatrix}\right]e^{-8t}$. A general solution is

$$\begin{pmatrix}x(t)\\y(t)\end{pmatrix}=c_1\begin{pmatrix}-1\\1\end{pmatrix}e^{-8t}+c_2\left[\begin{pmatrix}-1\\1\end{pmatrix}t+\begin{pmatrix}-1/2\\0\end{pmatrix}\right]e^{-8t}=\begin{pmatrix}-c_1e^{-8t}-c_2te^{-8t}-1/2c_2e^{-8t}\\c_1e^{-8t}+c_2te^{-8t}\end{pmatrix}.$$

27. Eigenvalues: $\lambda_1=-2$, $\lambda_2=-1$, $\lambda_3=4$; Eigenvectors: $\mathbf{v}_1=\begin{pmatrix}0\\1\\0\end{pmatrix}$, $\mathbf{v}_2=\begin{pmatrix}-1\\0\\5\end{pmatrix}$, $\mathbf{v}_3=\begin{pmatrix}1\\0\\0\end{pmatrix}$;

$$\begin{cases}x(t)=c_1e^{4t}+c_2e^{-t}\\ y(t)=c_3e^{-2t}\\ z(t)=-5c_2e^{-t}\end{cases}$$

31. In matrix form, the system is $\mathbf{X}'=\mathbf{A}\,\mathbf{X}$, where

$$\mathbf{A}=\begin{pmatrix}1&2&3\\0&1&2\\0&-2&1\end{pmatrix}.$$

The eigenvalues of $\mathbf{A}$ are $\lambda_1=1$ and $\lambda_{2,3}=1\pm 2\mathrm{i}$ with corresponding eigenvectors $\mathbf{v}_1=\begin{pmatrix}1\\0\\0\end{pmatrix}$ and

$\mathbf{v}_{2,3}=\begin{pmatrix}-1\mp\frac{3}{2}\mathrm{i}\\ \mp\mathrm{i}\\ 1\end{pmatrix}=\begin{pmatrix}-1\\0\\1\end{pmatrix}\pm\begin{pmatrix}-3/2\\-1\\0\end{pmatrix}\mathrm{i}$, respectively. A general solution to the system is

$$\mathbf{X}=c_1e^t\begin{pmatrix}1\\0\\0\end{pmatrix}+c_2e^t\left[\begin{pmatrix}-1\\0\\1\end{pmatrix}\cos 2t-\begin{pmatrix}-3/2\\-1\\0\end{pmatrix}\sin 2t\right]+c_3e^t\left[\begin{pmatrix}-1\\0\\1\end{pmatrix}\sin 2t+\begin{pmatrix}-3/2\\-1\\0\end{pmatrix}\cos 2t\right]$$

$$=\begin{pmatrix}c_1e^t+e^t\left[\left(-c_2-\frac{3}{2}c_3\right)\cos 2t+\left(\frac{3}{2}c_2-c_3\right)\sin 2t\right]\\ e^t\left(-c_3\cos 2t+c_2\sin 2t\right)\\ e^t\left(c_2\cos 2t+c_3\sin 2t\right)\end{pmatrix}$$

or

$$\begin{cases}x(t)=c_1e^t+e^t\left[\left(-c_2-\frac{3}{2}c_3\right)\cos 2t+\left(\frac{3}{2}c_2-c_3\right)\sin 2t\right]\\ y(t)=e^t\left(-c_3\cos 2t+c_2\sin 2t\right)\\ z(t)=e^t\left(c_2\cos 2t+c_3\sin 2t\right)\end{cases}$$

35. $\begin{pmatrix}4&0\\2&4\end{pmatrix}$ has the repeated eigenvalues $\lambda_1=\lambda_2=4$ and corresponding linearly independent eigenvector $\mathbf{v}_1=\begin{pmatrix}0\\1\end{pmatrix}$. One solution is $\begin{pmatrix}0\\1\end{pmatrix}e^{4t}$; a second linearly independent solution is $\left[\begin{pmatrix}0\\1\end{pmatrix}t+\mathbf{w}_2\right]e^{4t}$ where $(A-4I)\mathbf{w}_2=\begin{pmatrix}0&0\\2&0\end{pmatrix}\begin{pmatrix}x_2\\y_2\end{pmatrix}=\begin{pmatrix}0\\1\end{pmatrix}$. All solutions to this system satisfy $x_2=1/2$, so we choose $\mathbf{w}_2=\begin{pmatrix}1/2\\0\end{pmatrix}$. Then, a general solution is

$$\begin{pmatrix}x(t)\\y(t)\end{pmatrix}=c_1\begin{pmatrix}0\\1\end{pmatrix}e^{4t}+c_2\left[\begin{pmatrix}0\\1\end{pmatrix}t+\begin{pmatrix}1/2\\0\end{pmatrix}\right]e^{4t}=\begin{pmatrix}1/2c_2e^{4t}\\ c_1e^{4t}+c_2te^{4t}\end{pmatrix}.$$

$\begin{pmatrix}x(0)\\y(0)\end{pmatrix}=\begin{pmatrix}1/2c_2\\c_1\end{pmatrix}=\begin{pmatrix}8\\0\end{pmatrix}$ implies that $c_1=0$ and $c_2=16$. Therefore, $\begin{pmatrix}x(t)\\y(t)\end{pmatrix}=\begin{pmatrix}8e^{4t}\\16te^{4t}\end{pmatrix}$.

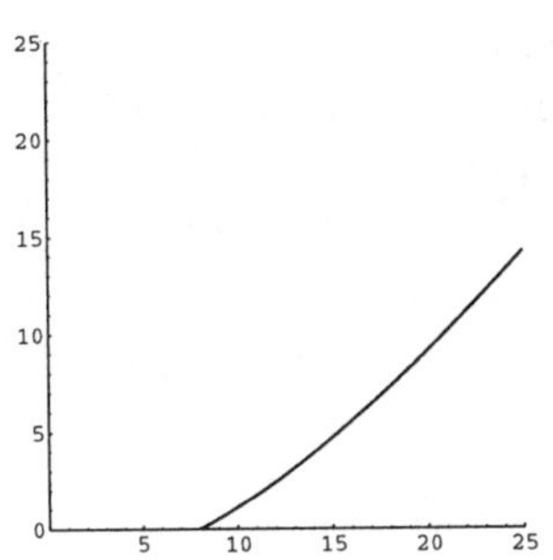

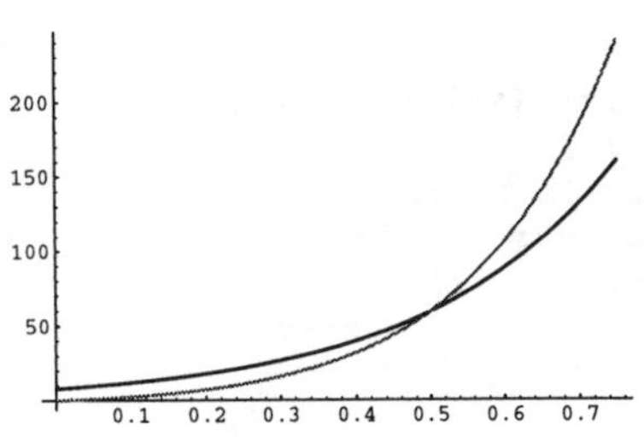

39. The eigenvalues and corresponding eigenvectors of $\begin{pmatrix} 4 & 0 & 1 \\ -2 & 1 & 0 \\ -2 & - & 1 \end{pmatrix}$ are $\lambda_1 = 2, \mathbf{v}_1 = \begin{pmatrix} 1 \\ -2 \\ -2 \end{pmatrix}$; $\lambda_2 = 3, \mathbf{v}_2 = \begin{pmatrix} -1 \\ 1 \\ 1 \end{pmatrix}$; and $\lambda_3 = 1, \mathbf{v}_3 = \begin{pmatrix} 0 \\ 1 \\ 0 \end{pmatrix}$, so a general solution is $\begin{pmatrix} x(t) \\ y(t) \\ z(t) \end{pmatrix} = c_1 \begin{pmatrix} 1 \\ -2 \\ -2 \end{pmatrix} e^{2t} + c_2 \begin{pmatrix} -1 \\ 1 \\ 1 \end{pmatrix} e^{3t} + c_3 \begin{pmatrix} 0 \\ 1 \\ 0 \end{pmatrix} e^{t}$. $\begin{pmatrix} x(0) \\ y(0) \\ z(0) \end{pmatrix} = \begin{pmatrix} -1 \\ 2 \\ 0 \end{pmatrix}$ yields $\{c_1 - c_2 = -1, -2c_1 + c_2 + c_3 = 2, -2c_1 + c_2 = 0\}$ with solution $\{c_1 = 1, c_2 = 2, c_3 = 2\}$. Therefore, $\begin{pmatrix} x(t) \\ y(t) \\ z(t) \end{pmatrix} = \begin{pmatrix} e^{2t} - 2e^{3t} \\ -2e^{2t} + 2e^{3t} + 2e^{t} \\ -2e^{2t} + 2e^{3t} \end{pmatrix}$.

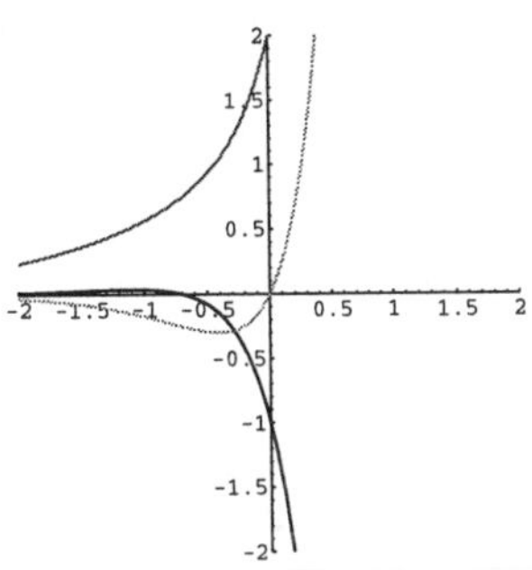

50.

```
Clear[x,y,t]
partsol=DSolve[{x'[t]==2x[t]-y[t],
    y'[t]==-x[t]+3y[t],x[0]==0,
    y[0]==1},{x[t],y[t]},t]
Plot[Evaluate[{x[t],y[t]} /.
    partsol],{t,-1,2},
    PlotStyle->{GrayLevel[0],
      GrayLevel[.4]}]
ParametricPlot[{x[t],y[t]} /.
    partsol,{t,-1,2},Compiled->False]
```

15
10
5
-1 -0.5 0.5 1 1.5 2
-5
-10

```
x:='x':y:='y':
sol:=dsolve({diff(x(t),t)=
     2*x(t)-y(t),
   diff(y(t),t)=-x(t)+3*y(t),
   x(0)=1,y(0)=1},{x(t),y(t)});
assign(sol):
plot({x(t),y(t)},t=-1..2);
plot([x(t),y(t),t=-1..2]);
```

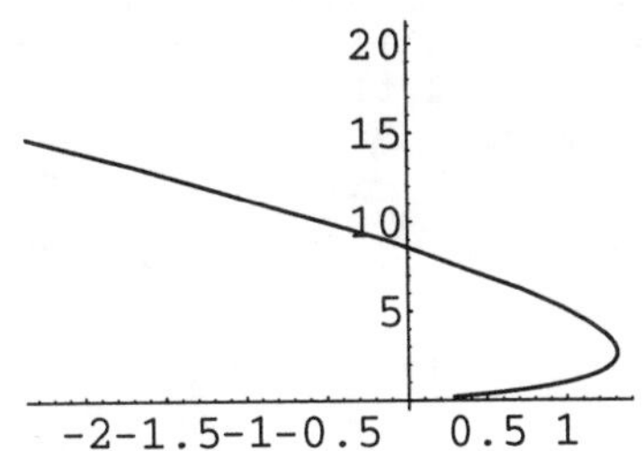

```
Clear[x,y,t]
partsol=DSolve[{x'[t]==2x[t],
    y'[t]==3x[t]+2y[t],x[0]==0,
    y[0]==1},{x[t],y[t]},t]
Plot[Evaluate[{x[t],y[t]} /.
    partsol],{t,-1,2},
    PlotStyle->{GrayLevel[0],
        GrayLevel[.4]}]
ParametricPlot[{x[t],y[t]} /.
    partsol,{t,-1,2},Compiled->False]
```

```
x:='x':y:='y':
sol:=dsolve({diff(x(t),t)=2*x(t),
    diff(y(t),t)=3*x(t)+2*y(t),
    x(0)=1,y(0)=1},{x(t),y(t)});
assign(sol):
plot({x(t),y(t)},t=-1..2);
plot([x(t),y(t),t=-1..2]);
```

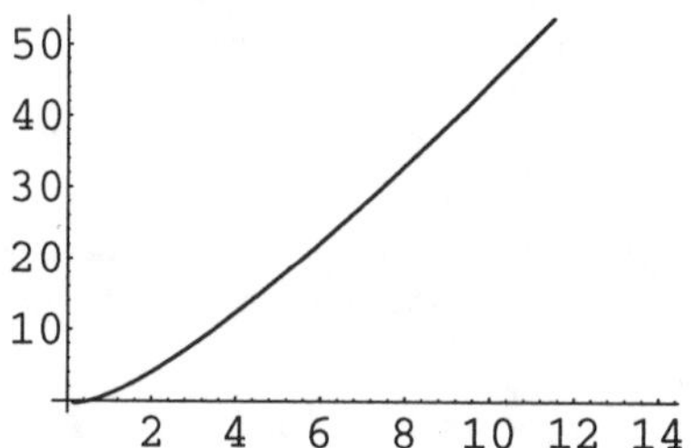


```
Clear[x,y,t]
partsol=DSolve[{x'[t]==x[t]+4y[t],
    y'[t]==-2x[t]-y[t],x[0]==1,
    y[0]==1},{x[t],y[t]},t]
<<Algebra`Trigonometry`
partsol=ComplexToTrig[partsol]//
    Simplify
Plot[Evaluate[{x[t],y[t]} /.
    partsol],{t,-1,5},
    PlotRange->{-3,3},AspectRatio->1,
    PlotStyle->{GrayLevel[0],
        GrayLevel[.4]}]
ParametricPlot[{x[t],y[t]} /.
    partsol,{t,-1,5},
    PlotRange->{{-3,3},{-3,3}},
    AspectRatio->1,Compiled->False]
```

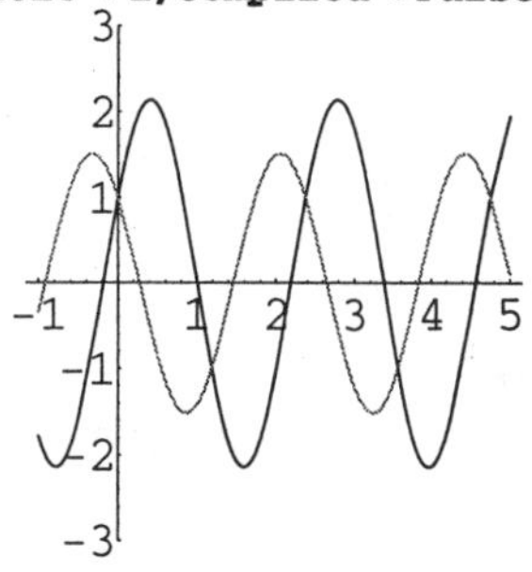


```
x:='x':y:='y':
sol:=dsolve({diff(x(t),t)=
        x(t)+4*y(t),
    diff(y(t),t)=-2*x(t)-y(t),
    x(0)=1,y(0)=1},{x(t),y(t)});
assign(sol):
plot({x(t),y(t)},t=0..2);
plot([x(t),y(t),t=0..2]);
```

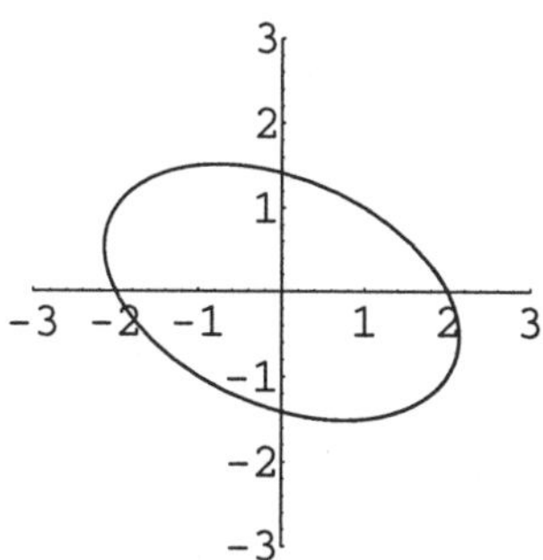

* 51.

```
gensol=DSolve[{x'[t]==3z[t],
    y'[t]==x[t]-4y[t]+2z[t],
    z'[t]==-4y[t]+z[t]},
    {x[t],y[t],z[t]},t]
<<Algebra`Trigonometry`
gensol=ComplexToTrig[gensol]//
    Simplify
```

```
gensol:=dsolve({diff(x(t),t)=3*z(t),
    diff(y(t),t)=x(t)-4*y(t)+2*z(t),
    diff(z(t),t)=-4*y(t)+z(t)},
    {x(t),y(t),z(t)});
assign(gensol):
```

```
Plot[Evaluate[{x[t],y[t],z[t]} /.
    gensol /.
        {C[1]->1,C[2]->1,
        C[3]->1}],
    {t,0,2Pi},
    PlotStyle->{GrayLevel[0],
        GrayLevel[.3],GrayLevel[.6]}]
Plot[Evaluate[{x[t],y[t],z[t]} /.
    gensol /.
        {C[1]->0,C[2]->-1,
        C[3]->1}],
    {t,0,2Pi},
    PlotStyle->{GrayLevel[0],
        GrayLevel[.3],GrayLevel[.6]}]
Plot[Evaluate[{x[t],y[t],z[t]} /.
    gensol /.
        {C[1]->2,C[2]->0,
        C[3]->-3}],
    {t,0,2Pi},
    PlotStyle->{GrayLevel[0],
        GrayLevel[.3],GrayLevel[.6]}]
```

```
plot(subs({_C1=1,_C2=1,_C3=1},
        {x(t),y(t),z(t)}),
    t=0..2*Pi);
plot(subs({_C1=-1,_C2=0,_C3=1},
        {x(t),y(t),z(t)}),
    t=0..2*Pi);
plot(subs({_C1=2,_C2=0,_C3=-3},
        {x(t),y(t),z(t)}),
    t=0..2*Pi);
```

52.

```
<<Graphics`PlotField`
Clear[x,y,lambda]
Do[
    PlotVectorField[{2x+lambda y,x},
        {x,-1,1},{y,-1,1},
        Axes->Automatic,
        AxesOrigin->{0,0},
        PlotPoints->20,
        Ticks->{{-1,1},{-1,1}},
        ScaleFunction->(1&)],
    {lambda,-2+2/7,0,2/7}]
```

```
x:='x':y:='y':
lambda:='lambda':
with(DEtools);
for lambda from -2 by 2/7 to 0 do
    dfieldplot([2*x+lambda*y,x],[x,y],
        -1..1,y=-1..1)
    od;
```

$\lambda = -2$ $\lambda = -12/7$ $\lambda = -10/7$ $\lambda = -8/7$

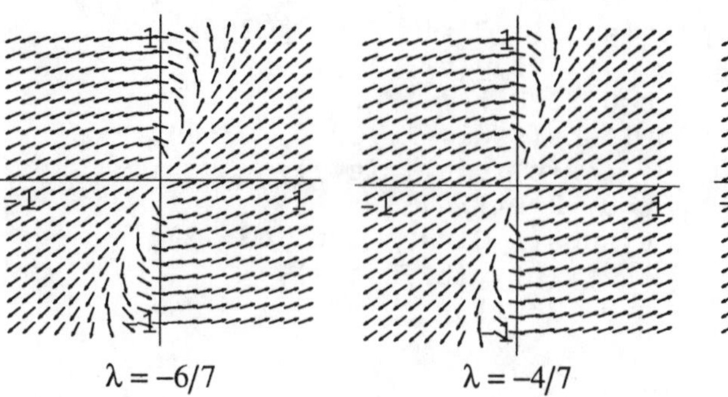

λ = −6/7 λ = −4/7

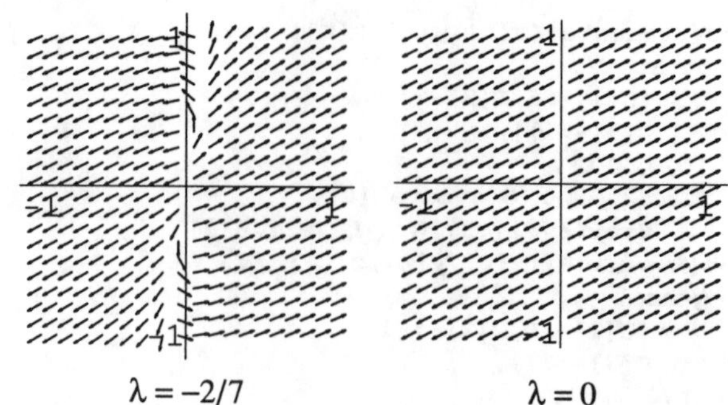

λ = −2/7 λ = 0

```
gensol=DSolve[{x'[t]==
    2x[t]+lambda y[t],y'[t]==x[t]},
    {x[t],y[t]},t]
Simplify[gensol]
lvals=Table[lambda,{lambda,-2,0,2/7}]
f[lambda_]:=DSolve[{x'[t]==
    2x[t]+lambda y[t],y'[t]==x[t]},
    {x[t],y[t]},t]//
    ComplexToTrig//Simplify
Map[f,lvals]
```

```
lambda:='lambda':
gensol:=dsolve({diff(x(t),t)=
   2*x(t)+lambda*y(t),
   diff(y(t),t)=x(t)},{x(t),y(t)});
lambdavals:=seq(-2+2/7*i,i=0..7);
seq(eval(subs(lambda=j,gensol)),
   j=lambdavals);
```

54.

```
Clear[capx,capa,x1,x2,x3,x4]
capa={{lambda,0,0,0},{0,lambda,0,0},
    {0,0,lambda,0},{0,0,0,lambda}};
capx[t_]={x1[t],x2[t],x3[t],x4[t]}
sys=Thread[capx'[t]==capa.capx[t]]
sol1=DSolve[sys,
    {x1[t],x2[t],x3[t],x4[t]},t]
Clear[capa,x1,x2,x3,x4]
capa={{lambda,1,0,0},{0,lambda,0,0},
    {0,0,lambda,0},{0,0,0,lambda}};
sys=Thread[capx'[t]==capa.capx[t]]
sol2=DSolve[sys,
    {x1[t],x2[t],x3[t],x4[t]},t]
Clear[capa,x1,x2,x3,x4]
capa={{lambda,1,0,0},{0,lambda,1,0},
    {0,0,lambda,0},{0,0,0,lambda}};
sys=Thread[capx'[t]==capa.capx[t]]
sol3=DSolve[sys,
    {x1[t],x2[t],x3[t],x4[t]},t]
Clear[capa,x1,x2,x3,x4]
capa={{lambda,1,0,0},{0,lambda,1,0},
    {0,0,lambda,1},{0,0,0,lambda}};
sys=Thread[capx'[t]==capa.capx[t]]
sol4=DSolve[sys,
    {x1[t],x2[t],x3[t],x4[t]},t]
```

```
with(linalg);
with(student);
capa:='capa':capx:='capx':x1:='x1':
x2:='x2':x3:='x3':x4:='x4':
capx:=vector([x1(t),x2(t),
   x3(t),x4(t)]);
capa:=array([[lambda,0,0,0],
   [0,lambda,0,0],
   [0,0,lambda,0],[0,0,0,lambda]]);
dcapx:=map(diff,capx,t);
rhseq:=evalm(capa &* capx);
sys:=equate(dcapx,rhseq);
sol1:=dsolve(sys,
   {x1(t),x2(t),x3(t),x4(t)});
capa:=array([[lambda,1,0,0],
   [0,lambda,0,0],
   [0,0,lambda,0],[0,0,0,lambda]]):
rhseq:=evalm(capa &* capx):
sys:=equate(dcapx,rhseq);
sol2:=dsolve(sys,
   {x1(t),x2(t),x3(t),x4(t)});
capa:=array([[lambda,1,0,0],
   [0,lambda,1,0],
   [0,0,lambda,0],[0,0,0,lambda]]):
rhseq:=evalm(capa &* capx):
sys:=equate(dcapx,rhseq);
sol3:=dsolve(sys,
   {x1(t),x2(t),x3(t),x4(t)});
capa:=array([[lambda,1,0,0],
   [0,lambda,1,0],
   [0,0,lambda,1],[0,0,0,lambda]]):
rhseq:=evalm(capa &* capx):
sys:=equate(dcapx,rhseq);
sol4:=dsolve(sys,
   {x1(t),x2(t),x3(t),x4(t)});
```

```
sol1=sol1 /. lambda->-1/2
sol2=sol2 /. lambda->-1/2
sol3=sol3 /. lambda->-1/2
sol4=sol4 /. lambda->-1/2
sys1={x1[t]==-1,x2[t]==0,x3[t]==1,
    x4[t]==2} /. sol1[[1]] /. t->0
cvals1=Solve[sys1]
sol1=sol1 /. cvals1[[1]]
sys2={x1[t]==-1,x2[t]==0,x3[t]==1,
    x4[t]==2} /. sol2[[1]] /. t->0
cvals2=Solve[sys2]
sol2=sol2 /. cvals2[[1]]
sys3={x1[t]==-1,x2[t]==0,x3[t]==1,
    x4[t]==2} /. sol3[[1]] /. t->0
cvals3=Solve[sys3]
sol3=sol3 /. cvals3[[1]]
sys4={x1[t]==-1,x2[t]==0,x3[t]==1,
    x4[t]==2} /. sol4[[1]] /. t->0
cvals4=Solve[sys4]
sol4=sol4 /. cvals4[[1]]
grays=Table[GrayLevel[i],{i,0,.7,.7/3}];
Plot[Evaluate[{x1[t],x2[t],x3[t],
    x4[t]} /. sol1],
    {t,0,10},PlotStyle->grays,
    PlotRange->All]
Plot[Evaluate[{x1[t],x2[t],x3[t],
    x4[t]} /. sol2],
    {t,0,10},PlotStyle->grays,
    PlotRange->All]
Plot[Evaluate[{x1[t],x2[t],x3[t],
    x4[t]} /. sol3],
    {t,0,10},PlotStyle->grays,
    PlotRange->All]
Plot[Evaluate[{x1[t],x2[t],x3[t],
    x4[t]} /. sol4],
    {t,0,10},PlotStyle->grays,
    PlotRange->All]
```

```
sol1:=subs(lambda=-1/2,sol1);
sol2:=subs(lambda=-1/2,sol2);
sol3:=subs(lambda=-1/2,sol3);
sol4:=subs(lambda=-1/2,sol4);
sys1:=eval(subs(t=0,subs(sol1,
   {x1(t)=-1,x2(t)=0,
   x3(t)=1,x4(t)=2})));
cvals1:=solve(sys1);
sol1:=subs(cvals1,sol1);
sys2:=eval(subs(t=0,subs(sol2,
   {x1(t)=-1,x2(t)=0,
   x3(t)=1,x4(t)=2})));
   cvals2:=solve(sys2);
sol2:=subs(cvals2,sol2);
sys3:=eval(subs(t=0,subs(sol3,
   {x1(t)=-1,x2(t)=0,
   x3(t)=1,x4(t)=2})));
   cvals3:=solve(sys3);
sol3:=subs(cvals3,sol3);
sys4:=eval(subs(t=0,subs(sol4,
   {x1(t)=-1,x2(t)=0,
   x3(t)=1,x4(t)=2})));
   cvals4:=solve(sys4);
sol4:=subs(cvals4,sol4);
assign(sol1):
plot({x1(t),x2(t),x3(t),x4(t)},
   t=0..10);
x1:='x1':x2:='x2':x3:='x3':x4:='x4':
assign(sol2):
plot({x1(t),x2(t),x3(t),x4(t)},
   t=0..10);
x1:='x1':x2:='x2':x3:='x3':x4:='x4':
assign(sol3):
plot({x1(t),x2(t),x3(t),x4(t)},
   t=0..10);
x1:='x1':x2:='x2':x3:='x3':x4:='x4':
assign(sol4):
plot({x1(t),x2(t),x3(t),x4(t)},
   t=0..10);
```

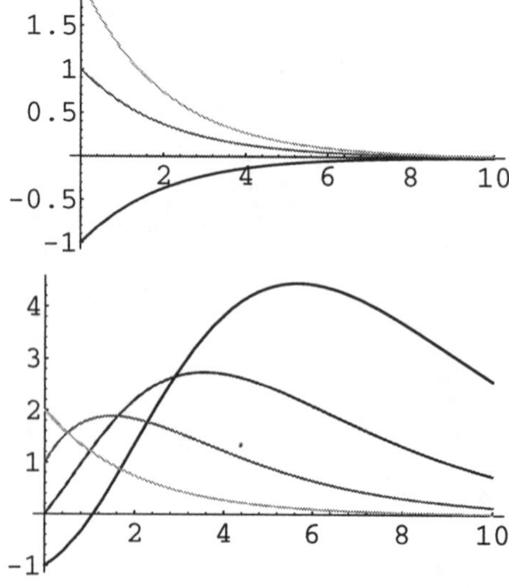

EXERCISES 6.5

3. The corresponding homogeneous system is $\mathbf{X}'=\mathbf{A}\,\mathbf{X}$, where $\mathbf{A}=\begin{pmatrix}-6 & 0\\ -1 & -6\end{pmatrix}$, with general solution $\begin{cases}x(t)=-c_2e^{-6t}\\ y(t)=c_1e^{-6t}+c_2te^{-6t}\end{cases}$. The nonhomogeneous system $\mathbf{X}'=\begin{pmatrix}-6 & 0\\ -1 & -6\end{pmatrix}\mathbf{X}+\begin{pmatrix}t\\ t^2\end{pmatrix}$ has a particular solution of the form $\mathbf{X}_p(t)=\mathbf{a}\,t^2+\mathbf{b}\,t+\mathbf{c}$, where $\mathbf{a}=\begin{pmatrix}a_1\\ a_2\end{pmatrix}$, $\mathbf{b}=\begin{pmatrix}b_1\\ b_2\end{pmatrix}$, and $\mathbf{c}=\begin{pmatrix}c_1\\ c_2\end{pmatrix}$. . Substitution into the nonhomogeneous system yields $\mathbf{a}=\begin{pmatrix}0\\ 1/6\end{pmatrix}$, $\mathbf{b}=\begin{pmatrix}1/6\\ -1/12\end{pmatrix}$, and $\mathbf{c}=\begin{pmatrix}-1/36\\ 1/54\end{pmatrix}$ so a general solution is

$$\begin{cases}x(t)=\dfrac{1}{6}t-\dfrac{1}{36}-c_2e^{-6t}\\ y(t)=\dfrac{1}{6}t^2-\dfrac{1}{12}t+\dfrac{1}{54}+c_1e^{-6t}+c_2te^{-6t}\end{cases}.$$

7. A solution to the corresponding homogeneous system is $\mathbf{X}_h(t)=\begin{pmatrix}c_1e^{4t}+7c_2e^{-2t}\\ c_1e^{4t}+c_2e^{-2t}\end{pmatrix}$. Assuming that $\mathbf{X}_p(t)=\mathbf{A}\cos t+\mathbf{B}\sin t+\mathbf{C}e^{-4t}$, we find that $\mathbf{C}=\begin{pmatrix}7/16\\ -1/16\end{pmatrix}$, $\mathbf{A}=\begin{pmatrix}43/85\\ 9/85\end{pmatrix}$, and $\mathbf{B}=\begin{pmatrix}19/85\\ 2/85\end{pmatrix}$. Therefore,

$$\begin{cases}x(t)=\dfrac{7}{16}e^{-4t}+\dfrac{43}{85}\cos t+\dfrac{19}{85}\sin t+c_1e^{4t}+7c_2e^{-2t}\\ y(t)=-\dfrac{1}{16}e^{-4t}+\dfrac{9}{85}\cos t+\dfrac{2}{85}\sin t+c_1e^{4t}+c_2e^{-2t}\end{cases}.$$

11. The eigenvalues of $\begin{pmatrix}9 & 5\\ -8 & -3\end{pmatrix}$ are $\lambda_{1,2}=3\pm 2i$ with corresponding eigenvectors $\mathbf{v}_{1,2}=\begin{pmatrix}-3/4\\ 1\end{pmatrix}\pm\begin{pmatrix}-1/4\\ 0\end{pmatrix}$. A general solution to the corresponding homogeneous system is

$$\mathbf{X}_h(t)=\begin{pmatrix}e^{3t}\left(-\frac{1}{4}\cos 2t-\frac{3}{4}\sin 2t\right) & e^{3t}\left(-\frac{3}{4}\cos 2t+\frac{1}{4}\sin 2t\right)\\ e^{3t}\cos 2t & e^{3t}\sin 2t\end{pmatrix}\begin{pmatrix}c_1\\ c_2\end{pmatrix}.$$

We use the method of variation of parameters

$$\mathbf{X}_h(t)=\begin{pmatrix}e^{3t}\left(-\frac{1}{4}\cos 2t-\frac{3}{4}\sin 2t\right) & e^{3t}\left(-\frac{3}{4}\cos 2t+\frac{1}{4}\sin 2t\right)\\ e^{3t}\cos 2t & e^{3t}\sin 2t\end{pmatrix}\begin{pmatrix}c_1\\ c_2\end{pmatrix}$$

$$\begin{cases}x(t)=e^{3t}\left[c_1\left(-\frac{1}{4}\cos 2t-\frac{3}{4}\sin 2t\right)+c_2\left(-\frac{3}{4}\cos 2t+\frac{1}{4}\sin 2t\right)\right]\\ \qquad +\frac{1}{8}e^{3t}(10-\cos 2t-12t\cos 2t+3\sin 2t+4t\sin 2t)\\ y(t)=e^{3t}\left[c_1\cos 2t+c_2\sin 2t\right]+\frac{1}{2}e^{3t}(-3+4t\cos 2t-\sin 2t)\end{cases}$$

15. A fundamental matrix for the corresponding homogeneous system is $\Phi(t)=\begin{pmatrix}\sin t & \cos t\\ \cos t & -\sin t\end{pmatrix}$ where $\Phi^{-1}(t)=\dfrac{1}{-1}\begin{pmatrix}-\sin t & -\cos t\\ -\cos t & \sin t\end{pmatrix}=\begin{pmatrix}\sin t & \cos t\\ \cos t & -\sin t\end{pmatrix}$. Then,

$$\mathbf{X}_p(t)=\Phi(t)\int\begin{pmatrix}\sin t & \cos t\\ \cos t & -\sin t\end{pmatrix}\begin{pmatrix}0\\ -2\cot t\end{pmatrix}dt=\Phi(t)\int\begin{pmatrix}-2\cos^2 t/\sin t\\ 2\cos t\end{pmatrix}dt$$

$$=\Phi(t)\begin{pmatrix}-2\ln|\csc t-\cot t|-\cos t\\ 2\sin t\end{pmatrix}=\begin{pmatrix}-2\sin t\ln|\csc t-\cot t|-2\sin t\cos t+2\sin t\cos t\\ -2\cos t\ln|\csc t-\cot t|-2\cos^2 t-2\sin^2 t\end{pmatrix}.$$

$$=\begin{pmatrix}-2\sin t\ln|\csc t-\cot t|\\ -2\cos t\ln|\csc t-\cot t|-2\end{pmatrix}$$

Therefore, $\mathbf{X}(t)=\begin{pmatrix}c_1\sin t+c_2\cos t-2\sin t\ln|\csc t-\cot t|\\ c_1\cos t-c_2\sin t-2\cos t\ln|\csc t-\cot t|-2\end{pmatrix}.$

19. $\Phi(t)=\begin{pmatrix}0 & 2e^{-2t} & -8e^{4t}\\ e^{-4t} & -e^{-2t} & -29e^{4t}\\ e^{-4t} & e^{-2t} & 11e^{4t}\end{pmatrix}$, $\Phi^{-1}(t)=\begin{pmatrix}\frac{-3}{16}e^{4t} & \frac{5}{16}e^{4t} & \frac{11}{16}e^{4t}\\ \frac{5}{12}e^{2t} & \frac{-1}{12}e^{2t} & \frac{1}{12}e^{2t}\\ \frac{-1}{48}e^{-4t} & \frac{-1}{48}e^{-4t} & \frac{1}{48}e^{-4t}\end{pmatrix}$,

$$\Phi^{-1}(t)\mathbf{F}(t)=\begin{pmatrix}\frac{-3}{16}+\frac{5}{16}te^{2t}+\frac{11}{16}te^{4t}\\ \frac{5}{12}e^{-2t}-\frac{t}{12}+\frac{1}{12}te^{2t}\\ \frac{-1}{48}e^{-8t}-\frac{1}{48}te^{-6t}+\frac{1}{48}te^{-4t}\end{pmatrix};$$

$$\int\Phi^{-1}(t)\mathbf{F}(t)dt=\begin{pmatrix}\left(\frac{5t}{32}-\frac{5}{64}\right)e^{2t}+\left(\frac{11t}{64}-\frac{11}{256}\right)e^{4t}-\frac{3t}{16}\\ -\frac{5}{24}e^{-2t}+\left(\frac{t}{24}-\frac{1}{48}\right)e^{2t}-\frac{1}{24}t^2\\ \frac{1}{384}e^{-8t}-\left(\frac{1}{768}+\frac{t}{192}\right)e^{-4t}+\left(\frac{1}{1728}+\frac{t}{288}\right)e^{-6t}\end{pmatrix}$$

$$\begin{cases}x(t)=-\frac{1}{36}te^{-2t}+\frac{1}{8}t-\frac{1}{32}-\frac{1}{12}t^2e^{-2t}-\frac{1}{216}e^{-2t}-\frac{7}{16}e^{-4t}+2c_2e^{-2t}-\frac{8}{11}c_3e^{4t}\\ y(t)=\frac{1}{18}te^{-2t}-\frac{5}{432}e^{-2t}+\frac{9}{32}t+\frac{1}{64}-\frac{3}{16}te^{-4t}+\frac{1}{24}t^2e^{-2t}-\frac{15}{128}e^{-4t}+\left(c_1+\frac{5}{16}\right)e^{-4t}\\ \quad+\left(-\frac{1}{12}-c_2\right)e^{-2t}-\frac{29}{11}c_3e^{4t}\\ z(t)=-\frac{3}{16}te^{-4t}+\frac{5}{32}t-\frac{5}{64}-\frac{31}{432}e^{-2t}+\frac{7}{36}te^{-2t}-\frac{15}{128}e^{-4t}-\frac{1}{24}t^2e^{-2t}+c_1e^{-4t}+c_2e^{-2t}+c_3e^{4t}\end{cases}$$

23. The eigenvalues of $\begin{pmatrix}3 & 2 & -3\\ 1 & 1 & 1\\ 0 & -4 & -4\end{pmatrix}$ are $\lambda_1=4$ and $\lambda_{2,3}=-2\pm i$ with corresponding eigenvectors $\mathbf{v}_1=\begin{pmatrix}-7\\ -2\\ 1\end{pmatrix}$

and $\mathbf{v}_{2,3}=\begin{pmatrix}\frac{3}{4}\pm\frac{1}{4}i\\ -\frac{1}{2}\mp\frac{1}{4}i\\ 1\end{pmatrix}=\begin{pmatrix}3/4\\ -1/2\\ 1\end{pmatrix}\pm\begin{pmatrix}1/4\\ -1/4\\ 0\end{pmatrix}i$, respectively. A general solution of the corresponding homogeneous

system is given by

$$\mathbf{X}_h(t) = c_1 e^{4t}\begin{pmatrix}-7\\-2\\1\end{pmatrix} + e^{-2t}\left\{c_1\left[\begin{pmatrix}3/4\\-1/2\\1\end{pmatrix}\cos t - \begin{pmatrix}1/4\\-1/4\\0\end{pmatrix}\sin t\right] + c_3\left[\begin{pmatrix}1/4\\-1/4\\0\end{pmatrix}\cos t + \begin{pmatrix}3/4\\-1/2\\1\end{pmatrix}\sin t\right]\right\}$$

$$= \begin{pmatrix}-7e^{4t} & e^{-2t}\left(\frac{3}{4}\cos t - \frac{1}{4}\sin t\right) & e^{-2t}\left(\frac{1}{4}\cos t + \frac{3}{4}\sin t\right)\\ -2e^{4t} & e^{-2t}\left(-\frac{1}{2}\cos t + \frac{1}{4}\sin t\right) & e^{-2t}\left(-\frac{1}{4}\cos t - \frac{1}{2}\sin t\right)\\ e^{4t} & e^{-2t}\cos t & e^{-2t}\sin t\end{pmatrix}\begin{pmatrix}c_1\\c_2\\c_3\end{pmatrix}.$$

Using the method of variation of parameters a particular solution to the nonhomogeneous system is given by

$$\mathbf{X}_p(t) = \begin{pmatrix}-7e^{4t} & e^{-2t}\left(\frac{3}{4}\cos t - \frac{1}{4}\sin t\right) & e^{-2t}\left(\frac{1}{4}\cos t + \frac{3}{4}\sin t\right)\\ -2e^{4t} & e^{-2t}\left(-\frac{1}{2}\cos t + \frac{1}{4}\sin t\right) & e^{-2t}\left(-\frac{1}{4}\cos t - \frac{1}{2}\sin t\right)\\ e^{4t} & e^{-2t}\cos t & e^{-2t}\sin t\end{pmatrix}\cdot$$

$$\int\begin{pmatrix}-7e^{4t} & e^{-2t}\left(\frac{3}{4}\cos t - \frac{1}{4}\sin t\right) & e^{-2t}\left(\frac{1}{4}\cos t + \frac{3}{4}\sin t\right)\\ -2e^{4t} & e^{-2t}\left(-\frac{1}{2}\cos t + \frac{1}{4}\sin t\right) & e^{-2t}\left(-\frac{1}{4}\cos t - \frac{1}{2}\sin t\right)\\ e^{4t} & e^{-2t}\cos t & e^{-2t}\sin t\end{pmatrix}^{-1}\begin{pmatrix}e^{4t}\\0\\t\end{pmatrix}dt$$

$$= \begin{pmatrix}-7e^{4t} & e^{-2t}\left(\frac{3}{4}\cos t - \frac{1}{4}\sin t\right) & e^{-2t}\left(\frac{1}{4}\cos t + \frac{3}{4}\sin t\right)\\ -2e^{4t} & e^{-2t}\left(-\frac{1}{2}\cos t + \frac{1}{4}\sin t\right) & e^{-2t}\left(-\frac{1}{4}\cos t - \frac{1}{2}\sin t\right)\\ e^{4t} & e^{-2t}\cos t & e^{-2t}\sin t\end{pmatrix}\cdot$$

$$\int\begin{pmatrix}-\frac{4}{37}e^{-4t} & -\frac{4}{37}e^{-4t} & \frac{1}{37}e^{-4t}\\ \frac{4}{37}e^{2t}(\cos t - 6\sin t) & \frac{4}{37}e^{2t}(\cos t + 31\sin t) & \frac{4}{37}e^{2t}(9\cos t + 20\sin t)\\ \frac{4}{37}e^{2t}(6\cos t + \sin t) & \frac{4}{37}e^{2t}(-31\cos t + \sin t) & \frac{4}{37}e^{2t}(-20\cos t + 9\sin t)\end{pmatrix}\begin{pmatrix}e^{4t}\\0\\t\end{pmatrix}dt$$

$$= \begin{pmatrix} -7e^{4t} & e^{-2t}\left(\frac{3}{4}\cos t - \frac{1}{4}\sin t\right) & e^{-2t}\left(\frac{1}{4}\cos t + \frac{3}{4}\sin t\right) \\ -2e^{4t} & e^{-2t}\left(-\frac{1}{2}\cos t + \frac{1}{4}\sin t\right) & e^{-2t}\left(-\frac{1}{4}\cos t - \frac{1}{2}\sin t\right) \\ e^{4t} & e^{-2t}\cos t & e^{-2t}\sin t \end{pmatrix} \cdot$$

$$\int \begin{pmatrix} -\frac{4}{37} + \frac{1}{37}te^{-4t} \\ \frac{4}{37}e^{6t}(\cos t - 6\sin t) + \frac{4}{37}te^{2t}(9\cos t + 20\sin t) \\ \frac{4}{37}e^{6t}(6\cos t + \sin t) + \frac{4}{37}te^{2t}(-20\cos t + 9\sin t) \end{pmatrix} dt$$

$$= \begin{pmatrix} -7e^{4t} & e^{-2t}\left(\frac{3}{4}\cos t - \frac{1}{4}\sin t\right) & e^{-2t}\left(\frac{1}{4}\cos t + \frac{3}{4}\sin t\right) \\ -2e^{4t} & e^{-2t}\left(-\frac{1}{2}\cos t + \frac{1}{4}\sin t\right) & e^{-2t}\left(-\frac{1}{4}\cos t - \frac{1}{2}\sin t\right) \\ e^{4t} & e^{-2t}\cos t & e^{-2t}\sin t \end{pmatrix} \cdot$$

$$\begin{pmatrix} -\frac{1}{592}e^{-4t}\left(1 + 4t + 64te^{4t}\right) \\ \frac{4}{34225}e^{2t}\left(1961\cos t + 300e^{4t}\cos t - 370t\cos t - 3552\sin t - 875e^{4t}\sin t + 9065t\sin t\right) \\ \frac{4}{34225}e^{2t}\left(3552\cos t + 875e^{4t}\cos t - 9065t\cos t + 1961\sin t + 300e^{4t}\sin t - 370t\sin t\right) \end{pmatrix}$$

$$= \begin{pmatrix} \frac{23}{80} + \frac{71}{1369}e^{4t} - \frac{1}{4}t + \frac{28}{37}te^{4t} \\ -\frac{43}{200} - \frac{59}{1369}e^{4t} + \frac{3}{10}t + \frac{8}{37}te^{4t} \\ \frac{91}{400} + \frac{48}{1369}e^{4t} - \frac{1}{20}t - \frac{4}{37}te^{4t} \end{pmatrix}.$$

Thus, a general solution to the nonhomogeneous system is

$$\mathbf{X}(t) = \begin{pmatrix} -7e^{4t} & e^{-2t}\left(\frac{3}{4}\cos t - \frac{1}{4}\sin t\right) & e^{-2t}\left(\frac{1}{4}\cos t + \frac{3}{4}\sin t\right) \\ -2e^{4t} & e^{-2t}\left(-\frac{1}{2}\cos t + \frac{1}{4}\sin t\right) & e^{-2t}\left(-\frac{1}{4}\cos t - \frac{1}{2}\sin t\right) \\ e^{4t} & e^{-2t}\cos t & e^{-2t}\sin t \end{pmatrix}\begin{pmatrix} c_1 \\ c_2 \\ c_3 \end{pmatrix} + \begin{pmatrix} \frac{23}{80} + \frac{71}{1369}e^{4t} - \frac{1}{4}t + \frac{28}{37}te^{4t} \\ -\frac{43}{200} - \frac{59}{1369}e^{4t} + \frac{3}{10}t + \frac{8}{37}te^{4t} \\ \frac{91}{400} + \frac{48}{1369}e^{4t} - \frac{1}{20}t - \frac{4}{37}te^{4t} \end{pmatrix}$$

27. A solution to the corresponding homogeneous system is $\mathbf{X}_h(t) = \begin{pmatrix} 2c_1e^{5t} - c_2e^{-t} \\ c_1e^{5t} + c_2e^{-t} \end{pmatrix}$. A particular solution has the form $\mathbf{X}_p(t) = \mathbf{A}e^t$. Substitution into the nonhomogeneous system indicates that $\mathbf{A} = \begin{pmatrix} -1/2 \\ 1/4 \end{pmatrix}$ so that $\mathbf{X}(t) = \begin{pmatrix} 2c_1e^{5t} - c_2e^{-t} - \frac{1}{2}e^t \\ c_1e^{5t} + c_2e^{-t} + \frac{1}{4}e^t \end{pmatrix}$. $\mathbf{X}(0) = \begin{pmatrix} 2c_1 - c_2 - \frac{1}{2} \\ c_1 + c_2 + \frac{1}{4} \end{pmatrix} = \begin{pmatrix} -1 \\ 1 \end{pmatrix}$ implies that $c_1 = \frac{1}{12}$ and $c_2 = \frac{2}{3}$. Therefore, $\mathbf{X}(t) = \begin{pmatrix} \frac{1}{6}e^{5t} - \frac{2}{3}e^{-t} - \frac{1}{2}e^t \\ \frac{1}{12}e^{5t} + \frac{2}{3}e^{-t} + \frac{1}{4}e^t \end{pmatrix}$.

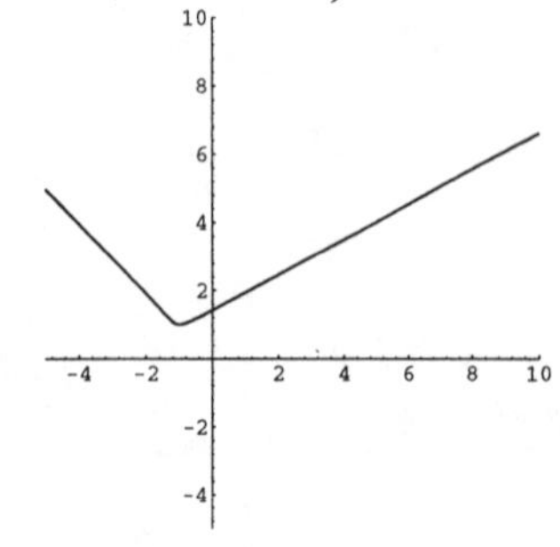

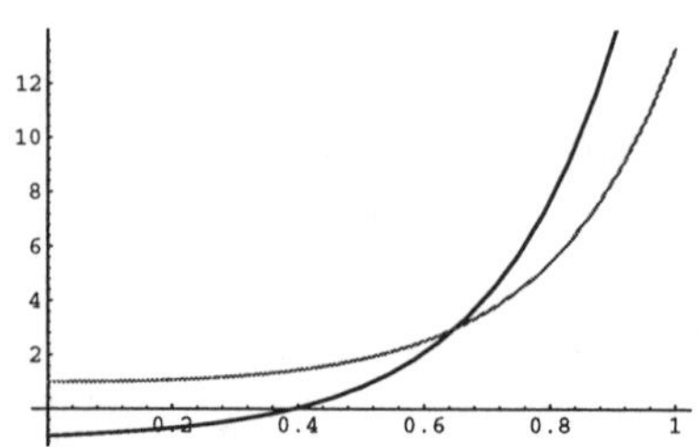

36. (a) $(\mathbf{X}_1 - \mathbf{X}_2)' = \mathbf{X}_1' - \mathbf{X}_2' = (\mathbf{A}\mathbf{X}_1 + \mathbf{F}(t)) - (\mathbf{A}\mathbf{X}_2 + \mathbf{F}(t)) = \mathbf{A}(\mathbf{X}_1 - \mathbf{X}_2)$; (b) Let $\mathbf{Y} = \mathbf{X}_{any} - \mathbf{X}_p$. Then, $\mathbf{Y}' = \mathbf{X}_{any}' - \mathbf{X}_p' = (\mathbf{A}\mathbf{X}_{any} + \mathbf{F}(t)) - (\mathbf{A}\mathbf{X}_p + \mathbf{F}(t)) = \mathbf{A}(\mathbf{X}_{any} - \mathbf{X}_p)$, so $\mathbf{Y}$ satisfies $\mathbf{X}' = \mathbf{A}\mathbf{X}$. There is a constant vector $\mathbf{C}$ so that $\mathbf{Y} = \Phi(t)\mathbf{C}$. Therefore, $\mathbf{X}_{any} - \mathbf{X}_p = \Phi(t)\mathbf{C}$ which implies that $\mathbf{X}_{any} = \Phi(t)\mathbf{C} + \mathbf{X}_p$.

37. (a)-(c)

```
sola=DSolve[{x'[t]==-7x[t]-3y[t]+1,
   y'[t]==-2x[t]-2y[t]+t  Exp[-t],
      x[0]==0,y[0]==1},{x[t],y[t]},t]
Plot[Evaluate[{x[t],y[t]}  /.  sola],
   {t,0,10},PlotStyle->{GrayLevel[0],
      GrayLevel[.5]}]
ParametricPlot[{x[t],y[t]}  /.  sola,
   {t,0,10},Compiled->False]
```

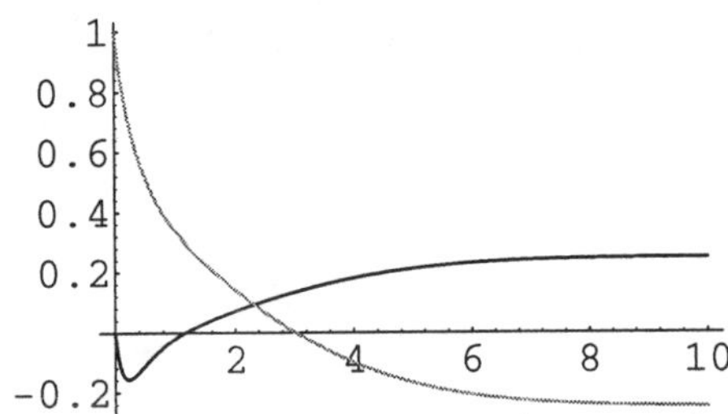


```
x:='z':y:='y':
partsol:=dsolve({diff(x(t),t)=
      -7*x(t)-3*y(t),
   diff(y(t),t)=-2*x(t)-
      2*y(t),
   x(0)=0,y(0)=1},{x(t),y(t)});
assign(partsol):
plot({x(t),y(t)},t=0..10);
plot([x(t),y(t),t=0..10]);
```

```
solb=DSolve[{x'[t]==8x[t]+10y[t]+
      t^2Exp[-2t],
   y'[t]==-7x[t]-9y[t]-t  Exp[t],
      x[0]==1,y[0]==0},{x[t],y[t]},t]
Plot[Evaluate[{x[t],y[t]}  /.  solb],
   {t,0,3},PlotStyle->{GrayLevel[0],
      GrayLevel[.5]}]
ParametricPlot[{x[t],y[t]}  /.  solb,
   {t,0,3},Compiled->False]
```

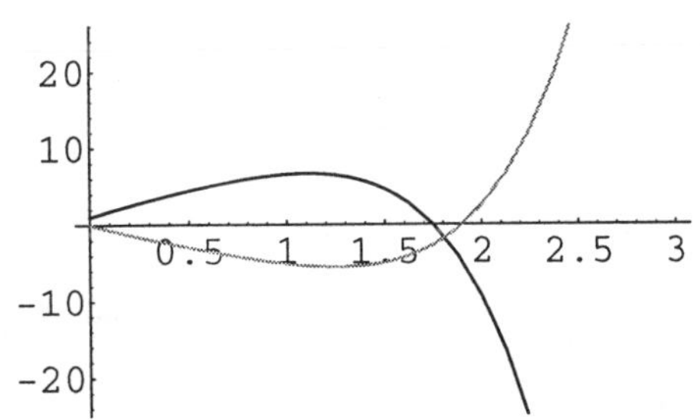


```
x:='x':y:='y':
partsol:=dsolve({diff(x(t),t)=
      8*x(t)+10*y(t)+t^2*exp(-2*t),
   diff(y(t),t)=-7*x(t)-9*y(t)-
      t*exp(t),
   x(0)=1,y(0)=0},{x(t),y(t)});
assign(partsol):
plot({x(t),y(t)},t=0..3);
plot([x(t),y(t),t=0..3]);
```

```
solc=DSolve[{x'[t]==2x[t]-
5y[t]+Sin[4t],
   y'[t]==4x[t]-2y[t]-t  Exp[-t],
      x[0]==1,y[0]==1},{x[t],y[t]},t]
<<Algebra`Trigonometry`
solc=ComplexToTrig[solc]//Simplify
Plot[Evaluate[{x[t],y[t]}  /.  solc],
   {t,0,2Pi},PlotStyle-
>{GrayLevel[0],
      GrayLevel[.5]}]
ParametricPlot[{x[t],y[t]}  /.  solc,
   {t,0,2Pi},Compiled->False]
```

```
x:='x':y:='y':
partsol:=dsolve({diff(x(t),t)=
      2*x(t)-5*y(t)+sin(4*t),
   diff(y(t),t)=4*x(t)-2*y(t)-
      t*exp(-t),
   x(0)=1,y(0)=1},{x(t),y(t)});
assign(partsol):
plot({x(t),y(t)},t=0..2*Pi);
plot([x(t),y(t),t=0..2*Pi]);
```

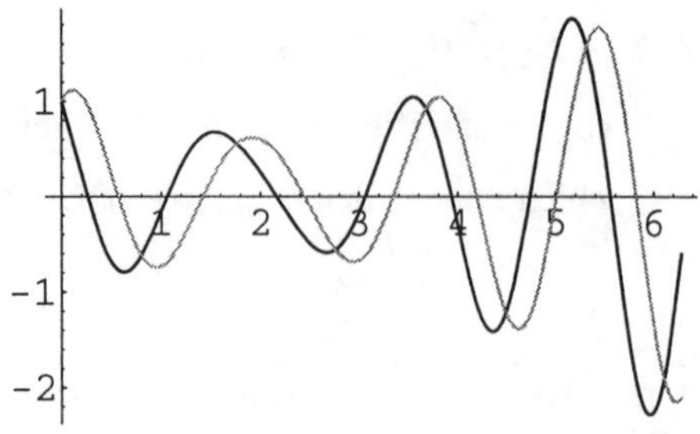

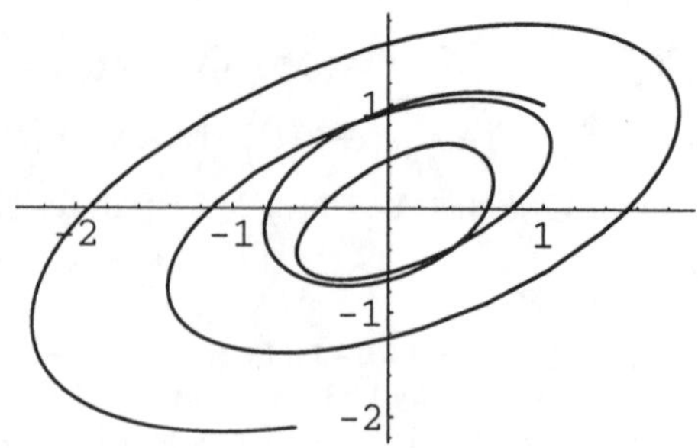

38.

```
Clear[x,y,f,t]
sol=DSolve[{x'[t]==-7x[t]+6y[t]+f[t],
   y'[t]==-8x[t]+12y[t],
   x[0]==0,y[0]==1},{x[t],y[t]},t]
```

```
x:='x':y:='y':f:='f':
partsol:=dsolve({diff(x(t),t)=
       -7*x(t)+6*y(t)+f(t),
    diff(y(t),t)=-8*x(t)+12*y(t),
    x(0)=0,y(0)=1},{x(t),y(t)});
```

EXERCISES 6.6

3. $\begin{vmatrix} 8-\lambda & -4 \\ 9 & -4-\lambda \end{vmatrix} = \lambda^2 - 4\lambda + 4$, $\lambda_1 = \lambda_2 = 2$, $\mathbf{v}_1 = \begin{pmatrix} 2 \\ 3 \end{pmatrix}$; Deficient Node, Unstable

7. $\begin{vmatrix} -1-\lambda & 0 \\ 0 & -1-\lambda \end{vmatrix} = (-1-\lambda)^2$, $\lambda_1 = \lambda_2 = -1$, $\mathbf{v}_1 = \begin{pmatrix} 1 \\ 0 \end{pmatrix}$, $\mathbf{v}_2 = \begin{pmatrix} 0 \\ 1 \end{pmatrix}$; Star Node, Asymptotically Stable

11. $\begin{vmatrix} -3-\lambda & 5 \\ -10 & 3-\lambda \end{vmatrix} = \lambda^2 + 41$, $\lambda_1 = i\sqrt{41}, \lambda_2 = -i\sqrt{41}$; Center, Stable

19. $(\lambda+3)(\lambda+3) = \lambda^2 + 6\lambda + 9$; $x'' + 6x' + 9x = 0$; $\begin{cases} x' = y \\ y' = -9x - 6y \end{cases}$; Deficient Node, Asymptotically Stable or $\begin{cases} x' = -3x \\ y' = -3y \end{cases}$; Star Node, Asymptotically Stable

23. $(-4,-2)$, $\lambda_1 = -3, \lambda_2 = -1$, Improper Node, Asymptotically Stable

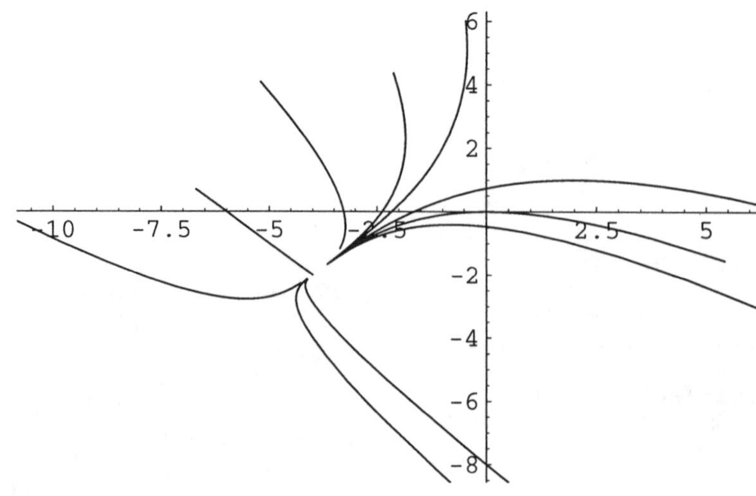

EXERCISES 6.7

3. Solve: $\begin{cases} y+y^2=0 \\ x+y=0 \end{cases}$; $y(1+y)=0$, so $y=0$ or $y=-1$; $x=-y$, so $x=0$ or $x=1$, respectively;

$J(x,y)=\begin{pmatrix} 0 & 1+2y \\ 1 & 1 \end{pmatrix}$; $J(0,0)=\begin{pmatrix} 0 & 1 \\ 1 & 1 \end{pmatrix}$, $\lambda_1=\left(1-\sqrt{5}\right)/2<0$, $\lambda_2=\left(1+\sqrt{5}\right)/2>0$; $(0,0)$, Saddle Point, Unstable; $J(1,-1)=\begin{pmatrix} 0 & -1 \\ 1 & 1 \end{pmatrix}$, $\lambda_{1,2}=\left(1\pm i\sqrt{3}\right)/2$, $(1,-1)$, Spiral Point, Unstable

7. Solve: $\begin{cases} x+y+y^2=0 \\ y=0 \end{cases}$, $y=0$, so $x=0$; $\lambda_1=\lambda_2=1$, $J(0,0)=\begin{pmatrix} 1 & 1 \\ 0 & 1 \end{pmatrix}$; $(0,0)$, Node or Spiral Point, Unstable

11. Solve: $\begin{cases} -3y-xy-4=0 \\ y^2-x^2=0 \end{cases}$, $y^2-x^2=(y+x)(y-x)=0$, $y=x$ or $y=-x$; $y=x$: $x^2+3x+4=0$, No real roots; $y=-x$: $x=-4$ or $x=1$; $J(x,y)=\begin{pmatrix} -y & -3-x \\ -2x & 2y \end{pmatrix}$, $J(-4,4)=\begin{pmatrix} -4 & 1 \\ 8 & 8 \end{pmatrix}$, $\lambda_1=2\left(1-\sqrt{11}\right)<0$, $\lambda_2=2\left(1+\sqrt{11}\right)>0$, $(-4,4)$, Saddle Point, Unstable; $J(1,-1)=\begin{pmatrix} 1 & -4 \\ -2 & -2 \end{pmatrix}$, $\lambda_1=\left(-1-\sqrt{41}\right)/2<0$, $\lambda_2=\left(-1+\sqrt{41}\right)/2>0$, $(1,-1)$, Saddle Point, Unstable

15. We find the equilibrium points by solving the system $\begin{cases} y-x=0 \\ x+y-2xy=0 \end{cases}$. From the first equation we have that $y=x$ and substitution into the second equation yields

$$\begin{aligned} x+x-2x^2&=0 \\ -2x^2+2x&=0 \\ x(x-1)&=0 \end{aligned}$$

so the equilibrium points are $(0,0)$ and $(1,1)$. For this problem,

$$J(x,y)=\begin{pmatrix} \frac{\partial}{\partial x}(y-x) & \frac{\partial}{\partial y}(y-x) \\ \frac{\partial}{\partial x}(x+y-2xy) & \frac{\partial}{\partial y}(x+y-2xy) \end{pmatrix}=\begin{pmatrix} -1 & 1 \\ 1-2y & 1-2x \end{pmatrix}.$$

The eigenvalues of $J(0,0)=\begin{pmatrix} -1 & 1 \\ 1 & 1 \end{pmatrix}$ are $\lambda_{1,2}=\pm\sqrt{2}$ so $(0,0)$ is a saddle point, unstable; the eigenvalues of $J(1,1)=\begin{pmatrix} -1 & 1 \\ -1 & -1 \end{pmatrix}$ are $\lambda_{1,2}=-1\pm i$ so $(1,1)$ is an asymptotically stable spiral point.

19. To find the equilibrium points we solve $\begin{cases} 1-xy=0 \\ y-x^3=0 \end{cases}$. Using the second equation we have that $y=x^3$ and substitution into the first equation gives us

$$\begin{aligned} 1-x^4&=0 \\ \left(1-x^2\right)\left(1+x^2\right)&=0 \\ (1-x)(1+x)\left(1+x^2\right)&=0 \end{aligned}$$

so the equilibrium points are $(1,1)$ and $(-1,-1)$. The Jacobian matrix is

$$J(x,y)=\begin{pmatrix}\frac{\partial}{\partial x}(1-xy) & \frac{\partial}{\partial y}(1-xy)\\ \frac{\partial}{\partial x}\left(y-x^3\right) & \frac{\partial}{\partial y}\left(y-x^3\right)\end{pmatrix}=\begin{pmatrix}-y & -x\\ -3x^2 & 1\end{pmatrix}.$$

The eigenvalues of $J(1,1)=\begin{pmatrix}-1 & -1\\ -3 & 1\end{pmatrix}$ are $\lambda_1=-2$ and $\lambda_2=2$ so $(1,1)$ is a saddle point, unstable; the eigenvalues of $J(-1,-1)=\begin{pmatrix}1 & 1\\ -3 & 1\end{pmatrix}$ are $\lambda_{1,2}=1\pm i\sqrt{3}$ so $(-1,-1)$ is an unstable spiral point.

25. The equilibrium points are found to be $(-5/9,-5/3)$, $(-1/5,1)$, and $(0,0)$ by solving $\begin{cases}-5xy-y^3=0\\ 3xy+y+2x=0\end{cases}$.
$-y\left(5x+y^2\right)=0\Rightarrow y=0$ or $x=-y^2/5$; $y=0$: $x=0$; $x=-y^2/5$: $-3y^3/5+y-2y^2/5=0$, $-y\left(3y^2/5+2y/5-1\right)=0$, $y=0$, $y=1$, $y=-5/3$; The only direction field with these equilibrium points is (d).

28. (a) We solve the system $\begin{cases}y+\mu x\left(x^2+y^2\right)=0\\ -x+\mu y\left(x^2+y^2\right)=0\end{cases}$ by multiplying the first equation by y and the second equation by x

$$\begin{cases}y^2+\mu xy\left(x^2+y^2\right)=0\\ -x^2+\mu xy\left(x^2+y^2\right)=0\end{cases}$$

and then subtracting the second equation from the first to obtain
$x^2+y^2=0$. The solution to this equation is $(x,y)=(0,0)$. The Jacobian matrix is

$$J(x,y)=\begin{pmatrix}\frac{\partial}{\partial x}\left[y+\mu x\left(x^2+y^2\right)\right] & \frac{\partial}{\partial y}\left[y+\mu x\left(x^2+y^2\right)\right]\\ \frac{\partial}{\partial x}\left[-x+\mu y\left(x^2+y^2\right)\right] & \frac{\partial}{\partial y}\left[-x+\mu y\left(x^2+y^2\right)\right]\end{pmatrix}=\begin{pmatrix}\mu\left(3x^2+y^2\right) & 1+2\mu xy\\ -1+2\mu xy & \mu\left(x^2+3y^2\right)\end{pmatrix}.$$

The eigenvalues of $J(0,0)=\begin{pmatrix}0 & 1\\ -1 & 0\end{pmatrix}$ are $\lambda_{1,2}=\pm i$ so $(0,0)$ is a center; the linearization about $(0,0)$ is

$$\begin{pmatrix}x'\\ y'\end{pmatrix}=\begin{pmatrix}0 & 1\\ -1 & 0\end{pmatrix}\begin{pmatrix}x\\ y\end{pmatrix}$$

or

$$\begin{cases}\dfrac{dx}{dt}=y\\ \dfrac{dy}{dt}=-x\end{cases}$$

(b) First, we differentiate:

$$x=r\cos\theta \underset{\text{product rule}}{\Rightarrow} \frac{dx}{dt}=\frac{d}{dt}(r)\cos\theta+r\frac{d}{dt}(\cos\theta) \underset{\text{chain rule}}{=} \frac{dr}{dt}\cos\theta-r\sin\theta\frac{d\theta}{dt}$$

and

$$y=r\sin\theta \underset{\text{product rule}}{\Rightarrow} \frac{dy}{dt}=\frac{d}{dt}(r)\sin\theta+r\frac{d}{dt}(\sin\theta) \underset{\text{chain rule}}{=} \frac{dr}{dt}\sin\theta+r\cos\theta\frac{d\theta}{dt}.$$

Substituting into the (original) system gives us

$$\begin{cases} \dfrac{dr}{dt}\cos\theta - r\sin\theta\dfrac{d\theta}{dt} = r\sin\theta + \mu r\cos\theta\left(r^2\cos^2\theta + r^2\sin^2\theta\right) \\ \dfrac{dr}{dt}\sin\theta + r\cos\theta\dfrac{d\theta}{dt} = -r\cos\theta + \mu r\sin\theta\left(r^2\cos^2\theta + r^2\sin^2\theta\right) \end{cases}$$

or

$$\begin{cases} \dfrac{dr}{dt}\cos\theta - r\sin\theta\dfrac{d\theta}{dt} = r\sin\theta + \mu r^3\cos\theta \\ \dfrac{dr}{dt}\sin\theta + r\cos\theta\dfrac{d\theta}{dt} = -r\cos\theta + \mu r^3\sin\theta \end{cases}$$

Multiplying the first equation by $\cos\theta$ and the second by $\sin\theta$

$$\begin{cases} \dfrac{dr}{dt}\cos^2\theta - r\cos\theta\sin\theta\dfrac{d\theta}{dt} = r\cos\theta\sin\theta + \mu r^3\cos^2\theta \\ \dfrac{dr}{dt}\sin^2\theta + r\cos\theta\sin\theta\dfrac{d\theta}{dt} = -r\cos\theta\sin\theta + \mu r^3\sin^2\theta \end{cases}$$

and then adding gives us

$$\frac{dr}{dt} = \mu r^3\cos^2\theta + \mu r^3\sin^2\theta = \mu r^3.$$

On the other hand, multiplying the first equation by $-\sin\theta$ and the second by $\cos\theta$

$$\begin{cases} -\dfrac{dr}{dt}\cos\theta\sin\theta + r\sin^2\theta\dfrac{d\theta}{dt} = -r\sin^2\theta - \mu r^3\cos\theta\sin\theta \\ \dfrac{dr}{dt}\cos\theta\sin\theta + r\cos^2\theta\dfrac{d\theta}{dt} = -r\cos^2\theta + \mu r^3\cos\theta\sin\theta \end{cases}$$

and then adding gives us

$$r\frac{d\theta}{dt} = -r$$

$$\frac{d\theta}{dt} = -1.$$

(c) $\dfrac{d\theta}{dt} = -1$ shows that $\theta(t) = -t + C$ so the solutions rotate clockwise about the origin. The equation indicates that for positive *r*, *r* increases. These observations indicate that the equilibrium point might also be classified as an unstable spiral point.

31. Regardless of the value of μ, the only equilbrium point of the system is (0,0).

```
rhseq1=mu  x+y-x(x^2+y^2);
rhseq2=mu  y-x-y(x^2+y^2);
Solve[{rhseq1==0,rhseq2==0},{x,y}]
```

```
rhseq1:=mu*x+y-x*(x^2+y^2):
rhseq2:=mu*y-x-y*(x^2+y^2):
solve({rhseq1=0,rhseq2=0},{x,y});
```

However, we see that the classification of the equilibrium point depends on the value of μ.

```
Eigenvalues[{{mu,1},{-1,mu}}]
```

```
eigenvals([[mu,1],[-1,mu]]);
```

```
Do[
   PlotVectorField[{rhseq1,rhseq2},
      {x,-1,1},{y,-1,1},
      Axes->Automatic,
      AxesOrigin->{0,0},
      PlotPoints->20,
      Ticks->{{-1,1},{-1,1}},
      ScaleFunction->(1&)],
   {mu,-1,1,1/4}]
```

```
mu:='mu':
for mu from -1 to 1 by 1/4 do
   DEplot2([rhseq1,rhseq2],[x,y],
      -10..10,x=-1..1,y=-1..1)
   od;
```

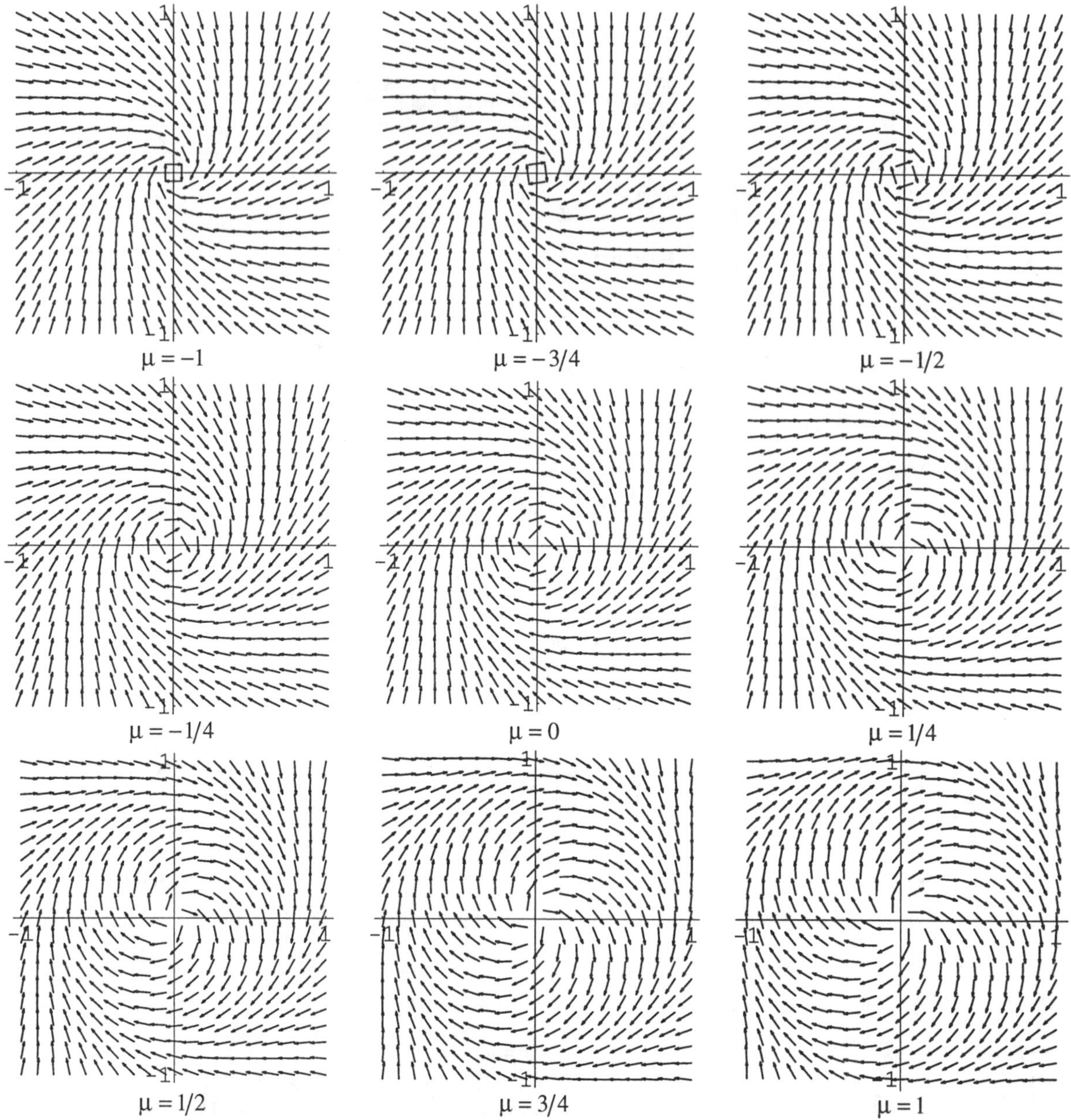

EXERCISES 6.8

We state the following algorithms that can be used to generate numerical solutions of some initial-value problems of the form $\begin{cases} \frac{dx}{dt} = f(x,y,t) \\ \frac{dy}{dt} = g(x,y,t) \\ x(0) = x_0,\ y(0) = y_0 \end{cases}$. For each algorithm, be sure to define f, g, the desired stepsize, h, and the initial conditions by replacing each question mark (?) with the appropriate value and/or definition. The following algorithm implements Euler's method.

Mathematica

```
Clear[f,g,t,h,xe,ye]
f[t_,x_,y_]=?;
g[t_,x_,y_]=?;
h=?;
t[n_]:=t0+n h;
t0=?;
xe[n_]:=xe[n]=xe[n-1]+h  f[t[n-1],
   xe[n-1],ye[n-1]];
ye[n_]:=ye[n]=ye[n-1]+h  g[t[n-1],
   xe[n-1],ye[n-1]];
xe[0]=?;
ye[0]=?;
```

Maple V

```
f:='f':g:='g':t:='t':h:='h':x:='x':
y:='y':
f:=(t,x,y)->?:
g:=(t,x,y)->?:
h:=?:
t0:=?:
t:=n->t0+n*h:
x:=proc(n) option remember;
   x(n-1)+h*f(t(n-1),x(n-1),y(n-1))
   end:
x(0):=?:
y:=proc(n) option remember;
   y(n-1)+h*g(t(n-1),x(n-1),y(n-1))
   end:
y(0):=?:
```

The following implements the Runge-Kutta method of order four.

```
Clear[t0,f,g,x,y,t,k1,k2,k3,k4,
   m1,m2,m3,m4,xr,yr]
f[t_,x_,y_]=?;
g[t_,x_,y_]=?;
t0=?;
h=?;
t[n_]:=t0+n h
x[n_]:=x[n]=x[n-1]+h/6(k1[n-1]+
   2 k2[n-1]+2 k3[n-1]+k4[n-1]);
x[0]=?;
y[n_]:=y[n]=y[n-1]+h/6(m1[n-1]+
   2 m2[n-1]+2 m3[n-1]+m4[n-1]);
y[0]=?;
k1[n_]:=k1[n]=f[t[n],x[n],y[n]];
k2[n_]:=k2[n]=f[t[n]+h/2,x[n]+
   h*k1[n]/2,y[n]+h  m1[n]/2];
k3[n_]:=k3[n]=f[t[n]+h/2,x[n]+
   h*k2[n]/2,y[n]+h  m2[n]/2];
k4[n_]:=k4[n]=f[t[n]+h,x[n]+
   h*k3[n],y[n]+h  m3[n]];
m1[n_]:=m1[n]=g[t[n],x[n],y[n]];
m2[n_]:=m2[n]=g[t[n]+h/2,x[n]+
   h*k1[n]/2,y[n]+h  m1[n]/2];
m3[n_]:=m3[n]=g[t[n]+h/2,x[n]+
   h*k2[n]/2,y[n]+h  m2[n]/2];
m4[n_]:=m4[n]=g[t[n]+h,x[n]+
   h*k3[n],y[n]+h  m3[n]];
```

```
f:=(t,x,y)->?:
g:=(t,x,y)->?:
h:=?:
t0:=?:
t:=n->t0+n*h:

xrk:='xrk':yrk:='yrk':
xrk:=proc(n)
   local k1,m1,k2,m2,k3,m3,k4,m4;
   option remember;
   k1:=f(t(n-1),xrk(n-1),yrk(n-1));
   m1:=g(t(n-1),xrk(n-1),yrk(n-1));
   k2:=f(t(n-1)+h/2,xrk(n-1)+
      h*k1/2,yrk(n-1)+h*m1/2);
   m2:=g(t(n-1)+h/2,xrk(n-1)+
      h*k1/2,yrk(n-1)+h*m1/2);
   k3:=f(t(n-1)+h/2,xrk(n-1)+
      h*k2/2,yrk(n-1)+h*m2/2);
   m3:=g(t(n-1)+h/2,xrk(n-1)+
      h*k2/2,yrk(n-1)+h*m2/2);
   k4:=f(t(n-1)+h,xrk(n-1)+
      h*k3,yrk(n-1)+h*m3);
   m4:=g(t(n-1)+h,xrk(n-1)+
      h*k3,yrk(n-1)+h*m3);
   xrk(n-1)+h/6*(k1+2*k2+2*k3+k4)
   end:
xrk(0):=?:
```

```
yrk:=proc(n)
   local k1,m1,k2,m2,k3,m3,k4,m4;
   option remember;
   k1:=f(t(n-1),xrk(n-1),yrk(n-1));
   m1:=g(t(n-1),xrk(n-1),yrk(n-1));
   k2:=f(t(n-1)+h/2,xrk(n-1)+
      h*k1/2,yrk(n-1)+h*m1/2);
   m2:=g(t(n-1)+h/2,xrk(n-1)+
      h*k1/2,yrk(n-1)+h*m1/2);
   k3:=f(t(n-1)+h/2,xrk(n-1)+
      h*k2/2,yrk(n-1)+h*m2/2);
   m3:=g(t(n-1)+h/2,xrk(n-1)+
      h*k2/2,yrk(n-1)+h*m2/2);
   k4:=f(t(n-1)+h,xrk(n-1)+
      h*k3,yrk(n-1)+h*m3);
   m4:=g(t(n-1)+h,xrk(n-1)+
      h*k3,yrk(n-1)+h*m3);
   yrk(n-1)+h/6*(m1+2*m2+2*m3+m4)
   end:
yrk(0):=?:
```

3. $x_{n+1} = x_n + h(3x_n - 5y_n)$,

$y_{n+1} = y_n + h\left(x_n - 2y_n + t_n^2\right)$,

$x_1 = x_0 + 0.1(3x_0 - 5y_0) = -1 + 0.1(3(-1) - 5(0)) = -1.3$,

$y_1 = y_0 + 0.1\left(x_0 - 2y_0 + t_0^2\right) = -0.1$

t_n	x_n	y_n
0	-1	0
0.1	-1.3	-0.1
0.2	-1.64	-0.209
0.3	-2.0275	-0.3272
0.4	-2.47215	-0.45551
0.5	-2.98604	-0.595623
0.6	-3.58404	-0.750102
0.7	-4.2842	-0.922486
0.8	-5.10822	-1.11741
0.9	-6.08198	-1.34075
1.	-7.2362	-1.5998

9.

t_n	x_n	y_n
0	0	0
0.1	-0.0404605	0.0954745
0.2	-0.156449	0.165045
0.3	-0.325008	0.188022
0.4	-0.507199	0.153632
0.5	-0.656017	0.0636206
0.6	-0.726213	-0.0679046
0.7	-0.683857	-0.217351
0.8	-0.513564	-0.356601
0.9	-0.221824	-0.458932
1.	0.164251	-0.504587

15 and 21. $\begin{cases} x' = y \\ y' = -9x \\ x(0) = 0,\ y(0) = 3 \end{cases}$; Euler's method yields $x(1) \approx 0.346313,\ y(1) \approx -4.49743$; Exact solution is $x(t) = \sin 3t$ so $x(1) = \sin 3 \approx 0.14112$; Runge-Kutta yields $x(1) \approx 0.141307,\ y(1) \approx -2.96975$.

	(Euler's Method)		**(Runge-Kutta**		**Exact**	
t_n	x_n	y_n	x_n	y_n	x_n	y_n
0	0	3	0	3	0	3.
0.1	0.3	3	0.2955	2.86601	0.29552	2.86601
0.2	0.6	2.73	0.564604	2.47605	0.564642	2.47601
0.3	0.873	2.19	0.783279	1.86494	0.783327	1.86483
0.4	1.092	1.4043	0.931992	1.08727	0.932039	1.08707
0.5	1.23243	0.4215	0.997463	0.2125	0.997495	0.212212
0.6	1.27458	-0.687687	0.973845	-0.681242	0.973848	-0.681606
0.7	1.20581	-1.83481	0.863248	-1.51413	0.863209	-1.51454
0.8	1.02233	-2.92004	0.675552	-2.21177	0.675463	-2.21218
0.9	0.730326	-3.84014	0.42752	-2.71187	0.42738	-2.71222
1.	0.346313	-4.49743	0.141307	-2.96975	0.14112	-2.96998

26. $$\begin{cases} x(t) = -\cos 3t \cosh 2t - \frac{1}{3}\cosh 2t \sin 3t + \cos 3t \sinh 2t + \frac{1}{3}\sin 3t \sinh 2t \\ y(t) = 13\left(\frac{1}{13}\cos 3t \cosh 2t + \frac{11}{39}\cosh 2t \sin 3t - \frac{1}{13}\cos 3t \sinh 2t - \frac{11}{39}\sin 3t \sinh 2t\right) \end{cases}$$

```
Clear[t0,f,g,x,y,t,k1,k2,k3,k4,
   m1,m2,m3,m4,xr,yr]
f[t_,x_,y_]=y;
g[t_,x_,y_]=-4x-13y;
t0=0;
h=0.1;
t[n_]:=t0+n h
x[n_]:=x[n]=x[n-1]+h/6(k1[n-1]+
   2 k2[n-1]+2 k3[n-1]+k4[n-1]);
x[0]=-1;
y[n_]:=y[n]=y[n-1]+h/6(m1[n-1]+
   2 m2[n-1]+2 m3[n-1]+m4[n-1]);
y[0]=1;
k1[n_]:=k1[n]=f[t[n],x[n],y[n]];
k2[n_]:=k2[n]=f[t[n]+h/2,x[n]+
   h*k1[n]/2,y[n]+h m1[n]/2];
k3[n_]:=k3[n]=f[t[n]+h/2,x[n]+
   h*k2[n]/2,y[n]+h m2[n]/2];
k4[n_]:=k4[n]=f[t[n]+h,x[n]+
   h*k3[n],y[n]+h m3[n]];
m1[n_]:=m1[n]=g[t[n],x[n],y[n]];
m2[n_]:=m2[n]=g[t[n]+h/2,x[n]+
   h*k1[n]/2,y[n]+h m1[n]/2];
m3[n_]:=m3[n]=g[t[n]+h/2,x[n]+
   h*k2[n]/2,y[n]+h m2[n]/2];
m4[n_]:=m4[n]=g[t[n]+h,x[n]+
   h*k3[n],y[n]+h m3[n]];

t2=Table[{x[n],y[n]},{n,0,10}]
lp2=ListPlot[t2,
   DisplayFunction->Identity];
```

```
f:=(t,x,y)->y:
g:=(t,x,y)->-4*x-13*y:
h:=0.1:
t0:=0:
t:=n->t0+n*h:
xrk:='xrk':yrk:='yrk':
xrk:=proc(n)
   local k1,m1,k2,m2,k3,m3,k4,m4;
   option remember;
   k1:=f(t(n-1),xrk(n-1),yrk(n-1));
   m1:=g(t(n-1),xrk(n-1),yrk(n-1));
   k2:=f(t(n-1)+h/2,xrk(n-1)+
      h*k1/2,yrk(n-1)+h*m1/2);
   m2:=g(t(n-1)+h/2,xrk(n-1)+
      h*k1/2,yrk(n-1)+h*m1/2);
   k3:=f(t(n-1)+h/2,xrk(n-1)+
      h*k2/2,yrk(n-1)+h*m2/2);
   m3:=g(t(n-1)+h/2,xrk(n-1)+
      h*k2/2,yrk(n-1)+h*m2/2);
   k4:=f(t(n-1)+h,xrk(n-1)+
      h*k3,yrk(n-1)+h*m3);
   m4:=g(t(n-1)+h,xrk(n-1)+
      h*k3,yrk(n-1)+h*m3);
   xrk(n-1)+h/6*(k1+2*k2+2*k3+k4)
   end:
xrk(0):=-1:
yrk:=proc(n)
   local k1,m1,k2,m2,k3,m3,k4,m4;
   option remember;
   k1:=f(t(n-1),xrk(n-1),yrk(n-1));
   m1:=g(t(n-1),xrk(n-1),yrk(n-1));
   k2:=f(t(n-1)+h/2,xrk(n-1)+
      h*k1/2,yrk(n-1)+h*m1/2);
   m2:=g(t(n-1)+h/2,xrk(n-1)+
      h*k1/2,yrk(n-1)+h*m1/2);
   k3:=f(t(n-1)+h/2,xrk(n-1)+
      h*k2/2,yrk(n-1)+h*m2/2);
   m3:=g(t(n-1)+h/2,xrk(n-1)+
      h*k2/2,yrk(n-1)+h*m2/2);
   k4:=f(t(n-1)+h,xrk(n-1)+
      h*k3,yrk(n-1)+h*m3);
   m4:=g(t(n-1)+h,xrk(n-1)+
      h*k3,yrk(n-1)+h*m3);
   yrk(n-1)+h/6*(m1+2*m2+2*m3+m4)
   end:
yrk(0):=1:
t1:={seq([xrk(n),yrk(n)],n=0..10)};
plot(t1,style=POINT);
```

```
Clear[x,y,t]
exactsol=DSolve[{x'[t]==y[t],
   y'[t]==-4x[t]-13y[t],
   x[0]==-1,y[0]==1},{x[t],y[t]},t]
p2=ParametricPlot[{x[t],y[t]} /.
   exactsol,{t,0,1},
   PlotStyle->GrayLevel[.6],
   DisplayFunction->Identity]
Show[p2,lp2,
   DisplayFunction->$DisplayFunction]
```

```
exactsol:=dsolve({diff(x(t),t)=y(t),
   diff(y(t),t)=-4*x(t)-13*y(t),
   x(0)=-1,y(0)=1},{x(t),y(t)});
plot(subs(exactsol,[x(t),y(t),
   t=0..1]));
```

27. $\begin{cases} x(t) = e^{-t}(2+2t) \\ y(t) = -2te^{-t} \end{cases}$

```
Clear[t0,f,g,x,y,t,k1,k2,k3,k4,
   m1,m2,m3,m4,xr,yr]
f[t_,x_,y_]=y;
g[t_,x_,y_]=-2y-x;
t0=0;
h=0.1;
t[n_]:=t0+n h
x[n_]:=x[n]=x[n-1]+h/6(k1[n-1]+
   2 k2[n-1]+2 k3[n-1]+k4[n-1]);
x[0]=2;
y[n_]:=y[n]=y[n-1]+h/6(m1[n-1]+
   2 m2[n-1]+2 m3[n-1]+m4[n-1]);
y[0]=0;
k1[n_]:=k1[n]=f[t[n],x[n],y[n]];
k2[n_]:=k2[n]=f[t[n]+h/2,x[n]+
   h*k1[n]/2,y[n]+h m1[n]/2];
k3[n_]:=k3[n]=f[t[n]+h/2,x[n]+
   h*k2[n]/2,y[n]+h m2[n]/2];
k4[n_]:=k4[n]=f[t[n]+h,x[n]+
   h*k3[n],y[n]+h m3[n]];
m1[n_]:=m1[n]=g[t[n],x[n],y[n]];
m2[n_]:=m2[n]=g[t[n]+h/2,x[n]+
   h*k1[n]/2,y[n]+h m1[n]/2];
m3[n_]:=m3[n]=g[t[n]+h/2,x[n]+
   h*k2[n]/2,y[n]+h m2[n]/2];
m4[n_]:=m4[n]=g[t[n]+h,x[n]+
   h*k3[n],y[n]+h m3[n]];
```

```
f:=(t,x,y)->y:
g:=(t,x,y)->-2*y-x:
h:=0.1:
t0:=0:
t:=n->t0+n*h:
xrk:='xrk':yrk:='yrk':
xrk:=proc(n)
   local k1,m1,k2,m2,k3,m3,k4,m4;
   option remember;
   k1:=f(t(n-1),xrk(n-1),yrk(n-1));
   m1:=g(t(n-1),xrk(n-1),yrk(n-1));
   k2:=f(t(n-1)+h/2,xrk(n-1)+
      h*k1/2,yrk(n-1)+h*m1/2);
   m2:=g(t(n-1)+h/2,xrk(n-1)+
      h*k1/2,yrk(n-1)+h*m1/2);
   k3:=f(t(n-1)+h/2,xrk(n-1)+
      h*k2/2,yrk(n-1)+h*m2/2);
   m3:=g(t(n-1)+h/2,xrk(n-1)+
      h*k2/2,yrk(n-1)+h*m2/2);
   k4:=f(t(n-1)+h,xrk(n-1)+
      h*k3,yrk(n-1)+h*m3);
   m4:=g(t(n-1)+h,xrk(n-1)+
      h*k3,yrk(n-1)+h*m3);
   xrk(n-1)+h/6*(k1+2*k2+2*k3+k4)
   end:
xrk(0):=2:
```

```
t3=Table[{x[n],y[n]},{n,0,20}]
lp3=ListPlot[t3,
   DisplayFunction->Identity];
Clear[x,y,t]
exactsol=DSolve[{x'[t]==y[t],
   y'[t]==-2y[t]-x[t],
   x[0]==2,y[0]==0},{x[t],y[t]},t]
p3=ParametricPlot[{x[t],y[t]}  /.
   exactsol,{t,0,2},
   PlotStyle->GrayLevel[.6],
   DisplayFunction->Identity]
Show[p3,lp3,
   DisplayFunction->$DisplayFunction]
```

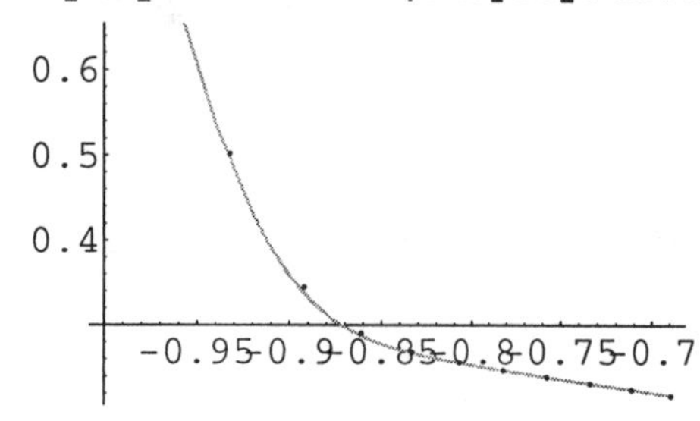

Graph for Problem 26

```
yrk:=proc(n)
   local k1,m1,k2,m2,k3,m3,k4,m4;
   option remember;
   k1:=f(t(n-1),xrk(n-1),yrk(n-1));
   m1:=g(t(n-1),xrk(n-1),yrk(n-1));
   k2:=f(t(n-1)+h/2,xrk(n-1)+
      h*k1/2,yrk(n-1)+h*m1/2);
   m2:=g(t(n-1)+h/2,xrk(n-1)+
      h*k1/2,yrk(n-1)+h*m1/2);
   k3:=f(t(n-1)+h/2,xrk(n-1)+
      h*k2/2,yrk(n-1)+h*m2/2);
   m3:=g(t(n-1)+h/2,xrk(n-1)+
      h*k2/2,yrk(n-1)+h*m2/2);
   k4:=f(t(n-1)+h,xrk(n-1)+
      h*k3,yrk(n-1)+h*m3);
   m4:=g(t(n-1)+h,xrk(n-1)+
      h*k3,yrk(n-1)+h*m3);
   yrk(n-1)+h/6*(m1+2*m2+2*m3+m4)
   end:
yrk(0):=0:
t1:={seq([xrk(n),yrk(n)],n=0..20)};
plot(t1,style=POINT);

exactsol:=dsolve({diff(x(t),t)=y(t),
   diff(y(t),t)=-2*y(t)-x(t),
   x(0)=2,y(0)=0},{x(t),y(t)});
plot(subs(exactsol,[x(t),y(t),
   t=0..2]));
```

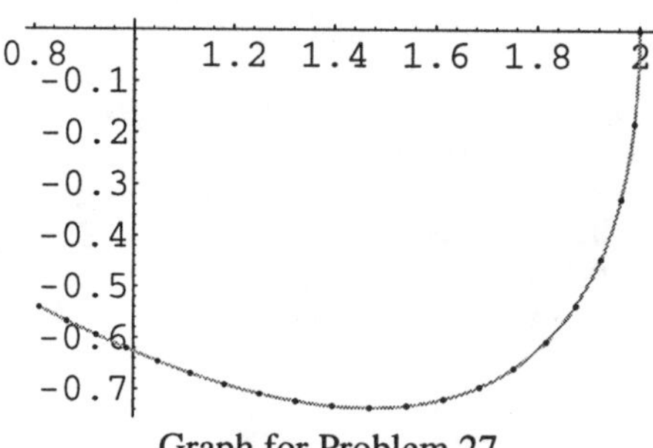

Graph for Problem 27

28. Necessary Code:
(a)

```
<<Graphics`PlotField`
pvf1=PlotVectorField[{y,-Sin[x]},
   {x,-7,7},{y,-4,4},
   DefaultColor->GrayLevel[.5],
   ScaleFunction->(1&),
   PlotPoints->25,
   DisplayFunction->Identity]
Show[pvf1,Axes->Automatic,
   AxesOrigin->{0,0},
   AxesStyle->GrayLevel[0],
   DisplayFunction->$DisplayFunction]
```

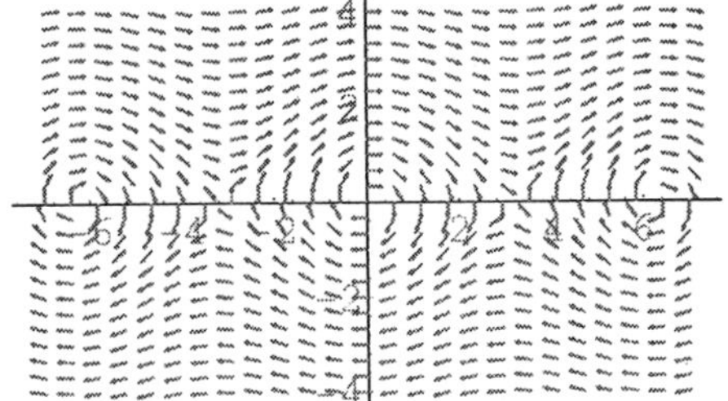

(b)

```
numsol1=NDSolve[{x1'[t]==y1[t],
   y1'[t]==-Sin[x1[t]],
   x1[0]==0,y1[0]==1},
   {x1[t],y1[t]},{t,0,7}]
p1=ParametricPlot[{x1[t],y1[t]}  /.
   numsol1,{t,0,7},
   Compiled->False,
   PlotStyle->{{GrayLevel[0],
      Thickness[0.0075]}},
   DisplayFunction->Identity]
Show[pvf1,p1,Axes->Automatic,
   AxesOrigin->{0,0},
   AxesStyle->GrayLevel[0],
   DisplayFunction->
      $DisplayFunction,
   AspectRatio->8/14]
numsol2=NDSolve[{x2'[t]==y2[t],
   y2'[t]==-Sin[x2[t]],
   x2[0]==0,y2[0]==2},
   {x2[t],y2[t]},{t,0,7}]
p2=ParametricPlot[{x2[t],y2[t]}  /.
   numsol2,{t,0,7},
   Compiled->False,
   PlotStyle->{{GrayLevel[.3],
      Thickness[0.0075]}},
   DisplayFunction->Identity]
Show[pvf1,p2,Axes->Automatic,
   AxesOrigin->{0,0},
   AxesStyle->GrayLevel[0],
   DisplayFunction->
      $DisplayFunction,
   AspectRatio->8/14]
```

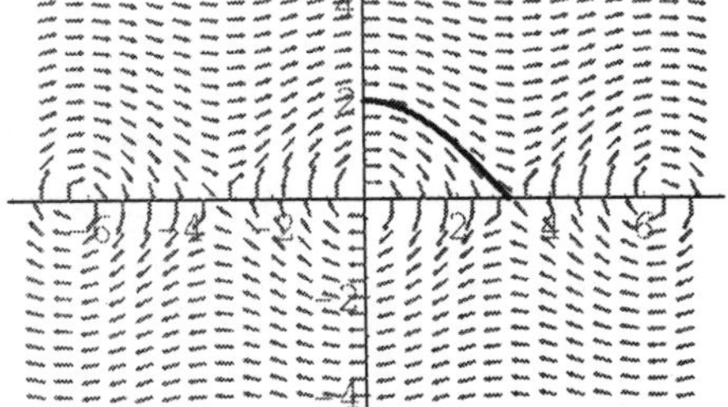

```
p3=ParametricPlot[
    {numsol1[[1,1,2]]+
       numsol2[[1,1,2]],
    numsol1[[1,2,2]]+
       numsol2[[1,2,2]]},{t,0,7},
    PlotStyle->{{GrayLevel[.15],
       Thickness[0.0075]}},
    DisplayFunction->Identity]
Show[pvf1,p3,Axes->Automatic,
    AxesOrigin->{0,0},
    AxesStyle->GrayLevel[0],
    DisplayFunction->
       $DisplayFunction,
    AspectRatio->8/14]
```

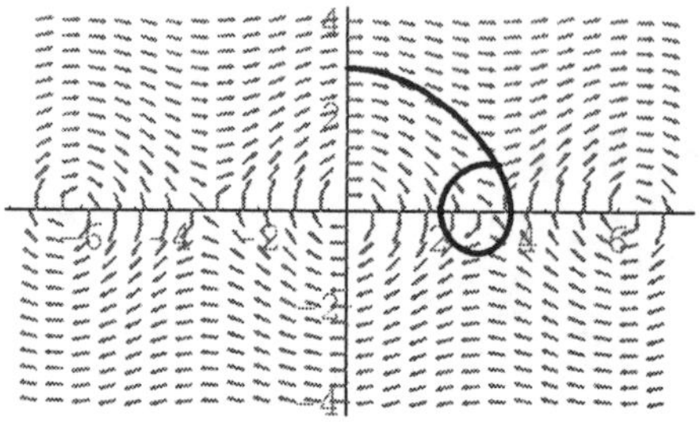

```
numsol3=NDSolve[{x'[t]==y[t],
    y'[t]==-Sin[x[t]],
    x[0]==0,y[0]==3},{x[t],y[t]},
    {t,0,7}]
p4=ParametricPlot[{x[t],y[t]}  /.
    numsol3,{t,0,7},
    Compiled->False,
    PlotStyle->{{GrayLevel[0],
       Thickness[0.0075]}},
    DisplayFunction->Identity]
Show[pvf1,p4,Axes->Automatic,
    AxesOrigin->{0,0},
    AxesStyle->GrayLevel[0],
    DisplayFunction->
       $DisplayFunction,
    AspectRatio->8/14,
    PlotRange->{{-7,7},{-4,4}}]
```

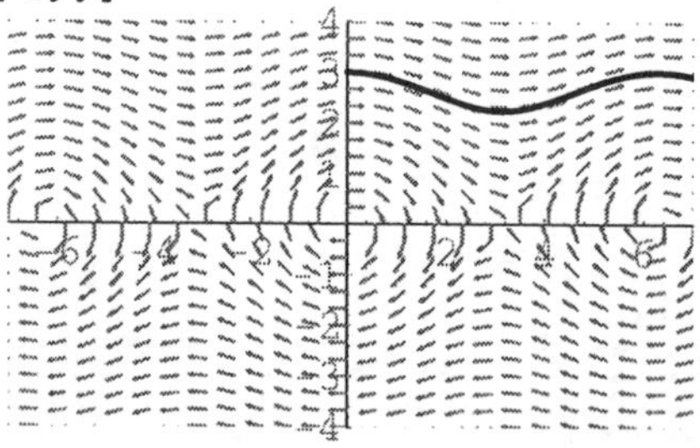

CHAPTER 6 REVIEW EXERCISES

3. First, we find the eigenvalues of $\mathbf{A}$:

$$\begin{vmatrix} -1-\lambda & 2 \\ -1 & -4-\lambda \end{vmatrix} = \lambda^2 + 5\lambda + 6 = (\lambda+3)(\lambda+2) = 0 \Rightarrow \lambda_1 = -3; \lambda_2 = -2.$$

We find an eigenvector $\mathbf{v}_1 = \begin{pmatrix} x_1 \\ y_1 \end{pmatrix}$ corresponding to $\lambda_1 = -3$ by solving

$$\left[\begin{pmatrix}-1 & 2\\-1 & -4\end{pmatrix}-(-3)\begin{pmatrix}1 & 0\\0 & 1\end{pmatrix}\right]\begin{pmatrix}x_1\\y_1\end{pmatrix}=\begin{pmatrix}0\\0\end{pmatrix}$$

$$\begin{pmatrix}2 & 2\\-1 & -1\end{pmatrix}\begin{pmatrix}x_1\\y_1\end{pmatrix}=\begin{pmatrix}0\\0\end{pmatrix}$$

$$\begin{pmatrix}1 & 1\\0 & 0\end{pmatrix}\begin{pmatrix}x_1\\y_1\end{pmatrix}=\begin{pmatrix}0\\0\end{pmatrix}\Rightarrow x_1=-y_1.$$

Choosing $y_1=1$, we obtain $\mathbf{v}_1=\begin{pmatrix}-1\\1\end{pmatrix}$. Similarly, we find an eigenvector $\mathbf{v}_2=\begin{pmatrix}x_2\\y_2\end{pmatrix}$ corresponding to $\lambda_2=-2$ by solving

$$\left[\begin{pmatrix}-1 & 2\\-1 & -4\end{pmatrix}-(-2)\begin{pmatrix}1 & 0\\0 & 1\end{pmatrix}\right]\begin{pmatrix}x_2\\y_2\end{pmatrix}=\begin{pmatrix}0\\0\end{pmatrix}$$

$$\begin{pmatrix}1 & 2\\-1 & -2\end{pmatrix}\begin{pmatrix}x_2\\y_2\end{pmatrix}=\begin{pmatrix}0\\0\end{pmatrix}$$

$$\begin{pmatrix}1 & 2\\0 & 0\end{pmatrix}\begin{pmatrix}x_2\\y_2\end{pmatrix}=\begin{pmatrix}0\\0\end{pmatrix}\Rightarrow x_2=-2y_2.$$

Choosing $y_2=1$ yields $\mathbf{v}_2=\begin{pmatrix}-2\\1\end{pmatrix}$.

7. First we find the eigenvalues of **A**:

$$\begin{vmatrix}3-\lambda & 1 & 1\\-1 & 3-\lambda & 0\\2 & 0 & 3-\lambda\end{vmatrix}=(3-\lambda)\begin{vmatrix}3-\lambda & 0\\0 & 3-\lambda\end{vmatrix}-\begin{vmatrix}-1 & 0\\2 & -3-\lambda\end{vmatrix}+\begin{vmatrix}-1 & 3-\lambda\\2 & 0\end{vmatrix}$$

$$=(3-\lambda)(3-\lambda)^2-(3+\lambda)-2(3-\lambda)$$

$$=-\lambda^3+9\lambda^2-26\lambda+24$$

$$=-(\lambda-2)(\lambda-3)(\lambda-4)=0\Rightarrow\lambda_1=2;\lambda_2=3;\lambda_3=4.$$

Now, we find corresponding eigenvectors. For $\lambda_1=2$,

$$\left[\begin{pmatrix}3 & 1 & 1\\-1 & 3 & 0\\2 & 0 & 3\end{pmatrix}-(2)\begin{pmatrix}1 & 0 & 0\\0 & 1 & 0\\0 & 0 & 1\end{pmatrix}\right]\begin{pmatrix}x_1\\y_1\\z_1\end{pmatrix}=\begin{pmatrix}0\\0\\0\end{pmatrix}$$

$$\begin{pmatrix}1 & 1 & 1\\-1 & 1 & 0\\2 & 0 & 1\end{pmatrix}\begin{pmatrix}x_1\\y_1\\z_1\end{pmatrix}=\begin{pmatrix}0\\0\\0\end{pmatrix}$$

$$\begin{pmatrix}1 & 0 & \frac{1}{2}\\0 & 1 & \frac{1}{2}\\0 & 0 & 0\end{pmatrix}\begin{pmatrix}x_1\\y_1\\z_1\end{pmatrix}=\begin{pmatrix}0\\0\\0\end{pmatrix}\Rightarrow x_1=y_1=-\tfrac{1}{2}z_1.$$

Choosing $z_1=2$ yields $\mathbf{v}_1=\begin{pmatrix}-1\\-1\\2\end{pmatrix}$. For $\lambda_2=3$ we have

$$\left[\begin{pmatrix} 3 & 1 & 1 \\ -1 & 3 & 0 \\ 2 & 0 & 3 \end{pmatrix} - (3)\begin{pmatrix} 1 & 0 & 0 \\ 0 & 1 & 0 \\ 0 & 0 & 1 \end{pmatrix}\right]\begin{pmatrix} x_2 \\ y_2 \\ z_2 \end{pmatrix} = \begin{pmatrix} 0 \\ 0 \\ 0 \end{pmatrix}$$

$$\begin{pmatrix} 0 & 1 & 1 \\ -1 & 0 & 0 \\ 2 & 0 & 0 \end{pmatrix}\begin{pmatrix} x_2 \\ y_2 \\ z_2 \end{pmatrix} = \begin{pmatrix} 0 \\ 0 \\ 0 \end{pmatrix}$$

$$\begin{pmatrix} 1 & 0 & 0 \\ 0 & 1 & 1 \\ 0 & 0 & 0 \end{pmatrix}\begin{pmatrix} x_2 \\ y_2 \\ z_2 \end{pmatrix} = \begin{pmatrix} 0 \\ 0 \\ 0 \end{pmatrix} \Rightarrow x_2 = 0 \textit{ and } y_2 = -z_2$$

and choosing $z_2 = 1$ yields $\mathbf{v}_2 = \begin{pmatrix} 0 \\ -1 \\ 1 \end{pmatrix}$. Finally, for $\lambda_3 = 4$ we have

$$\left[\begin{pmatrix} 3 & 1 & 1 \\ -1 & 3 & 0 \\ 2 & 0 & 3 \end{pmatrix} - (4)\begin{pmatrix} 1 & 0 & 0 \\ 0 & 1 & 0 \\ 0 & 0 & 1 \end{pmatrix}\right]\begin{pmatrix} x_3 \\ y_3 \\ z_3 \end{pmatrix} = \begin{pmatrix} 0 \\ 0 \\ 0 \end{pmatrix}$$

$$\begin{pmatrix} -1 & 1 & 1 \\ -1 & -1 & 0 \\ 2 & 0 & -1 \end{pmatrix}\begin{pmatrix} x_3 \\ y_3 \\ z_3 \end{pmatrix} = \begin{pmatrix} 0 \\ 0 \\ 0 \end{pmatrix}$$

$$\begin{pmatrix} 1 & 0 & -\frac{1}{2} \\ 0 & 1 & \frac{1}{2} \\ 0 & 0 & 0 \end{pmatrix}\begin{pmatrix} x_3 \\ y_3 \\ z_3 \end{pmatrix} = \begin{pmatrix} 0 \\ 0 \\ 0 \end{pmatrix} \Rightarrow x_3 = \frac{1}{2}z_3 \textit{ and } y_3 = -\frac{1}{2}z_3$$

and choosing $z_3 = 2$ yields $\mathbf{v}_3 = \begin{pmatrix} 1 \\ -1 \\ 2 \end{pmatrix}$.

13. First, we find the eigenvalues of $\begin{pmatrix} 2 & 0 \\ 1 & -1 \end{pmatrix}$:

$$\begin{vmatrix} 2-\lambda & 0 \\ 1 & -1-\lambda \end{vmatrix} = \lambda^2 - \lambda - 2 = (\lambda+1)(\lambda-2) = 0 \Rightarrow \lambda_1 = -1 \text{ and } \lambda_2 = 2.$$

Now, we find corresponding eigenvectors. For $\lambda_1 = -1$, we have

$$\left[\begin{pmatrix} 2 & 0 \\ 1 & -1 \end{pmatrix} - (-1)\begin{pmatrix} 1 & 0 \\ 0 & 1 \end{pmatrix}\right]\begin{pmatrix} x_1 \\ y_1 \end{pmatrix} = \begin{pmatrix} 0 \\ 0 \end{pmatrix}$$

$$\begin{pmatrix} 3 & 0 \\ 1 & 0 \end{pmatrix}\begin{pmatrix} x_1 \\ y_1 \end{pmatrix} = \begin{pmatrix} 0 \\ 0 \end{pmatrix}$$

$$\begin{pmatrix} 1 & 0 \\ 0 & 0 \end{pmatrix}\begin{pmatrix} x_1 \\ y_1 \end{pmatrix} = \begin{pmatrix} 0 \\ 0 \end{pmatrix} \Rightarrow x_1 = 0 \text{ so } \mathbf{v}_1 = \begin{pmatrix} 0 \\ 1 \end{pmatrix};$$

for $\lambda_2 = 2$ we have

$$\left[\begin{pmatrix}2 & 0\\ 1 & -1\end{pmatrix}-(2)\begin{pmatrix}1 & 0\\ 0 & 1\end{pmatrix}\right]\begin{pmatrix}x_2\\ y_2\end{pmatrix}=\begin{pmatrix}0\\ 0\end{pmatrix}$$

$$\begin{pmatrix}0 & 0\\ 1 & -3\end{pmatrix}\begin{pmatrix}x_2\\ y_2\end{pmatrix}=\begin{pmatrix}0\\ 0\end{pmatrix}$$

$$\begin{pmatrix}1 & -3\\ 0 & 0\end{pmatrix}\begin{pmatrix}x_2\\ y_2\end{pmatrix}=\begin{pmatrix}0\\ 0\end{pmatrix}\Rightarrow x_2=3y_2 \text{ so } \mathbf{v}_2=\begin{pmatrix}3\\ 1\end{pmatrix}.$$

Thus, a general solution to the system is $\begin{cases}x=3c_2e^{2t}\\ y=c_1e^{-t}+c_2e^{2t}\end{cases}$. Application of the initial conditions yields the system of equations $\begin{cases}3c_2=3\\ c_1+c_2=3\end{cases}$ which has solution $c_1=2$ and $c_2=1$ so the solution to the initial-value problem is $\begin{cases}x=3e^{2t}\\ y=2e^{-t}+e^{2t}\end{cases}$.

17. The eigenvalues of $\begin{pmatrix}-4 & -1\\ 5 & -2\end{pmatrix}$ are $\lambda_{1,2}=-3\pm 2i$. An eigenvector corresponding to $\lambda_1=-3+2i$ is found with

$$\left(\begin{pmatrix}-4 & -1\\ 5 & -2\end{pmatrix}-(-3+2i)\begin{pmatrix}1 & 0\\ 0 & 1\end{pmatrix}\right)\begin{pmatrix}x_1\\ y_1\end{pmatrix}=\begin{pmatrix}0\\ 0\end{pmatrix}$$

$$\begin{pmatrix}-1-2i & -1\\ 5 & 1-2i\end{pmatrix}\begin{pmatrix}x_1\\ y_1\end{pmatrix}=\begin{pmatrix}0\\ 0\end{pmatrix}$$

$$\begin{pmatrix}1 & \frac{1}{5}-\frac{2}{5}i\\ 0 & 0\end{pmatrix}\begin{pmatrix}x_1\\ y_1\end{pmatrix}=\begin{pmatrix}0\\ 0\end{pmatrix}\Rightarrow x_1=\left(-\frac{1}{5}+\frac{2}{5}i\right)y_1 \text{ so } \mathbf{v}_1=\begin{pmatrix}-1+2i\\ 5\end{pmatrix}=\underbrace{\begin{pmatrix}-1\\ 5\end{pmatrix}}_{\mathbf{a}}+\underbrace{\begin{pmatrix}2\\ 0\end{pmatrix}}_{\mathbf{b}}i.$$

A general solution is then

$$\mathbf{X}=e^{-3t}\begin{pmatrix}-\cos 2t-2\sin 2t & 2\cos 2t-\sin 2t\\ 5\cos 2t & 5\sin 2t\end{pmatrix}\begin{pmatrix}c_1\\ c_2\end{pmatrix}$$

or

$$\mathbf{X}=e^{-3t}\begin{pmatrix}\sin 2t-2\cos 2t & -\cos 2t\\ -5\sin 2t & \cos 2t-2\sin 2t\end{pmatrix}\begin{pmatrix}c_1\\ c_2\end{pmatrix}.$$

21. The eigenvalues of $\begin{pmatrix}2 & -1\\ 18 & -4\end{pmatrix}$ are $\lambda_{1,2}=-1\pm 3i$. An eigenvector corresponding to $\lambda_1=-1+3i$ satisfies

$$\left(\begin{pmatrix}2 & -1\\ 18 & -4\end{pmatrix}-(-1+3i)\right)\begin{pmatrix}x_1\\ y_1\end{pmatrix}=\begin{pmatrix}0\\ 0\end{pmatrix}$$

$$\begin{pmatrix}3-3i & -1\\ 18 & -3-3i\end{pmatrix}\begin{pmatrix}x_1\\ y_1\end{pmatrix}=\begin{pmatrix}0\\ 0\end{pmatrix}$$

$$\begin{pmatrix}1 & -\frac{1}{6}-\frac{1}{6}i\\ 0 & 0\end{pmatrix}\begin{pmatrix}x_1\\ y_1\end{pmatrix}=\begin{pmatrix}0\\ 0\end{pmatrix}$$

which indicates that $x_1=\left(\frac{1}{6}+\frac{1}{6}i\right)y_1$. Choosing $y_1=1$ yields $\mathbf{v}_1=\begin{pmatrix}\frac{1}{6}+\frac{1}{6}i\\ 1\end{pmatrix}=\begin{pmatrix}1/6\\ 1\end{pmatrix}+\begin{pmatrix}1/6\\ 0\end{pmatrix}i$. Thus, a general solution of the system is given by

$$\mathbf{X}(t)=e^{-t}\left\{c_1\left[\begin{pmatrix}1/6\\1\end{pmatrix}\cos 3t-\begin{pmatrix}1/6\\0\end{pmatrix}\sin 3t\right]+c_2\left[\begin{pmatrix}1/6\\0\end{pmatrix}\cos 3t+\begin{pmatrix}1/6\\1\end{pmatrix}\sin 3t\right]\right\}$$
$$=e^{-t}\begin{pmatrix}c_1\left(\frac{1}{6}\cos 3t-\frac{1}{6}\sin 3t\right)+c_2\left(\frac{1}{6}\cos 3t+\frac{1}{6}\sin 3t\right)\\ c_1\cos 3t+c_2\sin 3t\end{pmatrix}$$

or

$$\begin{cases}x(t)=c_1e^{-t}\left(\frac{1}{6}\cos 3t-\frac{1}{6}\sin 3t\right)+c_2e^{-t}\left(\frac{1}{6}\cos 3t+\frac{1}{6}\sin 3t\right)\\ y(t)=e^{-t}\left(c_1\cos 3t+c_2\sin 3t\right)\end{cases}.$$

Application of the initial conditions yields the system of equations

$$\begin{cases}\frac{1}{6}c_1+\frac{1}{6}c_2=2\\ c_1=3\end{cases},$$

which has solution $c_1=3$ and $c_2=9$ so the solution to the initial-value problem is

$$\begin{cases}x(t)=e^{-t}(2\cos 3t+\sin 3t)\\ y(t)=3e^{-t}(\cos 3t+3\sin 3t)\end{cases}.$$

25. The eigenvalues of $\begin{pmatrix}8&-9\\1&-2\end{pmatrix}$ are $\lambda_1=-1$ and $\lambda_2=7$ with corresponding eigenvectors $\mathbf{v}_1=\begin{pmatrix}1\\1\end{pmatrix}$ and $\mathbf{v}_2=\begin{pmatrix}9\\1\end{pmatrix}$, respectively, so a general solution of the corresponding homogeneous system $\mathbf{X}'=\begin{pmatrix}8&-9\\1&-2\end{pmatrix}\mathbf{X}$ is

$$\mathbf{X}_h=c_1e^{-t}\begin{pmatrix}1\\1\end{pmatrix}+c_2e^{7t}\begin{pmatrix}9\\1\end{pmatrix}=\begin{pmatrix}e^{-t}&9e^{7t}\\e^{-t}&e^{7t}\end{pmatrix}\begin{pmatrix}c_1\\c_1\end{pmatrix}.$$

Using the method of variation of parameters, a particular solution to the nonhomogeneous equation is

$$\mathbf{X}_p=\begin{pmatrix}e^{-t}&9e^{7t}\\e^{-t}&e^{7t}\end{pmatrix}\int\begin{pmatrix}e^{-t}&9e^{7t}\\e^{-t}&e^{7t}\end{pmatrix}^{-1}\begin{pmatrix}-8e^{7t}\\8e^{7t}\end{pmatrix}dt$$
$$=\begin{pmatrix}e^{-t}&9e^{7t}\\e^{-t}&e^{7t}\end{pmatrix}\int\begin{pmatrix}-\frac{1}{8}e^{t}&\frac{9}{8}e^{t}\\\frac{1}{8}e^{-7t}&-\frac{1}{8}e^{-7t}\end{pmatrix}\begin{pmatrix}-8e^{7t}\\8e^{7t}\end{pmatrix}dt$$
$$=\begin{pmatrix}e^{-t}&9e^{7t}\\e^{-t}&e^{7t}\end{pmatrix}\int\begin{pmatrix}10e^{8t}\\-2\end{pmatrix}dt=\begin{pmatrix}e^{-t}&9e^{7t}\\e^{-t}&e^{7t}\end{pmatrix}\begin{pmatrix}\frac{5}{4}e^{8t}\\-2t\end{pmatrix}=\begin{pmatrix}\frac{5}{4}e^{7t}-18t\,e^{7t}\\\frac{5}{4}e^{7t}-2t\,e^{7t}\end{pmatrix}.$$

A general solution of the nonhomogeneous system is

$$\mathbf{X}=\begin{pmatrix}e^{-t}&9e^{7t}\\e^{-t}&e^{7t}\end{pmatrix}\begin{pmatrix}c_1\\c_1\end{pmatrix}+\begin{pmatrix}\frac{5}{4}e^{7t}-18t\,e^{7t}\\\frac{5}{4}e^{7t}-2t\,e^{7t}\end{pmatrix}$$

or

$$\begin{cases}x(t)=c_1e^{-7t}+9c_2e^{7t}+\frac{5}{4}e^{7t}-18t\,e^{7t}\\ y(t)=c_1e^{-7t}+c_2e^{7t}+\frac{5}{4}e^{7t}-2t\,e^{7t}\end{cases}$$

29. $\mathbf{X}_h(t)=c_1\begin{pmatrix}\cos t\\\sin t\end{pmatrix}+c_2\begin{pmatrix}-\sin t\\\cos t\end{pmatrix}$, $\mathbf{X}_p(t)=\mathbf{a}t\cos t+\mathbf{b}t\sin t$,

$\mathbf{X}_p'(t) = \mathbf{a}\cos t - \mathbf{a}t\sin t + \mathbf{b}\sin t + \mathbf{b}t\cos t$, $\mathbf{AX}_p(t) + \begin{pmatrix} -\sin t \\ \cos t \end{pmatrix} = \mathbf{Aa}t\cos t + \mathbf{Ab}t\sin t + \begin{pmatrix} 0 \\ 1 \end{pmatrix}\cos t + \begin{pmatrix} -1 \\ 0 \end{pmatrix}\sin t$,

$\mathbf{a} = \begin{pmatrix} -1 \\ 0 \end{pmatrix}$, $\mathbf{b} = \begin{pmatrix} 0 \\ 1 \end{pmatrix}$, $\mathbf{X}(t) = \mathbf{X}_h(t) + \mathbf{X}_p(t) = c_1\begin{pmatrix} \cos t \\ \sin t \end{pmatrix} + c_2\begin{pmatrix} -\sin t \\ \cos t \end{pmatrix} + \begin{pmatrix} -1 \\ 0 \end{pmatrix}t\cos t + \begin{pmatrix} 0 \\ 1 \end{pmatrix}t\sin t$,

$\mathbf{X}(0) = c_1\begin{pmatrix} 1 \\ 0 \end{pmatrix} + c_2\begin{pmatrix} 0 \\ 1 \end{pmatrix} = \begin{pmatrix} 1 \\ 0 \end{pmatrix}$, $c_1 = 1$, $c_2 = 0$, $x(t) = \cos t - t\sin t$, $y(t) = \sin t + t\cos t$

Differential Equations at Work B: Controlling the Spread of a Disease

5 and 6.

```
jacmat={{D[eq1,s],D[eq1,i]},
   {D[eq2,s],D[eq2,i]}};
MatrixForm[jacmat]
s1=jacmat /. eqpts[[2]] // Simplify
s2=Eigenvalues[s1]
```

```
with(linalg):
jac_mat:=jacobian([rhs(Eq_1),
   rhs(Eq_2)],[s,i]);
s1:=subs(eq_pts[2],eval(jac_mat));
s2:=eigenvals(s1);
```

9.

```
<<Graphics`PlotField`
p1=Plot[1-x,{x,0,1},
   PlotStyle->Thickness[0.0075],
   DisplayFunction->Identity];
mu=?;
gamma=?;
sigma=?;
lambda=sigma*(gamma+mu);
pvf1=PlotVectorField[{eq1,eq2},{s,0,1
},
   {i,0,1},ScaleFunction->(1&),
   PlotPoints->20,
   DefaultColor->GrayLevel[.5],
   DisplayFunction->Identity];
initconds1=Table[{i/10,0.01},{i,1,9}]
;
initconds2=Table[{1-
i/10,i/10},{i,1,9}];
initconds=Union[initconds1,initconds2
]
numgraph[{s0_,i0_}]:=Module[{numsol},
   numsol=NDSolve[{s'[t]==
          -lambda  s[t]  i[t]+mu  -mu
s[t],
      i'[t]==lambda  s[t]  i[t]-gamma
i[t]-
          mu  i[t],
      s[0]==s0,i[0]==i0},
      {s[t],i[t]},{t,0,20}];
   ParametricPlot[{s[t],i[t]}/.
      numsol,{t,0,20},
      PlotStyle->GrayLevel[0],
      Compiled->False,
      DisplayFunction->Identity]
      ]
toshow=Map[numgraph,initconds];
Show[pvf1,toshow,p1,
   PlotRange->{{0,1},{0,1}},
   AspectRatio->1,
   DisplayFunction->$DisplayFunction,
   Axes->Automatic]
```

```
mu=0.0142857;
gamma=0.0526316;
sigma=8.1;
lambda=sigma*(gamma+mu);
pvf1=PlotVectorField[{eq1,eq2},{s,0,1
},
   {i,0,1},
   ScaleFunction->(1&),PlotPoints-
>20,
   DefaultColor->GrayLevel[.5],
   DisplayFunction->Identity];
numgraph[{s0_,i0_}]:=Module[{numsol},
   numsol=NDSolve[{s'[t]==
      -lambda s[t] i[t]+
         mu -mu s[t],
      i'[t]==lambda s[t] i[t]-
         gamma i[t]-mu i[t],
      s[0]==s0,i[0]==i0},
      {s[t],i[t]},{t,0,30}];
   ParametricPlot[{s[t],i[t]}/.
      numsol,{t,0,30},
      PlotStyle->GrayLevel[0],
      Compiled->False,
      DisplayFunction->Identity]
      ]
toshow=Map[numgraph,initconds];
Show[pvf1,toshow,p1,
   PlotRange->{{0,1},{0,1}},
   AspectRatio->1,
   DisplayFunction->$DisplayFunction,
   Axes->Automatic]
```

Applications of Systems of Ordinary Differential Equations

7

EXERCISES 7.1

1. (a) We solve the system

$$\begin{cases} \dfrac{dQ}{dt} = I \\ \dfrac{dI}{dt} = -\dfrac{1}{3\cdot 0.1}Q - \dfrac{10}{3}I \\ Q(0)=0,\ I(0)=1 \end{cases} \Leftrightarrow \begin{cases} \dfrac{dQ}{dt} = I \\ \dfrac{dI}{dt} = -\dfrac{10}{3}Q - \dfrac{10}{3}I \\ Q(0)=0,\ I(0)=1 \end{cases} \Leftrightarrow \begin{cases} \begin{pmatrix} Q' \\ I' \end{pmatrix} = \begin{pmatrix} 0 & 1 \\ -10/3 & -10/3 \end{pmatrix}\begin{pmatrix} Q \\ I \end{pmatrix} \\ \begin{pmatrix} Q(0) \\ I(0) \end{pmatrix} = \begin{pmatrix} 0 \\ 1 \end{pmatrix} \end{cases} :$$

$$\begin{cases} Q(t) = \dfrac{3\sqrt{5}}{5}e^{-5t/3}\sin\left(\dfrac{\sqrt{5}}{3}t\right) \\ I(t) = -\sqrt{5}e^{-5t/3}\sin\left(\dfrac{\sqrt{5}}{3}t\right) + e^{-5t/3}\cos\left(\dfrac{\sqrt{5}}{3}t\right) \end{cases}$$

(b) We solve the system

$$\begin{cases} \dfrac{dQ}{dt} = I \\ \dfrac{dI}{dt} = -\dfrac{1}{3\cdot 0.1}Q - \dfrac{10}{3}I + \dfrac{1}{3}e^{-t} \\ Q(0)=0,\ I(0)=1 \end{cases} \Leftrightarrow \begin{cases} \dfrac{dQ}{dt} = I \\ \dfrac{dI}{dt} = -\dfrac{10}{3}Q - \dfrac{10}{3}I + \dfrac{1}{3}e^{-t} \\ Q(0)=0,\ I(0)=1 \end{cases} \Leftrightarrow \begin{cases} \begin{pmatrix} Q' \\ I' \end{pmatrix} = \begin{pmatrix} 0 & 1 \\ -10/3 & -10/3 \end{pmatrix}\begin{pmatrix} Q \\ I \end{pmatrix} + \begin{pmatrix} 0 \\ 1/3 \end{pmatrix}e^{-t} \\ \begin{pmatrix} Q(0) \\ I(0) \end{pmatrix} = \begin{pmatrix} 0 \\ 1 \end{pmatrix} \end{cases} :$$

$$\begin{cases} Q(t) = \dfrac{1}{3}e^{-t} + \dfrac{7\sqrt{5}}{15}e^{-5t/3}\sin\left(\dfrac{\sqrt{5}}{3}t\right) - \dfrac{1}{3}e^{-5t/3}\cos\left(\dfrac{\sqrt{5}}{3}t\right) \\ I(t) = -\dfrac{1}{3}e^{-t} - \dfrac{2\sqrt{5}}{3}e^{-5t/3}\sin\left(\dfrac{\sqrt{5}}{3}t\right) + \dfrac{4}{3}e^{-5t/3}\cos\left(\dfrac{\sqrt{5}}{3}t\right) \end{cases}$$

5. We solve

$$\begin{cases} \dfrac{dQ}{dt} = -Q - I_2 \\ \dfrac{dI}{dt} = Q - 3I_2 \\ Q(0)=10^{-6},\ I_2(0)=0 \end{cases} :$$

(a) $\begin{cases} Q(t) = 10^{-6}e^{-2t}(1+t) \\ I_2(t) = 10^{-6}te^{-2t} \end{cases} \Rightarrow I(t) = \dfrac{dQ}{dt} = -10^{-6}e^{-2t}(1+2t) \Rightarrow I_1(t) = I(t) + I_2(t) = -10^{-6}e^{-2t}(1+t)$

(b) We solve

$$\begin{cases} \dfrac{dQ}{dt} = -Q - I_2 + 90 \\ \dfrac{dI}{dt} = Q - 3I_2 \\ Q(0)=0,\ I_2(0)=0 \end{cases} :$$

$$\begin{cases} Q(t) = \frac{135}{2} - \frac{135}{2}e^{-2t} - 45te^{-2t} \\ I_2(t) = \frac{45}{2} - \frac{45}{2}e^{-2t} - 45te^{-2t} \end{cases} \Rightarrow I(t) = 90e^{-2t}(1+t) \Rightarrow I_1(t) = \frac{45}{2}e^{-2t}\left(3 + e^{2t} + 2t\right)$$

9. (a) We solve $\begin{cases} \frac{dQ}{dt} = -Q - I_2 + 90 \\ \frac{dI_2}{dt} = Q - I_2 + I_3 \\ \frac{dI_3}{dt} = I_2 - I_3 - 90 \\ Q(0) = 0,\ I_2(0) = 1,\ I_3(0) = 0 \end{cases}$:

$$\begin{cases} Q(t) = 90 - te^{-t} - 90e^{-t} \\ I_2(t) = e^{-t} \\ I_3(t) = -90 + te^{-t} + 90e^{-t} \end{cases};$$

(b) We solve $\begin{cases} \frac{dQ}{dt} = -Q - I_2 + 90\sin t \\ \frac{dI_2}{dt} = Q - I_2 + I_3 \\ \frac{dI_3}{dt} = I_2 - I_3 - 90\sin t \\ Q(0) = 0,\ I_2(0) = 1,\ I_3(0) = 0 \end{cases}$:

$$\begin{cases} Q(t) = -te^{-t} + 45e^{-t} + 45\sin t - 45\cos t \\ I_2(t) = e^{-t} \\ I_3(t) = te^{-t} - 45e^{-t} - 45\sin t + 45\cos t \end{cases}$$

13. $\begin{cases} \frac{dx}{dt} = y \\ \frac{dy}{dt} = -9x \end{cases}$, $\begin{vmatrix} -\lambda & 1 \\ -9 & -\lambda \end{vmatrix} = \lambda^2 + 9 = 0 \Rightarrow \lambda = \pm 3i$; undamped; center

17. We let $y = \frac{dx}{dt}$. Then, $x'' + 10x' + 50x = 0 \Rightarrow \frac{dy}{dt} = \frac{d^2x}{dt^2} = -50x - 10x' = -50x - 10y$ so $\begin{cases} \frac{dx}{dt} = y \\ \frac{dy}{dt} = -50x - 10y \end{cases}$, $\begin{vmatrix} -\lambda & 1 \\ -50 & -10-\lambda \end{vmatrix} = \lambda^2 + 10\lambda + 50 = 0 \Rightarrow \lambda_1 = -5 + 5i, \lambda_2 = -5 - 5i$; underdamped, stable spiral

21. (13) $\begin{cases} x(t) = \cos 3t \\ y(t) = -3\sin 3t \end{cases}$; (15) $\begin{cases} x(t) = \frac{9}{8}e^{-t} - \frac{1}{8}e^{-9t} \\ y(t) = -\frac{9}{8}e^{-t} + \frac{9}{8}e^{-9t} \end{cases}$; (19) Because $\lambda_1 = \lambda_2 = -5$ with corresponding eigenvector $\mathbf{v} = \begin{pmatrix} 1 \\ -5 \end{pmatrix}$, $\mathbf{X}_1(t) = \begin{pmatrix} 1 \\ -5 \end{pmatrix}e^{-5t}$ and $\mathbf{X}_2(t) = \left[\begin{pmatrix} 1 \\ -5 \end{pmatrix}t + \begin{pmatrix} 0 \\ 1 \end{pmatrix}\right]e^{-5t}$. A general solution is $\mathbf{X}(t) = c_1\begin{pmatrix} 1 \\ -5 \end{pmatrix}e^{-5t} + c_2\left[\begin{pmatrix} 1 \\ -5 \end{pmatrix}t + \begin{pmatrix} 0 \\ 1 \end{pmatrix}\right]e^{-5t}$. $\mathbf{X}(0) = \begin{pmatrix} c_1 \\ -5c_1 + c_2 \end{pmatrix} = \begin{pmatrix} 1 \\ 0 \end{pmatrix}$ implies that $c_1 = 1$ and $c_2 = 5$..

25.

```
sys={q'[t]==i[t],i'[t]==-1/(l c)q[t]-
   r/l i[t]+e[t]/l,
   q[0]==q0,i[0]==i0}
l=1;
r=40;
c=4/1000;
e[t_]=120Sin[t];
q0=0;
i0=0;
sol1=DSolve[sys,{q[t],i[t]},t]
<<Algebra`Trigonometry`
sol1=ComplexToTrig[sol1]//Simplify
sol1b=N[sol1]
Plot[Evaluate[{q[t],i[t]}/.sol1],
   {t,0,7},PlotStyle->{GrayLevel[0],
      GrayLevel[.4]}]
ParametricPlot[{q[t],i[t]}/.sol1,
   {t,0,7},AspectRatio->1,
   PlotRange->{{-1,1},{-1,1}},
   Compiled->False]
```

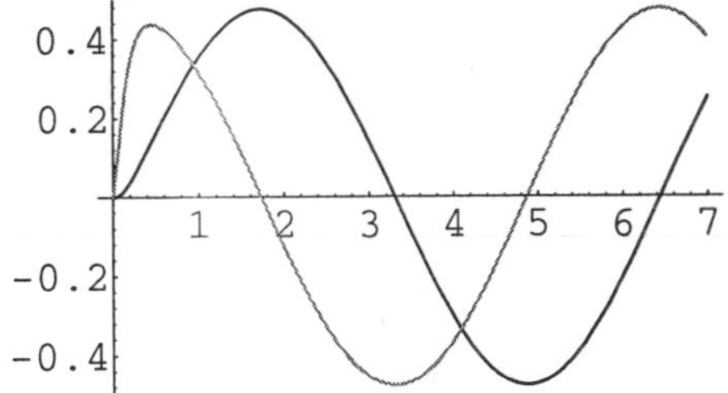


```
sys:='sys':q:='q':i:='i':
sys:={diff(q(t),t)=i(t),
   diff(i(t),t)=-1/(l*c)*q(t)-
      r/l*i(t)+e(t)/l,
   q(0)=q0,i(0)=i0};
l:=1;
r:=40;
c:=4/1000;
e:=t->120*sin(t);
q0:=0;
i0:=0;
sol1:=dsolve(sys,{q(t),i(t)});
plot(subs(sol1,{q(t),i(t)}),t=0..7);
plot(subs(sol1,[q(t),i(t),t=0..7]));
```

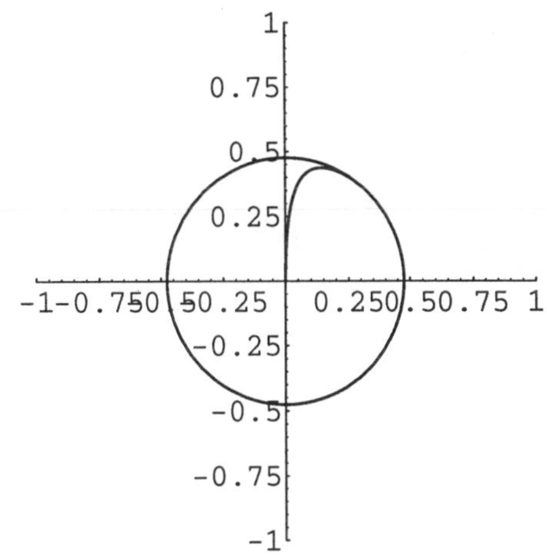

30.

```
Clear[x,y,k,m,c]
sys={x'[t]==y[t],y'[t]==-k/m x[t]-
      c/m y[t],
   x[0]==x0,y[0]==y0}
m=1;
c=1;
k=1/2;
sol=DSolve[sys,{x[t],y[t]},t]
<<Algebra`Trigonometry`
sol=ComplexToTrig[sol]//Simplify
toplot=Table[{sol[[1,1,2]],
   sol[[1,2,2]]}/.
   {x0->10 Cos[s],y0->10 Sin[s]},
      {s,0,2Pi-2Pi/9,2Pi/9}];
Short[toplot]
grays=Table[GrayLevel[i],
   {i,0,.5,.5/8}];
ParametricPlot[Evaluate[toplot],
   {t,0,10},PlotStyle->grays,
   PlotRange->{{-10,10},{-10,10}},
   AspectRatio->1]
```

```
Spr_2:=dsolve({diff(x(t),t)=y(t),
   diff(y(t),t)=-1/2*x(t)-
      y(t)},{x(t),y(t)});
to_plot:=proc(pair)
   subs({_C1=pair[1],_C2=pair[2]},
      [rhs(Spr_2[2]),rhs(Spr_2[1]),
      t=-Pi/2..2*Pi])
   end:
pairs:=[[0,0],[1,1],[-1,1],
   [1,-1],[1,2],[1,-2],
   [-1,2],[-1,-2],[2,1],
   [2,-1],[3,1],[3,-1],
   [3,2],[-2,-3],[-2,3],[-2,-1],
   [-3,1],[-3,-1],
   [-4,1],[-4,-2],[-4,3],[-4,-4],
   [-4,5]];
to_graph:=map(to_plot,pairs);
plot(convert(to_graph,set),
   view=[-5..5,-5..5]);
```

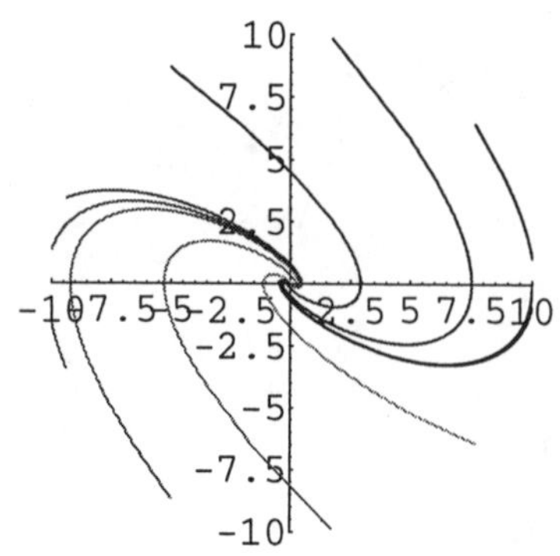

Stable Spiral Point.

31.

```
m=1;
c=2;
k=3/4;
sol=DSolve[sys,{x[t],y[t]},t]
toplot=Table[{sol[[1,1,2]],
   sol[[1,2,2]]}/.
   {x0->10 Cos[s],y0->10 Sin[s]},
      {s,0,2Pi-2Pi/9,2Pi/9}];
Short[toplot]
ParametricPlot[Evaluate[toplot],
   {t,-10,10},PlotStyle->grays,
   PlotRange->{{-10,10},{-10,10}},
   AspectRatio->1]
```

```
Spr_2:=dsolve({diff(x(t),t)=y(t),
   diff(y(t),t)=-3/4*x(t)-
     y(t)},{x(t),y(t)});
to_plot:=proc(pair)
   subs({_C1=pair[1],_C2=pair[2]},
      [rhs(Spr_2[2]),rhs(Spr_2[1]),
      t=-Pi/2..2*Pi])
   end:
pairs:=[[0,0],[1,1],[-1,1],
   [1,-1],[1,2],[1,-2],
   [-1,2],[-1,-2],[2,1],
   [2,-1],[3,1],[3,-1],
   [3,2],[-2,-3],[-2,3],[-2,-1],
   [-3,1],[-3,-1],
   [-4,1],[-4,-2],[-4,3],[-4,-4],
   [-4,5]];
to_graph:=map(to_plot,pairs);
plot(convert(to_graph,set),
   view=[-5..5,-5..5]);
```

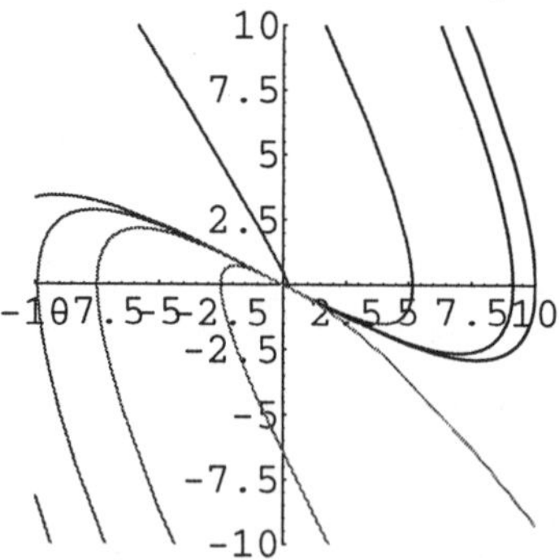

Stable Node.

EXERCISES 7.2

3. We solve $\begin{cases} x' = y - \frac{1}{2}x \\ y' = \frac{1}{2}x - y \\ x(0) = 0,\ y(0) = 4 \end{cases}$

$$\begin{cases} x(t) = \frac{8}{3} - \frac{8}{3}e^{-3t/2} \\ y(t) = \frac{4}{3} + \frac{8}{3}e^{-3t/2} \end{cases}, \quad \lim_{t\to\infty}\left(x(t), y(t)\right) = \left(\frac{8}{3}, \frac{4}{3}\right)$$

7. We solve $\begin{cases} x' = 2 - \frac{1}{5}x \\ y' = \frac{1}{5}x - \frac{1}{5}y \\ x(0) = 0,\ y(0) = 0 \end{cases}$

$\left\{x(t) = 10 - 10e^{-t/5}, y(t) = 10 - 10e^{-t/5} - 2te^{-t/5}\right\}$, $\lim_{t\to\infty} x(t) = \lim_{t\to\infty} y(t) = 10$, $x(t)$

11. We solve $\begin{cases} x' = 15 - \frac{1}{20}x \\ y' = \frac{1}{20}x - \frac{1}{10}y \\ x(0) = 0,\ y(0) = 0 \end{cases}$

$\left\{x(t) = 300 - 300e^{-t/20}, y(t) = 150 + 150e^{-t/10} - 300e^{-t/20}\right\}$, (b) $\lim_{t\to\infty} x(t) = 300$, $\lim_{t\to\infty} y(t) = 150$

15. System: $\begin{cases} \frac{dx}{dt} = 2y - 2x + 1 \\ \frac{dy}{dt} = -3y + x + 1 \end{cases}$

$\left\{x(t) = \frac{5}{4} - \frac{23}{12}e^{-4t} + \frac{8}{3}e^{-t}, y(t) = \frac{3}{4} + \frac{23}{12}e^{-4t} + \frac{4}{3}e^{-t}\right\}$, $\lim_{t\to\infty} x(t) = \frac{5}{4}$, $\lim_{t\to\infty} y(t) = \frac{3}{4}$; Max. $x = 4$, $y \approx 1$

21. We solve $\mathbf{X}' = \begin{pmatrix} 2 & 3 \\ 0 & -1 \end{pmatrix}\mathbf{X}$, $\mathbf{X}(0) = \begin{pmatrix} 5 \\ 10 \end{pmatrix}$: $\mathbf{X}(t) = \begin{pmatrix} x(t) \\ y(t) \end{pmatrix} = \begin{pmatrix} 15e^{2t} - 10e^{-t} \\ 10e^{-t} \end{pmatrix}$, $\lim_{t\to\infty} x(t) = 0$

25. We solve $\mathbf{X}' = \begin{pmatrix} 1 & 5 & 1 \\ 2 & -6 & 2 \\ 4 & 8 & 4 \end{pmatrix}\mathbf{X}$, $\mathbf{X}(0) = \begin{pmatrix} 8 \\ 2 \\ 0 \end{pmatrix}$: $\mathbf{X}(t) = \begin{pmatrix} x(t) \\ y(t) \\ z(t) \end{pmatrix} = \begin{pmatrix} \frac{43}{7} + \frac{46}{21}e^{7t} - \frac{1}{3}e^{-8t} \\ \frac{4}{3}e^{7t} + \frac{2}{3}e^{-8t} \\ -\frac{43}{7} + \frac{136}{21}e^{7t} - \frac{1}{3}e^{-8t} \end{pmatrix}$; $z(1) \approx 7095.86$

29. $\begin{vmatrix} -a-\lambda & 0 & 0 \\ a & -a-\lambda & 0 \\ 0 & a & -\lambda \end{vmatrix} = (-a-\lambda)(-a-\lambda)(-\lambda) = 0$; $\lambda_1 = \lambda_2 = -a, \lambda_3 = 0$; A general solution is

$\mathbf{X} = c_1\mathbf{X}_1(t) + c_2\mathbf{X}_2(t) + c_3\mathbf{X}_3(t)$ where $\mathbf{X}_1(t) = \begin{pmatrix} 0 \\ -1 \\ 1 \end{pmatrix}e^{-at}$, $\mathbf{X}_2(t) = \left[\begin{pmatrix} 0 \\ -1 \\ 1 \end{pmatrix}t + \begin{pmatrix} -1/a \\ 1/a \\ 0 \end{pmatrix}\right]e^{-at}$, $\mathbf{X}_3(t) = \begin{pmatrix} 0 \\ 0 \\ 1 \end{pmatrix}$.

$\mathbf{X}(0) = \begin{pmatrix} x_0 \\ y_0 \\ z_0 \end{pmatrix}$ yields $\left\{-\frac{c_2}{a} = x_0, -c_1 + \frac{c_2}{a} = y_0, c_1 + c_3 = z_0\right\}$ with solution $c_1 = -(x_0 + y_0)$, $c_2 = -ax_0$, and $c_3 = x_0 + y_0 + z_0$:

$$\begin{cases} x(t) = x_0e^{-at} \\ y(t) = y_0e^{-at} + ax_0te^{-at} \\ z(t) = x_0 + y_0 + z_0 - (x_0 + y_0)e^{-at} - ax_0te^{-at} \end{cases}; \quad \lim_{t\to\infty} x(t) = 0,\ \lim_{t\to\infty} y(t) = 0,\ \lim_{t\to\infty} z(t) = x_0 + y_0 + z_0$$

39. We solve $\begin{cases} x' = -x+10 \\ y' = x-y \\ z' = y \\ x(0)=8,\ y(0)=2,\ z(0)=2 \end{cases}$: $\begin{cases} x(t) = 10-2e^{-t} \\ y(t) = 10-8e^{-t}-2te^{-t} \\ z(t) = 10t-8+10e^{-t}+2te^{-t} \end{cases}$; $\lim_{t\to\infty} x(t)=10,\ \lim_{t\to\infty} y(t)=10,\ \lim_{t\to\infty} z(t)=\infty$

41.

```
Clear[x,y,a,b,p,v2,v1]
sol=DSolve[{x'[t]==p(y[t]/v2-
      x[t]/v1),
   y'[t]==p(x[t]/v1-y[t]/v2),
   x[0]==a,y[0]==b},{x[t],y[t]},t]
sol=Simplify[sol]
Apart[sol[[1,1,2]]]
Apart[sol[[1,2,2]]]
```

```
x:='x':y:='y':
sol:=dsolve({diff(x(t),t)=
   p*(y(t)/v2-x(t)/v1),
   diff(y(t),t)=p*(x(t)/v1-y(t)/v2),
   x(0)=a,y(0)=b},{x(t),y(t)});
```

(a) The results indicate that $\lim_{t\to\infty} x(t) = \dfrac{V_1(a+b)}{V_1+V_2}$ and $\lim_{t\to\infty} y(t) = \dfrac{V_2(a+b)}{V_1+V_2}$. (b) We must have $\dfrac{V_1(a+b)}{V_1+V_2} > \dfrac{V_2(a+b)}{V_1+V_2} \Rightarrow V_1 > V_2$. (c)

```
D[sol[[1,1,2]],t]//Apart
D[sol[[1,2,2]],t]//Apart
```

Thus, for *x(t)* to be an increasing function we must have $bV_1 - aV_2 > 0$ and for $y(t)$ to be an increasing function we must have $-bV_1 + aV_2 > 0$. Thus, it is not possible for both $x(t)$ and $y(t)$ to be increasing functions. (d)

```
Clear[x,y,a,b,p,v2,v1]
sol=DSolve[{x'[t]==p(y[t]/v1-
      x[t]/v1),
   y'[t]==p(x[t]/v1-y[t]/v1),
   x[0]==a,y[0]==a},{x[t],y[t]},t]
sol=Simplify[sol]
```

```
sol:=dsolve({diff(x(t),t)=
   p*(y(t)/v1-x(t)/v1),
   diff(y(t),t)=p*(x(t)/v1-y(t)/v1),
   x(0)=a,y(0)=a},{x(t),y(t)});
```

They are both constant.

42.

```
Clear[x,y,a,b,p,v2,v1]
sol=DSolve[{x'[t]==p(y[t]-x[t]),
   y'[t]==p(x[t]-y[t]),
   x[0]==1,y[0]==2},{x[t],y[t]},t]
sola=sol /. p->0.25
Plot[Evaluate[{x[t],y[t]} /.
   sola],{t,0,10},
   PlotStyle->{GrayLevel[0],
   GrayLevel[.4]},
   PlotRange->{0,2}]
ParametricPlot[{x[t],y[t]} /.
   sola,{t,0,10},
   PlotStyle->{GrayLevel[0],
   GrayLevel[.4]},Compiled->False,
   AxesOrigin->{0,0},
   PlotRange->{{0,5},{0,5}},
   AspectRatio->1]
```

```
sol:=dsolve({diff(x(t),t)=
   p*(y(t)-x(t)),
   diff(y(t),t)=p*(x(t)-y(t)),
   x(0)=1,y(0)=2},{x(t),y(t)});
sola:=subs(p=0.25,sol);
plot(subs(sola,{x(t),y(t)}),t=0..10);
plot(subs(sola,[x(t),y(t),t=0..10]));
```

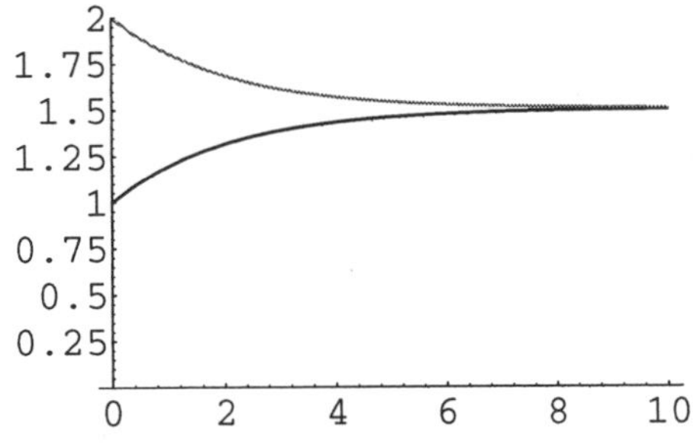

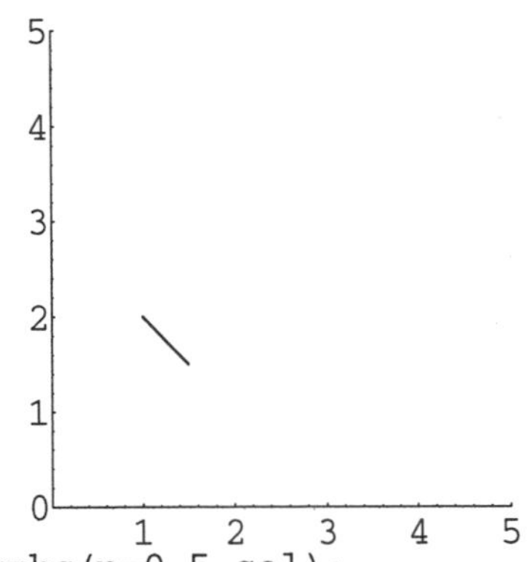

```
solb=sol /. p->0.5
Plot[Evaluate[{x[t],y[t]} /.
   solb],{t,0,10},
   PlotStyle->{GrayLevel[0],
      GrayLevel[.4]},
   AxesOrigin->{0,0},
   PlotRange->{0,2}]
ParametricPlot[{x[t],y[t]} /.
   solb,{t,0,10},
   PlotStyle->{GrayLevel[0],
   GrayLevel[.4]},Compiled->False,
   AxesOrigin->{0,0},
   PlotRange->{{0,5},{0,5}},
   AspectRatio->1]
```

```
solb:=subs(p=0.5,sol);
plot(subs(solb,{x(t),y(t)}),t=0..10);
plot(subs(solb,[x(t),y(t),t=0..10]));
```

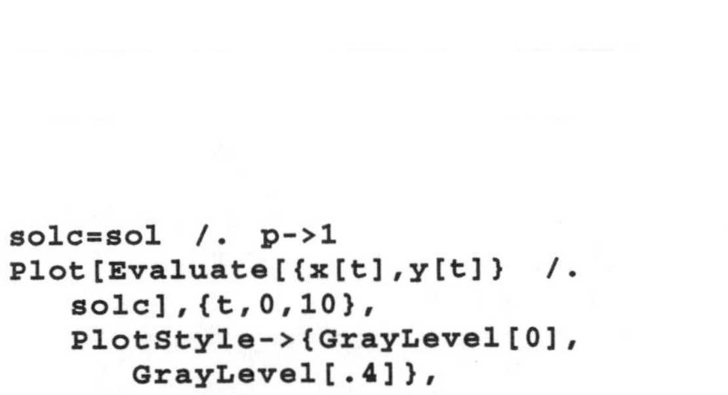

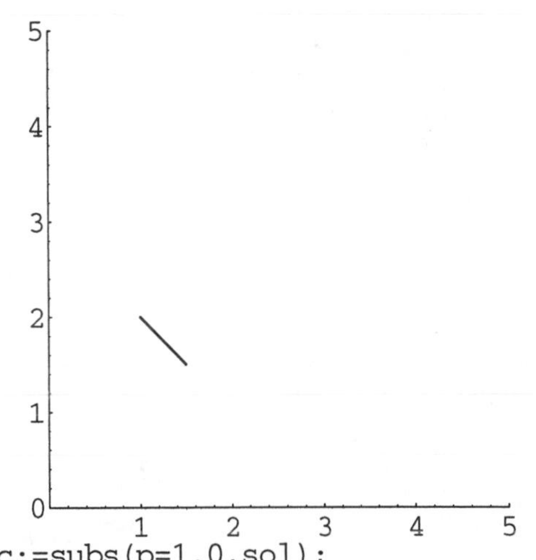

```
solc=sol /. p->1
Plot[Evaluate[{x[t],y[t]} /.
   solc],{t,0,10},
   PlotStyle->{GrayLevel[0],
      GrayLevel[.4]},
   AxesOrigin->{0,0},
   PlotRange->{0,2}]
ParametricPlot[{x[t],y[t]} /.
   solc,{t,0,10},
   PlotStyle->{GrayLevel[0],
      GrayLevel[.4]},Compiled->False,
   AxesOrigin->{0,0},
   PlotRange->{{0,5},{0,5}},
   AspectRatio->1]
```

```
solc:=subs(p=1.0,sol);
plot(subs(solc,{x(t),y(t)}),t=0..10);
plot(subs(solc,[x(t),y(t),t=0..10]));
```

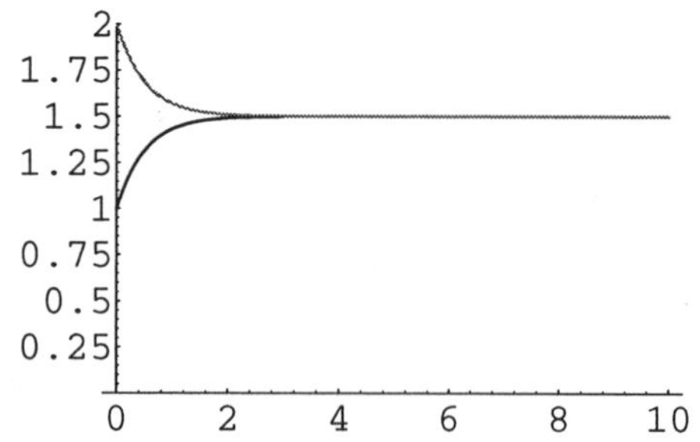


```
sold=sol /. p->2
Plot[Evaluate[{x[t],y[t]} /.
   sold],{t,0,10},
   PlotStyle->{GrayLevel[0],
      GrayLevel[.4]},
   AxesOrigin->{0,0},
   PlotRange->{0,2}]
ParametricPlot[{x[t],y[t]} /.
   sold,{t,0,10},
   PlotStyle->{GrayLevel[0],
      GrayLevel[.4]},
   Compiled->False,
   AxesOrigin->{0,0},
   PlotRange->{{0,5},{0,5}},
   AspectRatio->1]
```

```
sold:=subs(p=2.0,sol);
plot(subs(sold,{x(t),y(t)}),t=0..10);
plot(subs(sold,[x(t),y(t),t=0..10]));
```

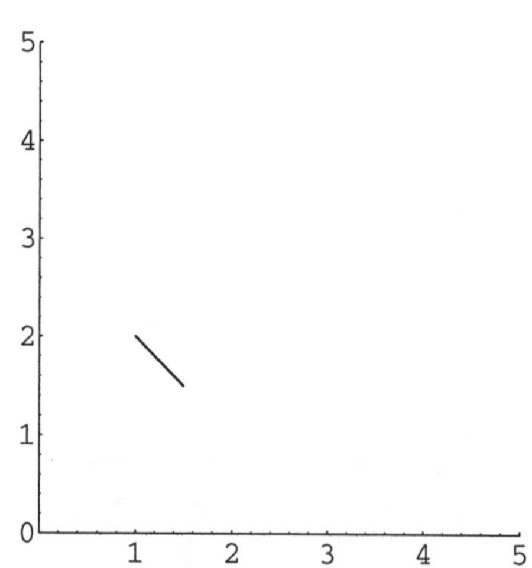

43. In matrix form the system is $\begin{pmatrix} dx/dt \\ dy/dt \end{pmatrix} = \begin{pmatrix} a_1 - a_2 & b_1 \\ a_2 & b_1 - b_2 \end{pmatrix} \begin{pmatrix} x \\ y \end{pmatrix}$. The eigenvalues of $\begin{pmatrix} a_1 - a_2 & b_1 \\ a_2 & b_1 - b_2 \end{pmatrix}$ are

$$\lambda_{1,2} = \frac{1}{2}\left(a_1 - a_2 + b_1 - b_2 \pm \sqrt{(a_1 - a_2 + b_1 - b_2)^2 - 4(a_1 b_1 - 2a_2 b_1 - a_1 b_2 + a_2 b_2)} \right).$$

```
capa={{a1-a2,b1},{a2,b1-b2}}
eigsa=Eigenvalues[capa]//Simplify
```

```
capa:=array([[a1-a2,b1],[a2,b1-b2]]);
linalg[eigenvals](capa);
```

```
DSolve[{x'[t]==(a1-a2)x[t]+b1  y[t],
   y'[t]==a2  x[t]+
      (b1-b2)y[t]},{x[t],y[t]},t]
```

```
dsolve({diff(x(t),t)=(a1-a2)*x(t)+
    b1*y(t),
    diff(y(t),t)=a2*x(t)+(b1-b2)*y(t),
    x(0)=x0,y(0)=y0},{x(t),y(t)});
```

Note: The problem may be solved as a system or the system may be written as a second order ODE where $x''=(a_1-a_2)x'+b_1y'$ (derivative of first eqn.) or $y'=\frac{1}{b_1}\left(x''-(a_1-a_2)x'\right)$. Substitution of y' and $y=\frac{1}{b_1}\left(x'-(a_1-a_2)x\right)$ into the second equation and simplification yields

$x''-\left[(a_1-a_2)+(b_1-b_2)\right]x'+\left[(b_1-b_2)(a_1-a_2)-a_2b_1\right]x=0.$

We consider the characteristics of this equation (same as the eigenvalues of the system).

Periodic if $a_1+b_1=a_2+b_2$ and $a_1b_1-2a_2b_1-a_1b_2+a_2b_2>0$. For example, if $a_1=a_2=-1$, $b_1=b_2=1$, $x_0=1$, and $y_0=0$, then $\{x(t)=\cos t,\ y(t)=-\sin t\}$.

Exponential Decay if $(b_2-b_1)>(a_1-a_2)$ and $(b_1-b_2)(a_1-a_2)-a_2b_1\geq 0$. For example, if $a_1=-1$, $a_2=0$, $b_1=b_2=1$, $x_0=1$, $y_0=0$, then $\left\{x(t)=e^{-t},\ y(t)=0\right\}$

Exponential Growth if $(a_1-a_2)>(b_2-b_1)$ (will have at least one positive eigenvalue). For example, if $a_1=2$, $a_2=0$, $b_1=b_2=1$, $x_0=1$, $y_0=0$, then $\left\{x(t)=e^{2t},\ y(t)=0\right\}$

EXERCISES 7.3

3. (a) $\begin{cases} I-2C=0 \\ 2(I-C-k)=0 \end{cases}\Rightarrow(I,C)=(2k,k)$ The eigenvalues of $\begin{pmatrix}1 & -2\\ 2 & -2\end{pmatrix}$ are $\lambda_{1,2}=-\frac{1}{2}\pm i\sqrt{7}$ so $(2k,k)$ is a stable spiral. (b) $\begin{cases} I-2C=0 \\ I-C-k=0 \end{cases}\Rightarrow(I,C)=(2k,k)$ The eigenvalues of $\begin{pmatrix}1 & -2\\ 1 & -1\end{pmatrix}$ are $\lambda_{1,2}=\pm i$ so $(2k,k)$ is a center.

7. $\begin{cases} \frac{dx}{dt}=y \\ \frac{dy}{dt}=-k^2x \end{cases}$; (b) $\frac{dy}{dx}=\frac{dy/dt}{dx/dt}=-\frac{k^2x}{y}$; (c) $\int y\,dy=-\int k^2x\,dx\Rightarrow\frac{1}{2}y^2=-\frac{k^2}{2}x^2+C$; (d) (0,0) Center; Solution is a family of ellipses as expected.

11. If $\frac{dx}{dt}=y-F(x)=y-\int_0^x f(u)du$ and $\frac{dy}{dt}=-g(x)$, then $\frac{d^2x}{dt^2}=\frac{dy}{dt}-f(x)\frac{dx}{dt}=-g(x)-f(x)\frac{dx}{dt}$ which is equivalent to $\frac{d^2x}{dt^2}+f(x)\frac{dx}{dt}+g(x)=0$.

15. $\begin{cases} \dfrac{du}{d\theta} = y \\ \dfrac{dy}{d\theta} = \alpha - u - ku^2 \end{cases}$; $y = 0$ and $\alpha - u - ku^2 = 0 \Rightarrow u = \dfrac{1 \pm \sqrt{1-4k\alpha}}{2k}$. We choose $u = \dfrac{1-\sqrt{1-4k\alpha}}{2k}$, because it is closer to $u = 0$ than $u = \dfrac{1+\sqrt{1-4k\alpha}}{2k}$, and we are considering at small change in the orbit. $J\left(\dfrac{1+\sqrt{1-4k\alpha}}{2k}, 0\right) = \begin{pmatrix} 0 & 1 \\ -\sqrt{1-4k\alpha} & 0 \end{pmatrix} \Rightarrow \lambda^2 + \sqrt{1-4k\alpha} = 0$; Center.

19. If $V(x) = \dfrac{1}{2}x^2$, then paths in phase plane are $\dfrac{1}{2}x^2 + \dfrac{1}{2}y^2 = C$.

Equilibrium point of $\begin{cases} \dfrac{dx}{dt} = y \\ \dfrac{dy}{dt} = -V'(x) = -x \end{cases}$ is (0,0) and is classified as a center.

23. $\begin{cases} \dfrac{d\theta}{dt} = y \\ \dfrac{dy}{dt} = -\dfrac{FR}{I} sgn(y) \end{cases}$; Equilibrium Points: $(\theta, 0)$; If $\dfrac{d\theta}{dt} > 0$, then we integrate $I\dfrac{d\theta}{dt}\dfrac{d}{d\theta}\left(\dfrac{d\theta}{dt}\right) = -FR$ with respect θ to obtain the parabolas $\dfrac{1}{2}I\left(\dfrac{d\theta}{dt}\right)^2 = -FR\theta + C$, $\dfrac{d\theta}{dt} > 0$. Similar calculations follow for $\dfrac{d\theta}{dt} < 0$.

26.

```
Clear[f,g]
f[x_,y_]=x-x^3/3-y;
g[x_,y_]=x+1-y;
Solve[{f[x,y]==0,g[x,y]==0}]
N[%]
<<Graphics`PlotField`
pvf=PlotVectorField[{f[x,y],g[x,y]},
   {x,-2,1},{y,-2,1},
   DefaultColor->GrayLevel[.4],
   PlotPoints->30,Axes->Automatic,
   AxesOrigin->{0,0},
   ScaleFunction->(1&)]
Clear[numsol]
numsol[{x0_,y0_}]:=NDSolve[{
   x'[t]==x[t]-x[t]^3/3-y[t],
   y'[t]==x[t]+1-y[t],
   x[0]==x0,y[0]==y0},{x[t],y[t]},
   {t,0,15}]
inits={{-2,0},{0,-2},{-0.5,-2},
   {-2,-2},{0,.75},{-1,.75},{-2,-1}};
toplot=Map[numsol,inits];
paremplots=ParametricPlot[
   Evaluate[{x[t],y[t]} /. toplot],
   {t,0,15},PlotStyle->GrayLevel[0],
   Compiled->False,
   DisplayFunction->Identity];
Show[pvf,paremplots,
   PlotRange->{{-2,1},{-2,1}}]
```

```
with(DEtools):
eqpts:=solve({x-x^3/3-y=0,
   x+1-y=0});
evalf(eqpts);
DEplot2([x-x^3/3-y,x+1-y],[x,y],
   0..15,x=-2..1,y=-2..1);
DEplot2([x-x^3/3-y,x+1-y],[x,y],
   0..15,{[0,-2,0],[0,0,-2],
   [0,-0.5,-2],[0,-2,-2],
   [0,0,0.75],[0,-1,0.75],
   [0,-2,-1]},x=-2..1,y=-2..1,
   stepsize=0.1);
```

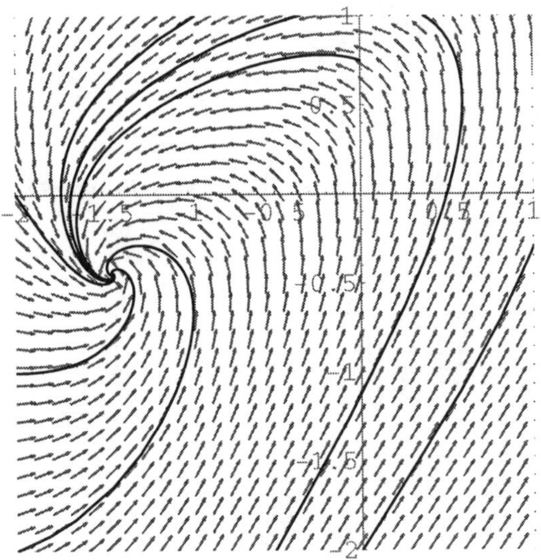

27. We guess $x_0 = 1$, graph the resulting solution parametrically, and then use the graph to obtain $x_0 = 2$.

```
n1=NDSolve[{x'[t]==y[t],
    y'[t]==-x[t]-(x[t]^2-1)y[t],
    x[0]==1,y[0]==0},{x[t],y[t]},
    {t,0,20}]
ParametricPlot[{x[t],y[t]} /. n1,
    {t,0,20},
    Compiled->False,
    PlotRange->{{-3,3},{-3,3}},
    AspectRatio->1]
n2=NDSolve[{x'[t]==y[t],
    y'[t]==-x[t]-(x[t]^2-1)y[t],
    x[0]==2,y[0]==0},{x[t],y[t]},
    {t,0,20}]
ParametricPlot[{x[t],y[t]} /. n2,
    {t,0,20},
    Compiled->False,
    PlotRange->{{-3,3},{-3,3}},
    AspectRatio->1]
```

```
DEplot2([y,-x-(x^2-1)*y],
    [x,y],0..20,{[0,1,0]},x=-3..3,
    y=-3..3,arrows='NONE',
    stepsize=0.1);
DEplot2([y,-x-(x^2-1)*y],
    [x,y],0..20,{[0,2,0]},x=-3..3,
    y=-3..3,arrows='NONE',
    stepsize=0.1);
```

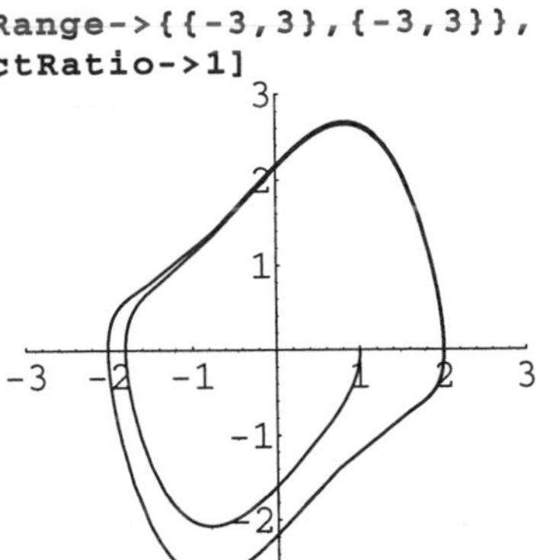

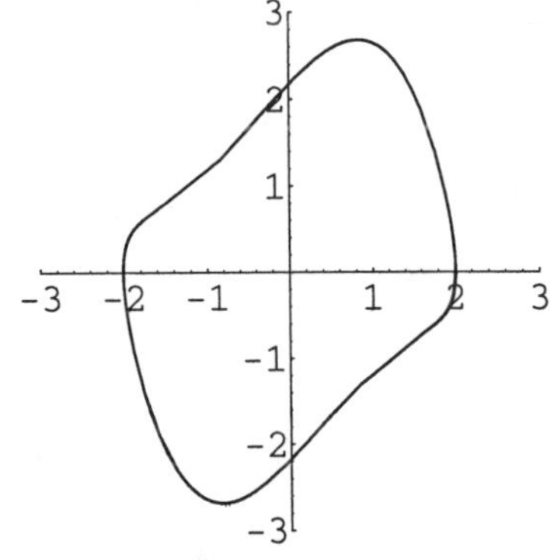

CHAPTER 7 REVIEW EXERCISES

3. (a) We solve $\begin{cases} \dfrac{dQ}{dt} = -Q - I_2 \\ \dfrac{dI_2}{dt} = Q - I_2 \\ Q(0) = 10^{-6}, I_2(0) = 0 \end{cases}$: $\begin{pmatrix} -1 & -1 \\ 1 & -1 \end{pmatrix}$ has eigenvalues and corresponding eigenvectors $\lambda_1 = -1 + i, \mathbf{v}_1 = \begin{pmatrix} i \\ 1 \end{pmatrix}$ and $\lambda_2 = -1 - i, \mathbf{v}_2 = \begin{pmatrix} -i \\ 1 \end{pmatrix}$ which results in $\begin{cases} Q(t) = \dfrac{1}{1000000} e^{-t} \cos t \\ I_2(t) = \dfrac{1}{1000000} e^{-t} \sin t \end{cases}$; (b) We solve $\begin{cases} \dfrac{dQ}{dt} = -Q - I_2 + 120 \\ \dfrac{dI_2}{dt} = Q - I_2 \\ Q(0) = 10^{-6}, I_2(0) = 0 \end{cases}$: $\begin{cases} Q(t) = 60 + 60e^{-t} \sin t - \dfrac{59999999}{1000000} e^{-t} \cos t \\ I_2(t) = 60 - \dfrac{59999999}{1000000} e^{-t} \sin t - 60e^{-t} \cos t \end{cases}$

7. We solve $\begin{cases} \dfrac{dx}{dt} = \dfrac{1}{2}(y - x) \\ \dfrac{dy}{dt} = \dfrac{1}{2}(x - y) \\ x(0) = 5, y(0) = 10 \end{cases}$: $\begin{pmatrix} -1/2 & 1/2 \\ 1/2 & -1/2 \end{pmatrix}$ has eigenvalues and corresponding eigenvectors $\lambda_1 = 0, \mathbf{v}_1 = \begin{pmatrix} 1 \\ 1 \end{pmatrix}$ and $\lambda_2 = -1, \mathbf{v}_2 = \begin{pmatrix} -1 \\ 1 \end{pmatrix}$; $\left\{ x(t) = \dfrac{15}{2} - \dfrac{5}{2} e^{-t}, y(t) = \dfrac{15}{2} + \dfrac{5}{2} e^{-t} \right\}$.

11. We solve $\begin{cases} \dfrac{dx}{dt} = -3x + 4y + 8z \\ \dfrac{dy}{dt} = x - 3y + 4z \\ \dfrac{dz}{dt} = 4x + 5y - 10z \\ x(0) = 4, y(0) = 4, z(0) = 8 \end{cases}$: $\begin{pmatrix} -3 & 4 & 8 \\ 1 & -3 & 4 \\ 4 & 5 & -10 \end{pmatrix}$ has eigenvalues and corresponding eigenvectors $\lambda_1 = 3, \mathbf{v}_1 = \begin{pmatrix} 2 \\ 1 \\ 1 \end{pmatrix}$, $\lambda_2 = -14, \mathbf{v}_2 = \begin{pmatrix} 8 \\ 4 \\ -13 \end{pmatrix}$, $\lambda_3 = -5, \mathbf{v}_3 = \begin{pmatrix} 10 \\ -7 \\ 1 \end{pmatrix}$; $\begin{cases} x(t) = -\frac{5}{3} e^{-5t} - \frac{128}{51} e^{-14t} + \frac{139}{17} e^{3t} \\ y(t) = \frac{7}{6} e^{-5t} - \frac{64}{51} e^{-14t} + \frac{139}{34} e^{3t} \\ z(t) = -\frac{1}{6} e^{-5t} + \frac{208}{51} e^{-14t} + \frac{139}{34} e^{3t} \end{cases}$

13. $(0,0)$, Saddle if $a > k$; Stable Node (Star) if $a \le k$. $\left(\dfrac{c}{d}, \dfrac{a-k}{b} \right)$, Center if $a > k$; Saddle if $a \le k$.

Differential Equations at Work B: Food Chains

```
<<Graphics`PlotField`
PlotVectorField[{36  x-4x^2-2  x  y,-y+1/4  x  y},{x,2,6},{y,8,12},
   ScaleFunction  ->  (0.5&),Axes->Automatic,AxesOrigin->{2,8},
   PlotPoints->30]

numsol=NDSolve[{x1'[t]==a*x1[t]-b*x1[t]^2-c*x1[t]*x2[t],
        x2'[t]==-d*x2[t]+e*x1[t]*x2[t],
```

```
       x1[0]==3,x2[0]==2},{x1[t],x2[t]},
       {t,0,15}]
pxy=Plot[Evaluate[{x1[t],x2[t]}/.numsol],{t,0,8},
       PlotStyle->{GrayLevel[0],GrayLevel[0.5]},AspectRatio->1,
     PlotRange->{0,11}]

pp=ParametricPlot[{x1[t],x2[t]}/.numsol,{t,0,8},Compiled->False,
     AspectRatio->1]
```

Differential Equations at Work C: Chemical Reactor

(1) and (2) are direct. (3)

```
cval=Solve[cin-c==v/q  c  kt,c]
step2=h  v/(q  cp)c  kt  /.  cval[[1]]
```

```
cval:=solve(cin-c=v/q*c*kt,c);
step2:=subs(c=cval,h*v/(q*cp)*c*kt);
```

(4) **FindRoot** (Mathematica) or **fsolve** (Maple V) may be useful.

Introduction to the Laplace Transform

8

EXERCISES 8.1

3.
$$\int_0^\infty e^{-st}\left(2e^t\right)dt = \lim_{M\to\infty}\left[\frac{2}{s-1}-\frac{2e^{M(1-s)}}{s-1}\right]$$
$$=\frac{2}{s-1},\ s>1$$

7.
$$\mathscr{L}\{f(t)\} = \int_0^\infty e^{-st}f(t)dt = \int_1^\infty e^{-st}dt$$
$$= \lim_{M\to\infty}\frac{-1}{s}e^{-st}\Big]_0^M = \frac{e^{-s}}{s},\ s>0$$

11. We use integration by parts to evaluate the resulting integral:
$$\mathscr{L}\{f(t)\} = \int_0^\infty e^{-st}f(t)dt = \int_0^3 (1-t)e^{-st}dt$$
$$= \int_0^3 e^{-st}dt - \underbrace{\int_0^3 te^{-st}dt}_{\substack{\text{use integration by parts}\\ \text{with } u=t\Rightarrow du=dt \text{ and}\\ dv=e^{-st}dt\Rightarrow v=-\frac{1}{s}e^{-st}dt}} = \frac{-1}{s}e^{-st}\Big]_0^3 - \left(\frac{-1}{s^2}e^{-st}(st+1)\right)\Big]_0^3$$
$$= \frac{1}{s}\left(1-e^{-3s}\right) - \frac{1}{s^2}\left(1-e^{-3s}(3s+1)\right) = \frac{1}{s^2}\left(e^{-3s}(2s+1)+s-1\right).$$

15.
$$\mathscr{L}\{f(t)\} = \int_0^\infty e^{-st}f(t)dt = \int_0^\infty e^{-st}\sin kt\,dt = \lim_{M\to\infty}\frac{-1}{s^2+k^2}e^{-st}(k\cos kt + s\sin kt)\Big]_0^M = \frac{k}{s^2+k^2}$$

19.
$$\mathscr{L}\left\{e^{at}f(t)\right\} = \int_0^\infty e^{-st}e^{at}f(t)dt = \int_0^\infty e^{-(s-a)t}f(t)dt = F(s-a),\ s>a$$

29. $\mathscr{L}\{\cos 5t\} = \dfrac{s}{s^2+5^2} = \dfrac{s}{s^2+25}$

37. $\mathscr{L}\left\{t^2\right\} = \dfrac{2}{s^3},\ \mathscr{L}\left\{t^2e^{-3t}\right\} = \dfrac{2}{(s+3)^3}$

45. $\mathscr{L}\{\sinh 7t\} = \dfrac{7}{s^2-49}$,
$$\mathscr{L}\{t\sinh 7t\} = -\frac{d}{ds}\left(\frac{7}{s^2-49}\right) = \frac{14s}{\left(s^2-49\right)^2}$$

51. $L\{\cos 7t\} = \dfrac{s}{s^2+49}$,
$$\mathscr{L}\left\{e^{5t}\cos 7t\right\} = \frac{s-5}{(s-5)^2+49}$$

61. (a) $\mathscr{L}\left\{\sin\left(t+\frac{\pi}{4}\right)\right\} = \mathscr{L}\left\{\frac{\sqrt{2}}{2}\cos t + \frac{\sqrt{2}}{2}\sin t\right\} = \frac{\sqrt{2}}{2}\left(\frac{s}{s^2+1}+\frac{1}{s^2+1}\right) = \frac{\sqrt{2}}{2}\left(\frac{s+1}{s^2+1}\right)$; (b) $\frac{\sqrt{2}}{2}\left(\frac{s-1}{s^2+1}\right)$; (c) $\mathscr{L}\left\{\cos\left(t+\frac{\pi}{6}\right)\right\} = \mathscr{L}\left\{\frac{\sqrt{3}}{2}\cos t - \frac{1}{2}\sin t\right\} = \frac{\sqrt{3}}{2}\frac{s}{s^2+1} - \frac{1}{2}\frac{1}{s^2+1} = \frac{1}{2}\left(\frac{\sqrt{3}s-1}{s^2+1}\right)$; (d) $\frac{1}{2}\left(\frac{s+\sqrt{3}}{s^2+1}\right)$

63. (a) $\alpha = -\frac{1}{2}$, $\mathscr{L}\left\{t^{-1/2}\right\} = \frac{1}{s^{-1/2+1}}\Gamma(-1/2+1) = \frac{1}{s^{1/2}}\Gamma\left(\frac{1}{2}\right) = \sqrt{\frac{\pi}{s}}$ because $\Gamma\left(\frac{1}{2}\right) = \sqrt{\pi}$. (b) $\alpha = \frac{1}{2}$, $\mathscr{L}\left\{t^{1/2}\right\} = \frac{1}{s^{1/2+1}}\Gamma(1/2+1) = \frac{1}{s^{3/2}}\Gamma\left(\frac{3}{2}\right) = \frac{\sqrt{\pi}}{2s^{3/2}}$ because $\Gamma\left(\frac{3}{2}\right) = \frac{1}{2}\Gamma\left(\frac{1}{2}\right)$.

69. $\mathscr{L}^{-1}\left\{\frac{1}{s^2}\right\}=t,\ te^{2t}$

75. $\mathscr{L}^{-1}\left\{\frac{s}{s^2+k^2}\right\}=\cos kt$ with $k=4$; $\cos 4t$

79. Use partial fractions to obtain $\frac{s-2}{s^2-4s}=\frac{1}{2s}+\frac{1}{2(s-4)}$. The inverse Laplace transform is $\frac{1}{2}+\frac{1}{2}e^{4t}$.

83. $\frac{1}{s^2-4s-12}=-\frac{1}{8}\frac{1}{s+2}+\frac{1}{8}\frac{1}{s-6}\Rightarrow\mathscr{L}^{-1}\left\{\frac{1}{s^2-4s-12}\right\}=-\frac{1}{8}e^{-2t}+\frac{1}{8}e^{6t}$

87. Note that

$$\frac{s-1}{s^2-2s+50}=\frac{s-1}{s^2-2s+1+49}=\frac{s-1}{(s-1)^2+49}$$

so the inverse Laplace transform is $e^t\cos 7t$.

89. Note that

$$\frac{s-7}{s^2-14s+50}=\frac{s-7}{s^2-14s+49+1}=\frac{s-7}{(s-7)^2+1}$$

so the inverse Laplace transform is $e^{7t}\cos t$.

95. Suppose that we let $C=1/2$. Then, with $b=1/2$, we compare $f(t)=\frac{1}{t}\sin t\cos 2t$ with $\frac{1}{2}e^{t/2}$ by graphing both functions.

```
Plot[{e^(t/2)/2, Sin[t] Cos[2 t]/t},
  {t, 0, 2 π},
  PlotStyle →
   {GrayLevel[.5], GrayLevel[0]}]

FindRoot[e^(t/2)/2 == Sin[t] Cos[2 t]/t,
  {t, 0.5}]

{t → 0.436697}
```

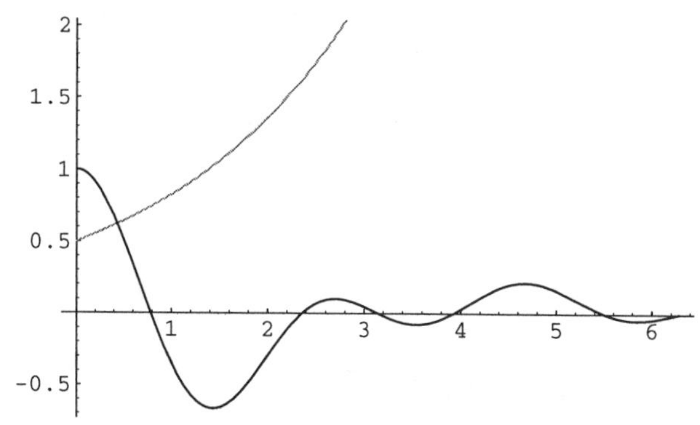

Notice that these curves intersect near $t=0.5$. We obtain a better approximation of $t\approx 0.436697..$ With $T=0.44$, $\left|\frac{1}{t}\sin t\cos 2t\right|<\frac{1}{2}e^{t/2}$ for $t>T$.

EXERCISES 8.2

3.

$$\left(s^2+3s-10\right)Y(s)+s+2=0$$

$$Y(s)=-\frac{2+s}{s^2+3s-10}$$

$$=\frac{-4}{7(s-2)}-\frac{3}{7(s+5)}$$

$$y(t)=-\frac{4}{7}e^{2t}-\frac{3}{7}e^{-5t}$$

```
sol =
 DSolve[
  {y''[t] + 3 y'[t] - 10 y[t] == 0,
   y[0] == -1, y'[0] == 1}, y[t], t]
```

$$\left\{\left\{y[t]\to e^{-5t}\left(-\frac{3}{7}-\frac{4e^{7t}}{7}\right)\right\}\right\}$$

```
Plot[y[t] /. sol, {t, -1 / 2, 3 / 2}]
```

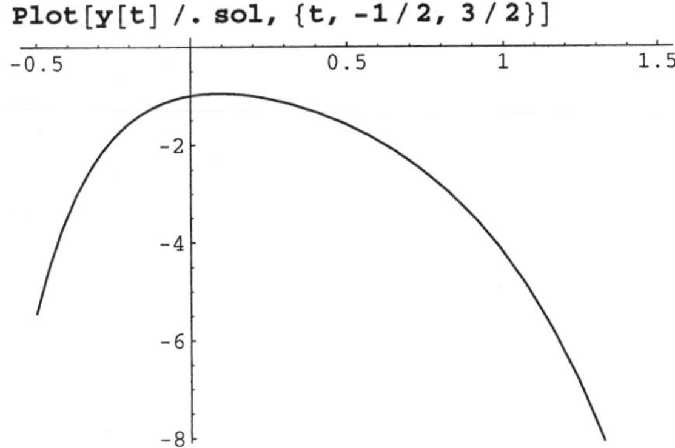

7. Taking the Laplace transform of the equation results in

$$\left(s^3-4s^2-9s+36\right)Y(s)-s^2+4s+10=0$$

and solving for $Y(s)$ yields

$$Y(s)=-\frac{-s^2+4s+10}{s^3-4s^2-9s+36}$$

$$=\frac{-10}{7(s-4)}+\frac{13}{6(s-3)}+\frac{11}{42(s+3)}$$

so $y(t)=-\frac{10}{7}e^{4t}+\frac{13}{6}e^{3t}+\frac{11}{42}e^{-3t}$.

```
sol =
 DSolve[
  {y'''[t] - 4 y''[t] - 9 y'[t] +
     36 y[t] == 0, y[0] == 1,
    y'[0] == 0, y''[0] == -1},
  y[t], t]
```

$$\left\{\left\{y[t]\to e^{-3t}\left(\frac{11}{42}+\frac{13e^{6t}}{6}-\frac{10e^{7t}}{7}\right)\right\}\right\}$$

```
Plot[y[t] /. sol, {t, -1, 1}]
```

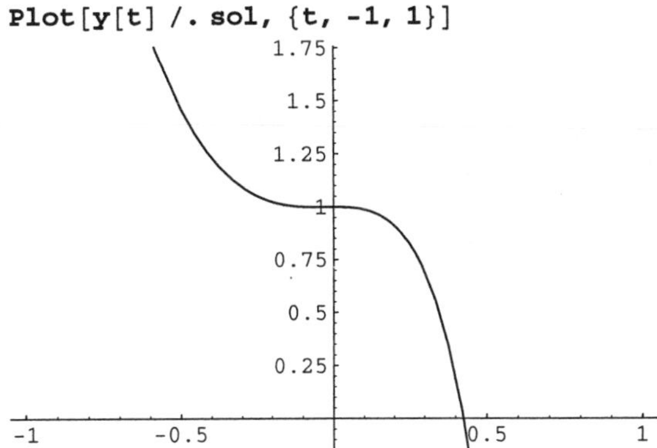

11.

$$\left(s^2-12s+40\right)Y(s)-s+12=\frac{2}{s^2+4}$$

$$Y(s)=-\frac{46-4s+12s^2-s^3}{\left(s^2+4\right)\left(s^2-12s+40\right)}$$

$$=\frac{s+3}{78\left(s^2+4\right)}+\frac{77s-927}{78\left(s^2-12s+40\right)}$$

$$y(t)=\frac{1}{156}\left(2\cos 2t+154e^{6t}\cos 2t+3\sin 2t-465e^{6t}\sin 2t\right)$$

```
sol =
 DSolve[
  {y''[t] - 12 y'[t] + 40 y[t] ==
    Sin[2 t], y[0] == 1, y'[0] == 0},
  y[t], t]
```

$$\{\{y[t] \to \frac{1}{156}(2\,\mathrm{Cos}[2\,t] + 154\,e^{6\,t}\,\mathrm{Cos}[2\,t] + 3\,\mathrm{Sin}[2\,t] - 465\,e^{6\,t}\,\mathrm{Sin}[2\,t])\}\}$$

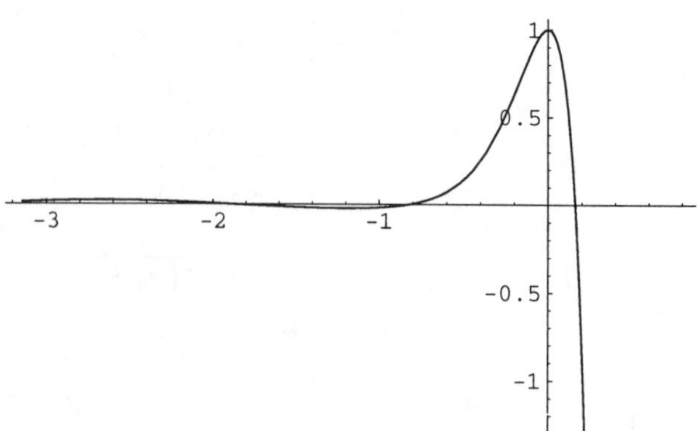


```
Plot[y[t] /. sol, {t, -Pi, Pi / 4}]
```

13.

$$\left(s^2 - s - 12\right)Y(s) + s - 2 = \frac{1}{s-2} - \frac{1}{s^2+1}$$

$$Y(s) = \frac{-53}{238(s-4)} - \frac{1}{10(s-2)} - \frac{47}{70(s+3)} + \frac{13-s}{170\left(1+s^2\right)}$$

$$y(t) = \frac{-53}{238}e^{4t} - \frac{1}{10}e^{2t} - \frac{47}{70}e^{-3t} - \frac{1}{170}\cos t + \frac{13}{170}\sin t$$

```
sol =
 DSolve[
  {y''[t] - y'[t] - 12 y[t] ==
    Exp[2 t] - Sin[t], y[0] == -1,
   y'[0] == 1}, y[t], t]
```

$$\{\{y[t] \to \frac{1}{170}e^{-3\,t}\left(-\frac{799}{7} - 17\,e^{5\,t} - \frac{265\,e^{7\,t}}{7} - e^{3\,t}\,\mathrm{Cos}[t] + 13\,e^{3\,t}\,\mathrm{Sin}[t]\right)\}\}$$

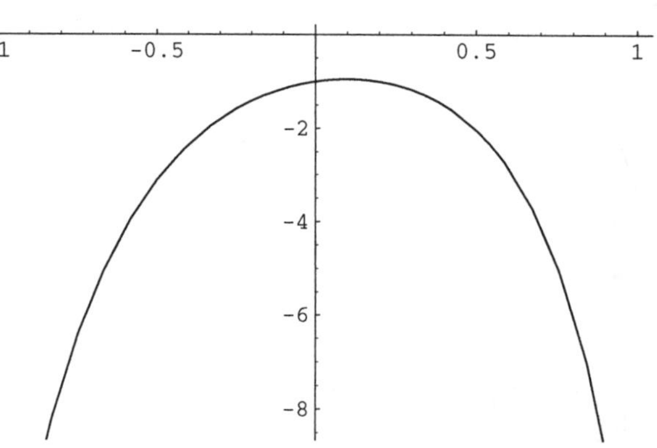


```
Plot[y[t] /. sol, {t, -1, 1}]
```

15. Note that $\mathcal{L}\{ty'(t)\} = -\frac{d}{ds}\left[\mathcal{L}\{y'(t)\}\right]$. Then, because $\mathcal{L}\{y'(t)\} = sY(s) - y(0)$,

$$\mathcal{L}\{ty'(t)\} = -\frac{d}{ds}\left[\mathcal{L}\{y'(t)\}\right] = -\frac{d}{ds}\left[sY(s) - y(0)\right] = -sY'(s) - Y(s).$$

Now we take the Laplace transform of both sides of the equation:

$$\mathcal{L}\{y'' + 4ty' - 8y\} = \mathcal{L}\{4\}$$

$$s^2Y(s) - sy(0) - y'(0) + 4\left[-sY'(s) - Y(s)\right] - 8Y(s) = \frac{4}{s}$$

$$\left(s^2 - 12\right)Y(s) - 4sY'(s) = \frac{4}{s}.$$

This is a first-order differential equation that can be solved for $Y(s)$ by using an integrating factor. Rewriting the equation as

$$Y'(s) + \left(\frac{3}{s} - \frac{s}{4}\right)Y(s) = -\frac{1}{s^2},$$

we see that an integrating factor is

$$\mu = e^{\int (3/s - s/4)ds} = e^{3\ln s - s^2/8} = s^3 e^{-s^2/8}.$$

Then,

$$\frac{d}{ds}\left[s^3 e^{-s^2/8} Y(s)\right] = -\frac{1}{s^2}\cdot s^3 e^{-s^2/8} = -se^{-s^2/8},$$

so integrating yields

$$s^3 e^{-s^2/8} Y(s) = 4e^{-s^2/8} + C$$
$$Y(s) = \frac{4}{s^3} + C\frac{1}{s^3}e^{s^2/8}.$$

Notice that in order for $y(t) = \mathcal{L}^{-1}\{Y(s)\}$ to exist, $\lim_{s\to\infty} Y(s) = 0$. This is not the case with $Y(s) = \frac{4}{s^3} + C\frac{1}{s^3}e^{s^2/8}$ unless $C = 0$. Hence,

$$Y(s) = \frac{4}{s^3},$$

so the solution to the initial-value problem is

$$y(t) = \mathcal{L}^{-1}\left\{\frac{4}{s^3}\right\} = 2\mathcal{L}^{-1}\left\{\frac{2}{s^3}\right\} = 2t^2.$$

19. First, we compute the Laplace transform of the equation which yields a first-order linear differential equation:

$$\left[s^2Y(s) - sy(0) - y'(0)\right] + 2\left[-sY'(s) - Y(s)\right] - 4Y(s) = \frac{4}{s}$$
$$-2sY'(s) + \left(s^2 - 6\right)Y(s) = \frac{4}{s}$$
$$Y'(s) - \frac{s^2 - 6}{2s}Y(s) = -\frac{2}{s^2}$$
$$\frac{d}{ds}\left[s^3 e^{-s^2/4} Y(s)\right] = -2se^{-s^2/4}$$

This equation has general solution $Y(s) = \frac{4}{s^3} + \frac{K}{s^3}e^{s^2/4}$ and because we must have that $\lim_{s\to\infty} Y(s) = 0$, it follows that $K = 0$ so $Y(s) = \frac{4}{s^3}$. Therefore, $y = 2t^2$.

24. First, we illustrate the method of Laplace transforms.

```
Clear[y,t]
s1=y''[t]+2y'[t]+4y[t]-t+Exp[-t]
s2=LaplaceTransform[s1,t,s]
s3=s2 /. {y[0]->1,y'[0]->-1}
s4=Solve[s3==0,LaplaceTransform[y[t],
t,s]]
sol=InverseLaplaceTransform[s4[[1,1,2
]],
   s,t]
Plot[sol,{t,0,7}],PlotRange->{-1,6},
   AspectRatio->1
```

```
step1:=laplace(diff(y(t),t$2)+
   2*diff(y(t),t)+4*y(t)=
      t-exp(-t),t,s);
step2:=subs(
   {y(0)=1,D(y)(0)=-1},step1);
step3:=solve(step2,
   laplace(y(t),t,s));
step4:=invlaplace(step3,s,t);
plot(step4,t=0..7,-1..6);
```

EXERCISES 8.3

3. $\frac{3}{s}e^{-8s} - e^{-4s} = \frac{3 - e^{4s}}{se^{8s}}$

7. $-42e^{-4s}\mathcal{L}\left\{e^t\right\} = \frac{-42e^{-4s}}{s-1}$

11. $\mathscr{L}\{\sin t\} = \dfrac{1}{s^2+1} \cdot \dfrac{-14}{e^{2\pi s/3}\left(s^2+1\right)}$

13.
$$\mathscr{L}\{f(t)\} = \frac{1}{1-e^{-2s}} \int_0^1 te^{-st}\,dt$$
$$= \frac{1}{1-e^{-2s}}\left[\frac{1}{s^2} - \frac{s+1}{s^2e^s}\right] = \frac{e^s - s - 1}{s^2e^s\left(1-e^{-2s}\right)}$$

19. $\mathscr{L}\{\delta(t-1)\} + \mathscr{L}\{\delta(t-2)\} = e^{-s} + e^{-2s}$

21.
$$e^{-2\pi s}\mathscr{L}\{\sin(t+2\pi)\} + e^{-\pi s/2}$$
$$= e^{-2\pi s}\mathscr{L}\{\sin t\} + e^{-\pi s/2}$$
$$= \frac{e^{-2\pi s}}{s^2+1} + e^{-\pi s/2}$$

25.
$$\mathscr{L}^{-1}\left\{\frac{2}{s}e^{-s} - \frac{3}{s}e^{-4s}\right\} =$$
$$2\mathscr{U}(t-1) - 3\mathscr{U}(t-4)$$

29. $f(t) = \mathscr{L}^{-1}\left\{\dfrac{1}{s-1}\right\} = e^t,$

$f(t-4)\mathscr{U}(t-4) = e^{t-4}\mathscr{U}(t-4)$

33. $f(t) = \mathscr{L}^{-1}\left\{\dfrac{1}{(s-2)(s-5)}\right\} = \mathscr{L}^{-1}\left\{\dfrac{1}{3(s-5)} - \dfrac{1}{3(s-2)}\right\} = \dfrac{1}{3}e^{5t} - \dfrac{1}{3}e^{2t};$

$f(t-5)\mathscr{U}(t-5) = \dfrac{1}{3}\left(e^{5(t-5)} - e^{2(t-5)}\right)\mathscr{U}(t-5) = \dfrac{1}{3}\left(e^{5t-25} - e^{2t-10}\right)\mathscr{U}(t-5)$

37. $f(t) = \mathscr{L}^{-1}\left\{\dfrac{s}{s^2-1}\right\} = \cosh t,\ f(t-4)\mathscr{U}(t-4) = \cosh(t-4)\mathscr{U}(t-4)$

41. $\mathscr{L}^{-1}\left\{\dfrac{1}{s^2+16}\right\} = \sin 4t$ so

$$\mathscr{L}^{-1}\left\{\frac{1}{\left(1-\left(-e^{-3s}\right)\right)\left(s^2+16\right)}\right\} = \mathscr{L}^{-1}\left\{\frac{1}{s^2+16}\left[1-e^{-3s}+e^{-6s}-e^{-9s}+\cdots\right]\right\}$$
$$= \sin 4t - \sin(4t-12)\mathscr{U}(t-3) + \sin(4t-24)\mathscr{U}(t-6) + \sin(4t-36)\mathscr{U}(t-9)+\cdots$$

43. First, we rewrite the fraction:
$$\frac{1}{s^2\left(s^2+4\right)\left(3e^{-2s}-1\right)} = \frac{1}{1-3e^{-2s}}\frac{-1}{s^2\left(s^2+4\right)} = \frac{1}{1-3e^{-2s}}\left[\frac{1}{4}\frac{1}{s^2+4} - \frac{1}{4}\frac{1}{s^2}\right]$$
$$= \left(1+3e^{-2s}+9e^{-4s}+27e^{-6s}+81e^{-8s}+\cdots\right)\left[\frac{1}{4}\frac{1}{s^2+4} - \frac{1}{4}\frac{1}{s^2}\right].$$

Now, $\mathscr{L}^{-1}\left\{\frac{1}{4}\frac{1}{s^2+4} - \frac{1}{4}\frac{1}{s^2}\right\} = \frac{1}{8}\sin 2t - \frac{1}{4}t.$ Thus,

$$\mathscr{L}^{-1}\left\{\frac{1}{s^2\left(s^2+4\right)\left(3e^{-2s}-1\right)}\right\} = \mathscr{L}^{-1}\left\{\left(1+3e^{-2s}+9e^{-4s}+27e^{-6s}+81e^{-8s}+\cdots\right)\left[\frac{1}{4}\frac{1}{s^2+4} - \frac{1}{4}\frac{1}{s^2}\right]\right\}$$
$$= \left(\tfrac{1}{8}\sin 2t - \tfrac{1}{4}t\right) + 3\left(\tfrac{1}{2} - \tfrac{1}{4}t + \tfrac{1}{8}\sin(2t-4)\right)\mathscr{U}(t-2) +$$
$$9\left(1 - \tfrac{1}{4}t + \tfrac{1}{8}\sin(2t-8)\right)\mathscr{U}(t-4) + 27\left(\tfrac{3}{2} - \tfrac{1}{4}t + \tfrac{1}{8}\sin(2t-12)\right)\mathscr{U}(t-6)+\cdots$$

45. Note that $f(t) = 1 - \mathscr{U}(t-2)$ and $\mathscr{L}\{f(t)\} = \dfrac{1}{s} - \dfrac{1}{s}e^{-2s}$. Then,

$$sY(s)-y(0)+3Y(s)=\frac{1}{s}-\frac{1}{s}e^{-2s}$$

$$Y(s)=\frac{1}{s(s+3)}-\frac{e^{-2s}}{s(s+3)}$$

$$y(t)=\frac{1}{3}-\frac{1}{3}e^{-3t}-\left[\frac{1}{3}-\frac{1}{3}e^{-3(t-2)}\right]\mathcal{U}(t-2)$$

```
f[t_] = UnitStep[t] - U

sol =
 DSolve[{y'[t] + 3 y[t] == f[t],
    y[0] == 0}, y[t], t] //
  Simplify

{{y[t] → 1/3 e^(-3 t)
    ((e^6 - e^(3 t)) UnitStep[-2 + t] +
      (-1 + e^(3 t)) UnitStep[t])}}
```

```
Plot[y[t] /. sol, {t, 0, 4}]
```

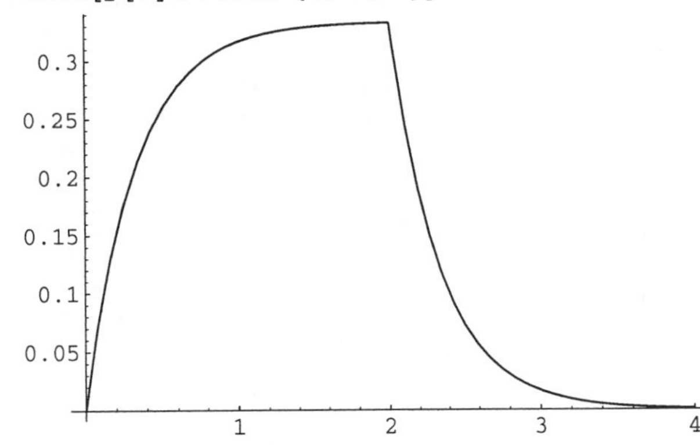

47. The Laplace transform of $f(t)=\begin{cases}\cos\left(\frac{\pi t}{2}\right), 0\le t<1\\ 0, t>1\end{cases}=\cos\left(\frac{\pi t}{2}\right)\{\mathcal{U}(t-0)-\mathcal{U}(t-1)\}$ is $\mathcal{L}\{f(t)\}=\dfrac{2\left(2s\,e^{s}+\pi\right)}{e^{s}\left(4s^{2}+\pi^{2}\right)}$. Applying the Laplace transform and initial conditions to the differential equation yields

$$s^2Y(s)-sy(0)-y'(0)-4(s\,Y(s)-y(0))+3Y(s)=\frac{2\left(2s\,e^{s}+\pi\right)}{e^{s}\left(4s^{2}+\pi^{2}\right)}$$

$$s^2Y(s)-4s\,Y(s)+3Y(s)-1=\frac{2\left(2s\,e^{s}+\pi\right)}{e^{s}\left(4s^{2}+\pi^{2}\right)}$$

$$Y(s)=\frac{2\pi+\pi^2e^s+4\,se^s+4s^2e^s}{e^s\left(s^2-4s+3\right)\left(4s^2+\pi^2\right)}.$$

To find $y(t)=\mathcal{L}^{-1}\{Y(s)\}$, we first rewrite $Y(s)$:

$$Y(s)=\frac{2\pi+\pi^2e^s+4\,se^s+4s^2e^s}{e^s\left(s^2-4s+3\right)\left(4s^2+\pi^2\right)}=\frac{2\pi}{e^s\left(s^2-4s+3\right)\left(4s^2+\pi^2\right)}+\frac{\pi^2+4\,s+4s^2}{\left(s^2-4s+3\right)\left(4s^2+\pi^2\right)}$$

$$=e^{-s}\left[\frac{\pi}{36+\pi^2}\frac{1}{s-3}-\frac{\pi}{4+\pi^2}\frac{1}{s-1}+\frac{8}{(4+\pi^2)(36+\pi^2)}\frac{12\pi-\pi^3+16\pi\,s}{4s^2+\pi^2}\right]+$$

$$\frac{48+\pi^2}{2(36+\pi^2)}\frac{1}{s-3}-\frac{8+\pi^2}{2(4+\pi^2)}\frac{1}{s-1}+\frac{16}{(4+\pi^2)(36+\pi^2)}\frac{-4\pi^2+12s-\pi^2s}{4s^2+\pi^2}.$$

Computing the inverse Laplace transform gives us

$$y(t)=\left[-\frac{\pi}{4+\pi^2}e^{t-1}+\frac{\pi}{36+\pi^2}e^{3t-3}+\frac{32\pi}{(4+\pi^2)(36+\pi^2)}\cos\left(\frac{\pi}{2}(t-1)\right)+\frac{48-4\pi^2}{(4+\pi^2)(36+\pi^2)}\sin\left(\frac{\pi}{2}(t-1)\right)\right]\mathcal{U}(t-1)-$$
$$\frac{8+\pi^2}{2(4+\pi^2)}e^t+\frac{48+\pi^2}{2(36+\pi^2)}e^{3t}+\frac{48-4\pi^2}{(4+\pi^2)(36+\pi^2)}\cos\left(\frac{\pi}{2}t\right)-\frac{32\pi}{(4+\pi^2)(36+\pi^2)}\sin\left(\frac{\pi}{2}t\right).$$

```
f[t_] = Cos[Pi t / 2]
   (UnitStep[t] -
     UnitStep[t - 1]);
sol =
 DSolve[
   {y''[t] - 4 y'[t] + 3 y[t] ==
     f[t], y[0] == 0, y'[0] == 1},
   y[t], t] // Simplify
```

```
{{y[t] →
   (2 (4 e^3 (-12 + π^2) Cos[π t / 2] +
        π (e^t (e^(2 t) (4 + π^2) - e^2
               (36 + π^2)) +
           32 e^3 Sin[π t / 2]))
       UnitStep[-1 + t] +
     e^3 (e^t (-1 + e^(2 t))
          (144 + 40 π^2 + π^4) -
        4 (e^t (36 + π^2 - 3 e^(2 t)
               (4 + π^2)) +
           2 (-12 + π^2) Cos[π t / 2] +
           16 π Sin[π t / 2])
         UnitStep[t])) /
   (2 e^3 (4 + π^2) (36 + π^2))}}
```

```
Plot[y[t] /. sol, {t, 0, 2}]
```

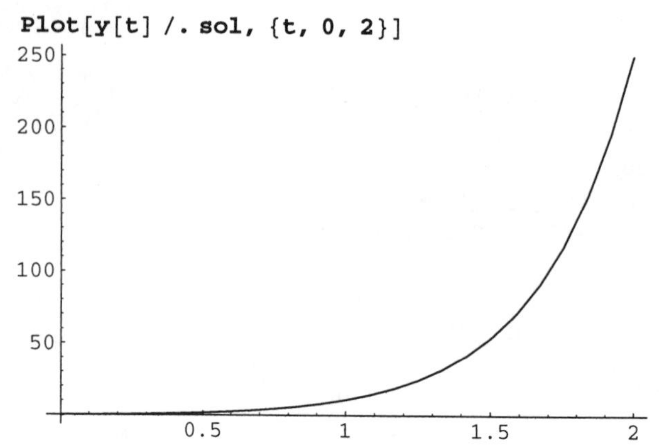

49.

$$s^2X(s) - sx(0) - x'(0) + X(s) = e^{-\pi s} + \frac{1}{s}$$

$$X(s) = \frac{e^{-\pi s}}{s^2+1} + \frac{1}{s\left(s^2+1\right)}$$

$$x(t) = \sin(t-\pi)\mathscr{U}(t-\pi) + 1 - \cos t = -\sin t\,\mathscr{U}(t-\pi) + 1 - \cos t$$

53.

$$s^2X(s) - sx(0) - x'(0) + 4\left(sX(s) - x(0)\right) + 13X(s) = e^{-\pi s} + \frac{1}{s^2+1}$$

$$X(s) = \frac{e^{-\pi s}}{s^2+4s+13} + \frac{1}{\left(s^2+1\right)\left(s^2+4s+13\right)}$$

Notice that $\frac{1}{\left(s^2+1\right)\left(s^2+4s+13\right)} = \frac{3-s}{40\left(s^2+1\right)} + \frac{1+s}{40\left(s^2+4s+13\right)}$. Therefore,

$$x(t) = \frac{1}{3}e^{-2(t-\pi)}\sin 3(t-\pi)\,\mathscr{U}(t-\pi) + \frac{3}{40}\sin t - \frac{1}{40}\cos t - \frac{1}{120}e^{-2t}\sin 3t + \frac{1}{40}e^{-2t}\cos 3t$$

$$= -\frac{1}{3}e^{2\pi-2t}\sin 3t\,\mathscr{U}(t-\pi) + \frac{3}{40}\sin t - \frac{1}{40}\cos t - \frac{1}{120}e^{-2t}\sin 3t + \frac{1}{40}e^{-2t}\cos 3t$$

```
Clear[x]
sol =
 DSolve[
  {x''[t] + 4 x'[t] + 13 x[t] ==
    DiracDelta[t - Pi] + Sin[t],
   x[0] == 0, x'[0] == 0}, x[t], t]
```

$$\left\{\left\{x[t] \to -\frac{1}{120}e^{-2t}\left(3e^{2t}\text{Cos}[t] - 3\,\text{Cos}[3t] - 9e^{2t}\text{Sin}[t] + \text{Sin}[3t] + 40e^{2\pi}\text{Sin}[3t]\,\text{UnitStep}[-\pi+t]\right)\right\}\right\}$$

```
Plot[x[t] /. sol, {t, 0, 2 Pi}]
```

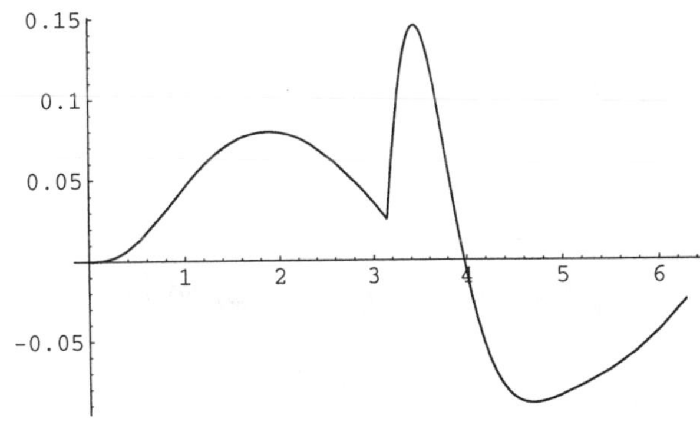

57.

$$s^2X(s) - sx(0) - x'(0) + 3\left(sX(s) - x(0)\right) + 2X(s) = e^{-s} + \frac{1}{s+1}$$

$$X(s) = \frac{e^{-s}}{(s+1)(s+2)} + \frac{1}{(s+1)^2(s+2)}$$

Notice that $\frac{1}{(s+1)(s+2)} = \frac{1}{s+1} - \frac{1}{s+2}$ and $\frac{1}{(s+1)^2(s+2)} = \frac{1}{(s+1)^2} - \frac{1}{s+1} - \frac{1}{s+2}$. Therefore,

$$x(t)=\left(-e^{-2(t-1)}+e^{-(t-1)}\right)\mathcal{U}(t-1)+te^{-t}-e^{-t}+e^{-2t}.$$

```
sol =
  DSolve[
   {x''[t] + 3 x'[t] + 2 x[t] ==
     DiracDelta[t - 1] + Exp[-t],
    x[0] == 0, x'[0] == 0}, x[t], t]

{{x[t] → e^(-2 t) (1 - e^t + e^t t -
      e^2 UnitStep[-1 + t] +
      e^(1+t) UnitStep[-1 + t])}}

Plot[x[t] /. sol, {t, 0, 2}]
```

61.

$$s^2X(s)-sx(0)-x'(0)+6(sX(s)-x(0))+10X=3e^{-\pi s}$$

$$X(s)=\frac{3e^{-\pi s}}{(s+3)^2+1}$$

$$x(t)=3e^{-3(t-\pi)}\sin(t-\pi)\mathcal{U}(t-\pi)=-3e^{-3(t-\pi)}\sin t\,\mathcal{U}(t-\pi)$$

```
sol =
 DSolve[
  {x''[t] + 6 x'[t] + 10 x[t] ==
    3 DiracDelta[t - Pi], x[0] == (
   x'[0] == 0}, x[t], t]

{{x[t] → -3 e^(3 π-3 t)
    Sin[t] UnitStep[-π + t]}}

Plot[x[t] /. sol, {t, 0, 2 Pi}]
```

67.

$$\mathcal{L}\{f(t)\}=\frac{1}{1-e^{-as}}\int_0^a e^{-st}f(t)dt=\frac{1}{1-e^{-as}}\int_0^a e^{-st}\cdot\frac{t}{a}dt$$

$$=\frac{1}{1-e^{-as}}\left(-\frac{1}{as^2}(st+1)e^{-st}\right)\Bigg]_0^a$$

$$=\frac{1}{1-e^{-as}}\cdot\frac{1}{as^2}\left(e^{as}-as-1\right)e^{-as}$$

$$=\frac{e^{as}-as-1}{as^2\left(e^{as}-1\right)}=\frac{1}{as^2}-\frac{1}{s\left(e^{as}-1\right)}=\frac{1}{as^2}-\frac{e^{-as}}{s\left(1-e^{-as}\right)}.$$

73. (Mathematica only) The limit of the solution to the first problem is 1; the limit of the solution to the second is 2.

```
Clear[y,t,sol]
sol=DSolve[{y''[t]+2y'[t]+y[t]==DiracDelta[t]+UnitStep[t-2Pi],
   y[0]==0,y'[0]==0},y[t],t]//Simplify
sol=sol[[1,1,2]]  /.  DiracDelta[2Pi]->0  //  Simplify
Plot[sol,{t,0,8Pi}]
Clear[y,t,sol]
sol2=DSolve[{y''[t]+2y'[t]+y[t]==100DiracDelta[t]+UnitStep[t-2Pi]+UnitStep[t-
4Pi],
   y[0]==0,y'[0]==0},y[t],t]//Simplify
sol2=sol2[[1,1,2]]  /.  {DiracDelta[2Pi]->0,DiracDelta[4Pi]->0}  //  Simplify
Plot[sol2,{t,0,8Pi}]
```

EXERCISES 8.4

3. $\int_0^t (t-v)e^{-v}dv = \left[(1-t+v)e^{-v}\right]_0^t = t-1+e^{-t}$

9. $h(t) = (f*g)(t)$ where $f(t)=t$ and $g(t)=\sin t$;

$$\mathscr{L}\{(f*g)(t)\} = \mathscr{L}\{f(t)\}\mathscr{L}\{g(t)\} = \frac{1}{s^2\left(s^2+1\right)}$$

15. $\dfrac{s}{\left(s^2+1\right)^2} = \dfrac{s}{s^2+1}\cdot\dfrac{1}{s^2+1}$;

$$\mathscr{L}^{-1}\left\{\frac{s}{\left(s^2+1\right)^2}\right\} = (f*g)(t),$$

where $f(t) = \mathscr{L}^{-1}\left\{\dfrac{s}{s^2+1}\right\} = \cos t$ and

$g(t) = \mathscr{L}^{-1}\left\{\dfrac{1}{s^2+1}\right\} = \sin t$. $(f*g)(t) = \dfrac{1}{2}t\sin t$

21. $\mathscr{L}^{-1}\left\{\dfrac{1}{\left(s^2+100\right)^2}\right\} = (f*g)(t)$ where $f(t) = g(t) = \mathscr{L}^{-1}\left\{\dfrac{1}{s^2+100}\right\} = \dfrac{1}{10}\sin 10t$;

25.

$$\mathscr{L}\{h(t)\} - \frac{4}{s+2} = \frac{s}{s^2+1} - \mathscr{L}\{\sin t * h(t)\}$$

$$H(s) - \frac{4}{s+2} = \frac{s}{s^2+1} - H(s)\frac{1}{s^2+1}$$

$$H(s) = \frac{s}{s^2+2} + \frac{4\left(s^2+1\right)}{\left(s^2+2\right)(s+2)} = \frac{s}{s^2+2} + \frac{10}{3(s+2)} + \frac{2(s-2)}{3\left(s^2+2\right)}$$

$$h(t) = \frac{10}{3}e^{-2t} + \frac{5}{3}\cos\left(\sqrt{2}t\right) - \frac{2\sqrt{2}}{3}\sin\left(\sqrt{2}t\right)$$

27.

$$sY(s) - y(0) - 4y(s) + \frac{4Y(s)}{s} = \frac{3!}{(s-2)^4}$$

$$\left(s^2-4s+4\right)Y(s) = \frac{6s}{(s-2)^4}$$

$$Y(s) = \frac{6s}{(s-2)^6} = \frac{12}{(s-2)^6} + \frac{6}{(s-2)^5}$$

$$y(t) = \left(\frac{1}{4}t^4 + \frac{1}{10}t^5\right)e^{2t}$$

33. The convolutions are

$$(\sin t)*(\cos kt) = \int_0^t \sin(t-v)\cos kv\, dv = \left(\frac{1}{2(k+1)}\cos(kv+v-t) - \frac{1}{2(k-1)}\cos(kv-v+t)\right)\Bigg]_{v=0}^{t}$$
$$= \frac{1}{(k-1)(k+1)}(\cos t - \cos kt)$$

and

$$(\cos t)*\left(\frac{1}{k}\sin kt\right) = \frac{1}{k}\int_0^t \cos(t-v)\sin kv\, dv = -\frac{1}{k}\left(\frac{k}{2(k+1)}\cos(kv+v-t) + \frac{k}{2(k-1)}\cos(kv-v+t)\right)\Bigg]_{v=0}^{t}$$
$$= \frac{1}{(k-1)(k+1)}(\cos t - \cos kt).$$

37.

```
Map[{#,InverseLaplaceTransform[#,s,t]
}//
      Simplify&,
   {4/((s-2)^2(s^2+16)),
    6!/s^7  (1/(2s)-s/(2  (s^2+4))),
    /s^4       (s/(4       (s^2+9))-
3s/(4(s^2+1)))}]
```

```
readlib(laplace):
map(invlaplace,
   {4/((s-2)^2*(s^2+16)),
   6!/s^7*(1/(2*s)-s/(2 (s^2+4))),
   6/s^4*(s/(4*(s^2+9))-
      3*s/(4*(s^2+1)))},
      s,t);
```

38.

```
conv[f_,g_]:=Integrate[f[t-v]g[v],
   {v,0,t}]//SimplifyClear[f,g]

Clear[f,g]
f[t_]=Sin[ k t]
g[t_]=Cos[ k t]
conv[f,g]

InverseLaplaceTransform[
   k  s/(s^2+k^2)^2,s,t]
```

```
convolution:=(f,g)->int(f(t-v)*g(v),
   v=0..t):

f:='f':g:='g':
f:=t->sin(k*t):
g:=t->cos(k*t):
convolution(f,g);

invlaplace(k*s/(s^2+k^2),s,t);
```

EXERCISES 8.5

3. Applying the Laplace transform to the system of equations yields the system of equations

$$\begin{cases} s\,X(s) - x(0) - 5X(s) - 5Y(s) = 0 \\ s\,Y(s) - y(0) + 4X(s) + 3Y(s) = 0 \end{cases}.$$

Next, we apply the initial conditions and rearrange the equations

$$\begin{cases} 4X(s) + (s+3)Y(s) = 0 \\ (s-5)X(s) - 5Y(s) = 2 \end{cases}.$$

We obtain $X(s)$ by eliminating $Y(s)$, which we do by multiplying the first equation by 5, the second equation by $s+3$

$$\begin{cases} 20X(s) + 5(s+3)Y(s) = 0 \\ (s+3)(s-5)X(s) - 5(s+3)Y(s) = 2(s+3) \end{cases}$$

and then adding

$$[(s+3)(s-5) + 20]X(s) = 2(s+3)$$

$$\left(s^2 - 2s + 5\right)X(s) = 2(s+3)$$

$$X(s) = \frac{2(s+3)}{(s-1)^2 + 4}.$$

Now, we use the inverse Laplace transform to find $x(t)$

$$x(t) = \mathcal{L}^{-1}\left\{\frac{2(s+3)}{(s-1)^2 + 4}\right\} = \mathcal{L}^{-1}\left\{\frac{2(s-1)}{(s-1)^2 + 4} + \frac{8}{(s-1)^2 + 4}\right\} = 2e^t(\cos 2t + 2\sin 2t).$$

We use the first differential equation to find $y(t)$

$$y(t) = \frac{1}{5}(x' - 5x) = -4e^t \sin 2t.$$

```
Clear[x, y]
sol =
 DSolve[
   {x'[t] - 5 x[t] - 5 y[t] == 0,
    y'[t] + 4 x[t] + 3 y[t] == 0,
    x[0] == 2, y[0] == 0},
   {x[t], y[t]}, t] //
  FullSimplify
```

```
{{x[t] ->
   2 e^t (Cos[2 t] + 2 Sin[2 t]),
  y[t] -> -4 e^t Sin[2 t]}}
```

```
ParametricPlot[
 Evaluate[{x[t], y[t]} /. sol],
 {t, 0, 5}]
```

```
Plot[Evaluate[
   {x[t], y[t]} /. sol], {
  PlotStyle ->
   {GrayLevel[0],
    GrayLevel[0.5]}}]
```

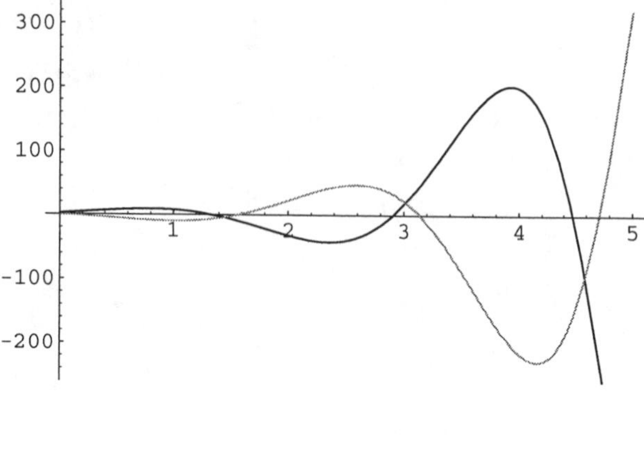

5. Applying the Laplace transform to the system yields the system

$$\begin{cases} s\,X(s) - x(0) - 5X(s) + 4Y(s) - 2Z(s) = 0 \\ s\,Y(s) - y(0) + 2X(s) + 2Y(s) + 2Z(s) = 0. \\ s\,Z(s) - z(0) - Z(s) = 0 \end{cases}$$

Now we apply the initial conditions and rearrange the equations

$$\begin{cases} (s-5)X(s) + 4Y(s) - 2Z(s) = 0 \\ 2X(s) + (s+2)Y(s) + 2Z(s) = 0 \\ (s-1)Z(s) = 15 \end{cases}$$

From the third equation we see that $Z(s) = \dfrac{15}{s-1}$ and substitution into the first and second equations yields the system

$$\begin{cases} 2\,X(s) + (s+2)Y(s) = -\dfrac{30}{s-1} \\ (s-5)X(s) + 4Y(s) = \dfrac{30}{s-1} \end{cases}$$

Note that by inspection of the system and initial conditions it is relatively easy to see that $z(t) = 15e^t$. We eliminate $Y(s)$ by applying -4 to the first equation, $s+2$ to the second equation

$$\begin{cases} -8\,X(s) - 4(s+2)Y(s) = \dfrac{120}{s-1} \\ (s+2)(s-5)X(s) + 4(s+2)Y(s) = \dfrac{30(s+2)}{s-1} \end{cases}$$

and then adding

$$\left[(s+2)(s-5)-8\right]X(s)=\frac{120+30(s+2)}{s-1}$$

$$\left(s^2-3s-18\right)X(s)=\frac{30(s+6)}{s-1}$$

$$X(s)=\frac{30(s+6)}{(s+3)(s-6)(s-1)}.$$

We use the inverse Laplace transform to find x(t):

$$x(t)=\mathscr{L}^{-1}\left\{\frac{30(s+6)}{(s+3)(s-6)(s-1)}\right\}=\mathscr{L}^{-1}\left\{\frac{5}{2}\frac{1}{s+3}+8\frac{1}{s-6}-\frac{21}{2}\frac{1}{s-1}\right\}=\frac{5}{2}e^{-3t}+8e^{6t}-\frac{21}{2}e^{t}.$$

We use the first equation to find $y(t)$:

$$y(t)=\frac{1}{4}(2z+5x-x')=-3e^{t}+5e^{-3t}-2e^{6t}.$$

```
sol =
 DSolve[
  {x'[t] - 5 x[t] + 4 y[t] -
      2 z[t] == 0,
   y'[t] + 2 x[t] + 2 y[t] +
      2 z[t] == 0,
   z'[t] - z[t] == 0, x[0] == 0,
   y[0] == 0, z[0] == 15},
  {x[t], y[t], z[t]}, t]
```

$\{\{x[t]\to\frac{1}{2}e^{-3t}(5-21e^{4t}+16e^{9t}),\ y[t]\to -e^{-3t}(-5+3e^{4t}+2e^{9t}),\ z[t]\to 15e^{t}\}\}$

```
Plot[Evaluate[
   {x[t], y[t], z[t]} /. sol],
  {t, 0, 2},
  PlotStyle ->
   {GrayLevel[0], GrayLevel[.3],
    GrayLevel[.6]},
  PlotRange -> {-100, 100}]
```

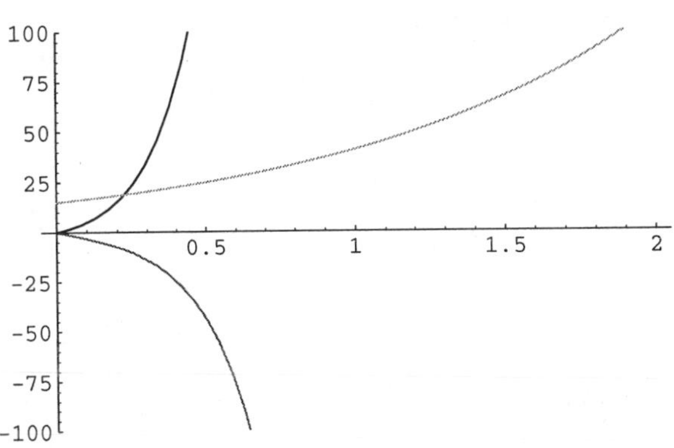

7. $\begin{cases}(s-1)X(s)-x(0)-3Y(s)=\dfrac{1}{s-4} \\ -5X(s)+(s+1)Y(s)-y(0)=0\end{cases}$; $X(s)=\dfrac{s+1}{s^3-4s^2-16s+64}=\dfrac{5}{8(s-4)^2}+\dfrac{3}{64(s-4)}-\dfrac{3}{64(s+4)}$;

$Y(s)=\dfrac{5}{s^3-4s^2-16s+64}=\dfrac{5}{8(s-4)^2}-\dfrac{5}{64(s-4)}+\dfrac{5}{64(s+4)}$; $\begin{cases}x(t)=\dfrac{5}{8}te^{4t}+\dfrac{3}{64}e^{4t}-\dfrac{3}{64}e^{-4t} \\ y(t)=\dfrac{5}{8}te^{4t}-\dfrac{5}{64}e^{4t}+\dfrac{5}{64}e^{-4t}\end{cases}$

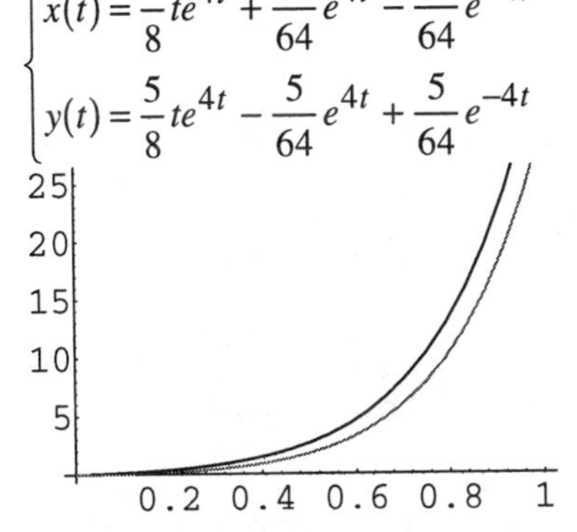

13. $\begin{cases} (s+2)X(s) - x(0) - 4Y(s) = \dfrac{s}{s^2+4} \\ 5X(s) + (s-2)Y(s) - y(0) = \dfrac{2}{s^2+4} \end{cases}$; $X(s) = \dfrac{2-s}{6\left(s^2+4\right)} + \dfrac{s-20}{6\left(s^2+16\right)}$; $Y(s) = \dfrac{4-3s}{12\left(s^2+4\right)} + \dfrac{-9s-28}{12\left(s^2+16\right)}$;

$$\begin{cases} x(t) = -\dfrac{1}{6}\cos 2t + \dfrac{1}{6}\cos 4t + \dfrac{1}{6}\sin 2t - \dfrac{5}{6}\sin 4t \\ y(t) = -\dfrac{1}{4}\cos 2t - \dfrac{3}{4}\cos 4t + \dfrac{1}{6}\sin 2t - \dfrac{7}{12}\sin 4t \end{cases}$$

```
sol =
 DSolve[
  {x'[t] + 2 x[t] - 4 y[t] ==
    Cos[2 t],
   y'[t] + 5 x[t] - 2 y[t] ==
    Sin[2 t], x[0] == 0,
   y[0] == -1}, {x[t], y[t]}, t]
```

```
{{x[t] →
   1
   - (-Cos[2 t] + Cos[4 t] + Sin[
   6
      2 t] - 5 Sin[4 t]), y[t] →
   1
   -- (-3 Cos[2 t] - 9 Cos[4 t] +
   12
     2 Sin[2 t] - 7 Sin[4 t])}}
```

```
Plot[Evaluate[
   {x[t], y[t]} /. sol],
  {t, 0, 3 Pi},
  PlotStyle ->
   {GrayLevel[0],
    GrayLevel[0.5]}]
```

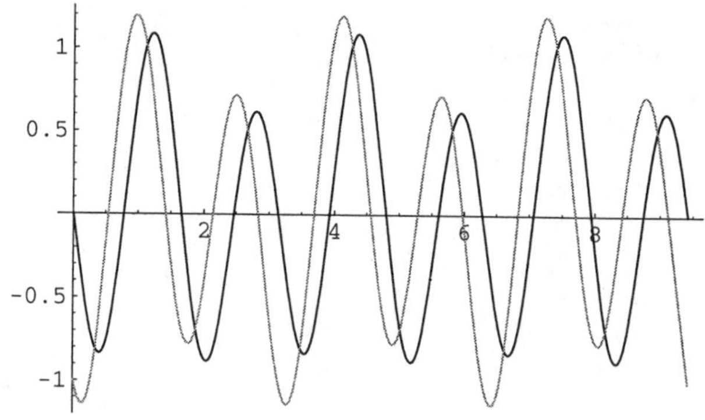


```
ParametricPlot[
 {x[t], y[t]} /. sol, {
```

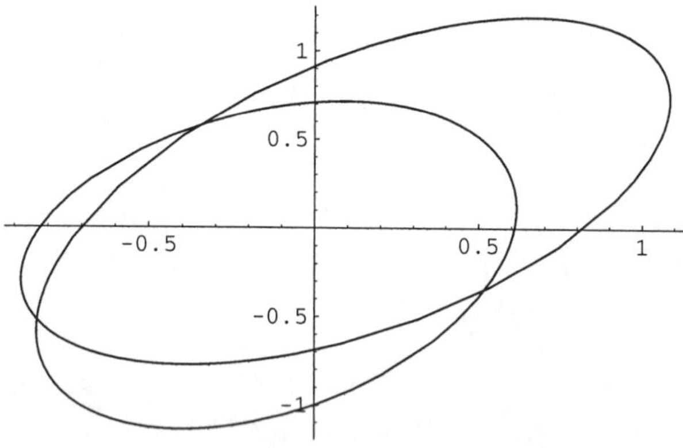

15. $\begin{cases} sX(s) - x(0) - Y(s) = 0 \\ X(s) + sY(s) - y(0) = \dfrac{1}{s^2+1} + \dfrac{e^{-\pi s}}{s^2+1} \end{cases}$; $X(s) = \dfrac{1}{\left(s^2+1\right)^2} + \dfrac{e^{-\pi s}}{\left(s^2+1\right)^2}$; $Y(s) = \dfrac{s}{\left(s^2+1\right)^2} + \dfrac{se^{-\pi s}}{\left(s^2+1\right)^2}$;

$$\begin{cases} x(t) = -\dfrac{t}{2}\cos t + \dfrac{1}{2}\sin t + \left(\dfrac{t}{2}\cos t - \dfrac{1}{2}\sin t - \dfrac{\pi}{2}\cos t\right)\mathcal{U}(t-\pi) \\ y(t) = \dfrac{t}{2}\sin t + \left(\dfrac{\pi}{2}\sin t - \dfrac{t}{2}\sin t\right)\mathcal{U}(t-\pi) \end{cases}$$

```
sol =
 DSolve[{x'[t] - y[t] == 0,
   y'[t] + x[t] ==
    Sin[t]
     (UnitStep[0] -
       UnitStep[t - Pi]),
   x[0] == 0, y[0] == 0},
  {x[t], y[t]}, t]
```

$$\left\{\left\{x[t] \to \frac{1}{2}\,(-t\,\mathrm{Cos}[t] + \mathrm{Sin}[t] - \pi\,\mathrm{Cos}[t]\,\mathrm{UnitStep}[-\pi + t] + t\,\mathrm{Cos}[t]\,\mathrm{UnitStep}[-\pi + t] - \mathrm{Sin}[t]\,\mathrm{UnitStep}[-\pi + t]),\; y[t] \to \frac{1}{2}\,(t\,\mathrm{Sin}[t] + \pi\,\mathrm{Sin}[t]\,\mathrm{UnitStep}[-\pi + t] - t\,\mathrm{Sin}[t]\,\mathrm{UnitStep}[-\pi + t])\right\}\right\}$$

```
Plot[Evaluate[
   {x[t], y[t]} /. sol],
  {t, 0, 4 Pi},
  PlotStyle ->
   {GrayLevel[0],
    GrayLevel[0.5]}]
```

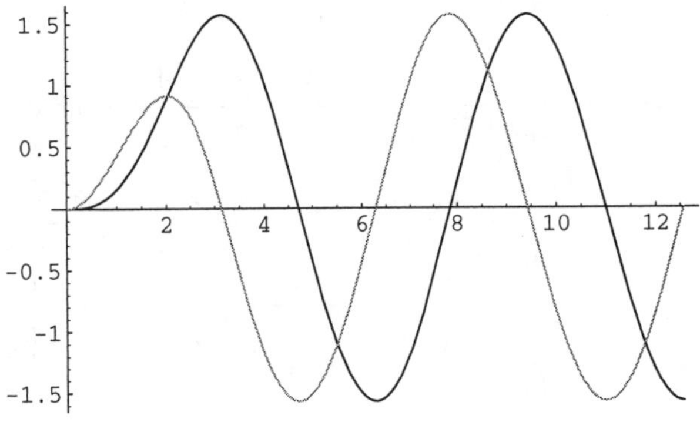

```
ParametricPlot[
 Evaluate[{x[t], y[t]} /. sol],
 {t, 0, 4 Pi},
 AspectRatio -> Automatic]
```

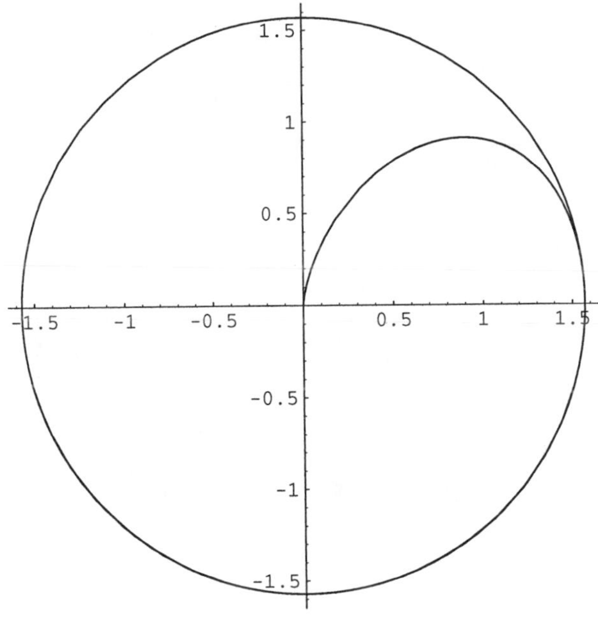

19. $\mathscr{L}\{f(t)\} = \dfrac{1}{1-e^{-3s}}\left(\int_1^2 e^{-st}dt + \int_2^3 2e^{-st}dt\right) = \dfrac{e^{2s}+e^{s}-1}{\left(1-e^{-3s}\right)se^{3s}} = \dfrac{e^{2s}+e^{s}-1}{\left(e^{3s}-1\right)s}$;

$$\begin{cases} (s+2)X(s) - x(0) + 3Y(s) = 0 \\ -X(s) + (6+s)Y(s) - y(0) = \dfrac{e^{2s}+e^{s}-2}{\left(1-e^{-3s}\right)se^{3s}} \end{cases}; \quad X(s) = \frac{-3\left(e^{2s}+e^{s}-2\right)}{\left(e^{3s}-1\right)s(s+3)(s+5)} + \frac{6+s}{(s+3)(s+5)};$$

$Y(s) = \dfrac{1}{(s+3)(s+5)} + \dfrac{(s+2)\left(e^{2s}+e^{s}-2\right)}{s\left(e^{3s}-1\right)(s+3)(s+5)}$; Consider $\dfrac{e^{2s}+e^{s}-2}{e^{3s}-1}$ in *X(s)* and *Y(s)*:

$$\frac{e^{2s}+e^{s}-2}{e^{3s}-1}\cdot\frac{e^{-3s}}{e^{-3s}} = \frac{e^{-s}+e^{-2s}-2e^{-3s}}{1-e^{-3s}} = \left(e^{-s}+e^{-2s}-2e^{-3s}\right)\left(1+e^{-3s}+e^{-6s}+\cdots\right)$$
$$= e^{-s}+e^{-2s}-2e^{-3s}+\cdots$$

With $\mathscr{L}^{-1}\left\{\dfrac{6+s}{(s+3)(s+5)}\right\} = -\dfrac{1}{2}e^{-5t}+\dfrac{3}{2}e^{-3t}$, $\mathscr{L}^{-1}\left\{\dfrac{-3}{s(s+3)(s+5)}\right\} = -\dfrac{1}{5}-\dfrac{3}{10}e^{-5t}+\dfrac{1}{2}e^{-3t}$,

$\mathscr{L}^{-1}\left\{\dfrac{1}{(s+3)(s+5)}\right\} = -\dfrac{1}{2}e^{-5t}+\dfrac{1}{2}e^{-3t}$, and $\mathscr{L}^{-1}\left\{\dfrac{s+2}{s(s+3)(s+5)}\right\} = \dfrac{2}{15}-\dfrac{3}{10}e^{-5t}+\dfrac{1}{6}e^{-3t}$, we have

$$\begin{cases} x(t) = \dfrac{1}{2}e^{-5t} - \dfrac{3}{2}e^{-3t} + \left[-\dfrac{2}{5}-\dfrac{3}{5}e^{-5(t-3)}+e^{-3(t-3)}\right]\mathscr{U}(t-3) + \\ \qquad \left[\dfrac{1}{5}+\dfrac{3}{10}e^{-5(t-2)}-e^{-3(t-2)}\right]\mathscr{U}(t-2) + \left[\dfrac{1}{5}+\dfrac{3}{10}e^{-5(t-1)}-e^{-3(t-1)}\right]\mathscr{U}(t-1)+\cdots \\ y(t) = \dfrac{1}{2}e^{-5t} - \dfrac{1}{2}e^{-3t} + \left[\dfrac{4}{15}-\dfrac{3}{5}e^{-5(t-3)}+\dfrac{1}{3}e^{-3(t-3)}\right]\mathscr{U}(t-3) - \\ \qquad \left[\dfrac{2}{15}-\dfrac{3}{10}e^{-5(t-2)}+\dfrac{1}{6}e^{-3(t-2)}\right]\mathscr{U}(t-2) - \left[\dfrac{2}{15}-\dfrac{3}{10}e^{-5(t-1)}+\dfrac{1}{6}e^{-3(t-1)}\right]\mathscr{U}(t-1)+\cdots \end{cases}$$

0.25
0.2
0.15
0.1
0.05
-0.3 -0.2 -0.1 0.1 0.2 0.3

1
0.8
0.6
0.4
0.2
-0.2
-0.4
1 2 3 4 5 6

21. $\begin{cases} -2s^2X(s)+2x(0)+2x'(0)-2sY(s)+2y(0)=0 \\ -sX(s)+x(0)+\left(1+s^2\right)Y(s)-sy(0)-y'(0)=\dfrac{s}{s^2+1} \end{cases}$; $X(s)=\dfrac{1}{s^2}+\dfrac{1}{s}-\dfrac{1}{s^2+1}+\dfrac{s+1}{s^2+2}$;

$Y(s)=\dfrac{1}{s}+\dfrac{s}{s^2+1}+\dfrac{2-s}{s^2+2}$; $\begin{cases} x(t)=t+1+\dfrac{\sqrt{2}}{2}\sin\sqrt{2}t+\cos\sqrt{2}t-\sin t \\ y(t)=1+\sqrt{2}\sin\sqrt{2}t-\cos\sqrt{2}t+\cos t \end{cases}$

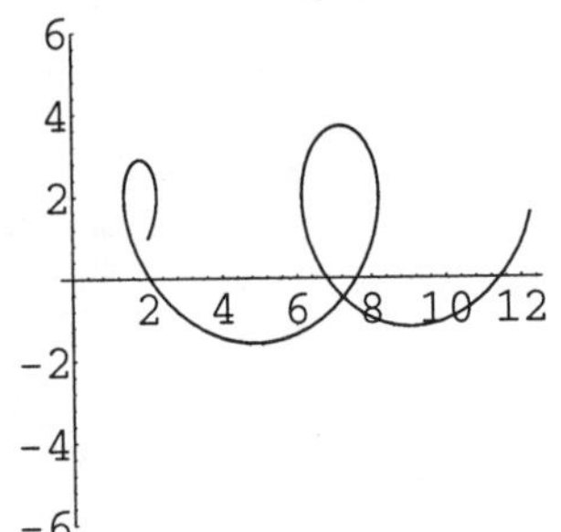

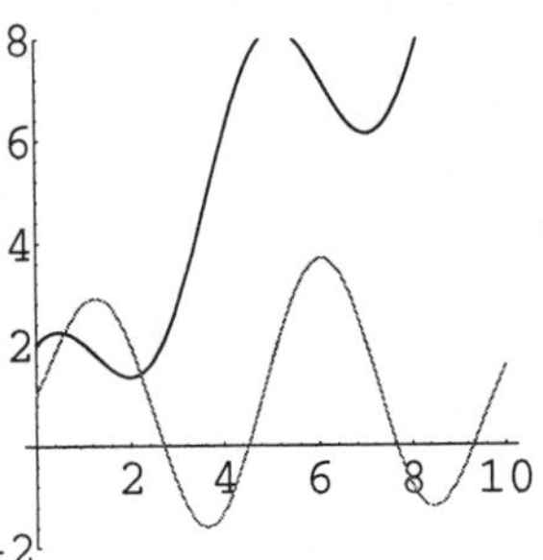

25. $\begin{cases} -\left(s^2+2\right)X(s)+sx(0)+x'(0)+2sY(s)-2y(0)=\dfrac{1}{s-1} \\ -\left(s^2+2\right)Y(s)+2sy(0)-y'(0)=\dfrac{1}{s^2} \end{cases}$; $X(s)=\dfrac{-1}{3(s-1)}-\dfrac{1}{2s}+\dfrac{2-3s}{s^2+1}+\dfrac{29s-16}{6\left(s^2+2\right)}$;

$Y(s)=\dfrac{-1}{2s^2}+\dfrac{-3-2s}{2\left(s^2+1\right)}$; $\begin{cases} x(t)=-\dfrac{1}{2}-\dfrac{1}{3}e^t+2\sin t-3\cos t-\dfrac{4\sqrt{2}}{3}\sin\sqrt{2}t+\dfrac{29}{6}\cos\sqrt{2}t \\ y(t)=-\dfrac{1}{2}t-\dfrac{3}{2}\sin t-\cos t \end{cases}$

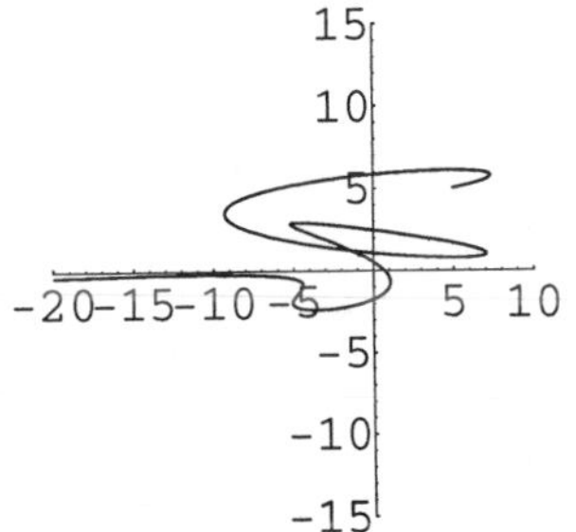

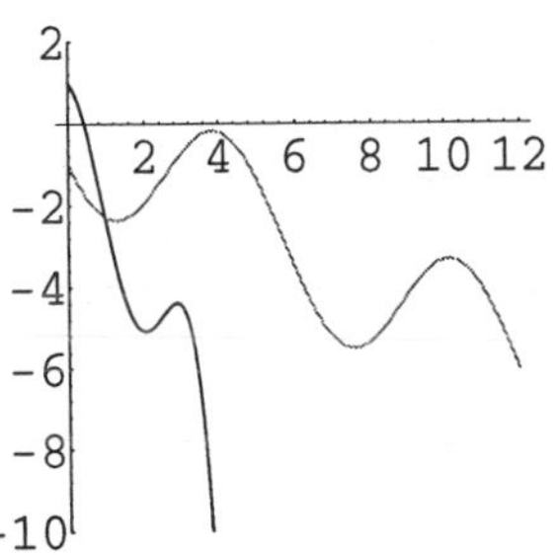

1. We illustrate how to implement the method of Laplace transforms.

```
<<Calculus`LaplaceTransform`
Clear[x,y,t]
deq1=x'[t]-y[t]==Exp[-t]
deq2=y'[t]+5x[t]+2y[t]==Sin[3t]
eq1=LaplaceTransform[deq1,t,s]
eq2=LaplaceTransform[deq2,t,s]

laps=Solve[{eq1,eq2},
    {LaplaceTransform[x[t],t,s],
    LaplaceTransform[y[t],t,s]}]
laps=laps /. {x[0]->x0,y[0]->y0}
laps[[1,1,2]]
laps[[1,2,2]]
x=InverseLaplaceTransform[
    laps[[1,1,2]],s,t]
y=InverseLaplaceTransform[
    laps[[1,2,2]],s,t]
```

```
readlib(laplace):
deq1:=diff(x(t),t)-y(t)=exp(-t);
deq2:=diff(y(t),t)+5*x(t)+2*y(t)=
    sin(3*t);
eq1:=laplace(deq1,t,s);
eq2:=laplace(deq2,t,s);
laps:=solve({eq1,eq2},
    {laplace(x(t),t,s),
        laplace(y(t),t,s)});
laps:=subs({x(0)=x0,y(0)=y0},laps);

x:=invlaplace(rhs(laps[1]),s,t);
y:=invlaplace(rhs(laps[2]),s,t);
```

```
Clear[toplot]
a=0.75;
toplot=Table[{x,y}  /.
   {x0->a  Cos[theta],
      y0->a  Sin[theta]},
   {theta,0,2Pi,2Pi/19}];
Short[toplot,3]
grays=Table[GrayLevel[i],
   {i,0,.6,.6/19}];
ParametricPlot[Evaluate[toplot],
   {t,-2,2},PlotRange->
      {{-3/2,3/2},{-3/2,3/2}},
   AspectRatio->1,
   PlotStyle->grays]
```

```
a:=0.75:
thetavals:=[seq(i*2*Pi/19,i=0..19)]:
toplot:={seq(subs({x0=a*cos(theta),
   y0=a*sin(theta)},[x,y,-2..2]),
   theta=thetavals)}:
plot(toplot,
   view=[-3/2..3/2,-3/2..3/2]);
```

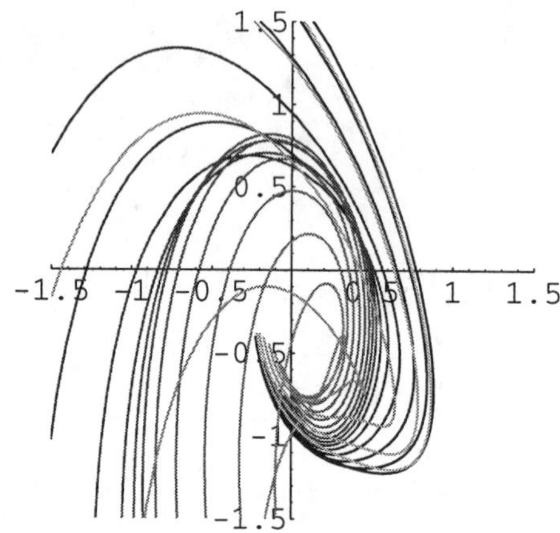

2.

```
Clear[x,y,t]
deq1=2x''[t]-5y'[t]==0
deq2=-y''[t]-3y[t]-x'[t]==Sin[t]
eq1=LaplaceTransform[deq1,t,s]
eq2=LaplaceTransform[deq2,t,s]
sysofeqs={eq1,eq2}  /.
   {x[0]->1,x'[0]->0,
   y[0]->1,y'[0]->1}
laps=Solve[sysofeqs,
   {LaplaceTransform[x[t],t,s],
   LaplaceTransform[y[t],t,s]}]
laps[[1,1,2]]
laps[[1,2,2]]
x=InverseLaplaceTransform[
   laps[[1,1,2]],s,t]
y=InverseLaplaceTransform[
   laps[[1,2,2]],s,t]
Plot[{x,y},{t,0,3Pi},
   PlotRange->{-5Pi/2,Pi/2},
   AspectRatio->1,
   PlotStyle->{GrayLevel[0],
      GrayLevel[.6]}]
ParametricPlot[{x,y},{t,0,3Pi},
   PlotRange->{{-10,2},{-6,6}},
   AspectRatio->1]
```

```
x:='x':y:='y':t:='t':
sol:=dsolve({2*diff(x(t),t$2)-
      5*diff(y(t),t)=0,
   -diff(y(t),t$2)-3*y(t)-
      diff(x(t),t)=sin(t),
   x(0)=1,D(x)(0)=0,y(0)=1,
   D(y)(0)=1},{x(t),y(t)},laplace);
assign(sol):
plot({x(t),y(t)},t=0..3*Pi,
   -5*Pi/2..Pi/2);
plot([x(t),y(t),t=0..3*Pi],
   view=[-10..2,-6..6]);
```

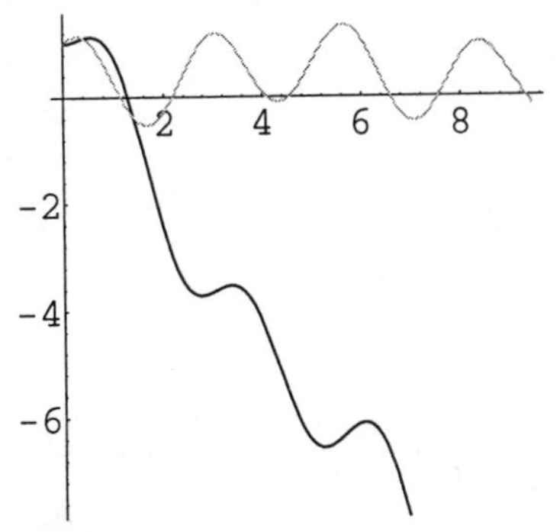
-2
-4
-6
2
4
6
8

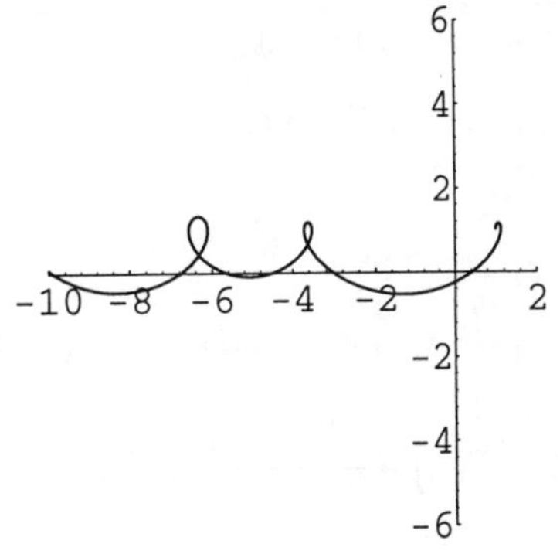
6
4
2
-2
-4
-6
-10
-8
-6
-4
-2
2

EXERCISES 8.6

3. We solve the initial-value problem

$$\begin{cases} 100\dfrac{dI}{dt}+100I=E(t) \\ I(0)=0 \end{cases}$$

The Laplace transform of the periodic function $E(t)$ is

$$\mathscr{L}\{E(t)\}=\frac{1}{1-e^{-2s}}\int_0^2 e^{-st}E(t)dt=\frac{50}{1-e^{-2s}}\int_0^1 e^{-st}dt=\frac{50}{1-e^{-2s}}\cdot\frac{1-e^{-s}}{s}=\frac{50\left(1-e^{-s}\right)}{s\left(1-e^{-2s}\right)}$$

so taking the Laplace transform of each side of the differential equation and applying the initial condition yields

$$100\,s\,\mathscr{L}\{I\}-100I(0)+100\mathscr{L}\{I\}=\frac{50\left(1-e^{-s}\right)}{s\left(1-e^{-2s}\right)}$$

$$100(s+1)\mathscr{L}\{I\}=\frac{50\left(1-e^{-s}\right)}{s\left(1-e^{-2s}\right)}$$

$$\mathscr{L}\{I\}=\frac{1-e^{-s}}{2s(s+1)\left(1-e^{-2s}\right)}.$$

To find $I(t)$, we first rewrite $\mathscr{L}\{I\}$:

$$\mathscr{L}\{I\}=\frac{1-e^{-s}}{2s(s+1)\left(1-e^{-2s}\right)}=\frac{1}{1-e^{-2s}}\frac{1}{2s(s+1)}-\frac{e^{-s}}{1-e^{-2s}}\frac{1}{2s(s+1)}$$

$$=\left(1+e^{-2s}+e^{-4s}+e^{-6s}+\cdots\right)\frac{1}{2s(s+1)}-\left(e^{-s}+e^{-3s}+e^{-5s}+e^{-7s}+\cdots\right)\frac{1}{2s(s+1)}.$$

Now, $\mathscr{L}^{-1}\left\{\dfrac{1}{2s(s+1)}\right\}=\dfrac{1}{2}-\dfrac{1}{2}e^{-t}$ so

$$I(t)=\frac{1}{2}\left(1-e^{-t}\right)-\frac{1}{2}\left(1-e^{-(t-1)}\right)\mathscr{U}(t-1)+\frac{1}{2}\left(1-e^{-(t-2)}\right)\mathscr{U}(t-2)-\frac{1}{2}\left(1-e^{-(t-3)}\right)\mathscr{U}(t-3)+\cdots.$$

7. We solve

$$\begin{cases} \dfrac{dI}{dt}+6I+9\displaystyle\int_0^t I(\alpha)d\alpha=100 \\ I(0)=0 \end{cases}$$

Applying the Laplace transform and initial condition results in

$$s\,\mathscr{L}\{I\}-I(0)+6\mathscr{L}\{I\}+9\frac{L\{I\}}{s}=\frac{100}{s}$$

$$\left(s+6+\frac{9}{s}\right)\mathscr{L}\{I\}=\frac{100}{s}$$

$$\mathscr{L}\{I\}=\frac{100}{s^2+6s+9}=\frac{100}{(s+3)^2}.$$

Thus, $I(t)=\mathscr{L}^{-1}\left\{\dfrac{100}{(s+3)^2}\right\}=100\,t\,e^{-3t}$.

11. (a) We solve

$$\begin{cases} \dfrac{dI}{dt} + 4I = \delta(t-1) \\ I(0) = 0 \end{cases}$$

Applying the Laplace transform and initial condition results in

$$s\,\mathscr{L}\{I\} - I(0) + 4\mathscr{L}\{I\} = e^{-s}$$

$$\mathscr{L}\{I\} = \frac{e^{-s}}{s+4} \Rightarrow I(t) = e^{-4t+4}\mathscr{U}(t-1).$$

(b) We solve

$$\begin{cases} \dfrac{dI}{dt} + 4I = \delta(t-1) \\ I(0) = 1 \end{cases}$$

We proceed as in (a) to obtain

$$s\,\mathscr{L}\{I\} - I(0) + 4\mathscr{L}\{I\} = e^{-s}$$

$$\mathscr{L}\{I\} = \frac{1+e^{-s}}{s+4} \Rightarrow I(t) = e^{-4t} + e^{-4t+4}\mathscr{U}(t-1).$$

15. We solve

$$\begin{cases} \dfrac{dI}{dt} + I = \delta(t-2) + \delta(t-6) \\ I(0) = 0 \end{cases}$$

Applying the Laplace transform and initial condition to the equation results in

$$s\,\mathscr{L}\{I\} - I(0) + \mathscr{L}\{I\} = e^{-2s} + e^{-6s}$$

$$\mathscr{L}\{I\} = \frac{e^{-2s} + e^{-6s}}{s+1} \Rightarrow I(t) = e^{-t+2}\mathscr{U}(t-2) + e^{-t+6}\mathscr{U}(t-6).$$

25. $\mathscr{L}\{x'' + 5x' + 4x\} = s^2X(s) - sx(0) - x'(0) + 5[sX(s) - x(0)] + 4X(s)$
$= \left(s^2 + 5s + 4\right)X(s) - s - 5,\ X(s) = \dfrac{s+5}{(s+1)(s+4)} = \dfrac{4/3}{s+1} - \dfrac{1/3}{s+4},\ x(t) = -\dfrac{1}{3}e^{-4t} + \dfrac{4}{3}e^{-t}$

29.

$$\left(s^2 + 5s + 4\right)X(s) = \frac{2}{s+1} + \frac{1}{s+4}$$

$$X(s) = \frac{3s+9}{(s+1)(s+4)\left(s^2 + 5s + 4\right)}$$

$$= \frac{2}{3(s+1)^2} - \frac{1}{9(s+1)} - \frac{1}{3(s+4)^2} + \frac{1}{9(s+4)}$$

$$x(t) = \frac{1}{9}e^{-4t} - \frac{1}{9}e^{-t} - \frac{1}{3}te^{-4t} + \frac{2}{3}te^{-t}$$

33. $f(t) = e^{-2t}[1 - \mathscr{U}(t-1)] + \mathscr{U}(t-1) - \mathscr{U}(t-2)$,
$\mathscr{L}\{f(t)\} = -e^{-2s}/s + e^{-s}/s + 1/(s+2) - e^{-(s+2)}/(s+2)$

$$x(t) = t^2e^{-t}\mathscr{U}(t) + \frac{1}{2}e^{-2t}\left(-2 + e^2 + e^{2t} + 4t - 2e^2t - 2t^2\right)\mathscr{U}(t-1) +$$
$$\frac{1}{2}e^{-2t}\left(-3e^4 - e^{2t} + 2e^4t\right)\mathscr{U}(t-2)$$

37. $\mathscr{L}\{f(t)\}=\frac{s}{s^2+\pi^2}\left(1+e^{-s}\right)$, $X(s)=\frac{s}{\left(s^2+\pi^2\right)\left(s^2+5s+4\right)}\left(1+e^{-s}\right)$

$$=\frac{1}{3\left(1+\pi^2\right)(s+1)}-\frac{1}{3\left(16+\pi^2\right)(s+4)}+\frac{4-\pi^2-5s}{\left(1+\pi^2\right)\left(16+\pi^2\right)\left(s^2+\pi^2\right)}$$

$$x(t)=\frac{-(1+e)}{3\left(\pi^2+1\right)}e^{-t}+\frac{4\left(1+e^4\right)}{3\left(\pi^2+16\right)}e^{-4t}+\frac{1}{3\pi^4+51\pi^2+48}e^{-4t}\left(-4e^4+16e^{3t+1}-4e^4\pi^2+\right.$$

$$\left.\pi^2e^{3t+1}+12e^{4t}\cos\pi t-3\pi^2e^{4t}\cos\pi t+15\pi e^{4t}\sin\pi t\right)\mathscr{U}(1-t)$$

41. We solve the equation $x''+4x'+13x=\delta(t-1)+\delta(t-3)$ to obtain
$x(t)=\frac{1}{3}\left[-\sin(3-3t)\mathscr{U}(t-1)-e^4e^{2-2t}\sin(9-3t)\mathscr{U}(t-3)\right]$.

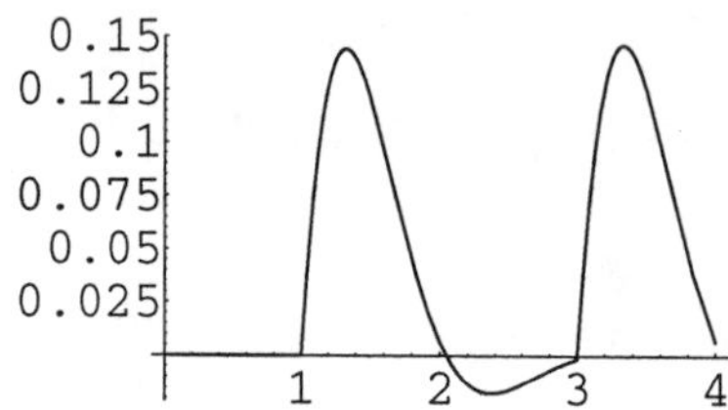

45.

$$x'+5x=500(2-\cos t)$$

$$s\,X(s)-x(0)+5X(s)=500\frac{s^2+2}{s\left(s^2+1\right)}$$

$$s\,X(s)-5000+5X(s)=500\frac{s^2+2}{s\left(s^2+1\right)}$$

$$X(s)=500\frac{10s^3+s^2+10s+2}{s(s+5)\left(s^2+1\right)}=200\frac{1}{s}+\frac{63650}{13}\frac{1}{s+5}-\frac{250}{13}\frac{5s+1}{s^2+1}$$

$$x(t)=200+\tfrac{63650}{13}e^{-5t}-\tfrac{1250}{13}\cos t-\tfrac{250}{13}\sin t;\ \textit{Bounded}$$

49. The Laplace transform of $f(t)=\begin{cases}5000(2-\sin t),\ 0\le t<5\\ 0,\ t\ge 5\end{cases}=5000(2-\sin t)\{\mathscr{U}(t-0)-\mathscr{U}(t-5)\}$ is

$$\mathscr{L}\{f(t)\}=5000\frac{-2+2e^{5s}-s\,e^{5s}-2s^2+2s^2e^{5s}+s\cos 5+s^2\sin 5}{s\,e^{5s}\left(s^2+1\right)}.$$

Then,

$$s\,X(s)-x(0)-X(s)=5000\frac{-2+2e^{5s}-s\,e^{5s}-2s^2+2s^2e^{5s}+s\cos 5+s^2\sin 5}{s\,e^{5s}\left(s^2+1\right)}$$

$$s\,X(s)-10000-X(s)=5000\frac{-2+2e^{5s}-s\,e^{5s}-2s^2+2s^2e^{5s}+s\cos 5+s^2\sin 5}{s\,e^{5s}\left(s^2+1\right)}$$

$$X(s)=-5000\frac{2-2e^{5s}-se^{5s}+2s^2-2s^2e^{5s}-2s^3e^{5s}-s\cos 5-s^2\sin 5}{s\,e^{5s}(s-1)\left(s^2+1\right)}.$$

Now,

$$X(s) = -5000\frac{2 - 2e^{5s} - se^{5s} + 2s^2 - 2s^2e^{5s} - 2s^3e^{5s} - s\cos 5 - s^2\sin 5}{s\,e^{5s}(s-1)(s^2+1)}$$

$$= -5000\left[\frac{2 + 2s^2 - s\cos 5 - s^2\sin 5}{s\,e^{5s}(s-1)(s^2+1)} + \frac{-2 - s - 2s^2 - 2s^3}{s\,(s-1)(s^2+1)}\right]$$

so

$$x(t) = -10000 + 17500e^t + 2500\cos t + 2500\sin t + \big(10000 - 10000e^{t-5} + 2500e^{t-5}\cos 5 -$$
$$2500\cos 5\cos(t-5) + 2500e^{t-5}\sin 5 - 2500\sin 5\cos(t-5) - 2500\cos 5\sin(t-5) +$$
$$2500\sin 5\sin(t-5)\big)\mathcal{U}(t-5)$$

55. First, we use Laplace transforms to solve the initial-value problem:

$$s\,X(s) - x(0) = c\frac{s^2 - s + 1}{s(s^2+1)} - k\,X(s)$$

$$s\,X(s) - x_0 = c\frac{s^2 - s + 1}{s(s^2+1)} - k\,X(s)$$

$$X(s) = \frac{x_0s^3 + c\,s^2 + (x_0 - c)s + c}{s(s+k)(s^2+1)} = \frac{c}{k}\frac{1}{s} - \frac{c}{k^2+1}\frac{k-s}{s^2+1} + \frac{k^3x_0 + kx_0 - ck^2 - c - kc}{k(k^2+1)}\frac{1}{s+k}$$

$$x(t) = \tfrac{c}{k} - \frac{ck}{k^2+1}\sin t + \frac{c}{k^2+1}\cos t + \frac{k^3x_0 + kx_0 - ck^2 - c - kc}{k^3+k}e^{-kt}.$$

For large values of t, $x(t) \approx \tfrac{c}{k} - \frac{ck}{k^2+1}\sin t + \frac{c}{k^2+1}\cos t$ is nearly periodic.

65. We solve $\begin{cases} 2x'' = -6x + 2y \\ y'' = 2x - 2y \\ x(0) = 0,\ x'(0) = 0,\ y(0) = 0,\ y'(0) = -1 \end{cases}$: $X(s) = \frac{s(s^2+2)}{s^4 + 5s^2 + 4} = \frac{s}{3(s^2+1)} + \frac{2s}{3(s^2+4)}$,

$$Y(s) = \frac{2s}{s^4 + 5s^2 + 4} = \frac{2s}{3(s^2+1)} - \frac{2s}{3(s^2+4)};\quad \begin{cases} x(t) = -\frac{1}{3}\sin t + \frac{1}{6}\sin 2t \\ y(t) = -\frac{2}{3}\sin t - \frac{1}{6}\sin 2t \end{cases}$$

71. We solve $\begin{cases} 2x'' = -6x + 2y + \cos t \\ y'' = 2x - 2y + \sin t \\ x(0) = 1,\ x'(0) = 0,\ y(0) = 0,\ y'(0) = 0 \end{cases}$:

$$X(s) = \frac{2s^5 + 7s^3 + 6s + 2}{2(s^2+1)^2(s^2+4)} = \frac{s+2}{6(s^2+1)^2} + \frac{4s-1}{9(s^2+1)} + \frac{5s+1}{9(s^2+4)},$$

$$Y(s) = \frac{2s^3 + s^2 + 3s + 3}{(s^2+1)^2(s^2+4)} = \frac{s+2}{3(s^2+1)^2} + \frac{5s-1}{9(s^2+1)} - \frac{5s+1}{9(s^2+4)},$$

$$\begin{cases} x(t) = \frac{1}{18}\sin 2t + \frac{5}{9}\cos 2t + \frac{1}{18}\sin t + \frac{4}{9}\cos t - \frac{t}{6}\cos t + \frac{t}{12}\sin t \\ y(t) = -\frac{1}{18}\sin 2t - \frac{5}{9}\cos 2t + \frac{4}{9}\sin t + \frac{5}{9}\cos t - \frac{t}{3}\cos t + \frac{t}{6}\sin t \end{cases}$$

75. We solve $\begin{cases} 1024\theta_1'' + 256\theta_2'' + 2048\theta_1 = 0 \\ 256\theta_2'' + 256\theta_1'' + 512\theta_2 = 0 \\ \theta_1(0) = 1, \theta_1'(0) = 1, \theta_2(0) = 0, \theta_2'(0) = 0 \end{cases}$:

$X(s) = \dfrac{s(3s^2+8)}{3s^4+16s^2+16} = \dfrac{s}{2(s^2+4)} + \dfrac{3s}{2(3s^2+4)}$, $Y(s) = \dfrac{8s}{3s^4+16s^2+16} = \dfrac{-s}{s^2+4} + \dfrac{3s}{3s^2+4}$;

$$\begin{cases} \theta_1(t) = \frac{1}{2}\cos 2t + \frac{1}{4}\sin 2t + \frac{1}{2}\cos\frac{2t}{\sqrt{3}} + \frac{\sqrt{3}}{4}\sin\frac{2t}{\sqrt{3}} \\ \theta_2(t) = -\cos 2t - \frac{1}{2}\sin 2t + \frac{\sqrt{3}}{2}\sin\frac{2t}{\sqrt{3}} + \cos\frac{2t}{\sqrt{3}} \end{cases}$$

79. We solve $\begin{cases} 1024\theta_1'' + 256\theta_2'' + 2048\theta_1 = 0 \\ 256\theta_2'' + 256\theta_1'' + 512\theta_2 = 0 \\ \theta_1(0) = 0, \theta_1'(0) = 0, \theta_2(0) = -1, \theta_2'(0) = -1 \end{cases}$: $X(s) = \dfrac{-2(s+1)}{3s^4+16s^2+16} = \dfrac{s+1}{4(s^2+4)} - \dfrac{3(s+1)}{4(3s^2+4)}$,

$Y(s) = -\dfrac{3s^3+3s^2+8s+8}{3s^4+16s^2+16} = \dfrac{-s-1}{2(s^2+4)} - \dfrac{3(s+1)}{2(3s^2+4)}$, $\begin{cases} \theta_1(t) = \frac{1}{4}\cos 2t + \frac{1}{8}\sin 2t - \frac{1}{4}\cos\frac{2t}{\sqrt{3}} - \frac{\sqrt{3}}{8}\sin\frac{2t}{\sqrt{3}} \\ \theta_2(t) = -\frac{1}{2}\cos 2t - \frac{1}{4}\sin 2t - \frac{\sqrt{3}}{4}\sin\frac{2t}{\sqrt{3}} - \frac{1}{2}\cos\frac{2t}{\sqrt{3}} \end{cases}$

85. $\begin{vmatrix} -\lambda & 1 & 0 & 0 \\ -\frac{(k_1+k_2)}{m_1} & -\lambda & \frac{k_2}{m_1} & 0 \\ 0 & 0 & -\lambda & 1 \\ \frac{k_2}{m_2} & 0 & -\frac{(k_2+k_3)}{m_2} & -\lambda \end{vmatrix} = \lambda^4 + B\lambda^2 + C$ where $B = \left(\dfrac{(k_2+k_3)}{m_2} + \dfrac{(k_1+k_2)}{m_1}\right)$ and

$C = \dfrac{(k_1+k_2)(k_2+k_3)}{m_1 m_2} - \dfrac{k_2^2}{m_1 m_2}$. $\lambda^4 + B\lambda^2 + C = 0 \Rightarrow \lambda^2 = \frac{1}{2}\left(-B \pm \sqrt{B^2-4C}\right)$.

$B^2 - 4C = \left(\dfrac{(k_2+k_3)}{m_2} - \dfrac{(k_1+k_2)}{m_1}\right)^2 + \dfrac{4k_2^2}{m_1 m_2} > 0$ and $B^2 - 4C < B$, so λ^2 equals two negative values, say $-\beta_1^2$ and $-\beta_2^2$. $\lambda^2 = -\beta_1^2 \Rightarrow \lambda = \pm\beta_1 i$ and $\lambda^2 = -\beta_2^2 \Rightarrow \lambda = \pm\beta_2 i$.

86.

(a). The Laplace transform of $f(t)$ is $\dfrac{e^s}{s(e^s+1)}$.

```
laprhs=1/(1-Exp[-2s])*
   Integrate[Exp[-s  t],{t,0,1}]//
     Together
```

```
laprh:=simplify(1/(1-exp(-2*s))*
   int(exp(-s*t),t=0..1));
```

Computing the Laplace transform of the left-hand side of the equation, setting the result equal to the Laplace transform of the right-hand side of the equation and solving the result for the Laplace transform of $x(t)$ yields

$$\mathscr{L}\{x(t)\} = \frac{e^s}{s(s^2+5s+6)(1+e^s)}.$$

```
Clear[m,k,c,x,t]
<<Calculus`LaplaceTransform`
eq=m  x''[t]+c  x'[t]+k  x[t]
m=1;k=6;c=5;
laplhs=LaplaceTransform[eq,t,s]  /.
   {x[0]->0,x'[0]->0}
Solve[laplhs==laprhs,
   LaplaceTransform[x[t],t,s]]
```

```
readlib(laplace):
laprhs:=simplify(1/(1-exp(-2*s))*
   int(exp(-s*t),t=0..1));
m:='m':k:='k':c:='c':x:='x':
eq:=m*diff(x(t),t$2)+c*diff(x(t),t)+
   k*x(t);
m:=1:k:=6:c:=5:
laplhs:=subs({D(x)(0)=0,x(0)=0},
   laplace(eq,t,s));
simplify(solve(laplhs=laprhs,
   laplace(x(t),t,s)));
```

Now,

$$\frac{e^s}{s\left(s^2+5s+6\right)\left(1+e^s\right)}=\frac{1}{s\left(s^2+5s+6\right)\left(1+e^{-s}\right)}=\frac{1}{1-\left(-e^{-s}\right)}\frac{1}{s\left(s^2+5s+6\right)}$$
$$=\left(1-e^{-s}+e^{-2s}-e^{-3s}+e^{-4s}-\cdots\right)\frac{1}{s\left(s^2+5s+6\right)}$$

and

$$\mathscr{L}^{-1}\left\{\frac{1}{s\left(s^2+5s+6\right)}\right\}=\frac{2-3e^t+e^{3t}}{6e^{3t}}$$

```
InverseLaplaceTransform[
   1/(s(s^2+5s+6)),s,t]//Together
```

```
invlaplace(1/(s*(s^2+5*s+6)),s,t);
```

so

$$x(t)=\frac{2-3e^t+e^{3t}}{6e^{3t}}-\frac{2-3e^{t-1}+e^{3(t-1)}}{6e^{3(t-1)}}\mathscr{U}(t-1)+$$
$$\frac{2-3e^{t-2}+e^{3(t-2)}}{6e^{3(t-2)}}\mathscr{U}(t-2)-\frac{2-3e^{t-3}+e^{3(t-3)}}{6e^{3(t-3)}}\mathscr{U}(t-3)+\cdots$$

(b) We proceed in the same manner as in 1.

```
laprhs=1/(1-Exp[-2s])*
   (Integrate[Exp[-s  t],{t,0,1}]+
      Integrate[-Exp[-s  t],
      {t,1,2}])//Together
Clear[m,k,c,x,t]
eq=m  x''[t]+c  x'[t]+k  x[t]
m=1;k=6;c=5;
laplhs=LaplaceTransform[eq,t,s]  /.
   {x[0]->0,x'[0]->0}
Solve[laplhs==laprhs,
   LaplaceTransform[x[t],t,s]]
```

```
laprhs:=simplify(1/(1-exp(-2*s))*
   (int(exp(-s*t),t=0..1)-
   int(exp(-s*t),t=1..2)));
m:='m':k:='k':c:='c':x:='x':
eq:=m*diff(x(t),t$2)+c*diff(x(t),t)+
   k*x(t);
m:=1:k:=6:c:=5:
laplhs:=subs({D(x)(0)=0,x(0)=0},
   laplace(eq,t,s));
simplify(solve(laplhs=laprhs,
   laplace(x(t),t,s)));
```

Thus,

$$\mathcal{L}\{x(t)\}=\frac{e^s-1}{s\left(1+e^s\right)\left(s^2+5s+6\right)}=\frac{1}{s\left(s^2+5s+6\right)}-\frac{2}{s\left(1+e^s\right)\left(s^2+5s+6\right)}$$

$$=\frac{1}{s\left(s^2+5s+6\right)}-\frac{2e^{-s}}{s\left(1+e^{-s}\right)\left(s^2+5s+6\right)}=\frac{1}{s\left(s^2+5s+6\right)}-\frac{2e^{-s}}{1-\left(-e^{-s}\right)}\frac{1}{s\left(s^2+5s+6\right)}$$

$$=\frac{1}{s\left(s^2+5s+6\right)}-2e^{-s}\left(1-e^{-s}+e^{-2s}-e^{-3s}+e^{-4s}-\dots\right)\frac{1}{s\left(s^2+5s+6\right)}$$

$$=\frac{1}{s\left(s^2+5s+6\right)}-2\left(e^{-s}-e^{-2s}+e^{-3s}-e^{-4s}+e^{-5s}-\dots\right)\frac{1}{s\left(s^2+5s+6\right)}.$$

Because $\mathcal{L}^{-1}\left\{\frac{1}{s\left(s^2+5s+6\right)}\right\}=\frac{2-3e^t+e^{3t}}{6e^{3t}}$,

```
InverseLaplaceTransform[
   1/(s(s^2+5s+6)),s,t]//Together
```

```
invlaplace(1/(s*(s^2+5*s+6)),s,t);
```

it follows that

$$x(t)=\frac{2-3e^t+e^{3t}}{6e^{3t}}-\frac{2-3e^{t-1}+e^{3(t-1)}}{3e^{3(t-1)}}\mathcal{U}(t-1)+\frac{2-3e^{t-2}+e^{3(t-2)}}{3e^{3(t-2)}}\mathcal{U}(t-2)-$$
$$\frac{2-3e^{t-3}+e^{3(t-3)}}{3e^{3(t-3)}}\mathcal{U}(t-3)+\frac{2-3e^{t-4}+e^{3(t-4)}}{3e^{3(t-4)}}\mathcal{U}(t-4)-\cdots$$

(c) (Mathematica only) We proceed in the same manner as in the previous problems.

```
laprhs=1/(1-Exp[-2  Pi  s])*Integrate[Sin[t]  Exp[-s  t],{t,0,Pi}]//Together
Clear[m,k,c,x,t]
eq=m   x''[t]+c  x'[t]+k  x[t]
m=1;k=13;c=4;
laplhs=LaplaceTransform[eq,t,s]  /.  {x[0]->0,x'[0]->0}
Solve[laplhs==laprhs,LaplaceTransform[x[t],t,s]]
```

Then,

$$\mathcal{L}\{x(t)\}=\frac{e^{\pi s}}{\left(s^2+1\right)\left(e^{\pi s}-1\right)\left(s^2+4s+13\right)}=\frac{1}{1-e^{-\pi s}}\frac{1}{\left(s^2+1\right)\left(s^2+4s+13\right)}$$

$$=\left(1+e^{-\pi s}+e^{-2\pi s}+e^{-3\pi s}+e^{-4\pi s}+\dots\right)\frac{1}{\left(s^2+1\right)\left(s^2+4s+13\right)}$$

and $\mathcal{L}^{-1}\left\{\frac{1}{\left(s^2+1\right)\left(s^2+4s+13\right)}\right\}=\frac{1}{120}e^{-2t}\left(-3e^{2t}\cos t+3\cos 3t+9e^{2t}\sin t-\sin 3t\right)$

```
InverseLaplaceTransform[1/((s^2+1)(s^2+4s+13)),s,t]//Together
```

so

$$x(t)=\frac{1}{120}e^{-2t}\left(-3e^{2t}\cos t+3\cos 3t+9e^{2t}\sin t-\sin 3t\right)+$$
$$\frac{1}{120}e^{-2(t-\pi)}\left(-3e^{2(t-\pi)}\cos(t-\pi)+3\cos(3(t-\pi))+9e^{2(t-\pi)}\sin(t-\pi)-\sin(3(t-\pi))\right)\mathcal{U}(t-\pi)+$$
$$\frac{1}{120}e^{-2(t-2\pi)}\left(-3e^{2(t-2\pi)}\cos(t-2\pi)+3\cos(3(t-2\pi))+\right.$$
$$\left.9e^{2(t-2\pi)}\sin(t-2\pi)-\sin(3(t-2\pi))\right)\mathcal{U}(t-2\pi)+\cdots$$

(d) (Mathematica only) We proceed in the same manner as in the previous problems.

```
laprhs=1/(1-Exp[-2s])(Integrate[Exp[-s  t],{t,0,1}]+
   Integrate[-Exp[-s  t],{t,1,2}])//Together
Clear[m,k,c,x,t]
eq=m  x''[t]+c  x'[t]+k  x[t]
m=1;k=13;c=4;
laplhs=LaplaceTransform[eq,t,s]  /.{x[0]->0,x'[0]->0}
Solve[laplhs==laprhs,LaplaceTransform[x[t],t,s]]
```

Thus,

$$\mathcal{L}\{x(t)\}=\frac{e^s-1}{s\left(1+e^s\right)\left(s^2+4s+13\right)}=\frac{1}{s\left(s^2+4s+13\right)}-\frac{2}{s\left(1+e^s\right)\left(s^2+4s+13\right)}$$
$$=\frac{1}{s\left(s^2+4s+13\right)}-\frac{2e^{-s}}{s\left(1+e^{-s}\right)\left(s^2+4s+13\right)}=\frac{1}{s\left(s^2+4s+13\right)}-2\frac{e^{-s}}{1-\left(-e^{-s}\right)}\frac{1}{s\left(s^2+4s+13\right)}$$
$$=\frac{1}{s\left(s^2+4s+13\right)}-2e^{-s}\left(1-e^{-s}+e^{-2s}-e^{-3s}+e^{-4s}-\cdots\right)\frac{1}{s\left(s^2+4s+13\right)}$$
$$=\frac{1}{s\left(s^2+4s+13\right)}-2\left(e^{-s}-e^{-2s}+e^{-3s}-e^{-4s}+e^{-5s}-\cdots\right)\frac{1}{s\left(s^2+4s+13\right)}.$$

Because $\mathcal{L}^{-1}\left\{\frac{1}{s\left(s^2+4s+13\right)}\right\}=\frac{1}{39}e^{-2t}\left(3e^{2t}-3\cos 3t-2\sin 3t\right)$,

```
InverseLaplaceTransform[1/(s(s^2+4s+13)),s,t]//Together
```

it follows that

$$x(t)=\frac{1}{39}e^{-2t}\left(3e^{2t}-3\cos 3t-2\sin 3t\right)-2\left\{\frac{1}{39}e^{-2(t-1)}\left(3e^{2(t-1)}-3\cos(3(t-1))-2\sin(3(t-1))\right)\mathcal{U}(t-1)-\right.$$
$$\left.\frac{1}{39}e^{-2(t-2)}\left(3e^{2(t-2)}-3\cos(3(t-2))-2\sin(3(t-2))\right)\mathcal{U}(t-2)+\cdots\right\}.$$

(e) For Parts (a), (b), and (c) we graph the solution on the interval [0,4]; for Part (c) we graph the solution on the interval [0,4π].

```
<<Calculus`DiracDelta`
g[t_]:=(2-3 Exp[t]+Exp[3 t])/(6 Exp[3
t])
x[t_]=g[t]+Sum[(-1)^n g[t-n]
   UnitStep[t-n],{n,1,5}]
Plot[x[t],{t,0,5}]
g[t_]:=(2-3 Exp[t]+Exp[3 t])/(6 Exp[3
t])
x[t_]=g[t]+Sum[2 (-1)^n g[t-n]
   UnitStep[t-n],{n,1,5}]
Plot[x[t],{t,0,5}]
Clear[g,x]
g[t_]:=1/120 Exp[-2 t] (-3 Exp[2t]
Cos[t]
   +3 Cos[3 t]+9 Exp[2t] Sin[t]-Sin[3
t])
x[t_]=g[t]+Sum[g[t-Pi n]
   UnitStep[t-Pi n],{n,1,5}]
Plot[x[t],{t,0,5 Pi}]
Clear[g,x]
g[t_]:=1/39 Exp[-2 t] (3 Exp[2t]
Cos[t]
   -3 Cos[3 t]-2 Sin[3 t])
x[t_]=g[t]+Sum[(-1)^n g[t-n]
   UnitStep[t-n],{n,1,5}]
Plot[x[t],{t,0,5}]
```

```
x:='x':f:='f':
m:=1:k:=6:c:=5:
f:=t->(Heaviside(t-0)-
   Heaviside(t-1))+(Heaviside(t-2)-
   Heaviside(t-3)):
sol1:=dsolve({m*diff(x(t),t$2)+
   c*diff(x(t),t)+k*x(t)=f(t),
   x(0)=0,D(x)(0)=0},x(t),laplace);
assign(sol1):
plot(x(t),t=0..4);
x:='x':f:='f':
m:=1:k:=6:c:=5:
f:=t->(Heaviside(t-0)-
   Heaviside(t-1))-
   (Heaviside(t-1)-Heaviside(t-2))+
   (Heaviside(t-2)-Heaviside(t-3))-
   (Heaviside(t-3)-Heaviside(t-4)):
sol2:=dsolve({m*diff(x(t),t$2)+
   c*diff(x(t),t)+k*x(t)=f(t),
   x(0)=0,D(x)(0)=0},x(t),laplace);
assign(sol2):
plot(x(t),t=0..4);
x:='x':f:='f':
m:=1:k:=13:c:=4:
f:=t->sin(t)*(Heaviside(t-0)-
   Heaviside(t-Pi))+
   sin(t)*(Heaviside(t-2*Pi)-
   Heaviside(t-3*Pi)):
sol3:=dsolve({m*diff(x(t),t$2)+
   c*diff(x(t),t)+k*x(t)=f(t),
   x(0)=0,D(x)(0)=0},x(t),laplace);
assign(sol3):
plot(x(t),t=0..4*Pi);
x:='x':f:='f':
m:=1:k:=13:c:=4:
f:=t->(Heaviside(t-0)-
   Heaviside(t-1))-
   (Heaviside(t-1)-Heaviside(t-2))+
   (Heaviside(t-2)-Heaviside(t-3))-
   (Heaviside(t-3)-Heaviside(t-4)):
sol4:=dsolve({m*diff(x(t),t$2)+
   c*diff(x(t),t)+k*x(t)=f(t),
   x(0)=0,D(x)(0)=0},x(t),laplace);
assign(sol4):
plot(x(t),t=0..4);
```

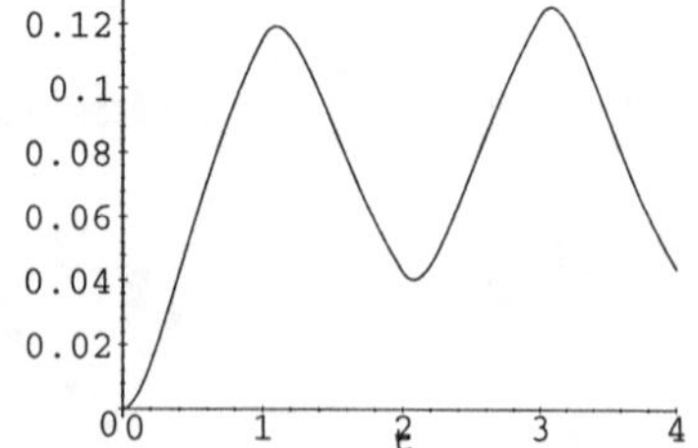

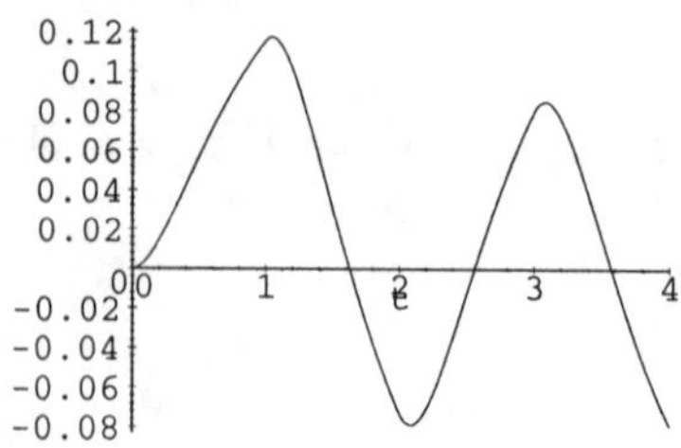

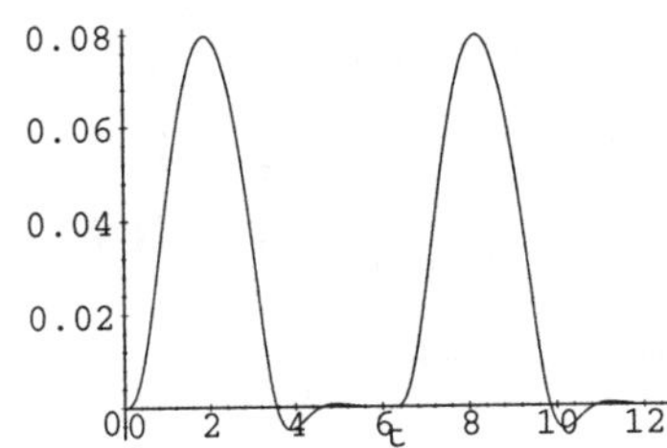

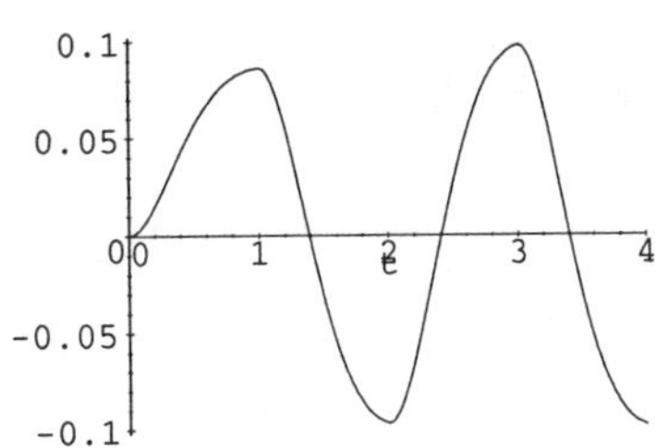

87. We solve $\frac{1}{10}\frac{d^2Q}{dt^2}+100\frac{dQ}{dt}+10^3Q=155\sin 377t,\ Q(0)=\frac{dQ}{dt}(0)=0$. (a)

```
<<Calculus`LaplaceTransform`
Clear[y,x,lhs,rhs,stepone,steptwo]
lhs=1/10  q''[t]+100  q'[t]+10^3  q[t]
rule={q[t]->lq,q'[t]->s  lq-q[0],
   q''[t]->s^2  lq-s  q[0]-q'[0]};
stepone=lhs/.rule//Simplify
e[t_]:=155  Sin[377  t]
rhs=LaplaceTransform[e[t],t,s]
steptwo=Solve[stepone==rhs,lq]
conds=steptwo/.{q[0]->0,q'[0]->0}
dcomp=Apart[conds[[1,1,2]]]
q[t_]=InverseLaplaceTransform[dcomp,s
,t]
i[t_]=q'[t]
Plot[i[t],{t,0,1}]
Plot[i[t],{t,0,.1}]
```

```
eq:=1/10*diff(q(t),t$2)+
   100*diff(q(t),t)+10^3*q(t)=
      155*sin(377*t);
sol:=dsolve({eq,q(0)=0,D(q)(0)=0},
   q(t));
assign(sol):
plot(diff(q(t),t),t=0..0.1);
```

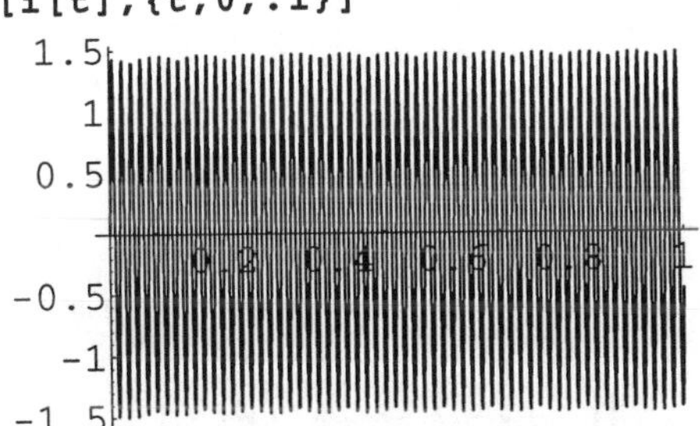

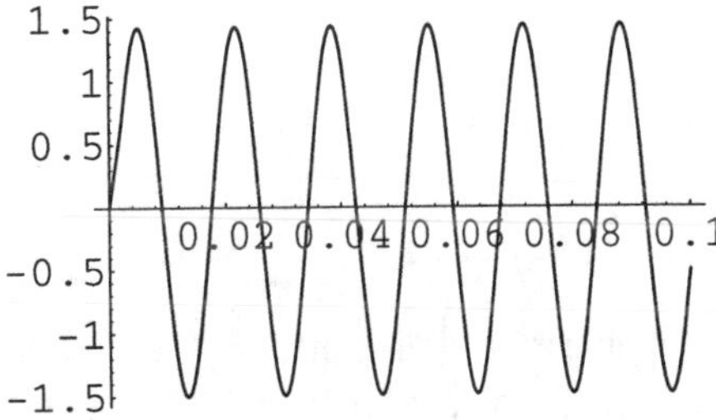

(b) and (c) The exponential terms cause the affect of damping to diminish quickly as we see by graphing *I(t)* over $0 \le t \le 0.1$. The maximum value is difficult to see from the graph. First, we find the first value of t where $I'(t)=0$ and evaluate $I(t)$ there. We then compute the period of the steady-state current, $\frac{584350}{159587072641}(132129\cos 377\text{t}-377000\sin 377\text{t})$ which is $\frac{2\pi}{377}\approx 0.0166663$, and we compute *I(t)* at approximate subsequent points where $I'(t)=0$. From the table, we see that the maximum value of *I(t)* is approximately 1.46 and first occurs at $t\approx 0.905025$. It also occurs periodically every $\frac{2\pi}{377}\approx 0.0166663$ units. We can also find the maximum value of I(t) by analyzing steady-state current only. (d) $\frac{377}{2\pi}\approx 60.0014$

```
sol=FindRoot[i'[t]==0,{t,.005}]
N[i[sol[[1,2]]]]
p=N[2 Pi/377]
Table[{sol[[1,2]]+j p,
   N[i[sol[[1,2]]+j p]]},{j,0,100}]
1/p
```

```
one:=fsolve(diff(q(t),t$2)=0,
   t=0..0.2);
p:=evalf(2*Pi/377);
seq([one+i*p,evalf(subs(t=one+i*p,
   diff(q(t),t)))],i=0..50);
```

88. We solve a similar initial-value problem as that in Problem 87. In this case, $\mathscr{L}\{E(t)\}=\dfrac{155s}{142129+s^2}+\dfrac{155(377\sin 377-s\cos 377)}{142129+s^2}e^{-s}$. Observing the graph, we find that the maximum value of $I(t)$ is approximately 1.46 and first occurs at $t\approx 0.0175$. It also occurs periodically every $\dfrac{2\pi}{377}\approx 0.0166663$ units until the forcing function becomes zero. The steady-state current consists of all terms in $I(t)$ on $0\le t\le 1$ that does not involve a negative exponential.

```
<<Calculus`LaplaceTransform`
Clear[q,i,y,x,lhs,rhs,stepone,steptwo
]
lhs=1/10 q''[t]+100 q'[t]+10^3 q[t]
rule={q[t]->lq,q'[t]->s lq-q[0],
   q''[t]->s^2 lq-s q[0]-q'[0]};
stepone=lhs/.rule//Simplify
e[t_]:=155 Cos[377 t]
rhs=Integrate[Exp[-s t] e[t],{t,0,1}]
steptwo=Solve[stepone==rhs,lq]
conds=steptwo/.{q[0]->0,q'[0]->0}
dcomp=Apart[conds[[1,1,2]]]
first=Apart[dcomp[[1]]]
q1[t_]=InverseLaplaceTransform[first,
s,t]
dcomp[[2]]
second=Apart[dcomp[[2]] Exp[s]]
q2[t_]=InverseLaplaceTransform[second
,s,t]
<<Calculus`DiracDelta`
q[t_]:=q1[t]+q2[t-1] UnitStep[t-1]
i[t_]=q'[t]
Plot[i[t],{t,0.8,1.05}]
Plot[i[t],{t,0,0.1}]
```

```
eq:=1/10*diff(q(t),t$2)+
   100*diff(q(t),t)+10^3*q(t)=
   155*cos(377*t)*(1-Heaviside(t-1));
sol:=dsolve({eq,q(0)=0,D(q)(0)=0},
   q(t),laplace);
assign(sol):
plot(diff(q(t),t),t=0.9..1.2,
   numpoints=200);
```

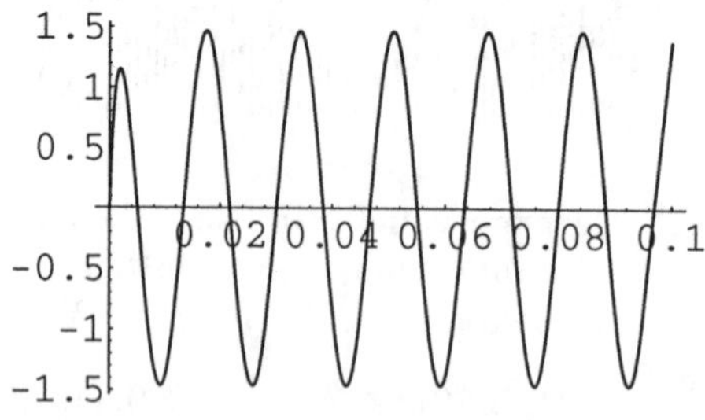

93. We use x and y to represent θ_1 and θ_2, respectively.

```
Clear[m1,m2,k1,k2,f1,f2,
   x0,y0,xp0,yp0]
sys={(m1+m2)l1^2  x''[t]+
   m2  l1  l2  y''[t]+(m1+m2)*
      l1  g  x[t]==0,
   m2  l2^2  y''[t]+m2  l1  l2  x''[t]+
      m2  l2  g  y[t]==0,
   x[0]==x0,x'[0]==xp0,y[0]==y0,
   y'[0]==yp0}
m1=3;
m2=1;
l1=16;
l2=16;
g=32;
x0=1;
xp0=0;
y0=0;
yp0=-1;
sys
sol=DSolve[sys,{x[t],y[t]},t]
<<Algebra`Trigonometry`
sol=ComplexToTrig[sol]//Simplify
Plot[Evaluate[{x[t],y[t]}  /.
   sol],{t,0,10},
   PlotStyle->{GrayLevel[0],
      GrayLevel[.4]}]
ParametricPlot[{x[t],y[t]}  /.
   sol,{t,0,10},
   PlotRange->{{-5/2,5/2},
      {-5/2,5/2}},AspectRatio->1]
```

```
m1:='m1':m2:='m2':k1:='k1':
k2:='k2':f1:='f1':f2:='f2':
x:='x':y:='y':x0:='x0':y0:='y0':
xp0:='xp0':yp0:='yp0':
sys:={(m1+m2)*l1^2*diff(x(t),t$2)+
   m2*l1*l2*diff(y(t),t$2)+
      (m1+m2)*l1*g*x(t)=0,
   m2*l2^2*diff(y(t),t$2)+
      m2*l1*l2*diff(x(t),t$2)+
         m2*l2*g*y(t)=0,
   x(0)=x0,D(x)(0)=xp0,D(y)(0)=y0,
   D(y)(0)=yp0};
m1:=3;
m2:=1;
l1:=16;
l2:=16;
g:=32;
x0:=1;
xp0:=0;
y0:=0;
yp0:=-1;
sys;
sol:=dsolve(sys,{x(t),y(t)},laplace);
plot(subs(sol,{x(t),y(t)}),t=0..10);
plot(subs(sol,[x(t),y(t),t=0..10]));
```

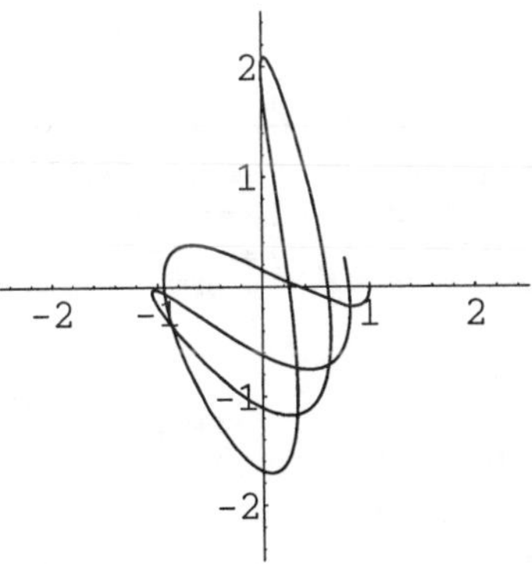

96. (a)

```
Clear[m1,m2,k1,k2,f1,f2,
   x0,y0,xp0,yp0]
sys={m1  q1''[t]+3k  q1[t]-k  q3[t]==0,
   ibar  q2''[t]+2k  l^2  q2[t]==0,
   m3  q3''[t]-k  q1[t]+k  q3[t]==0,
   q1[0]==q10,q1'[0]==q1p0,
   q2[0]==q20,q2'[0]==q2p0,
   q3[0]==q30,q3'[0]==q3p0}
k=2;
m1=1;
m3=1;
ibar=1;
l=1;
q10=0;
q1p0=1;
q20=0;
q2p0=0;
q30=0;
q3p0=0;
sol=DSolve[sys,{q1[t],q2[t],q3[t]},t]
sol=ComplexToTrig[sol]//Simplify
Plot[Evaluate[{q1[t],q2[t],q3[t]}  /.
   sol],{t,0,10},
   PlotStyle->{GrayLevel[0],
      GrayLevel[.3],GrayLevel[.6]}]
```

```
m1:='m1':m2:='m2':k1:='k1':
k2:='k2':f1:='f1':f2:='f2':
x:='x':y:='y':x0:='x0':y0:='y0':
xp0:='xp0':yp0:='yp0':
sys:={m1*diff(q1(t),t$2)+3*k*q1(t)-
      k*q3(t)=0,
   ibar*diff(q2(t),t$2)+
      2*k*l^2*q2(t)=0,
   m3*diff(q3(t),t$2)-
      k*q1(t)+k*q3(t)=0,
   q1(0)=q10,D(q1)(0)=q1p0,
   q2(0)=q20,D(q2)(0)=q2p0,
   q3(0)=q30,D(q3)(0)=q3p0};
k:=2;
m1:=1;
m3:=1;
ibar:=1;
l:=1;
q10:=0;
q1p0:=1;
q20:=0;
q2p0:=0;
q30:=0;
q3p0:=0;
sol:=dsolve(sys,{q1(t),q2(t),q3(t)},
   laplace);
plot(subs(sol,{q1(t),q2(t),q3(t)}),
   t=0..10);
```

97. (Mathematica only) Here we generate a numerical solution.

```
Clear[m1,m2,k1,k2,f1,f2,x0,y0,xp0,yp0]
sys={m1  x1''[t]==-k1  x1[t]+k2  (x2[t]-x1[t]),
   m2  x2''[t]==-k2(x2[t]-x1[t])+k3(x3[t]-x2[t]),
   m3  x3''[t]==-k3  (x3[t]-x2[t]),
   x1[0]==x10,x1'[0]==x1p0,x2[0]==x20,x2'[0]==x2p0,
   x3[0]==x30,x3'[0]==x3p0}
m1=2;
m2=1;
m3=1;
k1=1/2;
k2=1;
k3=2;
x10=0;
x1p0=-2;
x20=0;
x2p0=1;
x30=0;
x3p0=2;
sol=NDSolve[sys,{x1[t],x2[t],x3[t]},{t,0,20}]
Plot[Evaluate[{x1[t],x2[t],x3[t]}  /.  sol],{t,0,20},
   PlotStyle->{GrayLevel[0],GrayLevel[.3],GrayLevel[.6]}]
```

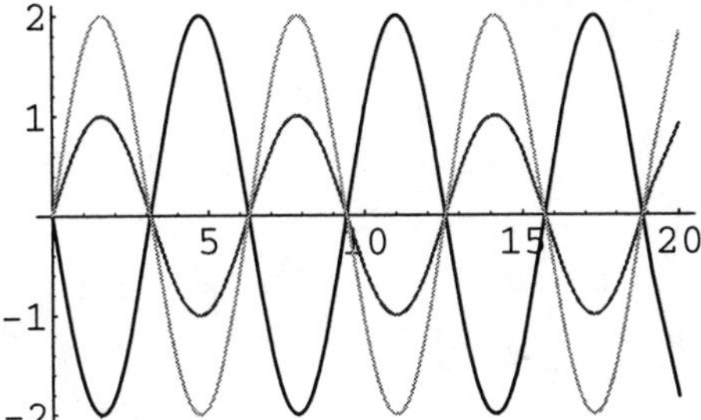
2
1
-1
-2
5
10
15
20

CHAPTER 8 REVIEW EXERCISES

3. $\int_0^5 e^{-st}\,dt = \frac{1}{s} - \frac{1}{s}e^{-5s} = \frac{e^{5s}-1}{se^{5s}}$

23.

$$\mathscr{L}\{f(t)\} = \frac{1}{1-e^{-2s}}\int_0^2 e^{-st} f(t)dt = \frac{1}{1-e^{-2s}}\left(\int_0^1 t\,e^{-st}dt + \int_1^2 (2-t)e^{-st}dt\right)$$

$$= \frac{1}{1-e^{-2s}}\left[-\frac{1}{s^2}e^{-st}(1+s\,t)\Big]_{t=0}^{t=1} + \frac{1}{s^2}e^{-st}(s\,t-2s+1)\Big]_{t=1}^{t=2}\right]$$

$$= \frac{1}{1-e^{-2s}}\left(-\frac{1}{s^2}\left(s\,e^{-s}+e^{-s}-s\right)+\frac{1}{s^2}\left(s\,e^{-s}+e^{-2s}-e^{-s}\right)\right)$$

$$= \frac{1}{1-e^{-2s}}\cdot -\frac{2e^{-s}-1-e^{-2s}}{s^2} = \frac{1+e^{-2s}-2e^{-s}}{s^2\left(1-e^{-2s}\right)} = \frac{e^{2s}-2e^{s}+1}{s^2\left(e^{2s}-1\right)}\frac{\left(e^s-1\right)\left(e^s-1\right)}{s^2\left(e^s+1\right)\left(e^s-1\right)} = \frac{e^s-1}{s^2\left(e^s+1\right)}$$

25.

$$\mathscr{L}\{f(t)\} = \frac{1}{1-e^{-2s}}\int_0^2 e^{-st} f(t)\,dt = \frac{1}{1-e^{-2s}}\int_0^1 e^{-st}\sin \pi t\,dt$$

$$= \frac{1}{1-e^{-2s}}\frac{-1}{s^2+\pi^2}e^{-st}(\pi\cos\pi t + s\sin\pi t)\Big]_{t=0}^{t=1} = \frac{1}{1-e^{-2s}}\frac{\pi}{s^2+\pi^2}\left(1+e^{-s}\right)$$

$$= \frac{\pi\,e^s+\pi}{e^s\left(s^2+\pi^2\right)\left(1-e^{-2s}\right)} = \frac{\pi e^s}{\left(e^s-1\right)\left(s^2+\pi^2\right)}$$

29. $\mathscr{L}^{-1}\left\{\frac{120}{s^6} - \frac{3}{s}\right\} = t^5 - 3$

31. $\mathscr{L}^{-1}\left\{\frac{s+5}{(s+5)^2+1} - \frac{2}{s}\right\} = e^{-5t}\cos t - 2$

43. The Laplace transform of $f(t)$ is

$$\mathscr{L}\{f(t)\} = \int_0^\infty e^{-st} f(t)dt = \int_1^\infty e^{-st}dt = \lim_{M\to\infty} -\frac{1}{s}e^{-st}\Big]_{t=0}^{M} = \frac{1}{s}e^{-s}.$$

Then,

$$y'' + 6y' + 8y = f(t)$$

$$s^2Y(s) - sy(0) - y'(0) + 6\,s\,Y(s) - 6y(0) + 8Y(s) = \frac{1}{s}e^{-s}$$

$$s^2Y(s) - s + 6\,s\,Y(s) - 6 + 8Y(s) = \frac{1}{s}e^{-s}$$

$$\left(s^2+6s+8\right)Y(s) = 6 + s + \frac{1}{s}e^{-s}$$

$$Y(s) = \frac{s^2+6s+e^{-s}}{s\left(s^2+6s+8\right)}.$$

We rewrite $Y(s)$:

$$Y(s) = \frac{s^2+6s+e^{-s}}{s\left(s^2+6s+8\right)} = 2\frac{1}{s+2} - \frac{1}{s+4} + e^{-s}\left(\frac{1}{8}\frac{1}{s} + \frac{1}{8}\frac{1}{s+4} - \frac{1}{4}\frac{1}{s+2}\right).$$

Computing the inverse Laplace transform yields

$$y(t) = -e^{-4t} + 2e^{-2t} + \left(\tfrac{1}{8} + \tfrac{1}{8}e^{4-4t} - \tfrac{1}{4}e^{2-2t}\right)u(t-1).$$

45. We first find the Laplace transform of $x(t)$:

$$x'' + 9x = \cos t + \delta(t-\pi)$$

$$s^2 X(s) - s\,x(0) - x'(0) + 9X(s) = \frac{s}{s^2+1} + e^{-\pi s}$$

$$\left(s^2+9\right)X(s) = \frac{s}{s^2+1} + e^{-\pi s}$$

$$X(s) = \frac{s + s^2 e^{-\pi s} + e^{-\pi s}}{\left(s^2+1\right)\left(s^2+9\right)}$$

After rewriting $X(s)$,

$$X(s) = \frac{s + s^2 e^{-\pi s} + e^{-\pi s}}{\left(s^2+1\right)\left(s^2+9\right)} = \frac{1}{8}\frac{s}{s^2+1} - \frac{1}{8}\frac{s}{s^2+9} + e^{-\pi s}\frac{1}{s^2+9}$$

we use the inverse Laplace transform to find that

$$x(t) = \tfrac{1}{8}\cos t - \tfrac{1}{8}\cos 3t - \tfrac{1}{3}\sin 3t\,\mathcal{U}(t-\pi).$$

47.

$$G(s) = \frac{1}{s^2+1} + \frac{G(s)}{s+1}$$

$$G(s) = \frac{s+1}{s\left(s^2+1\right)} = \frac{1}{s} + \frac{1-s}{s^2+1}$$

$$g(t) = 1 - \cos t + \sin t$$

67. We use the method of Laplace transforms. First, we compute

$$\mathcal{L}\{f(t)\} = \int_0^\infty e^{-st} f(t)\,dt = \int_0^2 e^{-st}\,dt = \frac{1-e^{-2s}}{s}.$$

Applying the Laplace transform to the system yields

$$\begin{cases} s\,X(s) - x(0) = X(s) - 2Y(s) \\ s\,Y(s) - y(0) = 3Y(s) + \dfrac{1-e^{-2s}}{s} \end{cases} \Rightarrow \begin{cases} (s-1)X(s) + 2Y(s) = 0 \\ (s-3)Y(s) = \dfrac{1-e^{-2s}}{s} \end{cases}$$

so

$$y(t) = -\tfrac{1}{3} + \tfrac{1}{3}e^{3t} - \left(-\tfrac{1}{3} + \tfrac{1}{3}e^{3t-6}\right)\mathcal{U}(t-2).$$

Substitution into the first equation then yields

$$x' = x - 2\left[-\tfrac{1}{3} + \tfrac{1}{3}e^{3t} - \left(-\tfrac{1}{3} + \tfrac{1}{3}e^{3t-6}\right)\mathcal{U}(t-2)\right],$$

which has solution

$$x = -\tfrac{2}{3} + e^{t} - \tfrac{1}{3}e^{3t} + 2\left(\tfrac{1}{3} + \tfrac{1}{6}e^{3t-6} - \tfrac{1}{2}e^{t-2}\right)\mathcal{U}(t-2).$$

72. We solve $\begin{cases} 1024\theta_1'' + 256\theta_2'' + 2048\theta_1 = 0 \\ 256\theta_2'' + 256\theta_1'' + 512\theta_2 = 0 \\ \theta_1(0) = 0,\ \theta_1'(0) = 0,\ \theta_2(0) = 1,\ \theta_2'(0) = -1 \end{cases}$: $X(s) = \dfrac{1-s}{4\left(s^2+4\right)} + \dfrac{3(s-1)}{4\left(3s^2+4\right)}$,

$$Y(s) = \frac{s-1}{2\left(s^2+4\right)} + \frac{3(s-1)}{2\left(3s^2+4\right)} \Rightarrow \begin{cases} \theta_1(t) = \dfrac{1}{8}\sin 2t - \dfrac{1}{4}\cos 2t - \dfrac{\sqrt{3}}{8}\sin\dfrac{2t}{\sqrt{3}} + \dfrac{1}{4}\cos\dfrac{2t}{\sqrt{3}} \\ \theta_2(t) = -\dfrac{1}{4}\sin 2t + \dfrac{1}{2}\cos 2t - \dfrac{\sqrt{3}}{4}\sin\dfrac{2t}{\sqrt{3}} + \dfrac{1}{2}\cos\dfrac{2t}{\sqrt{3}} \end{cases}$$

73. We solve $\begin{cases} 1024\theta_1'' + 256\theta_2'' + 2048\theta_1 = 0 \\ 256\theta_2'' + 256\theta_1'' + 512\theta_2 = 0 \\ \theta_1(0) = 0, \theta_1'(0) = -1, \theta_2(0) = 1, \theta_2'(0) = 1 \end{cases}$ $X(s) = \dfrac{-3}{4(4+s^2)} - \dfrac{3}{4(4+3s^2)}$,

$$Y(s) = \frac{3}{2(4+s^2)} - \frac{3}{2(4+3s^2)} \Rightarrow \begin{cases} \theta_1(t) = -\frac{3}{8}\sin 2t - \frac{\sqrt{3}}{8}\sin\frac{2t}{\sqrt{3}} \\ \theta_2(t) = \frac{3}{4}\sin 2t - \frac{\sqrt{3}}{4}\sin\frac{2t}{\sqrt{3}} \end{cases}$$

Differential Equations At Work D: Free Vibration of a Three-Story Building

2. If we attempt to find an exact solution with the method of Laplace transforms, we find that each denominator of $\mathcal{L}\{x_1(t)\}$, $\mathcal{L}\{x_2(t)\}$, and $\mathcal{L}\{x_3(t)\}$ is a positive function of s. Therefore, the roots are complex and solutions will involve sines and/or cosines. (Below, we use *x(t), y(t),* and *z(t)* in the place of $x_1(t)$, $x_2(t)$, and $x_3(t)$.) (Note: In *Mathematica* 4.0 and above, **LaplaceTransform** is a built-in function, so the package does not need to be loaded.)

```
<<Calculus`LaplaceTransform`

Clear[x,y,rule,eq1,eq2]
eq1=m1  x''[t]+(k1+k2)  x[t]-k2  y[t];
eq2=m2  y''[t]-k2  x[t]+(k2+k3)  y[t]-k3  z[t];
eq3=m3  z''[t]-k3  y[t]+k3  z[t];

rule={x[t]->lx,x'[t]->s  lx-x[0],x''[t]->s^2  lx-s  x[0]-x'[0],
    y[t]->ly,y'[t]->s  ly-y[0],y''[t]->s^2  ly-s  y[0]-y'[0],
    z[t]->lz,z'[t]->s  lz-z[0],z''[t]->s^2  lz-s  z[0]-z'[0]};

eqs={eq1,eq2,eq3}/.rule;
sols=Solve[eqs=={0,0,0},{lx,ly,lz}];

conds=sols/.{x[0]->a,x'[0]->b,y[0]->c,y'[0]->d,
    z[0]->e,z'[0]->f}

(**Note:  Be sure to notice the order in which lx, ly, and lz are given above.
This affects the definition of x[t],y[t], and z[t].**)

one=Apart[conds[[1,1,2]]]
two=Apart[conds[[1,2,2]]]
three=Apart[conds[[1,3,2]]]
```

3 and 5. Using the method of Laplace transforms, we approximate the factors of the denominator of $\mathcal{L}\{x_1(t)\}$, $\mathcal{L}\{x_2(t)\}$, and $\mathcal{L}\{x_3(t)\}$ and compute the partial fraction decomposition of each.

```
Clear[x,y,rule,eq1,eq2,eq3,k1,k2,k3]
k1=3;k2=2;k3=1;m1=1;m2=2;m3=3;
eq1=m1  x''[t]+(k1+k2)  x[t]-k2  y[t];
eq2=m2  y''[t]-k2  x[t]+(k2+k3)  y[t]-k3  z[t];
eq3=m3  z''[t]-k3  y[t]+k3  z[t];

rule={x[t]->lx,x'[t]->s  lx-x[0],x''[t]->s^2  lx-s  x[0]-x'[0],
    y[t]->ly,y'[t]->s  ly-y[0],y''[t]->s^2  ly-s  y[0]-y'[0],
    z[t]->lz,z'[t]->s  lz-z[0],z''[t]->s^2  lz-s  z[0]-z'[0]};

eqs={eq1,eq2,eq3}/.rule;

sols=Solve[eqs=={0,0,0},{lx,ly,lz}];
```

```
conds=sols/.{x[0]->0,x'[0]->1/4,y[0]->0,y'[0]->-1/2,
   z[0]->0,z'[0]->1}

(**Note:  Be sure to notice the order in which lx, ly, and lz are given above
in conds.  This affects the definition of x1[t],x2[t], and x3[t] below.**)

one=Apart[conds[[1,1,2]]]

vals=NRoots[Denominator[one]==0,s]

vals2=Chop[vals]

fracs=Apart[1/24  Numerator[one]/
   ((s^2-vals2[[1,2]]^2)(s^2-vals2[[3,2]]^2)(s^2-vals2[[5,2]]^2)

x1[t_]=InverseLaplaceTransform[fracs,s,t]

Plot[x1[t],{t,0,100}]

two=Apart[conds[[1,2,2]]]

fracs2=Apart[1/12  Numerator[two]/
   ((s^2-vals2[[1,2]]^2)(s^2-vals2[[3,2]]^2)(s^2-vals2[[5,2]]^2)

x3[t_]=InverseLaplaceTransform[fracs2,s,t]

Plot[x3[t],{t,0,300}]

three=Apart[conds[[1,3,2]]]

fracs3=Apart[1/12  Numerator[three]/
   ((s^2-vals2[[1,2]]^2)(s^2-vals2[[3,2]]^2)(s^2-vals2[[5,2]]^2)

x2[t_]=InverseLaplaceTransform[fracs3,s,t]

Plot[x2[t],{t,0,200}]
```

We can also use `NDSolve` with a 6×6 system of first-order ordinary differential equations obtained with the substitutions $x_1{}' = y_1$, $x_2{}' = y_2$, and $x_3{}' = y_3$ where $y_1{}' = x_1{}'' = \frac{1}{m_1}\left[-k_1x_1 + k_2(x_2 - x_1)\right]$, $y_2{}' = x_2{}'' = \frac{1}{m_2}\left[-k_2(x_2 - x_1) + k_3(x_3 - x_2)\right]$, and $y_3{}' = x_3{}'' = \frac{1}{m_3}\left[-k_3(x_3 - x_2)\right]$. Therefore, we have

$$\begin{cases} x_1{}' = y_1 \\ y_1{}' = \dfrac{1}{m_1}\left[-(k_1 + k_2)x_1 + k_2x_2\right] \\ x_2{}' = y_2 \\ y_2{}' = \dfrac{1}{m_2}\left[k_2x_1 - (k_2 + k_3)x_2 + k_3x_3\right] \\ x_3{}' = y_3 \\ y_3{}' = \dfrac{1}{m_3}\left[k_3x_2 - k_3x_3\right] \end{cases}$$

Unfortunately in many cases, `NDSolve` limits the interval of t over which the system is solved. Because of this, we are unable to approximate the period of $x_1(t)$, $x_2(t)$, and $x_3(t)$.

```
m1=1;m2=2;m3=3;k1=3;k2=2;k3=1;
eq1=x1'[t]==y1[t];
eq2=y1'[t]==1/m1  (-(k1+k2)  x1[t]+k2  x2[t]);
eq3=x2'[t]==y2[t];
eq4=y2'[t]==1/m2  (k1  x1[t]-(k2+k3)  x2[t]+k3  x3[t]);
eq5=x3'[t]==y3[t];
eq6=y3'[t]==1/m3  (k3  x2[t]-k3  x3[t]);
sol=NDSolve[{eq1,eq2,eq3,eq4,eq5,eq6,
    x1[0]==0,y1[0]==1/4,x2[0]==0,y2[0]==-1/2,
    x3[0]==0,y3[0]==1},
    {x1[t],y1[t],x2[t],y2[t],x3[t],y3[t]},
    {t,0,15}]

p1=Plot[x1[t]/.sol,{t,0,15}]

p2=Plot[x2[t]/.sol,{t,0,15},PlotStyle->GrayLevel[.3]]

p3=Plot[x3[t]/.sol,{t,0,15},PlotStyle->GrayLevel[.5]]

Show[p1,p2,p3]
```

4. Approximate the period graphically. This is difficult in many cases because the period is quite large.

6. After computing (or redefining) $x_1(t)$, $x_2(t)$, and $x_3(t)$ for a particular case, we can construct an outline of a three-story building and observe its vibration. Below, we have the code for $k_1=3$, $k_2=2$, $k_3=1$, $m_1=1$, $m_2=2$, and $m_3=3$. ($x_1(t)$, $x_2(t)$, and $x_3(t)$ were calculated above in Problem 3.) The width and height of the floors were selected arbitrarily to be 20 and 1, respectively.

```
Clear[bldg]
bldg[t_]:=
    Show[Graphics[{
        Line[{{0,0},{20,0}}],
        PointSize[.05],Point[{0,0}],
        PointSize[.05],Point[{20,0}],
        Line[{{0,0},{x1[t],1}}],
        PointSize[.05],Point[{x1[t],1}],
        Line[{{20,0},{20+x1[t],1}}],
        PointSize[.05],Point[{20+x1[t],1}],
        Line[{{x1[t],1},{x2[t],2}}],
        PointSize[.05],Point[{x2[t],2}],
        Line[{{20+x1[t],1},{20+x2[t],2}}],
        PointSize[.05],Point[{20+x2[t],2}],
        Line[{{x2[t],2},{x3[t],3}}],
        PointSize[.05],Point[{x3[t],3}],
        Line[{{20+x2[t],2},{20+x3[t],3}}],
        PointSize[.05],Point[{20+x3[t],3}],
        Line[{{x3[t],3},{20+x3[t],3}}]},
    Axes->None,Ticks->None,
    PlotRange->{{-2,22},{-1,4}}]]
```

The list of graphics produced in the following Do loop can be animated.

```
Do[bldg[t],{t,0,5,.25}]
```

9

Eigenvalue Problems and Fourier Series

EXERCISES 9.1

3. $y(x)=c_1x+c_2,\ y'(x)=c_1,\ y'(0)=c_1=0; y'(x)=0,\ y'(1)=0\neq 1$; No solution

7. $y(x)=c_1\cos 3x+c_2\sin 3x,\ y(\pi)=c_1=0; y(x)=c_2\sin 3x$;
$y'(x)=3c_2\cos 3x; y'(0)=3c_2=0, c_2=0; y(x)=0$

11. $y(x)=c_1\cos 2\pi x+c_2\sin 2\pi x,\ y(1)=c_1=2\pi; y(0)=2\pi;\ y'(x)=-2\pi c_1\sin 2\pi x+2\pi c_2\cos 2\pi x,\ y'(0)=2\pi c_2$,
$y(0)+y'(0)=2\pi+2\pi c_2=0,\ c_2=-1;\ y(x)=2\pi\cos 2\pi x-\sin 2\pi x$

17. Case I: $\lambda=0$: $y(x)=c_1x+c_2,\ y(0)=c_2=0,\ y'(x)=c_1,\ y'(1)=c_1=0$; $\lambda=0$ is not an eigenvalue. Case II: $\lambda=-k^2<0,\ y(x)=c_1e^{kx}+c_2e^{-kx},\ y(0)=c_1+c_2=0$ or $c_2=-c_1,\ y'(x)=c_1ke^{kx}-c_2ke^{-kx}$,
$y'(1)=c_1k\left(e^k+e^{-k}\right)=0,\ c_1=0,\ c_2=0$, No negative eigenvalues. Case III: $\lambda=k^2>0$,
$y(x)=c_1\cos kx+c_2\sin kx,\ y(0)=c_1=0,\ y'(x)=c_2k\cos kx,\ y'(1)=c_2k\cos k=0,\ k=(2n-1)\pi/2,\ n=1,2,\ldots$;
$y_n(x)=\sin[(2n-1)x\pi/2],\ n=1,2,\ldots$

21. Case I: $\lambda=1/2,\ y(x)=e^{-x/2}(c_1+c_2x),\ y(0)=c_1=0,\ y(x)=c_2xe^{-x/2},\ y'(x)=c_2\left(e^{-x/2}-1/2\,xe^{-x/2}\right)$,
$y'(1)=c_2e^{-1/2}=0,\ c_2=0,\ \lambda=1/2$ is not an eigenvalue.
Case II: $4-8\lambda=-k^2<0\ (\lambda>1/2),\ y(x)=e^{-x/2}\left(c_1e^{kx}+c_2e^{-kx}\right),\ y(0)=c_1+c_2=0$ or $c_2=-c_1$,
$y'(1)=c_1e^{-1/2}\left[e^k(k-1/2)+e^{-k}(k+1/2)\right]=0,\ c_1=0,\ c_2=0$, No eigenvalues $\lambda,\ \lambda>1/2$. Case III:
$4-8\lambda=k^2>0\ (\lambda<1/2),\ y(x)=e^{-x/2}(c_1\cos kx+c_2\sin kx),\ y(0)=c_1=0,\ y(x)=c_2e^{-x/2}\sin kx$,
$y'(x)=c_2e^{-x/2}(k\cos kx-1/2\sin kx),\ y'(1)=c_2e^{-1/2}(k\cos k-1/2\sin k),\ \tan k=2k,\ y_n(x)=e^{-x/2}\sin kx$ where k satisfies $\tan k=2k$.

27. $d/dx(dy/dx)+\lambda y=0,\ s(x)=1$. If $m\neq n$,

$$\int_0^1\sin[(2m-1)\pi x/2]\sin[(2n-1)\pi x/2]dx=\frac{(m+n-1)\sin[(m-n)\pi]+n\sin[(m+n-1)\pi]+m\sin[(m+n)\pi]}{2(m-n)(m+n-1)\pi}=0$$

30. $\frac{d}{dx}\left[\left(1-x^2\right)P_n'(x)\right]P_m(x)+n(n+1)P_n(x)P_m(x)=0,\ \frac{d}{dx}\left[\left(1-x^2\right)P_m'(x)\right]P_n(x)+m(m+1)P_m(x)P_n(x)=0$,
Subtract and integrate by parts where

$$\int_{-1}^1\frac{d}{dx}\left[\left(1-x^2\right)P_n'(x)\right]P_m(x)dx=\underbrace{\left[\left(1-x^2\right)P_n'(x)P_m(x)\right]_{-1}^1}_{=0}-\int_{-1}^1\left(1-x^2\right)P_n'(x)P_m'(x)dx,$$

and

$$\int_{-1}^{1} \frac{d}{dx}\left[\left(1-x^2\right)P_m'(x)\right]P_n(x)dx = \underbrace{\left[\left(1-x^2\right)P_m'(x)P_n(x)\right]_{-1}^{1}}_{=0} - \int_{-1}^{1}\left(1-x^2\right)P_m'(x)P_n'(x)dx.$$

36. (a) $\begin{vmatrix} \sin 0 & \cos 0 \\ \sin \pi & \cos \pi \end{vmatrix} = \begin{vmatrix} 0 & 1 \\ 0 & -1 \end{vmatrix} = 0$, Nontrivial solutions exist;

(c) $\begin{vmatrix} \sin 0 & \cos 0 \\ e^{-5}\sin 5\pi & e^{-5}\cos 5\pi \end{vmatrix} = \begin{vmatrix} 0 & 1 \\ 0 & -e^{-5} \end{vmatrix} = 0$, Nontrivial solutions exist;

41. $y_p(x) = A\sin 2x + B\cos 2x$, $y_p''(x) = -4A\sin x - 4B\cos x$, $-4A\sin x - 4B\cos x = \sin 2x$, $A = -1/3$, $B = 0$, $y(x) = c_1\cos x + c_2\sin x - 1/3\sin 2x$, $y(0) = c_1 = 0$, $y(\pi) = 0$ for any c_2; $y(x) = c_2\sin x - 1/3\sin 2x$.

45. (a) $y(x) = C_1x + C_2$, $y(0) = C_2 = 0$, $y(x) = C_1x$, $y(1) + y'(1) = 2C_1 = 0$, $C_1 = 0$.
(b) $\lambda = -k^2 < 0$, $y(x) = C_1e^{kx} + C_2e^{-kx}$, $y(0) = C_1 + C_2 = 0$, $C_2 = -C_1$, $y(1) + y'(1) = C_1\left[(1+k)e^k + (k-1)e^{-k}\right] > 0$ unless $C_1 = 0$. Therefore, $C_2 = 0$. (c) $\lambda = k^2 > 0$, $y(x) = C_1\cos kx + C_2\sin kx$, $y(0) = C_1 = 0$, $y(1) + y'(1) = C_2[\sin k + k\cos k] = 0$, $k = -\tan k$.
(d) Graph $y = x$ and $y = -\tan x$. Approximate points of intersection. $\lambda_1 \approx (2.02876)^2 \approx 4.11586$, $\lambda_2 \approx (4.91318)^2 \approx 24.1393$, $\lambda_3 \approx (7.97867)^2 \approx 63.6591$

```
Plot[{k, -Tan[k]}, {k, 0, 10}]
```

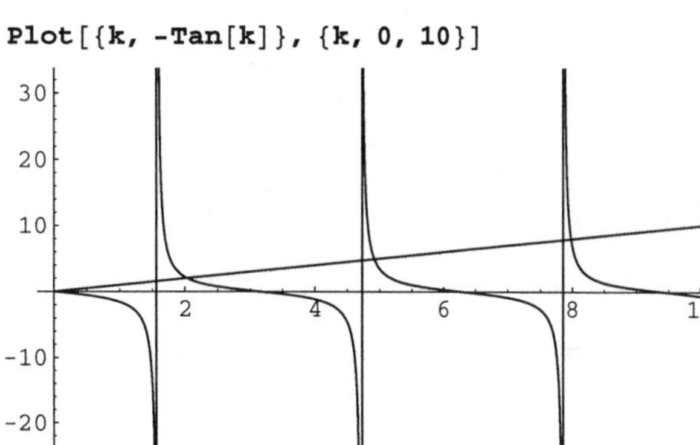


```
FindRoot[k == -Tan[k], {k, 2}]

{k → 2.02876}
FindRoot[k == -Tan[k], {k, 5}]

{k → 4.91318}
FindRoot[k == -Tan[k], {k, 8}]

{k → 7.97867}
```

EXERCISES 9.2

3. $a_0 = 2\int_0^1 x^2 dx = 2\left[\frac{x^3}{3}\right]_0^1 = \frac{2}{3}$

$$a_n = 2\int_0^1 x^2 \cos n\pi x dx = 2\left[\frac{2n\pi x\cos n\pi x - 2\sin n\pi x + n^2\pi^2 x^2 \sin n\pi x}{n^3\pi^3}\right]_0^1 = \frac{4n\pi\cos n\pi}{n^2\pi^2} = \frac{4}{n^2\pi^2}(-1)^n,\ n \ge 1$$

$$f(x) = \frac{1}{3} + \sum_{n=1}^{\infty}\frac{4(-1)^n}{n^2\pi^2}\cos n\pi x = \frac{1}{3} - \frac{4}{\pi^2}\cos\pi x + \frac{4}{2^2\pi^2}\cos 2\pi x + \cdots$$

$\int x^2\, dx$

$\frac{x^3}{3}$

$\int_0^1 x^2\, dx$

$\frac{1}{3}$

$\int x^2 \text{Cos}[n\,\pi\,x]\, dx$

$\frac{2\,n\,\pi\,x\,\text{Cos}[n\,\pi\,x] - 2\,\text{Sin}[n\,\pi\,x] + n^2\,\pi^2\,x^2\,\text{Sin}[n\,\pi x]}{n^3\,\pi^3}$

$\int_0^1 x^2 \text{Cos}[n\,\pi\,x]\, dx$

$\frac{2\,n\,\pi\,\text{Cos}[n\,\pi] - 2\,\text{Sin}[n\,\pi] + n^2\,\pi^2\,\text{Sin}[n\,\pi]}{n^3\,\pi^3}$

$\text{f}[\text{x\_},\ \text{n\_}] := \frac{1}{2} + \sum_{k=1}^{n} -\frac{4\,\text{Cos}[(2\,k-1)\,\pi\,x]}{(2\,k-1)^2\,\pi^2}$

```
Plot[f[x, 10], {x, 0, 1}]
```


7. $a_0 = \frac{2}{4}\int_0^2 x dx + \frac{2}{4}\int_2^4 dx = 2$

$$a_n = \frac{2}{4}\int_0^2 x\cos\frac{n\pi x}{4}dx + \frac{2}{4}\int_2^4 \cos\frac{n\pi x}{4}dx = \frac{2}{n^2\pi^2}\left[4\cos\frac{n\pi x}{4} + n\pi x\sin\frac{n\pi x}{4}\right]_0^2 + \frac{2}{n\pi}\left[\sin\frac{n\pi x}{4}\right]_2^4$$

$$= \frac{2}{n^2\pi^2}\left(-4 + 4\cos\frac{n\pi}{2} + n\pi\sin\frac{n\pi}{2}\right),\ n \ge 1$$

$$f(x) = 1 + \sum_{n=1}^{\infty}\frac{2}{n^2\pi^2}\left(-4 + 4\cos\frac{n\pi}{2} + n\pi\sin\frac{n\pi}{2}\right)\cos\frac{n\pi x}{4}$$

13.

$$b_n = 2\int_0^1 x^2 \sin n\pi x dx = 2\left[\frac{2\cos n\pi x - n^2\pi^2x^2\cos n\pi x + 2n\pi x\sin n\pi x}{n^3\pi^3}\right]_0^1$$

$$= 2\left[\frac{-2+2\cos n\pi - n^2\pi^2\cos n\pi + 2n\pi\sin n\pi}{n^3\pi^3}\right] = 2(-1)^n\left[\frac{2}{n^3\pi^3} - \frac{1}{n\pi}\right] - \frac{4}{n^3\pi^3},\ n \ge 1$$

$$f(x) = \sum_{n=1}^{\infty}\left[2(-1)^n\left[\frac{2}{n^3\pi^3} - \frac{1}{n\pi}\right] - \frac{4}{n^3\pi^3}\right]\sin n\pi x = 2\left(\frac{-4}{\pi^3} + \frac{1}{\pi}\right)\sin \pi x + 2\left(-\frac{1}{2\pi}\right)\sin 2\pi x + 2\left(\frac{-4}{3^3\pi^3} + \frac{1}{3\pi}\right)\sin 3\pi x + \cdots$$

17.

$$b_n = \frac{2}{4}\int_0^2 x\sin\frac{n\pi x}{4}dx + \frac{2}{4}\int_2^4 \sin\frac{n\pi x}{4}dx = \frac{-2}{n^2\pi^2}\left[n\pi x\cos\frac{n\pi x}{4} - 4\sin\frac{n\pi x}{4}\right]_0^2 - \frac{2}{n\pi}\left[\cos\frac{n\pi x}{4}\right]_2^4$$

$$= \frac{-2}{n^2\pi^2}\left(2n\pi\cos\frac{n\pi}{2} - 4\sin\frac{n\pi}{2}\right) + \frac{2}{n\pi}\left(\cos\frac{n\pi}{2} - \cos n\pi\right) = \frac{-2}{n^2\pi^2}\left(n\pi\left(\cos\frac{n\pi}{2} - \cos n\pi\right) - 4\sin\frac{n\pi}{2}\right),\ n \ge 1$$

$$f(x) = \sum_{n=1}^{\infty}\frac{-2}{n^2\pi^2}\left(n\pi\left(\cos\frac{n\pi}{2} - \cos n\pi\right) - 4\sin\frac{n\pi}{2}\right)\sin\frac{n\pi x}{4}$$

$$\frac{2}{4}\int x\,\mathrm{Sin}\left[\frac{n\pi x}{4}\right]dx$$

$$-\frac{2\left(n\pi x\,\mathrm{Cos}\left[\frac{n\pi x}{4}\right] - 4\,\mathrm{Sin}\left[\frac{n\pi x}{4}\right]\right)}{n^2\pi^2}$$

$$\frac{2}{4}\int \mathrm{Sin}\left[\frac{n\pi x}{4}\right]dx$$

$$-\frac{2\,\mathrm{Cos}\left[\frac{n\pi x}{4}\right]}{n\pi}$$

$$\frac{2}{4}\int_0^2 x\,\mathrm{Sin}\left[\frac{n\pi x}{4}\right]dx$$

$$-\frac{2\left(2n\pi\,\mathrm{Cos}\left[\frac{n\pi}{2}\right] - 4\,\mathrm{Sin}\left[\frac{n\pi}{2}\right]\right)}{n^2\pi^2}$$

$$\frac{2}{4}\int_2^4 \mathrm{Sin}\left[\frac{n\pi x}{4}\right]dx$$

$$\frac{1}{2}\left(\frac{4\,\mathrm{Cos}\left[\frac{n\pi}{2}\right]}{n\pi} - \frac{4\,\mathrm{Cos}[n\pi]}{n\pi}\right)$$

$$\mathrm{Simplify}\left[\frac{2}{4}\int_0^2 x\,\mathrm{Sin}\left[\frac{n\pi x}{4}\right]dx + \frac{2}{4}\int_2^4 \mathrm{Sin}\left[\frac{n\pi x}{4}\right]dx\right]$$

$$-\frac{4\,\mathrm{Cos}\left[\frac{n\pi}{4}\right]\left(n\pi\,\mathrm{Cos}\left[\frac{3n\pi}{4}\right] - 4\,\mathrm{Sin}\left[\frac{n\pi}{4}\right]\right)}{n^2\pi^2}$$

$$\frac{2}{4}\int_0^2 x\,\mathrm{Sin}\left[\frac{n\pi x}{4}\right]dx + \frac{2}{4}\int_2^4 \mathrm{Sin}\left[\frac{n\pi x}{4}\right]dx$$

$$\frac{1}{2}\left(\frac{4\,\mathrm{Cos}\left[\frac{n\pi}{2}\right]}{n\pi} - \frac{4\,\mathrm{Cos}[n\pi]}{n\pi}\right) - \frac{2\left(2n\pi\,\mathrm{Cos}\left[\frac{n\pi}{2}\right] - 4\,\mathrm{Sin}\left[\frac{n\pi}{2}\right]\right)}{n^2\pi^2}$$

$$\mathrm{b[n\_]} := -\frac{4\,\mathrm{Cos}\left[\frac{n\pi}{4}\right]\left(n\pi\,\mathrm{Cos}\left[\frac{3n\pi}{4}\right] - 4\,\mathrm{Sin}\left[\frac{n\pi}{4}\right]\right)}{n^2\pi^2}$$

$$\mathrm{f[x\_,\ k\_]} := \sum_{n=1}^{k} \mathrm{b[n]}\,\mathrm{Sin}\left[\frac{n\pi x}{4}\right]$$

```
Plot[f[x, 20], {x, 0, 4}]
```

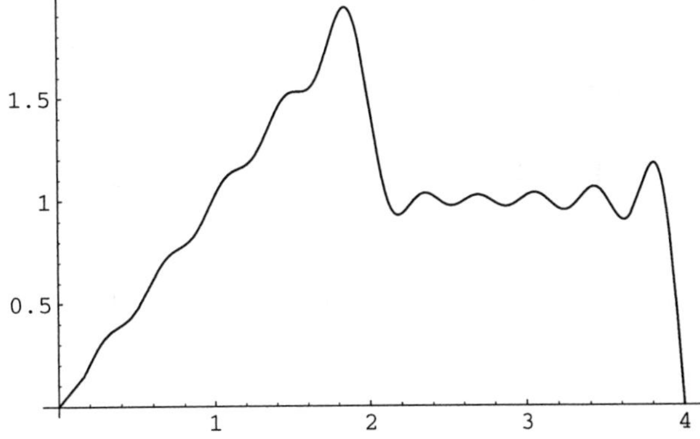

23. $\cos^3 x = \dfrac{3}{4}\cos x + \dfrac{1}{4}\cos 3x$

EXERCISES 9.3

* 3. f is odd; $a_0 = 0; a_n = 0, \ n \geq 1;$

$$b_n = \frac{1}{\pi}\int_{-\pi}^{\pi} x^3 \sin nx dx = \frac{2}{\pi}\left[\frac{-x\left(n^2x^2-6\right)\cos nx}{n^3} + \frac{3\left(n^2x^2-2\right)\sin nx}{n^4}\right]_0^{\pi} = -\frac{2\left(n^2\pi^2-6\right)\cos n\pi}{n^3}, \ n \geq 1;$$

$$f(x) = \sum_{n=1}^{\infty}\left(\frac{12}{n^3} - \frac{2\pi^2}{n}\right)(-1)^n \sin nx = \left(2\pi^2-12\right)\sin x + \left(\frac{3}{2} - \pi^2\right)\sin 2x + \left(\frac{2\pi^2}{3} - \frac{4}{9}\right)\sin 3x + \cdots$$

```
Clear[f]
f[x_] := x^3
p = Pi;
1/p Integrate[f[x] Sin[n Pi x/p], x]
```

$$\frac{1}{\pi}\left(-\frac{x\left(-6+n^2 x^2\right)\text{Cos}[n\,x]}{n^3} + \frac{3\left(-2+n^2 x^2\right)\text{Sin}[n\,x]}{n^4}\right)$$

```
b[n_] =
  1/p Integrate[f[x] Sin[n Pi x/p],
    {x, -p, p}]
```

$$\frac{1}{\pi}\left(-\frac{2\pi\left(-6+n^2\pi^2\right)\text{Cos}[n\,\pi]}{n^3} + \frac{6\left(-2+n^2\pi^2\right)\text{Sin}[n\,\pi]}{n^4}\right)$$

```
g[x_, k_] := Sum[b[n] Sin[n Pi x/p],
   {n, 1, k}]
g[x, 3]
```

$$2\left(-6+\pi^2\right)\text{Sin}[x] + \frac{1}{4}\left(6-4\pi^2\right)\text{Sin}[2\,x] + \frac{2}{27}\left(-6+9\pi^2\right)\text{Sin}[3\,x]$$

```
p1 = Plot[g[x, 10], {x, -p, p}]
```

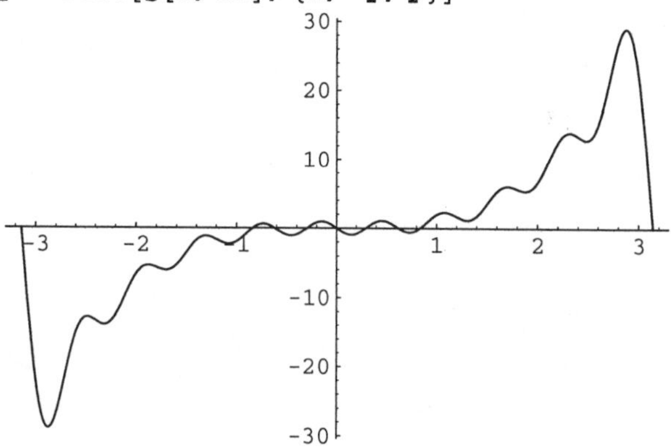

* 7. $$a_0 = \frac{1}{1}\left(\int_{-1}^{0} x dx + \int_{0}^{1} dx\right) = \left[\frac{x^2}{2}\right]_{-1}^{0} + [x]_0^1 = -\frac{1}{2} + 1 = \frac{1}{2};$$

$$a_n = \int_{-1}^{0} x\cos nx dx + \int_{0}^{1}\cos nx dx = \left[\frac{\cos n\pi x}{n^2\pi^2} + \frac{x\sin n\pi x}{n\pi}\right]_{-1}^{0} + \left[\frac{\sin n\pi x}{n\pi}\right]_0^1$$

$$= \frac{1}{n^2\pi^2}(1-\cos n\pi) = \frac{1}{n^2\pi^2}\left(1-(-1)^n\right), n \geq 1$$

$$b_n = \int_{-1}^{0} x\sin n\pi x dx + \int_{0}^{1}\sin n\pi x dx = \left[\frac{-x\cos n\pi x}{n\pi} + \frac{\sin n\pi x}{n^2\pi^2}\right]_{-1}^{0} + \left[\frac{-\cos n\pi x}{n\pi}\right]_0^1$$

$$= \frac{1}{n\pi}(1-2\cos n\pi) = \frac{1}{n\pi}\left(1-2(-1)^n\right), n \geq 1$$

$$f(x) = \frac{1}{4} + \sum_{n=1}^{\infty}\left[\frac{1}{n^2\pi^2}\left(1-(-1)^n\right)\cos n\pi x + \frac{1}{n\pi}\left(2(-1)^{n+1}+1\right)\sin n\pi x\right] = \frac{1}{4} + \frac{2}{\pi^2}\cos\pi x + \frac{3}{\pi}\sin\pi x - \frac{1}{2\pi}\sin 2\pi x + \cdots$$

```
p = 1;
a[0] =
 1 / p
  (Integrate[x, {x, -1, 0}] +
    Integrate[1, {x, 0, 1}])
```

$\frac{1}{2}$

```
a[n_] =
 1 / p
  (Integrate[x Cos[n Pi x / p], {x,
    Integrate[1 Cos[n Pi x / p], {x
```

$\frac{1}{n^2 \pi^2} - \frac{\text{Cos}[n\pi]}{n^2 \pi^2}$

```
Integrate[x Cos[n Pi x / p], x]
```

$\frac{\text{Cos}[n\pi x]}{n^2 \pi^2} + \frac{x \,\text{Sin}[n\pi x]}{n\pi}$

```
Integrate[Cos[n Pi x / p], x]
```

$\frac{\text{Sin}[n\pi x]}{n\pi}$

```
b[n_] =
 1 / p
  (Integrate[x Sin[n Pi x / p], {x,
    Integrate[1 Sin[n Pi x / p], {x
```

$\frac{1}{n\pi} - \frac{2\,\text{Cos}[n\pi]}{n\pi} + \frac{\text{Sin}[n\pi]}{n^2 \pi^2}$

```
Integrate[x Sin[n Pi x / p], x]
```

$-\frac{x\,\text{Cos}[n\pi x]}{n\pi} + \frac{\text{Sin}[n\pi x]}{n^2 \pi^2}$

```
Integrate[Sin[n Pi x / p], x]
```

$-\frac{\text{Cos}[n\pi x]}{n\pi}$

```
g[x_, k_] :=
 1 / 4 +
  Sum[1 / (n^2 Pi^2) (1 - (-1)^n) Cos[n Pi x
    1 / (n * Pi) (1 - 2 (-1)^n) Sin[n Pi x],
   {n, 1, k}]
```

```
g[x, 3]
```

$\frac{1}{4} + \frac{2\,\text{Cos}[\pi x]}{\pi^2} + \frac{2\,\text{Cos}[3\pi x]}{9\pi^2} +$
$\frac{3\,\text{Sin}[\pi x]}{\pi} - \frac{\text{Sin}[2\pi x]}{2\pi} + \frac{\text{Sin}[3\pi x}{\pi}$

```
p1 = Plot[g[x, 20], {x, -1, 1}]
```

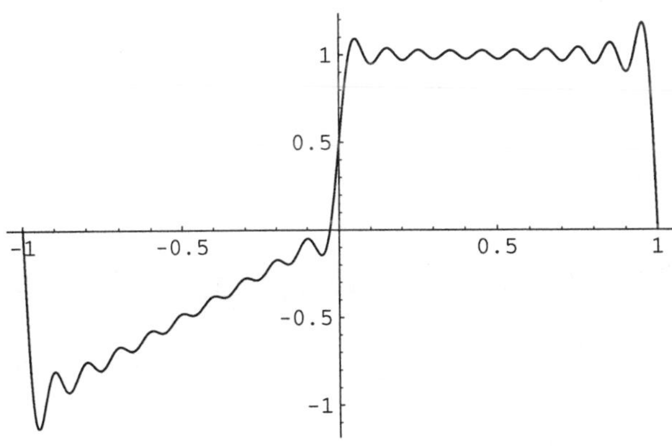

* 13. Neither; $f(-x) = (-x)^2 - (-x) = x^2 + x \neq -f(x) \; or \; f(x)$

* 17. Odd; $f(x) = \begin{cases} x^2, x \geq 0 \\ -x^2, x < 0 \end{cases}, \; f(-x) = -f(x)$

* 37. (a) 0; (b) $\frac{2+(-1/2)}{2} = \frac{3}{4}$; (c) $\frac{1/2+(-1)}{2} = -\frac{1}{4}$; (d) $\frac{2+(-1)}{2} = \frac{1}{2}$; (e) $\frac{2+(-1)}{2} = \frac{1}{2}$

41. $a_0 = \frac{2}{\pi}\int_0^{\pi} x(\pi - x)dx = \frac{2}{\pi}\left[\frac{\pi x^2}{2} - \frac{x^3}{3}\right]_0^{\pi} = \frac{\pi^2}{3},$

$$a_n = \frac{2}{\pi}\int_0^{\pi} x(\pi - x)\cos nx dx = \left[\frac{2}{\pi}\left(\frac{(\pi - 2x)\cos nx}{n^2} - \frac{\left(-2 - n^2\pi x + n^2x^2\right)\sin nx}{n^3}\right)\right]_0^{\pi} = \frac{-2}{n^2}(1 + \cos n\pi), \qquad n \ge 1;$$

$$f(x) = \frac{\pi^2}{6} + \sum_{n=1}^{\infty} -\frac{2}{n^2}\left[1 + (-1)^n\right]\cos nx = \frac{\pi^2}{6} + \sum_{n=1}^{\infty} -\frac{4}{(2n)^2}\cos 2nx = 1 - \cos 2x - \frac{1}{4}\cos 4x - \cdots$$

```
p = Pi;
a[0] = 2 / p Integrate[x (Pi - x), {x, 0, p}]
```

$\frac{\pi^2}{3}$

```
2 / p Integrate[x (Pi - x), x]
```

$\frac{2\left(\frac{\pi x^2}{2} - \frac{x^3}{3}\right)}{\pi}$

```
2 / p Integrate[x (Pi - x) Cos[n Pi x / p], x]
```

$\frac{1}{\pi}\left(2\left(\frac{(\pi - 2 x) \text{Cos}[n x]}{n^2} - \frac{\left(-2 - n^2 \pi x + n^2 x^2\right) \text{Sin}[n x]}{n^3}\right)\right)$

```
f[x_, k_] :=
  1 / 2 a[0] + Sum[a[n] Cos[n x], {n, 1, k}]
a[n_] =
  2 / p Integrate[x (Pi - x) Cos[n Pi x / p],
    {x, 0, p}]
```

$\frac{2\left(-\frac{\pi}{n^2} - \frac{\pi \text{Cos}[n \pi]}{n^2} + \frac{2 \text{Sin}[n \pi]}{n^3}\right)}{\pi}$

```
p1 = Plot[f[x, 6], {x, 0, Pi}]
```

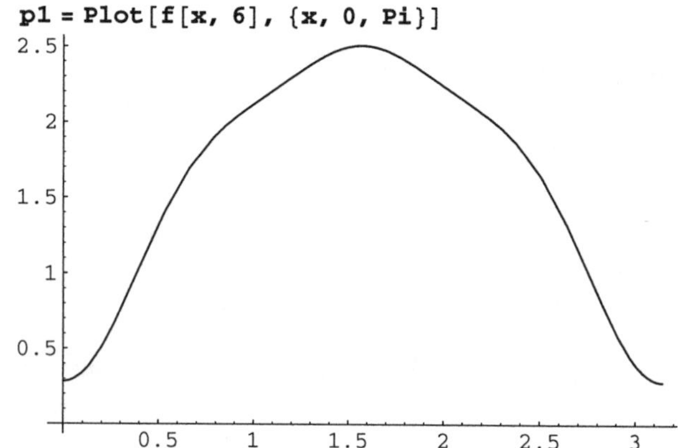

45. $b_n = \frac{4}{n^3\pi}[1 - \cos n\pi], n \ge 1,$

$$\sum_{n=1}^{\infty}\frac{4}{n^3\pi}\left[1 - (-1)^n\right]\sin nx = \sum_{n=1}^{\infty}\frac{8}{(2n-1)^3\pi}\sin(2n-1)x = \frac{8}{\pi}\sin x + \frac{8}{3^3\pi}\sin 3x + \frac{8}{5^3\pi}\sin 5x + \cdots$$

```
p = Pi;
2 / p Integrate[x (Pi - x) Sin[n Pi x / p], x]
```

$$\frac{1}{\pi}\left(2\left(\frac{(-2 - n^2 \pi x + n^2 x^2) \operatorname{Cos}[n x]}{n^3} + \frac{(\pi - 2 x) \operatorname{Sin}[n x]}{n^2}\right)\right)$$

```
b[n_] =
  2 / p Integrate[x (Pi - x) Sin[n Pi x / p],
    {x, 0, p}]
```

$$\frac{2\left(\frac{2}{n^3} - \frac{2 \operatorname{Cos}[n \pi]}{n^3} - \frac{\pi \operatorname{Sin}[n \pi]}{n^2}\right)}{\pi}$$

```
f[x_, k_] := Sum[b[n] Sin[n Pi x / p], {n

p1 = Plot[f[x, 6], {x, 0, p}]
p1 = Plot[f[x, 6], {x, 0, p}];
p2 = Plot[x (Pi - x), {x, 0, Pi}];
Show[p1, p2]
```

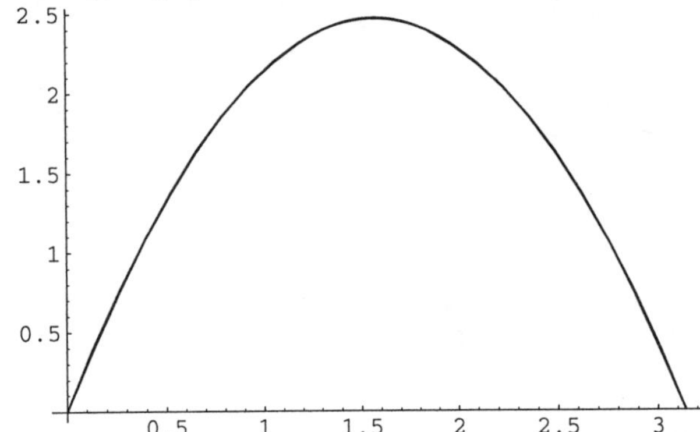

*49. $a_0 = \frac{\pi^2}{3}, a_n = -\frac{1}{n^2}, n \geq 1; b_n = 0, n \geq 1;$

$$f(x) = \frac{\pi^2}{6} - \sum_{n=1}^{\infty} \frac{1}{n^2} \cos 2nx = \frac{\pi^2}{6} - \cos 2x - \frac{1}{4}\cos 4x - \frac{1}{9}\cos 6x + \cdots$$

```
p = Pi;
f[x_] := x (Pi - x)
a[0] = 2 / p Integrate[f[x], {x, 0, p}]
```

$$\frac{\pi^2}{3}$$

```
a[n_] =
  2 / p Integrate[f[x] Cos[2 n Pi x / p],
    {x, 0, p}]
```

$$\frac{2\left(-\frac{\pi}{4 n^2} + \frac{-n \pi \operatorname{Cos}[2 n \pi] + \operatorname{Sin}[2 n \pi]}{4 n^3}\right)}{\pi}$$

```
Simplify[%]
```

$$\frac{\operatorname{Cos}[n \pi] (-n \pi \operatorname{Cos}[n \pi] + \operatorname{Sin}[n \pi])}{n^3 \pi}$$

```
b[n_] =
  2 / p Integrate[f[x] Sin[2 n Pi x / p],
    {x, 0, p}]
```

$$\frac{2\left(\frac{1}{4 n^3} + \frac{-\operatorname{Cos}[2 n \pi] - n \pi \operatorname{Sin}[2 n \pi]}{4 n^3}\right)}{\pi}$$

```
Simplify[b[n]]
```

$$\frac{\operatorname{Sin}[n \pi] (-n \pi \operatorname{Cos}[n \pi] + \operatorname{Sin}[n \pi])}{n^3 \pi}$$

61. (a) $a_n = 0,\ n \geq 0;\ b_n = \frac{1}{\pi}\int_0^{\pi} x^3 \sin nx dx = \frac{-2(-6 + n^2\pi^2)}{n^3}\cos n\pi,\ n \geq 1$

(c) $a_0 = \frac{1}{\pi}\int_0^{\pi} x^4 dx = \frac{2\pi^4}{5},\ a_n = \frac{1}{\pi}\int_0^{\pi} x^4 \cos nx dx = \frac{8(-6 + n^2\pi^2)}{n^4}\cos n\pi,\ n \geq 1;\ b_n = 0,\ n \geq 1$

EXERCISES 9.4

* 3. $p(x)=e^{2x}, s(x)=e^{2x}, \int_0^1 e^{2x}\left(e^{-x}\sin n\pi x\right)^2 dx=\frac{1}{2}, c_n=2\int_0^1 e^{2x}f(x)\left(e^{-x}\sin n\pi x\right)dx$;

(a) $c_n=\frac{2n\pi}{n^2\pi^2+1}(1-e\cos n\pi)=\frac{2n\pi}{n^2\pi^2+1}\left[1-e(-1)^n\right]$; (b) $c_n=\frac{2}{n^2\pi^2+1}\left(n\pi-e^{1/2}n\pi\cos\frac{n\pi}{2}+e^{1/2}\sin\frac{n\pi}{2}\right)$

```
c[n_] =
 2 Integrate[
   Exp[2 x] Exp[-x] Sin[n Pi x],
   {x, 0, 1}]
```

$$2\left(\frac{n\pi}{1+n^2\pi^2}+\frac{E\,(-n\pi\,\mathrm{Cos}[n\pi]+\mathrm{Sin}[n\pi])}{1+n^2\pi^2}\right)$$

```
Integrate[
 Exp[2 x] (Exp[-x] Sin[n Pi x])^2,
 {x, 0, 1}]
```

$$\frac{1}{2}-\frac{\mathrm{Sin}[2n\pi]}{4n\pi}$$

```
f[x_, k_] :=
 Sum[c[n] Exp[-x] Sin[n Pi x],
  {n, 1, k}]
Plot[f[x, 20], {x, 0, 1}]
```

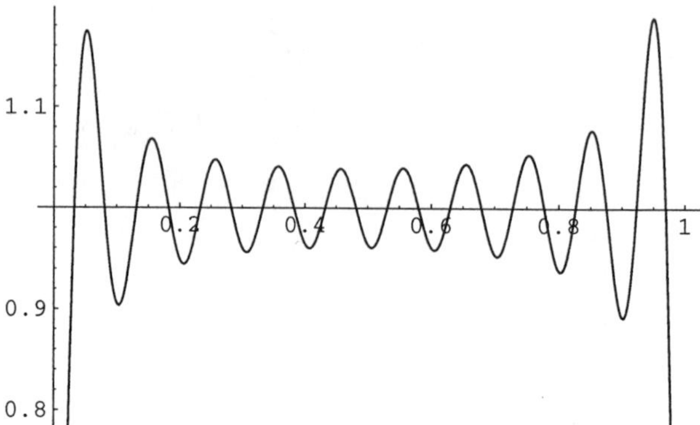


```
c[n_] =
 2 Integrate[
   Exp[2 x] Exp[-x] Sin[n Pi x],
   {x, 0, 1/2}]
```

$$2\left(\frac{n\pi}{1+n^2\pi^2}-\frac{\sqrt{E}\left(n\pi\,\mathrm{Cos}\left[\frac{n\pi}{2}\right]-\mathrm{Sin}\left[\frac{n\pi}{2}\right]\right)}{1+n^2\pi^2}\right)$$

```
f[x_, k_] :=
 Sum[c[n] Exp[-x] Sin[n Pi x],
  {n, 1, k}]
Plot[f[x, 20], {x, 0, 1}]
```

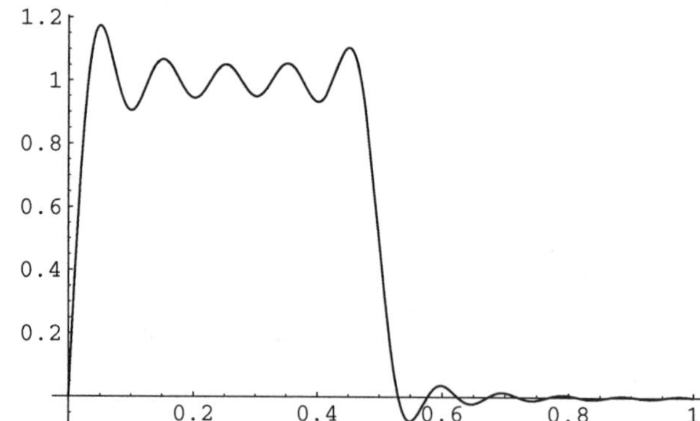

* 7. $r=\frac{-1\pm\sqrt{1-4(1-\lambda)}}{2}=\frac{-1\pm\sqrt{-3+4\lambda}}{2}$, Case I: $\lambda=3/4$, $y(x)=e^{-x/2}(c_1+c_2x)$, $y(0)=c_1=0$, $y(1)=e^{-1/2}c_2=0$, $c_2=0$, $\lambda=3/4$ is not an eigenvalue; Case II: $-3+4\lambda=k^2>0$ yields $y(x)=0$; Case III: $-3+4\lambda=-k^2<0$, $y(x)=e^{-x/2}\left(c_1\cos\frac{kx}{2}+c_2\sin\frac{kx}{2}\right)$, $y(0)=c_1=0$, $y(1)=e^{-1/2}c_2\sin\frac{k}{2}=0$, $\frac{k}{2}=n\pi$, $k=2n\pi$, $\lambda_n=\frac{3-n^2\pi^2}{4}, n=1,2,\ldots$, $y_n(x)=e^{-x/2}\sin\frac{kx}{2}$, $n=1,2,\ldots$; $\frac{d}{dx}\left(e^x y\right)+e^x(1-\lambda)y=0$, $s(x)=e^x$; $c_n=2\int_0^1 e^x g(x)\left(e^{-x/2}\sin n\pi x\right)dx$

Chapter 9 Review Exercises

* 5. $b_n = \int_0^1 x\sin\frac{n\pi x}{2}dx + \int_1^2 (2-x)\sin\frac{n\pi x}{2}dx = \frac{-2}{n^2\pi^2}\left(-n\pi\cos\frac{n\pi}{2} - 2\sin\frac{n\pi}{2}\right)$

$-\frac{2}{n^2\pi^2}\left(n\pi\cos\frac{n\pi}{2} - 2\sin\frac{n\pi}{2}\right) - \frac{4}{n^2\pi^2}\sin n\pi = \frac{8}{n^2\pi^2}\sin\frac{n\pi}{2},\ n \geq 1$

9. $a_0 = \int_0^1 x dx + \int_1^2 (2-x)dx = 1;$

$a_n = \int_0^1 x\cos\frac{n\pi x}{2}dx + \int_1^2 (2-x)\cos\frac{n\pi x}{2}dx = \frac{4}{n^2\pi^2}\left(-1 - \cos n\pi + 2\cos\frac{n\pi}{2}\right),\ n \geq 1$

```
p = 2;
a[0] =
 2 / p
  ( Integrate[x, {x, 0, 1}] +
    Integrate[2 - x, {x, 1, 2}])
```

1

```
a[n_] =
 2 / p
  ( Integrate[x Cos[n Pi x / p],
     {x, 0, 1}] +
    Integrate[
     (2 - x) Cos[n Pi x / p],
     {x, 1, 2}])
```

$-\frac{4}{n^2\pi^2} - \frac{4\,\mathrm{Cos}[n\pi]}{n^2\pi^2} + \frac{2\left(2\,\mathrm{Cos}[\frac{n\pi}{2}] - n\pi\,\mathrm{Sin}[\frac{n\pi}{2}]\right)}{n^2\pi^2} + \frac{2\left(2\,\mathrm{Cos}[\frac{n\pi}{2}] + n\pi\,\mathrm{Sin}[\frac{n\pi}{2}]\right)}{n^2\pi^2}$

```
f[x_, k_] :=
 1 / 2 a[0] +
  Sum[a[n] Cos[n Pi x / p],
   {n, 1, k}]
```

```
approx[x_] = f[x, 20]
```

$\frac{1}{2} - \frac{4\,\mathrm{Cos}[\pi x]}{\pi^2} - \frac{4\,\mathrm{Cos}[3\pi x]}{9\pi^2} - \frac{4\,\mathrm{Cos}[5\pi x]}{25\pi^2} - \frac{4\,\mathrm{Cos}[7\pi x]}{49\pi^2} - \frac{4\,\mathrm{Cos}[9\pi x]}{81\pi^2}$

```
Plot[approx[x], {x, 0, 2},
 PlotRange -> All]
```

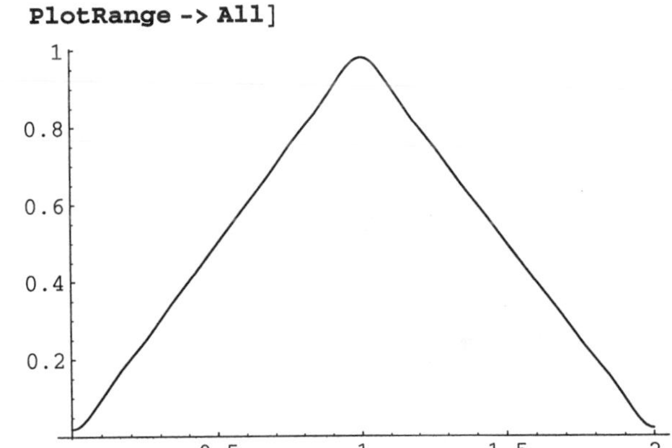

* 13. $a_0 = \frac{1}{\pi}\left(\int_{-\pi}^0 x dx + \int_0^\pi dx\right) = 1 - \frac{\pi}{2};\ a_n = \frac{1}{\pi}\left(\int_{-\pi}^0 x\cos nx dx + \int_0^\pi \cos nx dx\right) = \frac{-1}{n^2\pi}(\cos n\pi - 1);$

$b_n = \frac{1}{\pi}\left(\int_{-\pi}^0 x\sin nx dx + \int_0^\pi \sin nx dx\right) = \frac{-1}{n\pi}[(\pi+1)\cos n\pi - 1],\ n \geq 1$

* 21. (a) $\sin^3 x = b_1 \sin x + b_3 \sin 3x$, $b_1 = \frac{3}{4}$, $b_3 = -\frac{1}{4}$, $b_n = 0$, $n \neq 1,3$;

(b) $\cos^3 x = a_1 \cos x + a_3 \cos 3x$ $a_1 = \frac{3}{4}$, $a_3 = \frac{1}{4}$, $a_n = 0$, $n \neq 1,3$

10 Partial Differential Equations

EXERCISES 10.1

* 3. $X'Y - yXY' = 0$, $X' = kX$, $X(x) = C_1e^{kx}$, $Y'/Y = k/y$, $Y(y) = e^{C_2}y^k = C_3y^k$, $u(x,y) = C_1C_3e^{kx}y^k = Ce^{kx}y^k$

* 7. $u_y = f(y)$, $u(x,y) = F(y) + G(x)$ where $F'(y) = f(y)$; F and G are arbitrary functions.

* 13. $u_x = e^x \sin y$, $u_{xx} = e^x \sin y$, $u_y = e^x \cos y$, $u_{yy} = -e^x \sin y$

* 19. $u_x = 2\cos 2x \cos 2t$, $u_{xx} = -4\sin 2x \cos 2t$, $u_t = -2\sin 2x \sin 2t$, $u_{tt} = -4\sin 2x \cos 2t$

* 25. $u_x = 10 - 4e^{-16t}\sin 4x$, $u_{xx} = -16e^{-16t}\cos 4x$, $u_t = -16e^{-16t}\cos 4x$

* 31. $u_x = 4\cos ct \cos 4x$, $u_{xx} = -16\cos ct \sin 4x$, $u_t = -c\sin ct \sin 4x$, $u_{tt} = -c^2 \cos ct \sin 4x$, $4c^2 = 16$, $c = \pm 2$

```
u[x_, t_] := Cos[c t] Sin[4 x]
D[u[x, t], x]

4 Cos[c t] Cos[4 x
D[u[x, t], {x, 2}]

-16 Cos[c t] Sin[4 x
4 D[u[x, t], t]

-4 c Sin[c t] Sin[4 x
```

```
4 D[u[x, t], {t, 2}]

-4 c^2 Cos[c t] Sin[4 x
Solve[-16 == -4 c^2 , c]

{{c -> -2}, {c -> 2}}
```

* 35. Yes, $u_t(x,t) = -2\sin 2x \sin 2t$, $u_t(x,0) = -2\sin 2x \sin 0 = 0$

EXERCISES 10.2

* 3. $b_n = 2\int_0^1 x(1-x)\sin n\pi x\,dx = \dfrac{-4}{n^3\pi^3}(\cos n\pi - 1)$; $u(x,t) = \sum_{n=1}^{\infty} b_n \sin n\pi x\, e^{-n^2\pi^2 t/2}$

```
f[x_] := x (1 - x)
p = 1;
```

```
b[n_] =
  2 / p Integrate[f[x] Sin[n Pi x / p], {x, 0, 1}]
```

$$2\left(\frac{2}{n^3\,\pi^3} - \frac{2\,\mathrm{Cos}[n\,\pi]}{n^3\,\pi^3} - \frac{\mathrm{Sin}[n\,\pi]}{n^2\,\pi^2}\right)$$

* 9. $S(x) = 20 - 10x$, $u(x,t) = S(x) + \sum_{n=1}^{\infty} b_n \sin n\pi x\, e^{-n^2\pi^2 t}$

$$b_n = \frac{2}{1}\int_0^1 (20-10x)\sin n\pi x dx = \left[\frac{-20}{n^2\pi^2}\left((x-2)n\pi\cos n\pi x - \sin n\pi x\right)\right]_0^1 = \frac{20}{n\pi}(\cos n\pi - 2)$$

```
2 Integrate[(10 x - 20) Sin[n Pi x], x]

- 1/(n^2 π^2) (20
  (-2 n π Cos[n π x] + n π x Cos[n π x] - Sin[n π x
```

```
b[n_] =
 2 Integrate[(10 x - 20) Sin[n Pi x], {x, 0, 1}]

2 (-20/(n π) - (10 (-n π Cos[n π] - Sin[n π]))/(n^2 π^2))
```

* 13. $S''=0,\ S(0)=T_0,\ S'(1)=0,\ S(x)=c_1x+c_2,\ S(0)=c_2=T_0,\ S'(x)=c_1,\ S'(1)=c_1=0,\ S(x)=T_0$

```
DSolve[{s''[x] == 0, s[0] == T0, s'[1] == 0},
 s[x], x]

{{s[x] → T0}}
```

* 19. $u(x,0)=A_0+\sum_{n=1}^{\infty}a_n\cos nx = 4\sin^2 x = 4\left[\frac{1}{2}(1-\cos 2x)\right]=2-2\cos 2x,\ A_0=2,\ a_2=-2,\ a_n=0,\ n\neq 2,$
$u(x,t)=2-2\cos 2x\, e^{-4t}$

* 23. (Problem has homogeneous BC's) $u(x,t)=X(x)T(t),\ X''+\lambda X=0,\ X(-\pi)=X(\pi),\ X'(-\pi)=X'(\pi)$, (This eigenvalue problem was solved in Section 9.1 Ex. 35) $X_0(x)=1,\ X_n(x)=a_n\cos nx+b_n\sin nx,\ n\geq 1,$

$u(x,t)=\frac{1}{2}a_0+\sum_{n=1}^{\infty}(a_n\cos nx+b_n\sin nx)e^{-n^2t},\ u(x,0)=\frac{1}{2}a_0+\sum_{n=1}^{\infty}(a_n\cos nx+b_n\sin nx)=f(x)$, all coefficients are Fourier series coefficients for $f(x)$ with $p=\pi$.

* 29. f is even; $A_0=\frac{1}{2}a_0=\frac{1}{\pi}\int_{-\pi}^{\pi}|x|dx=\frac{2}{\pi}\int_0^{\pi}xdx=\pi,\ a_n=\frac{1}{\pi}\int_{-\pi}^{\pi}|x|\cos nxdx=\frac{2}{\pi}\int_0^{\pi}x\cos nxdx=\frac{2}{n^2\pi}(\cos n\pi-1),$

$b_n=0,\ n\geq 1;\ u(x,t)=\pi/2+\sum_{n=1}^{\infty}\frac{-4}{(2n-1)^2\pi}\cos(2n-1)x\, e^{-(2n-1)^2t}.$

```
p = Pi;
a[0] = 2 / Pi Integrate[x, {x, 0, Pi}]

π

a[n_] =
 2 / Pi Integrate[x Cos[n Pi x / p], {x, 0, Pi}]

2 (-1/n^2 + (Cos[n π] + n π Sin[n π])/n^2) / π
```

```
b[n_] =
 1 / Pi
  (Integrate[-x Sin[n Pi x / p], {x, -Pi, 0}] +
   Integrate[x Sin[n Pi x / p], {x, 0, Pi}])

((n π Cos[n π] - Sin[n π])/n^2 + (-n π Cos[n π] + Sin[n π])/n^2) / π
```

```
f[x_, k_] :=
 1/2 a[0] +
  Sum[a[n] Cos[n Pi x/p] + b[n] Sin[n Pi x/p],
   {n, 1, k}]
```

```
Plot[f[x, 10], {x, -p, p}]
```

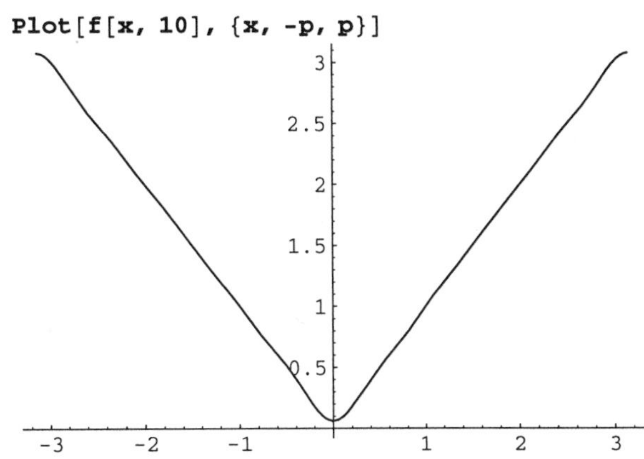

* 35. $S(x) = 50 + 50x$,

$$b_n = 2\int_0^1 \left[x\left(1 - x^3\right) - (50 + 50x)\right]\sin n\pi x\,dx = \frac{2\left(-24 - 50n^4\pi^4 + 4\left(6 - 3n^2\pi^2 + 25n^4\pi^4\right)\right)}{n^5\pi^5}\cos n\pi;$$

$$u(x,t) = \sum_{n=1}^{\infty} b_n \sin n\pi x\, e^{-n^2\pi^2 t}$$

```
p = 1;
b[n_] =
 2/p Integrate[(x (1 - x^3) - (50 + 50 x))
    Sin[n Pi x/p], {x, 0, 1}]
```

$$2\left(-\frac{24 + 50\,n^4\,\pi^4}{n^5\,\pi^5} + \frac{\left(24 - 12\,n^2\,\pi^2 + 100\,n^4\,\pi^4\right)\,\mathrm{Cos}[n\,\pi]}{n^5\,\pi^5} - \frac{\left(-24 + 53\,n^2\,\pi^2\right)\,\mathrm{Sin}[n\,\pi]}{n^4\,\pi^4}\right)$$

```
Simplify[%]
```

$$\frac{1}{n^5\,\pi^5}\left(2\left(-24 - 50\,n^4\,\pi^4 + 4\left(6 - 3\,n^2\,\pi^2 + 25\,n^4\,\pi^4\right)\mathrm{Cos}[n\,\pi] + \left(24\,n\,\pi - 53\,n^3\,\pi^3\right)\mathrm{Sin}[n\,\pi]\right)\right)$$

```
u[x_, t_, k_] :=
 Sum[b[n] Sin[n Pi x/p] Exp[-n^2 Pi^2 t],
   {n, 1, k}] + (50 + 50 x)
uTab = Table[u[x, t, 12], {t, 0, 1, 0.1}];
Plot[Evaluate[uTab], {x, 0, 1}, Pl‹
```

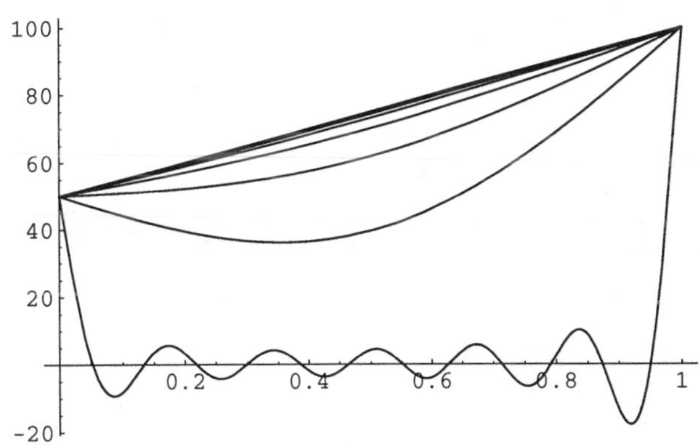

* 39. $u(x,t) = X(x)T(t)$, $X'' + \lambda X = 0$, $X(0) = 0, -X'(1) = X(1)$; $\lambda = 0$ is not an eigenvalue; No negative eigenvalues; $\lambda = k^2 > 0$: $X(x) = c_1\cos kx + c_2\sin kx$, $X(0) = c_1 = 0$, $X'(x) = kc_2\cos kx$, $kc_2\cos k = \sin k$ or $k = -\tan k$; eigenvalues are $\lambda_n = k_n^2$ where k_n satisfies $k_n = -\tan k_n$, $n \geq 1$; $u(x,t) = \sum_{n=1}^{\infty} c_n \sin k_n x\, e^{-k_n^2 t}$,

$$u(x,0)=\sum_{n=1}^{\infty} c_n \sin k_n x = f(x),\ c_n\int_0^1 \sin^2 k_n x dx = \int_0^1 f(x)\sin k_n x dx \text{ when } m=n,$$

$$\int_0^1 \sin^2 k_n x dx = \frac{1}{2}\int_0^1 (1-\cos 2k_n x)dx = \frac{1}{2}\left[x-\frac{1}{2k_n}\sin 2k_n x\right]_0^1 = \frac{1}{2}\left(1-\frac{1}{2k_n}\sin 2k_n\right),$$

$$\frac{1}{2}\left(1-\frac{1}{2k_n}\sin 2k_n\right)=\frac{1}{2}\left(1-\frac{1}{2k_n}2\sin k_n\cos k_n\right)=\frac{1}{2}\left(1-\frac{1}{k_n}(-k_n\cos k_n)\cos k_n\right)=\frac{1}{2}\left(1+\cos^2 k_n\right),$$

$$\frac{c_n}{2}\left(1+\cos^2 k_n\right)=\int_0^1 f(x)\sin k_n x dx,\ c_n=\frac{2}{1+\cos^2 k_n}\int_0^1 f(x)\sin k_n x dx$$

EXERCISES 10.3

* 3. $u_t(x,0)=\sum_{n=1}^{\infty} b_n n\pi \sin n\pi x = \sin 3\pi x,\ 3b_3\pi = 1,\ b_3 = 1/(3\pi),\ b_n = 0,\ n\neq 3,\ a_n = 0,\ n\geq 0,$
$u(x,t)=1/(3\pi)\sin 3\pi t\sin 3\pi x$

* 7. $u(x,t)=\sum_{n=1}^{\infty}(a_n\cos nt + b_n \sin nt)\sin nx;\ a_2=1, a_n=0, n\neq 2,\ b_n=0, n\geq 1;\ u(x,t)=\cos 2t\sin 2x$

* 13. $u_x = u_r;\ u_{xx}=u_{rr};\ u_{xy}=u_{rr}+u_{rs};\ u_{yx}-u_{xx}=u_{rs},\ u_{rs}=0,\ u(r,s)=F(r)+G(s),$
$u(x,y)=F(x+y)+G(y)$

* 17. $B^2-4AC=(-4)^2-4(1)(2)=8>0$, hyperbolic

* 27. $p=1,\ u(x,0)=\sum_{n=1}^{\infty} a_n \sin n\pi x = \sin\pi x,\ a_1=1,\ a_n=0,\ n\geq 2,\ b_n=0,\ n\geq 1,\ u(x,t)=\cos\pi t\sin\pi x,$
$u(x,t)=\frac{1}{2}[\sin\pi(x+t)+\sin\pi(x-t)]=\frac{1}{2}[\sin\pi x\cos\pi t+\sin\pi t\cos\pi x+\sin\pi x\cos\pi t-\sin\pi t\cos\pi x]$
$=\sin\pi x\cos\pi t$

EXERCISES 10.4

* 3. $u(x,y)=\sum_{n=1}^{\infty}\left(A_n\sinh\frac{n\pi(1-x)}{2}+B_n\sinh\frac{n\pi x}{2}\right)\sin\frac{n\pi y}{2},\ u(1,y)=\sum_{n=1}^{\infty}A_n\sinh\frac{n\pi x}{2}\sin\frac{n\pi y}{2}=0,\ A_n=0,\ n\geq 1,$

$$u(0,y)=\sum_{n=1}^{\infty}B_n\sinh\frac{n\pi}{2}\sin\frac{n\pi y}{2}=y,\ B_n=\frac{1}{\sinh(n\pi/2)}\int_0^2 y\sin\frac{n\pi y}{2}dy=\frac{-4\cos n\pi}{n\pi\sinh(n\pi/2)}$$

```
1 / Sinh[ n Pi / 2]
 Integrate[y Sin[n Pi y / 2], {y, 0, 2}]
```

$$-\frac{2\,\mathrm{Csch}\left[\frac{n\pi}{2}\right]\ (2\,n\,\pi\,\mathrm{Cos}[n\,\pi]-2\,\mathrm{Sin}[n\,\pi])}{n^2\,\pi^2}$$

* 7. $u(x,y)=\sum_{n=1}^{\infty}\left(A_n\cosh n\pi y+B_n\sinh n\pi y\right)\sin n\pi x,\ A_n=0,\ n\ge 1,\ u(x,2)=\sum_{n=1}^{\infty}B_n\sinh 2n\pi\sin n\pi x=\sin\pi x,$

$B_1=1/\sinh 2\pi,\ B_n=0, n\ge 2,\ u(x,y)=\sinh\pi y\sin\pi x/\sinh 2\pi$

13. $B_n=\frac{2}{1}\int_0^1 T_0\sin n\pi y\,dy=\frac{2T_0}{n\pi}(1-\cos n\pi)$

```
2 Integrate[T0 Sin[n Pi y], {y, 0, 1}]
```

$$2\left(\frac{\mathrm{T0}}{\mathrm{n}\,\pi}-\frac{\mathrm{T0}\,\mathrm{Cos}[\mathrm{n}\,\pi]}{\mathrm{n}\,\pi}\right)$$

17. $b_n=\frac{2}{2}\int_0^2 x\sin\frac{n\pi x}{2}dx=\frac{4}{n\pi}(-1)^{n+1}$

23. $B_{mn}=\frac{2}{m^2n^2\pi^3}\left[-4mn\pi\cos m\pi+mn\pi(1+\pi)\cos(m-n)\pi-2mn\pi^2\cos n\pi+mn\pi(1+\pi)\cos(m+n)\pi\right]$

27. $u(x,y,t)=1/(4\lambda_{12})\sin\lambda_{12}t\sin x\sin 2y,\ \lambda_{12}=c\sqrt{5}$

33. (c) $w(x,y)=\sum_{n=1}^{\infty}\left(\tilde{A}_n\sinh\frac{n\pi x}{b}+\tilde{B}_n\sinh\frac{n\pi(a-x)}{b}\right)\sin\frac{n\pi y}{b},\ \tilde{B}_n=\frac{2}{b\sinh\frac{n\pi a}{b}}\int_0^b h(y)\sin\frac{n\pi y}{b}dy,$

$\tilde{A}_n=\frac{2}{b\sinh\frac{n\pi a}{b}}\int_0^b k(y)\sin\frac{n\pi y}{b}dy$

37. $b_n=\frac{2}{\pi\sinh n}\int_0^{\pi}x(1-x)\sin nx\,dx=\frac{2\operatorname{csch}n\left(2+\left(-2+n^2(\pi-1)\pi\right)\cos n\pi\right)}{n^3\pi},$

$d_n=\frac{2}{\pi\sinh n\pi}\int_0^{\pi}\sin nx\,dx=\frac{-2(-1+\cos n\pi)}{n\pi}\operatorname{csch}n\pi,\ c_n=\frac{2}{\pi\sinh n\pi^2}\int_0^1\sin n\pi y\,dy=\frac{-20(-1+\cos n\pi)}{n\pi^2}\operatorname{csch}n\pi^2,$

$n\ge 1$

$u(x,y)=\sum_{n=1}^{\infty}\left(b_n\sinh ny\sin nx+c_n\sinh n\pi x\sin n\pi y+d_n\sinh n\pi(\pi-x)\sin n\pi y\right)$

41. $a_n = \frac{2}{\sinh 2n\pi}\int_0^1 x(1-x^2)\sin n\pi x dx = \frac{-12\operatorname{csch}2n\pi\cos n\pi}{n^3\pi^3}$, $b_n = 0$, $c_n = 0$,

$$d_n = \frac{1}{\sinh(n\pi/2)}\int_0^2 y\sin(n\pi y/2)dy = \frac{-4\cos n\pi}{n\pi}\operatorname{csch}(n\pi/2),$$

$$u(x,y) = \sum_{n=1}^{\infty}\left(a_n \sinh n\pi y \sin n\pi x + d_n \sinh n\pi(1-x)/2 \sin(n\pi y/2)\right)$$

```
Clear[a]
a[n_] = 1 / Sinh[2 n Pi] 2
  Integrate[x (1 - x^2) Sin[n Pi x], {x, 0, 1}]
```

$$\frac{1}{n^4\pi^4}\left(2\,\text{Csch}[2\,n\,\pi]\left(-6\,n\,\pi\,\text{Cos}[n\,\pi] + 6\,\text{Sin}[n\,\pi] - 2\,n^2\,\pi^2\,\text{Sin}[n\,\pi]\right)\right)$$

```
Clear[d]
d[n_] = 1 / Sinh[ n Pi / 2]
  Integrate[y Sin[n Pi y / 2], {y, 0, 2}]
```

$$-\frac{2\,\text{Csch}\left[\frac{n\pi}{2}\right]\left(2\,n\,\pi\,\text{Cos}[n\,\pi] - 2\,\text{Sin}[n\,\pi]\right)}{n^2\,\pi^2}$$

45. $u(x,y) = \frac{1}{\sinh\pi}\sinh y\sin x - \frac{1}{2\sinh 2\pi}\sinh 2y\sin 2x$

EXERCISES 10.5

* 3. $u(1,\theta) = A_0 + \sum_{n=1}^{\infty}(A_n\cos n\theta + B_n\sin n\theta) = 1/2 + 1/2\cos 2\theta$, $A_0 = 1/2$, $A_2 = 1/2$, $A_n = 0$, $n \neq 0,2$, $B_n = 0$, $n \geq 1$, $u(r,\theta) = 1/2 + 1/2\,r^2\cos 2\theta$

* 7. $B_n = \frac{2}{\pi}\int_0^{\pi}\sin n\theta d\theta = \frac{-2}{n\pi}(1-\cos n\pi), n \geq 1$; $A_n = 0, n \geq 0$, $u(r,\theta) = -\sum_{n=1}^{\infty} 4r^{2n-1}/((2n-1)\pi)\sin(2n-1)\theta$

```
2 / (Pi) Integrate[-Sin[n t], {t, 0, Pi}]
```

$$\frac{2\left(-\frac{1}{n} + \frac{\text{Cos}[n\,\pi]}{n}\right)}{\pi}$$

* 13. $u(r,0)=\sum_{n=1}^{\infty} A_n J_0(\alpha_n r)=0,\ A_n=0, n\ge 1,$

$u_t(r,0)=\sum_{n=1}^{\infty} c\alpha_n B_n J_0(\alpha_n r)=J_0(\alpha_1 r),\ B_1=1/(c\alpha_1), B_n=0, n\ge 2,\ u(r,t)=1/(c\,\alpha_1)\sin(c\,\alpha_1 t)J_0(\alpha_1 r)$

17. $1/\alpha_n\int_0^1 \alpha_n r J_0(\alpha_n r)dr=1/\alpha_n\int_0^1 d/dr\left[rJ_1(\alpha_n r)\right]dr=1/\alpha_n\left[rJ_1(\alpha_n r)\right]_0^1=1/\alpha_n\,J_1(\alpha_n)$

21. $A_n=\dfrac{2}{\left[J_1(\alpha_n)\right]^2}\left[\dfrac{J_1(\alpha_n)}{\alpha_n}-\dfrac{J_2(\alpha_n)}{\alpha_n{}^2}\right];\ B_n=0, n\ge 1$

23. Note that $r=\sqrt{x^2+y^2}$ and $\tan\theta=y/x$ (or $\theta=\arctan(y/x)$);

$r_x=1/2\left(x^2+y^2\right)^{-1/2}(2x)=x/\sqrt{x^2+y^2}=r\cos\theta/r=\cos\theta,\ r_y=y/r=r\sin\theta/r=\sin\theta,$

$\theta_x=\dfrac{1}{1+(y/x)^2}\left(-\dfrac{y}{x^2}\right)=\dfrac{-y}{x^2+y^2}=\dfrac{-r\cos\theta}{r^2}=\dfrac{-\cos\theta}{r},\ \theta_y=\dfrac{1}{1+(y/x)^2}\left(\dfrac{x}{x^2}\right)=\dfrac{x}{x^2+y^2}=\dfrac{r\sin\theta}{r^2}=\dfrac{\sin\theta}{r},$

$u_x=u_r r_x+u_\theta\theta_x=\cos\theta\,u_r-\dfrac{\sin\theta}{r}u_\theta,\ u_y=u_r r_y+u_\theta\theta_y=\sin\theta\,u_r+\dfrac{\cos\theta}{r}u_\theta,$

$u_{xx}=\sin^2\theta\,u_{rr}+\dfrac{2\sin\theta\cos\theta}{r}u_{r\theta}+\dfrac{\sin^2\theta}{r^2}u_{\theta\theta}+\dfrac{\sin^2\theta}{r}u_r+\dfrac{2\sin\theta\cos\theta}{r}u_\theta,$

$u_{yy}=\sin^2\theta\,u_{rr}+\dfrac{2\sin\theta\cos\theta}{r}u_{r\theta}+\dfrac{\cos^2\theta}{r^2}u_{\theta\theta}+\dfrac{\cos^2\theta}{r}u_r+\dfrac{2\sin\theta\cos\theta}{r}u_\theta$

29. $\cos 2\phi=2\cos^2\phi-1;\ \frac{4}{3}P_2(\cos\phi)-\frac{1}{3};\ u(r,\phi)=\frac{4}{3}r^2P_2(\cos\phi)-\frac{1}{3}$

33. (a) $A_0=\dfrac{1}{2\pi}\int_{-\pi}^{\pi}\theta^2\cos\theta\,d\theta=-2,\ A_n=\dfrac{1}{\pi}\int_{-\pi}^{\pi}\theta^2\cos\theta\cos n\theta\,d\theta=\dfrac{\left(4\pi-4n^4\pi\right)\cos n\pi}{\pi(n-1)^3(n+1)^3},\ B_n=0,\ n\ge 1$

(b) $A_0=0,\ A_n=0,\ B_n=\dfrac{1}{\pi}\int_{-\pi}^{\pi}\theta^2\sin\theta\cos n\theta\,d\theta=\dfrac{\left(8n-8n^3\right)\cos n\pi}{(n-1)^3(n+1)^3},\ n\ge 1$

(c) $A_0=\dfrac{1}{2\pi}\int_{-\pi}^{\pi}\theta^2\left(1+\theta^2\right)d\theta=\pi^2\left(5+3\pi^2\right)/15,\ A_n=\dfrac{1}{\pi}\int_{-\pi}^{\pi}\theta^2\left(1+\theta^2\right)\cos n\theta\,d\theta=\dfrac{4\left(-12+n^2\left(1+2\pi^2\right)\right)\cos n\pi}{n^4},$

$B_n=0,\ n\ge 1$

Chapter 10 Review Exercises

* 3. (a) $u_{xx}=0, u_{yy}=0$; (b) $u_{xx}=9e^{3x}\cos 3y, u_{yy}=-9e^{3x}\cos 3y$; (c) $u_{xx}=\dfrac{2\left(y^2-x^2\right)}{\left(x^2+y^2\right)^2}, u_{yy}=\dfrac{2\left(x^2-y^2\right)}{\left(x^2+y^2\right)^2}$

* 7. $u(0,t)=A_0$. u at the center of the disk equals the average value of f around the boundary of the disk.

11. $rR'' + R' + \lambda^2 rR = 0$, $R(a) = 0$, $Z'' - \lambda^2 Z = 0$, $Z(0) = 0$, $R(r) = c_1 J_0(\lambda r) + c_2 Y_0(\lambda r)$, $c_2 = 0$ (bounded), $J_0(\lambda a) = 0$, $\lambda_n = \alpha_n / a$, α_n - nth zero of J_0. $Z(z) = c_3 \cosh \lambda z + c_4 \sinh \lambda z$, $c_3 = 0$,

$$u(r,z) = \sum_{n=1}^{\infty} a_n \sin \lambda_n z \, J_0(\lambda_n r) \,, \quad a_n = \frac{2T_0}{\sinh b\lambda_n 2^2 J_1{}^2(a\lambda_n)} \int_0^a rJ_0(\lambda_n r)dr$$